Natalie Salter

Scanning Electron Microscopy and X-Ray Microanalysis

A Text for Biologists, Materials Scientists, and Geologists

Scanning Electron Microscopy and X-Ray Microanalysis

A Text for Biologists, Materials Scientists, and Geologists

Joseph I. Goldstein

Lehigh University
Bethlehem, Pennsylvania

Dale E. Newbury

National Bureau of Standards
Washington, D.C.

Patrick Echlin

University of Cambridge
Cambridge, England

David C. Joy

Bell Laboratories
Murray Hill, New Jersey

Charles Fiori

National Institutes of Health
Bethesda, Maryland

Eric Lifshin

General Electric Corporate Research and Development
Schenectady, New York

PLENUM PRESS • NEW YORK AND LONDON

Library of Congress Cataloging in Publication Data

Main entry under title:

Scanning electron microscopy and X-ray microanalysis.

Bibliography: p.
Includes index.
1. Scanning electron microscope. 2. X-ray microanalysis. I. Goldstein, Joseph, 1939-
QH212.S3S29   502 '8'25   81-13766
ISBN 0-306-40768-X   AACR2

First Printing — November 1981
Second Printing — April 1984

© 1981 Plenum Press, New York
A Division of Plenum Publishing Corporation
233 Spring Street, New York, N.Y. 10013

Printed in the United States of America

Preface

This book has evolved by processes of selection and expansion from its predecessor, *Practical Scanning Electron Microscopy* (PSEM), published by Plenum Press in 1975. The interaction of the authors with students at the Short Course on Scanning Electron Microscopy and X-Ray Microanalysis held annually at Lehigh University has helped greatly in developing this textbook. The material has been chosen to provide a student with a general introduction to the techniques of scanning electron microscopy and x-ray microanalysis suitable for application in such fields as biology, geology, solid state physics, and materials science. Following the format of PSEM, this book gives the student a basic knowledge of (1) the user-controlled functions of the electron optics of the scanning electron microscope and electron microprobe, (2) the characteristics of electron-beam–sample interactions, (3) image formation and interpretation, (4) x-ray spectrometry, and (5) quantitative x-ray microanalysis. Each of these topics has been updated and in most cases expanded over the material presented in PSEM in order to give the reader sufficient coverage to understand these topics and apply the information in the laboratory. Throughout the text, we have attempted to emphasize practical aspects of the techniques, describing those instrument parameters which the microscopist can and must manipulate to obtain optimum information from the specimen. Certain areas in particular have been expanded in response to their increasing importance in the SEM field. Thus energy-dispersive x-ray spectrometry, which has undergone a tremendous surge in growth, is treated in substantial detail. Moreover, we have come to realize the importance of developing careful procedures for qualitative x-ray microanalysis, that is, the identification of the elemental constituents present in a sample; suitable procedures for both energy-dispersive and wavelength-dispersive x-ray spectrometry are described.

The most conspicuous addition to the book is the material on biological specimen preparation and coating. Because of the great difficulties in properly preparing a biological sample for SEM examination and analysis, this topic has been considered in detail. It should be recognized that this material is of value not only to biologists, but also to many nonbiological disciplines in which fragile samples, often containing water or other fluids, must be prepared for the SEM. These include polymers, pigments, corrosion products, textiles, and many others.

For the convenience of readers who are confronted with a need for numerical information on important parameters for SEM and x-ray microanalysis calculations, we have included a data base of frequently used information, including x-ray energies of principal lines, mass absorption coefficients, backscattering factors, and others.

Some material from PSEM has been deleted in preparing this book. Some chapters, including "Contrast Mechanisms of Special Interest in Materials Science" and "Ion Microprobe Mass Analysis," have been removed. These topics and several others will be presented in a companion volume tentatively titled *Advanced Topics in Scanning Electron Microscopy and Microanalysis*, a specialist volume which is specifically intended for advanced workers who have completed an introductory course.

The authors wish to thank their many colleagues who have contributed to this volume by their kindness in allowing us to use material from their publications, by their criticism of PSEM and the present manuscript, and by their general support. One of the authors (J. I. Goldstein) wishes to acknowledge the research support and encouragement from the Planetary Materials Program of the National Aeronautics and Space Administration and from the Geochemistry Program of the Earth Sciences Division of the National Science Foundation. Special thanks go to Betty Fekete Zdinak and Louise Valkenburg of Lehigh for their extra efforts in the preparation of the original manuscript, to Carol Swyt of the National Institutes of Health for her constructive criticisms of many of the original chapters, to Roger Bolon and Mike Ciccarelli of General Electric, and Bob Myklebust and Harvey Yakowitz of the National Bureau of Standards, for specific contributions.

The Authors

Contents

Introduction

In our rapidly expanding technology, the scientist is required to observe, analyze, and correctly explain phenomena occurring on a micrometer (μm) or submicrometer scale. The scanning electron microscope and electron microprobe are two powerful instruments which permit the observation and characterization of heterogeneous organic and inorganic materials and surfaces on such a local scale. In both instruments, the area to be examined, or the microvolume to be analyzed, is irradiated with a finely focused electron beam, which may be static or swept in a raster across the surface of the specimen. The types of signals produced when the electron beam impinges on a specimen surface include secondary electrons, backscattered electrons, Auger electrons, characteristic x-rays, and photons of various energies. These signals are obtained from specific emission volumes within the sample and can be used to examine many characteristics of the sample (composition, surface topography, crystallography, etc.).

In the scanning electron microscope (SEM), the signals of greatest interest are the secondary and backscattered electrons, since these vary as a result of differences in surface topography as the electron beam is swept across the specimen. The secondary electron emission is confined to a volume near the beam impact area, permitting images to be obtained at relatively high resolution. The three dimensional appearance of the images is due to the large depth of field of the scanning electron microscope as well as to the shadow relief effect of the secondary electron contrast. Other signals are available which prove similarly useful in many cases.

In the electron probe microanalyzer (EPMA), frequently referred to as the electron microprobe, the primary radiation of interest is the characteristic x-rays which are emitted as a result of the electron bombardment. The analysis of the characteristic x-radiation can yield both qualitative and quantitative compositional information from regions of a specimen as small as a few micrometers in diameter.

Historically, the scanning electron microscope and electron microprobe evolved as separate instruments. It is obvious on inspection, however, that these two instruments are quite similar but differ mainly in the way in which they are utilized. The development of each of these instruments (SEM and EPMA) and the differences and similarities of modern commercial instruments are discussed in this chapter.

1.1. Evolution of the Scanning Electron Microscope

The scanning electron microscope (SEM) is one of the most versatile instruments available for the examination and analysis of the microstructural characteristics of solid objects. The primary reason for the SEM's usefulness is the high resolution which can be obtained when bulk objects are examined; values of the order of 5 nm (50 Å) are usually quoted for commercial instruments. Advanced research instruments have been described which have achieved resolutions of about 2.5 nm (25 Å) (Broers, 1974b). The high-resolution micrograph, shown in Figure 1.1, was taken with an advanced commercial SEM under typical operating conditions.

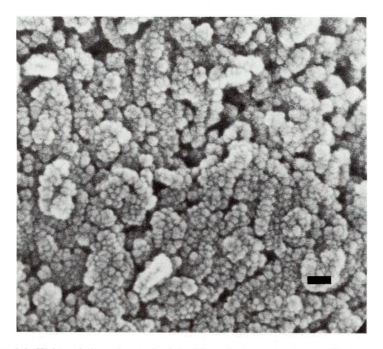

Figure 1.1. High-resolution micrograph of Au:Pd coating on magnetic tape. The image was taken with a JEOL 100CX equipped with a standard tungsten filament. Marker = 30 nm. (Micrograph courtesy of J. Geller, JEOL.)

Another important feature of the SEM is the three-dimensional appearance of the specimen image, which is a direct result of the large depth of field. Figure 1.2a shows the skeleton of a small marine organism (the radiolarian *Trochodiscus longispinus*) viewed optically, and Figure 1.2b as viewed with the SEM. The greater depth of field of the SEM provides much more information about the specimen. In fact, the SEM literature indicates that it is this feature which is of the most value to the SEM user. Most SEM micrographs have been produced with magnifications below 8000 diameters. At these magnifications the SEM is operating well within its resolution capabilities. Figure 1.3 is a micrograph of a pollen grain of *Ipomoea purpurea L.* (morning glory) and shows in one picture the complex surface topography of the wall of this single plant cell. The only other way to obtain this type of detailed information would be to painstakingly reconstruct the three-dimensional structure from planar serial sections observed in the transmission electron microscope. It would be difficult to make a faithful replica of such a detailed and irregular surface.

The SEM is also capable of examining objects at very low magnification. This feature is useful in forensic studies as well as other fields. An example of a low-magnification micrograph of an archeological subject is shown in Figure 1.4.

The basic components of the SEM are the lens system, electron gun, electron collector, visual and recording cathode ray tubes (CRTs), and the electronics associated with them. The first successful commercial packaging of these components (the Cambridge Scientific Instruments Mark I instrument) was offered in 1965. Considering the present popularity of the SEM, the fact that 23 years passed between the time Zworykin, Hillier, and Snyder (1942) published the basis for a modern SEM and this development seems incredible. The purpose of this brief historical introduction is to point out the pioneers of scanning electron microscopy and in the process trace the evolution of the instrument.

The earliest recognized work describing the construction of a scanning electron microscope is that of von Ardenne in 1938. In fact, von Ardenne added scan coils to a transmission electron microscope (TEM) and in so doing produced what amounts to the first scanning transmission electron microscope (STEM). Both the theoretical base and practical aspects of STEM were discussed in fairly complete detail. The first STEM micrograph was of a ZnO crystal imaged at an operating voltage of 23 kV, at a magnification of 8000 × and with a spatial resolution between 50 and 100 nm. The photograph contained 400 × 400 scan lines and took 20 min to record (von Ardenne, 1938a, b) because the film was mechanically scanned in synchronism with the beam. The instrument had two electrostatic condenser lenses, with the scan coils being placed between the lenses. The instrument possessed a CRT but it was not used to photograph the image (von Ardenne, 1938a, b).

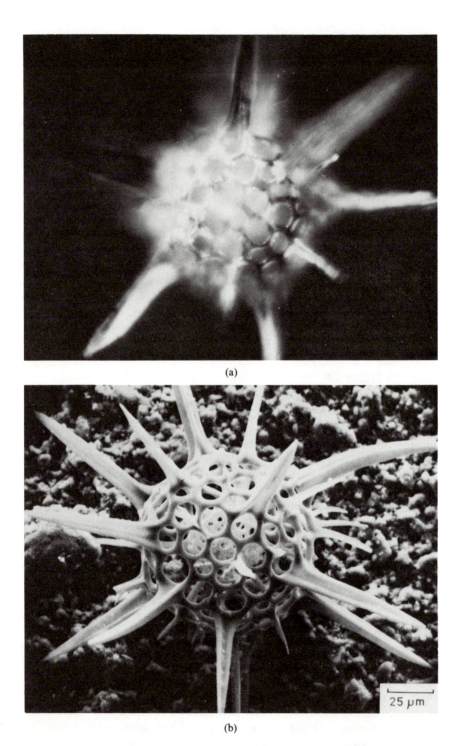

(a)

(b)

Figure 1.2. (a) Optical micrograph of the radiolarian *Trochodiscus longispinus*. (b) SEM micrograph of same radiolarian shown in (a). The depth of focus and superior resolving capability in this micrograph are apparent.

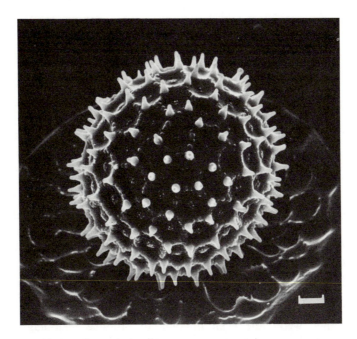

Figure 1.3. A whole pollen grain from *Ipomoea purpurea L.* (morning glory). Note the large spines, some of which are as long as 5 μm. Marker = 6 μm.

Figure 1.4. SEM image of facial area of a helmeted demonlike warrier from the rear of the handle of an 18th Century Japanese sword. The region of the helmet is gilt. (Courtesy of M. Notis, Lehigh University.) Marker = 1 mm.

The first SEM used to examine thick specimens was described by Zworykin *et al.* (1942). These authors recognized that secondary electron emission would be responsible for topographic contrast and accordingly constructed the design shown in Figure 1.5. The collector was biased positive to the specimen by 50 V and the secondary electron current collected on it produced a voltage drop across a resistor. This voltage drop was sent through a television set to produce the image; however, resolutions of only 1 μm were attained. This was considered unsatisfactory since they sought better resolution than that obtainable with optical microscopes (200 nm).

Therefore, Zworykin *et al.* (1942) decided to reduce spot size, as well as improve signal-to-noise ratio, and so produce a better instrument. A detailed analysis of the interrelationship of lens aberrations, gun brightness, and spot size resulted in a method for determining the minimum spot size as a function of beam current (Zworykin *et al.*, 1942). They sought to improve gun brightness by using a field emitter source, but instability in these cold cathode sources forced a return to thermionic emission sources, although they were able even in 1942 to produce high-magnification, high-resolution images (Zworykin *et al.*, 1942). The next contribution was the use of an electron multiplier tube as a preamplifier for the secondary emission current from the specimen. Useful but noisy (by today's standards) photomicrographs were obtained. The electron optics of the instrument consisted of three electrostatic lenses with scan coils placed between the second and third lenses. The electron gun was at the bottom so that the

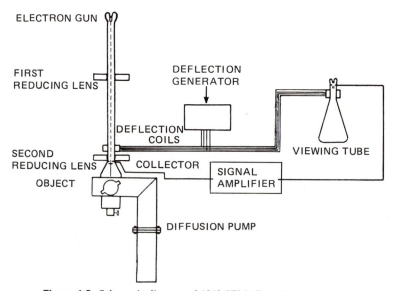

Figure 1.5. Schematic diagram of 1942 SEM (Zworykin *et al.*, 1942).

specimen chamber was at a comfortable height for the operator. Resolutions of at least 50 nm (500 Å) were achieved with this, the first modern SEM (Zworykin *et al.*, 1942). World War II caused a hiatus in the development of the SEM and so Zworykin *et al.* were unable to fully develop their instrument. In 1948, C. W. Oatley at the University of Cambridge in England became interested in building electron microscopes. He and McMullan built the first SEM at Cambridge, and by 1952 this unit had achieved a resolution of 50 nm (500 Å) (McMullan, 1952). McMullan was followed by K. C. A. Smith (1956), who, recognizing that signal processing could improve micrographs, introduced nonlinear signal amplification (γ processing). He also replaced electrostatic lenses with electromagnetic lenses, improved the scanning system by introducing double deflection scanning, and was also the first to insert a stigmator into the SEM (Smith, 1956).

The next step forward was the modification of the detector described by Zworykin *et al.* (1942). This was accomplished by Everhart and Thornley, who attached the scintillator to a light pipe directly at the face of the photomultiplier (Everhart and Thornley, 1960). This improvement increased the amount of signal collected and resulted in an improvement in signal-to-noise ratio. Hence, weak contrast mechanisms could be better investigated. Using this detector Oatley and Everhart (1957) were able to observe, for the first time, the phenomenon known as voltage contrast.

Pease (1963) built a system, known as SEM V, with three magnetic lenses, the gun at the bottom and the Everhart–Thornley detector system. This instrument led to the prototype of the Cambridge Scientific Instruments Mark I and, in many ways, was similar to the 1942 instrument (Pease, 1963; Pease and Nixon, 1965). A. D. G. Stewart and co-workers at the Cambridge Scientific Instrument Co. carried out the commercial design and packaging of the instrument. In the ensuing 16 years, more than 6000 SEM units have been sold by a dozen or more manufacturers, representing the U.S.A., U.K., France, Holland, Japan, and Germany, who are actively developing new, improved instruments. Yet, even today, the basic SEM is not far removed from the one described in 1942.

Since 1965, many advances have been made. One of these was the development of the lanthanum hexaboride (LaB_6) electron cathode by Broers (1969a). This source provides a high-brightness electron gun; hence more electron current can be concentrated into a smaller beam spot. In this way, an effective improvement in resolution can be obtained.

The field emission tip electron source first used in the SEM in 1942 was revived by Crewe (1969) and developed to the point where it can be used in obtaining high-resolution images. The reason that the field emission gun yields high-resolution probes is that the source is very small and the brightness is very high. Hence, current densities of thousands of amperes per square centimeter are available even when the beam current is on the order of 0.1 nA. Field emission sources have two drawbacks: (1) if one draws currents exceeding a few nanoamperes from them, resolution deteriorates rapidly, and (2) the source requires a clean, high vacuum of the order of 0.1 μPa (10^{-9} Torr) or lower to operate stably. Such pressures require special attention to vacuum technology and cannot be easily achieved with ordinary instrument designs. Even under the best conditions, a field emission tip provides a beam which is not very stable with time.

Other advances involve contrast mechanisms not readily available in other types of instrumentation. Crystallographic contrast, produced by crystal orientation and lattice interactions with the primary beam, was found by Coates (1967) and exploited initially by workers at Oxford University (Booker, 1970). Contrast from magnetic domains in uniaxial materials was observed by Joy and Jakubovics (1968). Magnetic contrast in cubic materials was first observed by Philibert and Tixier (1969); this contrast mechanism was explained later by Fathers *et al.*, (1973, 1974).

Often the contrast of the features one is examining is so low as to be invisible to the eye. Therefore contrast enhancement by processing of the signal was needed. Early signal processing included nonlinear amplification, as noted, and differential amplification (black level suppression), both incorporated into SEMs at Cambridge University. Derivative signal processing (differentiation) to enhance small details was introduced later (Lander *et al.*, 1963; Heinrich, Fiori, and Yakowitz, 1970; Fiori, Yakowitz, and Newbury, 1974). Most commercial SEM units today are provided with these signal processing capabilities.

The image itself can be processed either in analog or digital form. Image-storing circuits have been developed so that one can observe the image and/or operate on it off-line (Catto, 1972; Yew and Pease, 1974; Holburn and Smith, 1979; Harland and Venables, 1980). These devices are extremely useful and not prohibitively expensive, but not as versatile as full computer image processing. White and co-workers have developed a series of image processing computer programs known as CESEMI (Computer Evaluation, SEM Images) which can provide a great deal of information such as grain size, amoung of phase present, etc. (Lebiedzik *et al.*, 1973). Digital scanning in which the coordinates of the picture point and the signal intensity of the point are delivered to the computer is needed to take full advantage of such programs. In fact, computer interaction and control of the SEM has already been accomplished by Herzog *et al.* (1974).

Some of the first scanning micrographs were of biological material. Thornley (1960) showed SEM images of frozen-dried biological material examined at 1 keV to avoid charging. Boyde and Stewart (1962) published one of the first biological papers in which SEM was used to specifically study a biological problem.

Advances in the biological field have been controlled to some extent by advances in specimen preparation. Most biological specimens are wet, radiation sensitive, thermolabile samples of low contrast and weak emissivity and are invariably poor conductors. Much attention has been paid to stabilizing the delicate organic material, removing or immobilizing the cell fluids, and coating the samples with a thin layer of a conducting film. Ways have been devised to selectively increase the contrast of specimens and to reveal their contents at predetermined planes within the sample. The development of low-temperature stages has helped to reduce the mass loss and thermal damage in sensitive specimens (Echlin, 1978). The preparative techniques and associated instrumentation have now advanced to the point where any biological material may be examined in the SEM with the certainty that the images which are obtained are a reasonable representation of the once living state.

The large depth of field available in the SEM makes it possible to observe three-dimensional objects in stereo. This is particularly important in the examination of biological samples, which unlike many metals, rocks, polymers, and ceramics are morphologically very heterogeneous and must be examined in depth as well as at a given surface. The three-dimensional images allow different morphological features to be correctly interrelated and definitively measured. Equipment has been developed which allows quantitative evaluation of surface topography making use of this feature (Boyde, 1974a; Boyde and Howell, 1977; Boyde, 1979). Provisions for direct stereo viewing on the SEM have been described as well (Dinnis, 1973).

The addition of an energy-dispersive x-ray detector to an electron probe microanalyzer (Fitzgerald *et al.*, 1968) signaled the eventual coupling of such instrumentation to the SEM. Today, a majority of SEM facilities are equiped with x-ray analytical capabilities. Thus, topographic, crystallographic, and compositional information can be obtained rapidly, efficiently, and simultaneously from the same area.

1.2. Evolution of the Electron Probe Microanalyzer

The electron probe microanalyzer (EPMA) is one of the most powerful instruments for the microanalysis of inorganic and organic materials. The primary reason for the EPMA's usefulness is that compositional information, using characteristic x-ray lines, with a spatial resolution of the order of 1 μm can be obtained from a sample. The sample is analyzed nondestructively, and quantitative analysis can be obtained in many cases with an accuracy of the order of 1%–2% of the amount present for a given element (5%–10% in biological material). In the case of some labile biological and organic substances, there can be a substantial, i.e., up to 90%, mass loss during the period of analysis. Elemental losses can also occur, particularly in the case of light elements in an organic matrix. Both mass loss and

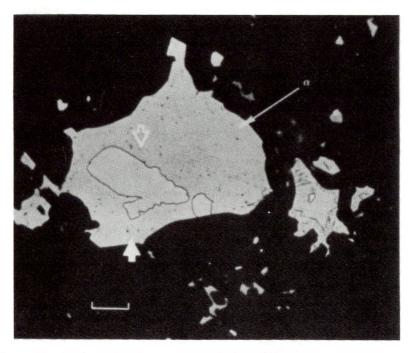

Figure 1.6a. Micrograph in reflected light of rock 73275,68. The α phase surrounds the cohenite and the arrows indicate the extent of the microprobe scan shown in Figure 1.6b. Scale bar = 16 μm. (Goldstein, *et al.*, 1976.)

elemental loss can be reduced to negligible properties by maintaining the specimen at low temperature.

Figure 1.6 shows a microprobe analysis of C, Fe, Ni, and Co across a carbide (cohenite) phase which has nucleated and grown by a diffusion-controlled process in a ferrite α-bcc phase matrix during cooling of a lunar metal particle on the moon's surface (Goldstein *et al.*, 1976). The microchemical data were taken with a commercial EPMA under typical operating conditions and illustrate the resolution of the quantitative x-ray analysis which can be obtained.

Another important feature of the EPMA is the capability for obtaining x-ray scanning pictures. The x-ray pictures show the elemental distribution in the area of interest. Figure 1.7 shows the distribution of Fe and P between the various phases in a metal grain from lunar solid 68501 (Hewins *et al.*, 1976). Magnifications up to 2500 $\times$ are possible without exceeding the resolution of the instrument. The attractiveness of this form of data gathering is that detailed microcompositional information can be directly correlated with optical metallography. In addition, reduced area scans can be used to compare the relative compositions of different areas of interest.

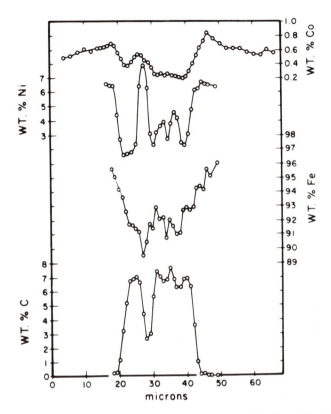

Figure 1.6b. Variation of C, Fe, Ni, and Co across α-cohenite (Fe₃C) particle in lunar rock 73275 (Figure 1.6a).

Figure 1.8 shows a simple biological application of EPMA in which differences in elemental ratios have been measured in developing plant roots. In this example chlorine is present in the nucleus but not detected in the cytoplasm. This analysis was carried out using a reduced raster measuring approximately 1 μm^2. The EPMA literature indicates that the x-ray scanning feature of the EPMA is of great value to the user. In addition, the variety of signals available in the EPMA (emitted electrons, photons, etc.) can provide useful information about surface topography and composition in small regions of the specimen. The early history of the development of the EPMA is reviewed in the following paragraphs.

In 1913, Moseley found that the frequency of emitted characteristic x-ray radiation is a function of the atomic number of the emitting element (Moseley, 1913). This discovery led to the technique of x-ray spectrochemical analysis by which the elements present in a specimen could be identified by the examination of the directly or indirectly excited x-ray spectra. The area analyzed, however, was quite large (> 1 mm²). The idea of the

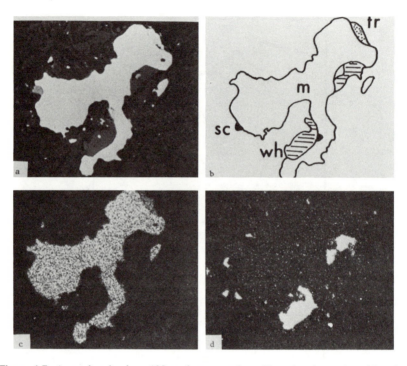

Figure 1.7. A metal grain about 105 μm long occurring with a phosphate mineral in a lunar melt rock from soil 68501 (Hewins *et al.*, 1976). (a) Backscattered electron image showing metal (white), phosphate and pyroxene (medium gray), and plagioclase (dark gray). (b) Sketch of (a) showing grains of metal (m), troilite, FeS (tr), phosphide $(FeNi)_3P$ (sc) and phosphate (wh). (c) Fe Kα image of (a). (d) P Kα image of (a).

electron microanalyzer, in which a focused electron beam was used to excite a small area on a specimen ($\sim$1 μm^2), and which included an optical microscope for locating the area of interest, was first patented in the 1940s (Hillier, 1947; Marton, 1941). It was not until 1949, however, that Castaing, under the direction of Guinier, described and built an instrument called the "microsonde electronique" or electron microprobe (Castaing and Guinier,

Figure 1.8. Freeze-dried vascular tissue of *Lemna minor L.* (duckweed) (marker = 5 μm). Energy-dispersive analysis carried out on nucleus (A) and cytoplasm (B) of one of the phloem parenchyma cells using a stationary reduced raster. Both compartments contain phosphorus, sulfur, and a small amount of magnesium. The nucleus contains a small amount of chloride and a much higher level of potassium than in the cytoplasm; the latter also contains a small amount of sodium. Computer-enhanced deconvolution of the two spectra shows the following elemental ratios:

	Na	Mg	P	S	Cl	K
Nucleus	0	6	35	6	2	42
Cytoplasm	2	4	31	5	0	29

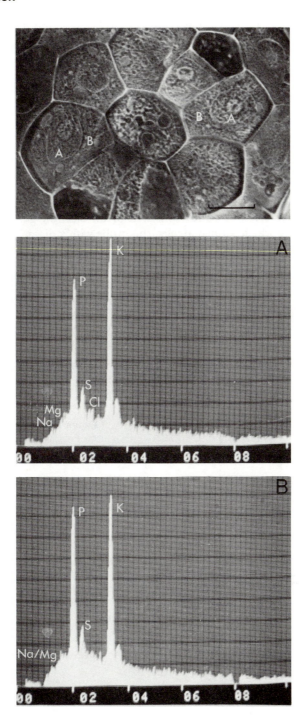

1950; Castaing, 1951; Castaing, 1960). In his doctoral thesis Castaing demonstrated that a localized chemical analysis could be made on the surface of a specimen. Concurrently with the work of Castaing in France, Borovskii (1953) in the U.S.S.R. also developed an EPMA quite dissimilar in design.

During the early 1950s several EPMA instruments were developed in laboratories both in Europe and the United States (Haine and Mulvey, 1959; Birks and Brooks, 1957; Fisher and Schwarts, 1957; Wittry, 1957; Cuthill, Wyman, and Yakowitz, 1963). The first EPMA commercial instrument, introduced by CAMECA in 1956, was based on the design of an EPMA built by Castaing in the laboratories of the Recherches Aeronautiques (Castaing, 1960). Figure 1.9 shows diagrammatically the design of the apparatus. The electron optics consisted of an electron gun followed by reducing lenses which formed an electron probe with a diameter of approximately 0.1 to 1 μm on the specimen. Since electrons produce x-rays from a volume often exceeding 1 μm in diameter and 1 μm in depth, it is usually unnecessary to use probes of very small diameter. An optical microscope for accurately choosing the point to be analyzed and a set of wavelength-dispersive spectrometers for analyzing the intensity of x-ray radiation emitted as a function of energy are also part of the instrument.

Cosslett and Duncumb (1956) designed and built the first scanning electron microprobe in 1956 at the Cavendish Laboratories in Cambridge,

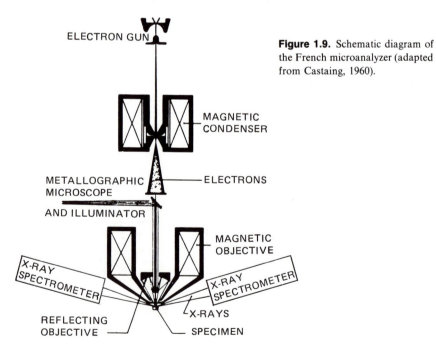

Figure 1.9. Schematic diagram of the French microanalyzer (adapted from Castaing, 1960).

ELECTRON GUN

MAGNETIC CONDENSER

METALLOGRAPHIC MICROSCOPE AND ILLUMINATOR

ELECTRONS

MAGNETIC OBJECTIVE

X-RAY SPECTROMETER

X-RAY SPECTROMETER

REFLECTING OBJECTIVE

X-RAYS

SPECIMEN

England. Whereas all previous electron microprobes had operated with a static electron probe, Cosslett and Duncumb swept the beam across the surface of a specimen in a raster, as is done in current SEMs. They used the backscattered electron signal to modulate the brightness of a cathode ray tube sweeping in synchronism with the electron probe. They also used the x-ray signal to modulate the brightness, permitting a scanned image to be obtained showing the lateral distribution of a particular element. Although the concept of a local x-ray analysis is in itself a strong incentive for the utilization of a microprobe, the addition of the scanning concept was an extremely significant contribution and probably accounted for the subsequent increased popularity of the electron microprobe.

In his doctoral thesis, Castaing developed the physical theory so that the analyst could convert measured x-ray intensities to chemical composition (Castaing, 1951). Castaing proposed a method of analysis based on comparisons of intensity, i.e., the comparison of the intensity I_i of a characteristic line of a given element emitted by a specimen under one set of conditions of electron bombardment and the intensity $I_{(i)}$ of the same characteristic radiation when emitted by a standard containing the same element under the same electron bombardment conditions. The ratio of the two readings is proportional to the mass concentration of a given element in the region analyzed. Recognition of the complexity of converting x-ray intensities to chemical composition has led numerous investigators (Wittry, 1963; Philibert, 1963; Duncumb and Shields, 1966) to expand the theoretical treatment of quantitative analysis proposed by Castaing.

In the years since the development of the first scanning EPMA instrument many advances have been made. Of particular importance was the development of diffracting crystals having large interplanar spacings (Henke, 1964; 1965). These crystals enable long-wavelength x-rays from the light elements to be measured with the wavelength dispersive spectrometers. The ability to detect fluorine, oxygen, nitrogen, carbon, and boron enabled users of the EPMA to investigate many new types of problems with the instrument (note Figure 1.6). Techniques of soft x-ray spectroscopy have now been applied in the EPMA to establish how each element is chemically combined in the sample. The chemical effect may be observed as changes in wavelength, shape or relative intensity of emission and absorption spectra.

The x-ray microanalysis of biological material is beset by the same problems associated with the examination of the sample in the SEM. The experimenter has to be very careful that the elements being measured remain in the specimen and are not removed or relocated by the preparative procedure. While specimen preparation for biological material is more exacting than the procedures used in the material sciences, the quantitation methods are somewhat simpler being based on the analysis of thin sections. However, such thin sections are all too easily damaged by the electron beam and much effort in the past decade has gone into devising instrumentation as well as preparative and analytical procedures to limit specimen damage in organic samples.

When the application of the EPMA was extended to nonmetallic specimens, it became apparent that other types of excitation phenomena might also be useful. For example, the color of visible light (cathodoluminescence) produced by the interaction of the electron probe, with the specimen has been associated with the presence of certain impurities in minerals (Long and Agrell, 1965). In addition photon radiation produced from the recombination of excess hole–electron pairs in a semiconductor can be studied (Kyser and Wittry, 1966). Measurement of cathodoluminescence in the EPMA has now been developed as another important use of this instrument.

Increased use of the computer in conjunction with the EPMA has greatly improved the quality and quantity of the data obtained. Many computer programs have been developed to convert the x-ray intensity ratios to chemical compositions, primarily because some of the correction parameters are functions of concentration and hence make successive approximations necessary. Most of these programs can be run on a minicomputer and compositions calculated directly from recorded digital data. The advantage of rapid calculation of chemical compositions is that the operator has greater flexibility in carrying out analyses. In addition, computer automation of the EPMA has been developed to varying degrees of complexity. Dedicated computers and the accompanying software to control the electron beam, specimen stage, and spectrometers are now commercially available. The advantages of automation are many, but in particular it greatly facilitates repetitive-type analysis, increases the amount of quantitative analysis performed, and leaves the operator free to concentrate on evaluating the analysis and designing experiments to be performed.

The development of the energy-resolving x-ray spectrometer based upon the silicon (lithium-drifted) Si(Li) solid state detector (Fitzgerald *et al.*, 1968), has revolutionized x-ray microanalysis. The energy-dispersive spectrometer (EDS) system is now the most common x-ray measurement system to be found in the SEM lab. Even in classical EPMA configurations, the Si(Li) detector plays an important part alongside the wavelength-dispersive spectrometer. The EDS system offers a means of rapidly evaluating the elemental constituents of a sample; major constituents (10 wt% or more) can often be identified in only 10 s, and a 100-s accumulation is often sufficient for identification of minor elements. In addition to rapid qualitative analysis, accurate quantitative analysis can also be achieved with EDS x-ray spectrometry. These great advantages are tempered by the relatively poor energy resolution of the EDS (150 eV at Mn $K\alpha$ compared with 5–10 eV for a wavelength spectrometer) which leads to frequent unresolved spectral interferences (e.g., S $K\alpha$, Mo $L\alpha$, and Pb $M\alpha$), poor peak-to-background values, and the resultant poor limit of detection (typi-cally ~0.5 wt% compared to ~0.1 wt% for a wavelength spectrometer).

1.3. Outline of This Book

The SEM and EPMA are in reality very similar instruments. Therefore several manufacturers have constructed instruments capable of being oper-ated as an electron microprobe and as a high-resolution scanning electron microscope. Figure 1.10 shows a schematic of the electron and x-ray optics of such a combination instrument. Both a secondary electron detector and

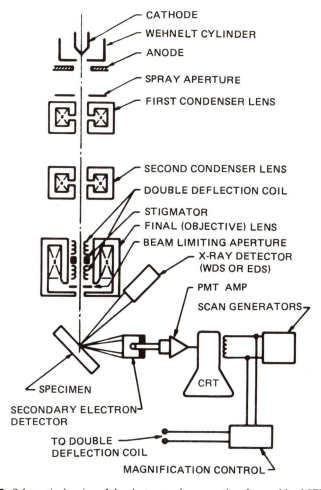

Figure 1.10. Schematic drawing of the electron and x-ray optics of a combined SEM–EPMA.

an x-ray detector(s) are placed below the final lens. For quantitative x-ray analysis and for the measurement of x-rays from the light elements, at least two wavelength-dispersive spectrometers and an energy-dispersive spectrometer are desirable. In addition, provision for a broad useful current range, 1 pA to 1 μA, must be made. The scanning unit allows both electron and x-ray signals to be measured and displayed on the CRT. It is logical therefore that the SEM–EPMA be considered as one instrument in this textbook as well.

The electron optics system and the signals produced during electron bombardment in the SEM–EPMA are discussed in Chapters 2 and 3. The remainder of the book is devoted to the details of measuring the available signals, to the techniques of using these signals to determine particular types of information about organic and inorganic samples, and to the preparation of the specimens. The emphasis is on the selection and utilization of techniques appropriate to the solution of problems often presented to the SEM and EPMA analysis staff by their clients.

Chapters 4 and 5 consider the detection and processing of secondary electron, backscattered electron, cathodoluminescence, and x-ray signals as obtained from the SEM–EPMA. Following this material Chapters 6, 7 and 8 discuss various methods for qualitative and quantitative x-ray analysis. The methods for preparation of solid materials such as rocks, metals, and ceramics for SEM and x-ray microanalysis are given in Chapter 9. Methods for specimen preparation are critical for most biological samples and other materials which contain water. The methods for preparation of biological samples for SEM are discussed in Chapter 11 and for x-ray microanalysis in Chapter 12. Coating techniques for biological and materials samples are considered in Chapter 10. Finally applications of the SEM and EPMA to biological and materials problems are considered in Chapter 13.

Electron Optics

The amount of current in a finely focused electron beam impinging on a specimen determines the magnitude of the signals (x-ray, secondary electrons, etc.) emitted, other parameters being equal. In addition, the size of the final probe determines the best possible resolution of the scanning electron microscope (SEM) and electron microprobe (EPMA) for many of the signals that are measured. Therefore the electron optical system in these instruments is designed so that the maximum possible current is obtained in the smallest possible electron probe. In order to use the instruments intelligently it is important to understand how the optical column is designed, how the various components of the optical system function, and how the final current and spot size are controlled. In this chapter we will discuss the various components of the electron optical system, develop the relationship between electron probe current and spot size, and discuss the factors which influence this relationship.

2.1. Electron Guns

2.1.1. Thermionic Emission

As shown in Chapter 1, electron optical columns for the EPMA and SEM consist of the electron gun and two or more electron lenses. The electron gun provides a stable source of electrons which is used to form the electron beam. These electrons are usually obtained from a source by a process called thermionic emission. In this process, at sufficiently high temperatures, a certain percentage of the electrons become sufficiently energetic to overcome the work function, E_w, of the cathode material and escape the source. Figure 2.1 illustrates the concept of the work function.

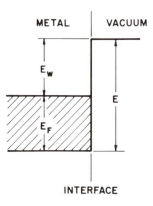

METAL | VACUUM

E_w

E

E_F

INTERFACE

Figure 2.1. Energy model for thermionic emission (adapted from Kittel, 1956).

The term E represents the work necessary to remove an electron to infinity from the lowest free energy state in the metal. If the electron is taken from the Fermi level, E_F, the highest free energy state in the metal, the work or work function E_w is

$$E_w = E - E_F \qquad (2.1)$$

The emission current density J_c obtained from the filament of the electron gun by the thermionic emission is expressed by the Richardson law,

$$J_c = A_c T^2 \exp(-E_w/kT) \quad \text{A/cm}^2 \qquad (2.2)$$

where A_c (A/cm^2 K^2) is a constant which is a function of the material and T (K) is the emission temperature.

   The filament or cathode has a V-shaped tip which is about 5 to 100 μm in radius. The filament materials that are used, W or LaB$_6$, have high values of A and low values of the work function E_w. Specific filament types are discussed in the following sections. The filament is heated directly or indirectly using a filament supply and is maintained at a high negative voltage (1–50 kV) during operation. At the operating filament temperature the emitted electrons leave the V-shaped tip and are accelerated to ground (anode) by a 1000 to 50,000 V potential between the cathode and anode. The configuration for a typical electron gun (Hall, 1953) is shown in Figure 2.2.

   Surrounding the filament is a grid cap or Wehnelt cylinder with a circular aperture centered at the filament apex. The grid cap is biased negatively between 0 and 2500 V with respect to the cathode. The effect of the electric field formed in such a gun configuration, the filament, Wehnelt cylinder and anode, causes the emitted electrons from the filament to converge to a crossover of dimension d_0. Figure 2.2 also shows the equipotential field or voltage lines which are produced between the filament grid cap and the anode. The constant field lines are plotted with respect to

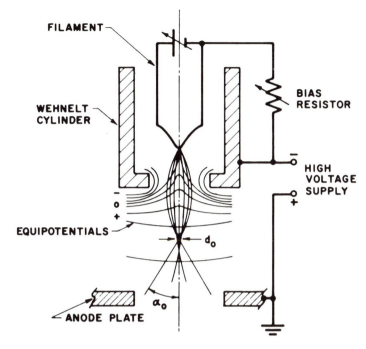

FILAMENT

WEHNELT
CYLINDER

BIAS
RESISTOR

HIGH
VOLTAGE
SUPPLY

−
∘
+

EQUIPOTENTIALS

d_0

α_0

ANODE PLATE

Figure 2.2. Configuration of self-biased electron gun (adapted from Hall, 1953).

the filament (cathode) voltage and vary between 0 at the filament to a negative potential (up to −2500 V) at the grid cap to the large positive potential (+1000 to 50,000 V) at the anode or anode plate.

The emitted electrons are accelerated through this voltage field and attempt to follow the maximum voltage gradient which is perpendicular to the field lines. Where the constant field lines represent a negative potential, however, the electrons are repelled. Figure 2.2 shows the paths of electrons through the field or constant voltage lines. Note the focusing action as the electrons approach and are repelled from the negatively biased grid cap. By the use of the grid cap, the electrons are focused to a crossover of dimension d_0 and divergence angle α_0 below the Wehnelt cylinder. The negative bias on the grid cap and its placement with respect to the tip of the filament control the focusing action. The intensity distribution of the electrons at crossover is usually assumed to be Gaussian. The condenser and probe-forming lenses produce a demagnified image of this crossover to yield the final electron probe.

The current density in the electron beam at crossover represents the current that could be concentrated into a focused spot on the specimen if no aberrations were present in the electron lenses. This current density, J_b

(A/cm^2), is the maximum intensity of electrons in the electron beam at crossover, and can be defined as

$$J_b = \frac{i_b}{\pi (d_0/2)^2} \qquad (2.3)$$

where i_b represents the total beam or emission current measured from the filament. The beam current usually varies between 100 and 200 μA. For comparison purposes, in a conventional sealed-off x-ray tube a considerably higher beam current, i_b, typically of the order of 15–25 mA, is produced.

It is desirable, in practice, to obtain the maximum current density in the final image. Since the maximum usable divergence angle of the focused electron beam is fixed by the aberrations of the final lens in the imaging system, the most important performance parameter of the electron gun is the current density per unit solid angle. This is called the electron beam brightness β and is defined as

$$\beta = \frac{\text{current}}{(\text{area})(\text{solid angle})} = \frac{4i}{\pi^2 d^2 \alpha^2} \quad \text{A/cm}^2 \, \text{sr} \qquad (2.4)$$

The steradian is defined as the solid angle subtended at the center of a sphere of unit radius by a unit area on the surface of the sphere; it is dimensionless. For the electron gun, $i = i_b$, $\alpha = \alpha_0$, and $d = d_0$. The electron beam brightness remains constant throughout the electron optical column even as i, d, and α change. As shown by Langmuir (1937) the brightness has a maximum value given by

$$\beta = \frac{J_c e E_0}{\pi k T} \quad \text{A/cm}^2 \, \text{sr} \qquad (2.5)$$

for high voltages, where J_c is the current density at the cathode surface, E_0 is the accelerating voltage, e is the electronic charge, and k is Boltzmann's constant. The brightness can be calculated as $\beta = 11,600 J_c E_0 / \pi T$, where the units are A/cm^2 for J_c and V for E_0. The current density can then be rewritten as

$$J_b = \pi \beta \alpha_0^2 \qquad (2.6)$$

and the maximum current density is

$$J_b = J_c \frac{e E_0 \alpha_0^2}{kT} \qquad (2.7)$$

The theoretical current density per unit solid angle (brightness) for a given gun configuration can be approached in practice provided an optimum bias voltage is applied between cathode and grid cap. In the electron gun the bias is produced by placing a variable "bias resistor" in series with the negative side of the high-voltage power supply and the filament (Figure 2.2). The bias resistor provides the bias voltage because the grid current of the high-voltage power supply flows through it. In this configuration, when the current is supplied to heat the filament a negative voltage will be applied across the grid cap. As the resistance of the "bias resistor" changes, the negative bias voltage changes in direct relation. The major effect of varying the bias voltage is to change the constant field lines near the cathode.

At low bias, the negative field gradient becomes weak (Figure 2.2) and the focusing action of the negative equipotentials is relatively ineffective. The electrons see only a positive field or voltage gradient towards the anode; and therefore, the emission current, i_b, is high. Since little or no focusing takes place, the crossover d_0 is large and the brightness β obtained is not optimum. If the bias is increased too much, however, the negative field lines around the filament will be so strong that the electrons which are emitted from the filament will observe only a negative field gradient and will return to the filament. In this case the emission current as well as the brightness decrease to zero. Figure 2.3 summarizes the relationship between emission current, brightness and bias voltage. As shown in Figure 2.3, there is an optimum bias setting for maximum brightness. This setting can be obtained for each electron gun. If the filament-to-Wehnelt cylinder distance can be changed, then the shape of the constant field lines can also be altered. In some instruments one may have the freedom to change the filament-to-grid spacing and/or the bias resistance. In other instruments these two factors have been properly adjusted to obtain the maximum brightness from the gun and need not be changed by the operator.

It is also important to obtain a stable well-regulated beam current. As the filament current, i_f, used to heat the cathode is increased from zero, the temperature of the filament is increased and electron emission occurs.

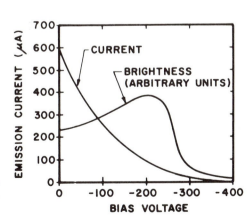

Figure 2.3. Relationship of emission current and brightness to bias voltage, schematic only.

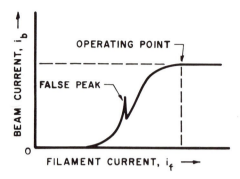

Figure 2.4. Emission characteristic of the self-biased electron gun. Schematic drawing of beam current i_b versus filament current i_f.

Figure 2.4 shows the emission characteristics of a self-biased gun in which the beam current i_b is plotted vs. the filament current i_f. If the bias resistor is correctly set for maximum brightness, the beam current i_b does not vary as the filament current is increased above a certain minimum value. This saturation condition occurs because as the filament current is increased above that necessary for emission, the bias voltage also increases, causing the negative field gradient around the filament to increase and therefore limiting the rise in i_b. This balancing condition is called saturation and produces a self-regulating gun and a stable beam current. In normal operation a false peak may be observed in the emission of a self-biased electron gun (Figure 2.4). The false peak may appear when emission occurs from a specific region of the filament which reaches the emission temperature before the filament tip. A false peak can be identified since electron emission will decrease as the filament current is increased further.

The current density at crossover J_b can be increased if the brightness of the gun is increased. Improved brightness would then provide either increased current for the same beam size, or a reduced beam diameter for the same current. The brightness of the filament can be increased according to Langmuir's formula [Equation (2.5)] by either increasing the high voltage E_0 or increasing the current density J_c at the cathode. Since gun voltages are held within relatively narrow limits, a major improvement in brightness will be made by an improvement in J_c such as increasing A or lowering E_w in the Richardson law, Equation (2.2).

2.1.2. Tungsten Cathode

The tungsten cathode is a wire filament approximately 0.01 cm in diameter, bent in the shape of a hairpin with a V-shaped tip which is about 100 μm in radius. The cathode is heated directly as the filament current, i_f, from the filament supply is passed through it. For tungsten, at a typical operating temperature of 2700 K, J_c is equal to 1.75 A/cm^2 as calculated from the Richardson expression, Equation (2.2), where $A_c = 60$ A/cm^2 K^2

and $E_w = 4.5$ eV. At normal operating temperatures the emitted electrons leave the V-shaped tip from an emission area of about 100×150 μm. The voltage of the emitted electrons is higher than that of the cathode itself. At 2900 K, for example, the maximum in the voltage distribution of the emitted electrons is about 0.25 V with respect to the cathode, while the range of electron potential is from 0 to about 2 V (Hall, 1953). The bias voltage is typically 0 to -500 V and a defined saturation is observed similar to that shown in Figure 2.4. The filament wire diameter decreases with time because of tungsten evaporation. Therefore the filament current necessary to reach the operating temperature and to obtain filament saturation decreases with the age of the filament. Filament life also decreases with increasing temperature. At an emission current of 1.75 A/cm^2 filament life should average around 40 to 80 h in a reasonably good vacuum. Although raising the operating temperature has the advantage of increasing J_c, this would be accomplished only at the loss of filament life.

Typical values of d_0 and α_0 for electron guns used in the SEM are $d_0 \sim 25$ to 100 μm, $\alpha_0 \sim 3 \times 10^{-3}$ to 8×10^{-3} rad. For a tungsten filament operated at 2700 K and a cathode current density J_c of 1.75 A/cm^2, the brightness at 25 kV, as calculated from Equation (2.5), is about 6×10^4 A/cm^2 sr. If the operating temperature is increased from 2700 to 3000 K, the emission current and brightness can be raised from $J_c = 1.75$ A/cm^2 and $\beta = 6 \times 10^4$ A/(cm^2 sr) at 25 kV to $J_c = 14.2$ A/cm^2 and $\beta = 4.4 \times 10^5$ A(cm^2 sr). However, although the brightness increases by over a factor of 7, the filament life decreases to an unacceptable low of 1 h, owing to W evaporation. Although it is of increasing interest to obtain cathodes of higher brightness, the W filament has served the TEM–SEM community well over the last 30 years. The W filament is reliable, its properties are well understood, and it is relatively inexpensive. Therefore for the majority of problems, where high brightness guns are not a necessity, the W filament will continue to play an important role.

2.1.3. The Lanthanum Hexaboride (LaB$_6$) Cathode

As new techniques in scanning microscopy are developed the need for sources of a higher brightness than the tungsten hairpin filament described in Section 2.1.2 becomes more evident. From the Richardson equation, Equation (2.2), it can be seen that the cathode current density, and hence the brightness, can be increased by lowering the work function E_w or by improving the value of the constant A_c. A considerable amount of work has therefore gone into searching for cathode materials which would have a lower value of E_w, a higher value of A_c, or both. The lower work function is usually of most significance because at the operating temperature of 2700 K each 0.1 eV reduction in E_w will increase J_c by about 1.5 times.

The most important possible cathode material with a low value of E_w so far developed has been lanthanum hexaboride, which was first investigated by Lafferty (1951). Lanthanum hexaboride, LaB_6, is a compound in which the lanthanum atoms are contained within the lattice formed by the boron atoms. When the material is heated the lanthanum can diffuse freely through the open boron lattice to replenish material evaporated from the surface. It is this action of the lanthanum that gives LaB_6 its low work function. Other rare earth hexaborides (praseodymium, neodymium, cerium, etc.) also have a similar property, and low work functions, but they have been less extensively investigated.

The measured work function (Fomenko, 1966; Swanson and Dickinson, 1976) for polycrystalline LaB_6 is of the order of 2.4 eV, with A_c typically 40 $A/cm^2 K^2$. This means that a current density equal to that produced by the conventional tungsten filament is available at an operating temperature of only about 1500 K and that at 2000 K nearly 100 A/cm^2 would be expected. This ability to produce useful cathode current densities at relatively low temperatures is important for two reasons. Firstly, the rate of evaporation will be low, so that a long operating life can be expected compared to the relatively short life of a tungsten hairpin. Secondly, from the Langmuir relation [Equation (2.5)] we see that for two sources both having the same cathode current density and accelerating voltage but operated at 1500 and 3000 K, respectively, the source at 1500 K will have twice the brightness of the otherwise equivalent source at 3000 K. LaB_6 therefore appears to offer substantial advantages over tungsten as an emitter material.

There are, however, compensating disadvantages which have slowed the acceptance of LaB_6. The material is extremely chemically reactive when hot, and readily forms compounds with all elements except carbon and rhenium. When this occurs the cathode is "poisoned" and it ceases to be an efficient emitter. This reactivity also means that the cathode can only be operated in a good vacuum, since in any pressure above 100 μPa an oxide, characteristically purple in color, forms, and this also impairs the performance. Finally LaB_6 is only available commercially as a fine-grain (5 μm diameter) powder, and considerable processing is therefore needed to form a useful cathode assembly.

The first practical LaB_6 gun was due to Broers (1969b) and is shown schematically in Figure 2.5. The cathode was made from powdered LaB_6 which was hot pressed and sintered to form a rod about 1 mm square in cross section and 1.6 cm long. One end of the rod was milled to a fine point with a radius of only a few microns. The other end was held in an oil-cooled heat sink. In this way the sharp emitting end of the cathode could be heated while maintaining the other end of the rod at a low temperature where its reactivity would not be a problem. The heat was

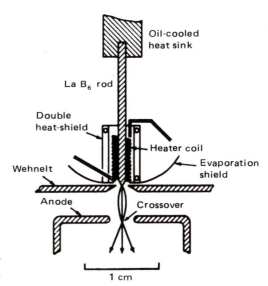

Figure 2.5. Gun configuration for LaB$_6$ cathode (from Broers, 1969).

applied from a coil of tungsten wire wrapped around, but not touching, the rod. A high current was passed through this coil, which was also held at a negative potential relative to the cathode rod. Heating then occurred through a combination of radiation and electron bombardment. To prevent lanthanum being evaporated onto other surfaces around the gun, heat and evaporation shields were placed around the heater. The temperature at the tip of the rod was in the range 1700 to 2100 K. In operation this gun was able to produce cathode current densities of the order of 65 A/cm^2 at a temperature of 1850 K with a lifetime of several hundred hours. At 25 kV this corresponds to a brightness of 3×10^6 A/cm^2 sr, which is at least five times greater than that of a tungsten filament operating at its maximum temperature.

The diameter of the end of the rod is such that the emitting area often consists of only one grain of LaB$_6$. During operation this grain will be gradually eroded by evaporation and an adjacent grain becomes the emitter. Because the work function of LaB$_6$ is a function of the crystallographic orientation of the emitting face, these random changes in the tip cause periodic fluctuations in the output of the gun.

Another effect can also occur to change the emission from the tip. Because of the small radius of the emitting cathode, and in the presence of a high bias voltage, there is locally a very high electric field present. This field, which may be 10^6 V/cm or more, will modify the potential barrier seen by an electron trying to leave the cathode by reducing the effective value of the work function E_w. The amount by which E_w is reduced will obviously depend on the strength of the field outside the cathode, but this "Schottky" effect can typically reduce E_w by 0.1 V or more (Broers, 1975) leading to increased emission from the tip. While this is a desirable thing, if the tip geometry changes, as will occur if the crystallite at the point is evaporated away, then the field at the cathode will change, thus altering the effective work function E_w and hence the emission.

More recently, directly heated LaB_6 emitters have been developed, usually being designed as direct plug-in replacements for the conventional tungsten hairpin. A variety of configurations have been described; Figure 2.6 shows one type of unit. The emitter is a tiny block of LaB_6, again formed from sintered material, about 100 μm square in cross section and 1/2 mm long, weighing only a few milligrams. The block is supported between two strips of graphite. The resistance of the graphite strips is chosen so that a current of one or two amperes creates sufficient joule heating to raise the temperature of the LaB_6 block to its required value. Typically only two or three watts of power are necessary. Assuming the vacuum in the gun chamber is adequate, a device of this kind can be used to replace the usual tungsten filament with a minimum of effort. It is usually only necessary to change the bias and the Wehnelt cylinder to achieve a good gain in brightness. Figures comparable with those from the Broers-type gun have been reported.

Because the LaB_6 block is so small it is possible to replace the sintered polycrystalline material by a single crystal, since small crystals are readily

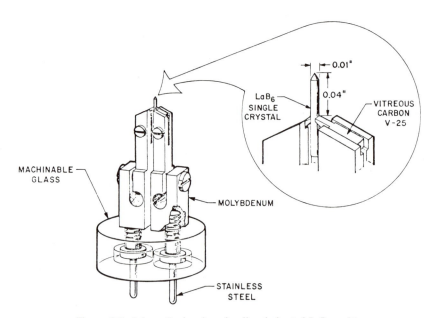

Figure 2.6. Schematic drawing of a directly heated LaB_6 emitter.

grown by the alumina-flux technique. This offers several advantages. Firstly since the work function is dependent on the crystallographic orientation, by selecting an optimum orientation, usually $\langle 110 \rangle$, the work function can be reduced to below 2 eV compared with the polycrystalline value of 2.4 eV. Secondly, because the material is homogeneous, the output is much more stable than that from the sintered cathodes. A general change to single crystals can thus be expected.

The high electric field present at the tip of the LaB_6 means that the type of saturation shown in Figure 2.4 will not usually occur. When correctly biased the emission will rise to a shallow "knee" but no clearly defined saturation will be found. Care must therefore be taken to avoid overheating the cathode in an attempt to reach saturation. It should also be noted that when first operated, or after exposure to the atmosphere, a LaB_6 emitter may require some time to activate itself. During this period, which may last from a few seconds to several minutes, contaminants on the surface will be removed by evaporation. When this is finished the output of the source will rapidly rise to its usual value. Any attempt to reach full output before activation has occurred can lead to the destruction of the cathode by overheating.

Several manufacturers have recently made available the LaB_6 gun for use on their SEMs. The necessity for vacuums of better than 100 μPa in the gun region requires improved gun pumping. This plus the greater difficulty in fabricating LaB_6 filaments leads to an increased cost for the LaB_6 gun. Nevertheless the increased brightness of the LaB_6 emitter yields a significantly smaller probe size at the same beam current or a larger beam current at the same probe size than a tungsten emitter. This improvement in performance will clearly justify the increased cost in many applications.

2.1.4. Field Emission Gun

The electron sources described so far have relied on the use of a high temperature to enable a fraction of the free electrons in the cathode material to overcome the barrier of the work function and leave. There is, however, another way of generating electrons that is free from some of the disadvantages of thermionic emission; this is the process of field emission. In field emission the cathode is in the form of a rod with a very sharp point at one end (typically of the order of 100 nm diameter or less). When the cathode is held at a negative potential relative to the anode, the electric field at the tip is so strong ($> 10^7$ V/cm) that the potential barrier discussed above becomes very narrow as well as reduced in height. As a result electrons can "tunnel" directly through the barrier and leave the cathode without needing any thermal energy to lift them over the barrier (Gomer, 1961). A cathode current density of between 1000 and 10^6 A/cm is obtained in this way, giving an effective brightness which is many hundreds of times higher than that of a thermionic source at the same operating voltage, even though the field emitter is at room temperature.

The usual cathode material is tungsten. This is because the field at the tip is so high that there is a very large mechanical stress on the cathode and only very strong materials can withstand this without failing. However, other substances, such as carbon fibers, have been used with some success. Because the work function is a function of the crystal orientation of the surface through which the electrons leave, the cathode must be a single crystal of a chosen orientation (usually $\langle 111 \rangle$ axial direction) to obtain the lowest work function and hence the highest emission. The expected work function of the cathode is only obtained on a clean material, that is when there are no foreign atoms of any kind on the surface. Even a single atom sitting on the surface will increase the work function and lower the emission. Since in a vacuum of 10 μPa a monolayer of atoms will form in less than 10 sec it is clear that field emission demands a very good vacuum if stable emission is to be obtained. Ideally the field emission tip is used in a vacuum of 10 nPa or better. Even in that condition, however, a few gas molecules will still land on the tip from time to time and these will cause fluctuations in the emission current. Eventually the whole tip will be covered and the output will become very unstable. The cathode must then be cleaned by rapidly heating it to a high temperature (2000°C) for a few seconds. Alternatively the tip can be kept warm (800–1000°C), in which case the majority of impinging molecules are immediately reevaporated. In this case acceptably stable emission is maintained even in a vacuum of 100 nPa or so.

The effective source or crossover size, d_0, of a field emitter is only about 10 nm as compared with 10 μm for LaB_6 and 50 μm for a tungsten hairpin. No further demagnifying lenses are therefore needed to produce an electron probe suitable for high-resolution SEM. However, the fact that the source size is so small can cause problems, because unless the gun, and the lenses following, are designed so as to minimize electron optical aberrations most of the benefits of the high brightness of the source will be lost.

Figure 2.7 shows one way of using a field emitter in a simple SEM (Crewe, 1968). A voltage V_1 of around 3 kV between the field emission tip and first anode controls the emission current while a second voltage V_0 between the tip and the second anode determines the accelerating voltage, which is up to 30 kV. The anodes act as a pair of electrostatic lenses and form a real image of the tip some distance (3–5 cm) beyond the second anode. The anode plates have a special profile which minimizes the aberrations, and this simple SEM can produce a probe of about 10 nm diameter at a beam current of 10^{-11} A.

In other applications where a wide range of probe sizes must be produced more elaborate gun designs are necessary (Troyon *et al.*, 1973). Even then the small source size and limited total current output from the

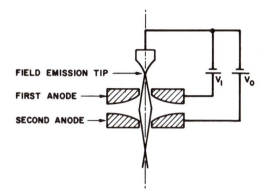

FIELD EMISSION TIP

FIRST ANODE

SECOND ANODE

V_1

V_0

Figure 2.7 Gun configuration for field emission (adapted from Crewe *et al.*, 1969).

tip restrict the useful range of probe sizes over which the field emitter is better than a thermionic source. Even with an optimized tip and gun configuration (Cleaver and Smith, 1973) more beam current is available on the specimen for probe sizes of 1 μm or more from a tungsten hairpin than from a field emitter. In addition, in most SEM instruments the requirement of a better than 100 nPa vacuum inhibits the use of the field emission gun, except in purpose-built ultrahigh-vacuum instruments.

2.2. Electron Lenses

2.2.1. General Properties of Magnetic Lenses

The condenser and objective lens systems are used to demagnify the electron image formed at crossover ($d_0 \sim 10$–50 μm) in the electron gun to the final spot size on the sample (5–200 nm). This represents a demagnification of as much as 10,000. The condenser lens system, which is composed of one or more lenses, determines the beam current which impinges on the sample. The final probe-forming lens, often called the objective lens, determines the final spot size of the electron beam. Conventional electromagnetic lenses are used and the electron beam is focused by the interaction of the electromagnetic field of the lens on the moving electrons. The vector equation which relates the force on the electron **F** to the velocity of the electron **v** and in a magnetic field of strength **H** is given by the functional equation in which

$$\mathbf{F} = -e(\mathbf{v} \times \mathbf{H}) \qquad (2.8)$$

where e is the charge on the electron and the multiplication operation is the vector cross product of **v** and **H**. In the electron optical instruments under discussion, the electron moves with a velocity **v** in a magnetic field **H** which is rotationally symmetric.

Figure 2.8 shows a schematic section of the cylindrical electromagnetic lens, commonly used as condenser lenses in these instruments. The lens is drawn so that the windings which are used to induce the magnetic field in the iron core may also be seen. The bore of the electromagnetic lens of diameter D is parallel to the direction in which the electrons are traveling. The gap located in the center of the iron core is the distance S between the north and south pole pieces of the lens. The condenser illuminating lenses in a typical 30-kV SEM are each 10 or 15 cm in height, a value which is a consequence of the low-current density, air-cooled windings. The strength of the magnetic lens, that is the intensity of the magnetic field in the gap, is proportional to NI, the number of turns, N, in the solenoid winding times the current, I, flowing through the lens. Since the velocity v of the electrons is directly proportional to the square root of the electron gun operating voltage, v_0, the effect of the lens on the electrons traveling down the column is inversely proportional to $v_0^{1/2}$.

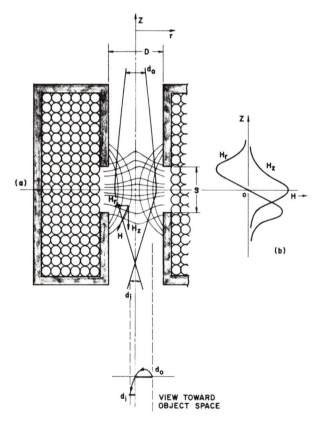

Figure 2.8. Schematic of an axially symmetric electromagnetic lens (adapted from Hall, 1953). The magnetic field lines are plotted along with the components of the magnetic field.

The focal length of a lens, f', is that point, on the z axis, where a ray initially parallel to the z axis crosses the axis after passing through the lens. Figure 2.9 shows the geometrical optics which define the focal length of a thin lens. In this illustration an image of diameter d_0 above the lens is brought to a focus and demagnified to a diameter d_i below the lens. The focal length of f' is inversely proportional to the strength of the lens. Therefore the greater the strength of the magnetic lens, the shorter will be its focal length. A beam of electrons which enters such a rotationally symmetric field will be forced to converge and may be brought to a focus. The shape and strength of the magnetic field determine the focal length and therefore the amount of demagnification. Figure 2.8 shows also that an electron beam which is diverging from an image of diameter d_0 above the lens is brought to a focus and demagnified to a diameter d_i outside the magnetic field of the lens.

A more detailed representation of the focusing action on the electromagnetic lens can also be seen in Figure 2.8. In Figure 2.8a the magnetic field lines (H) formed within the gap of the lens are depicted for a lens with axial symmetry. The magnetic field lines flow between two parallel iron pole pieces of axial symmetry. **H** is the vector which is parallel to the magnetic field, $\mathbf{H}_r$ is the vector which is perpendicular to the direction of the electron optical axis, and $\mathbf{H}_z$ is the magnetic field which is parallel to the electron optical axis. Since the magnetic field **H** varies as a function of position through the electromagnetic lens the values of $\mathbf{H}_r$ and $\mathbf{H}_z$ also vary. In Figure 2.8b a plot of the magnetic field intensity $\mathbf{H}_z$ parallel to the electron optical axis is given. The maximum magnetic field is obtained in the center of the gap. $\mathbf{H}_r$, which is the radial magnetic field, has a zero component when the magnetic field is parallel to the z axis, that is in the center of the gap of the electromagnetic lens, and has a maximum on either

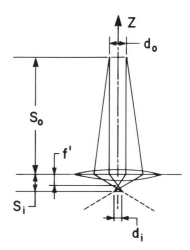

Figure 2.9. Schematic of the geometric optics for a thin lens system.

side of the center of the gap. The lines of constant magnetic field strength, the equipotential lines, which are normal to the magnetic field lines, stretch across the gap (Figure 2.8a) and show the converging field produced. Figures 2.8a and 2.8c also show the focusing action brought about by the combined magnetic fields, H_r and H_z in the electromagnetic lens. A ray initially parallel to the optical axis is made to converge towards the center of the lens (Figure 2.8a) and to rotate with respect to its original position (Figure 2.8c). The detailed motions of the electrons can be determined by applying Equation (2.8) using the two components (H_r and H_z) of the magnetic field.

The probe-forming lens is substantially different in design from the axially symmetric electromagnetic condenser lenses (Figure 2.10). In addition to the production of small electron beam spot size in the SEM–EPMA, other performance characteristics must also be considered. For the SEM, secondary electrons are emitted over a wide solid angle, have energies of only a few electron volts, and must reach the detector unimpeded in order to be measured. Therefore the magnetic field at the specimen must be low enough so as not to hinder the efficient collection of secondary electrons from the surface of the sample. The lens design must also allow a clear path for x-rays, produced in the specimen, to be collected without hitting parts of the final lens. In addition the bore of the lens should allow enough room to place the scanning coils, stigmator, and beam-limiting aperture and should be large enough to avoid obstructing those electrons which are deflected a considerable distance from the electron optical axis during scanning. Since aberrations of the lens increase rapidly with focal length, the focal length should also be as short as possible.

All these performance characteristics suggest, for the best resolution, that the specimen should be mounted immediately outside the bore of the lens. In order to keep the magnetic field at the specimen low enough to allow efficient collection of secondary electrons and minimize the focal length, the lens has a different configuration, Figure 2.10. In this design the diameter of the outer pole piece is much smaller than that of the inner pole piece. The lens is highly asymmetrical as opposed to the more conventional

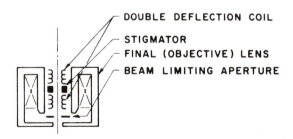

Figure 2.10. Final lens of the pinhole variety (adapted from Broers, 1973).

condenser lenses (Figure 2.8) and $\mathbf{H}_z$ reaches its maximum value quite close to the inner face of the pole piece containing the smaller bore (Oatley, 1972). The limiting aperture which controls the amount of the focused beam which enters the final lens is usually placed at this position, where $\mathbf{H}_z$ is a maximum. With the combination of the condenser lenses and the final probe-forming lens, the reduction of the crossover spot size d_0 from the electron gun to a minimum spot size focused on the sample can now be considered.

2.2.2. Production of Minimum Spot Size

Reduction of the electron beam at crossover to a focused electron probe is shown diagramatically in Figure 2.11. Ray traces of the electrons as they pass from the crossover point in the electron gun through the condenser lens system and the probe-forming lens are given for a typical scanning electron microscope column. The α angles drawn in Figure 2.11 and subsequent figures are highly exaggerated with respect to actual α

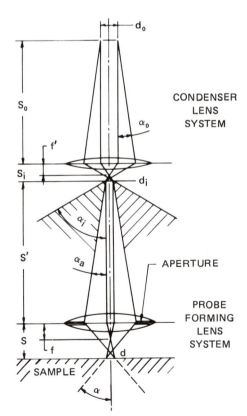

Figure 2.11. Schematic of ray traces in a typical scanning electron microscope column.

angles in the SEM. The electron image at crossover, of diameter d_0 and divergence angle α_0, passes through the condenser lens and is focused to a diameter d_i with a divergence angle α_i. In most instruments, the entrance aperture of the condenser lens system is greater than the divergence of the beam, enabling almost all of the beam to enter the lens system. The distance from the crossover point to the condenser lens gap S_0 is essentially constant (Figure 2.11). The distance at which the electrons are focused on the other side of the condenser lens is S_i and can be varied by changing the strength of the condenser lens system. As the current I in the condenser lens is increased, the magnetic field increases and the focal length of the lens, f', decreases. The demagnification (M) of such a lens is given by the equation $M = S_0/S_i$, and is greater than 1 in the SEM. The diameter, d_i, of the electron beam after passing through the condenser lens is equal to the crossover spot size d_0 divided by the demagnification M; the divergence angle α_i of the electrons from the focused image below the condenser lens at S_i is equal to α_0 times the demagnification M.

If, as is often the case, the lens thickness is negligible in comparison to S_0 and S_i, the Gaussian form of the thin lens equation from geometrical optics can be applied to the lens systems in the scanning electron microscope–electron microprobe; see Figure 2.9. This equation has the form (Hall, 1953)

$$\frac{1}{S_0} + \frac{1}{S_i} = \frac{1}{f'} \tag{2.9}$$

where f' is the focal length of the lens. We can apply this equation to the condenser lens system as shown in Figure 2.11. As discussed previously, as the current I in the condenser lens system increases, the strength of the lens increases and the focal length of the lens decreases. Therefore according to Equation (2.9), since S_0 is constant, S_i will decrease. Also as the strength of the lens increases the demagnification M will increase. However, α_i, the divergence of the electrons from the focused spot, will increase. For the two-lens system as shown in Figure 2.11, Equation (2.9) can be applied for each lens separately. The total demagnification is the product of the demagnification of each lens.

The distance from the intermediate image to the objective lens gap is S' and the final focused electron beam which impinges on the specimen is a distance S from the probe-forming lens (Figure 2.11). The distance from the bottom pole piece of the objective lens to the sample surface is called the working distance and is around 5 to 25 mm in most instruments. Such a long working distance is necessary so that low-energy secondary electrons and magnetic specimens remain outside the magnetic field of the lens. Also, in some instruments the longer working distance prevents the x-rays emitted from the sample from being absorbed by the lens pole pieces. An objective aperture (100 to 300 μm in diameter) is placed in or just above the gap in the probe-forming lens (Figures 2.10, 2.11). This aperture decreases the divergence angle of the electron beam α_i from the condenser lens system to a smaller divergence angle, α_a for the electrons entering the final lens. The electrons which enter the lens are then focused by the probe-forming lens to a resultant spot size d and a corresponding divergence angle α on the sample. The demagnification of the final lens is given by $M = S'/S$ and the final spot size on the sample is equal to the electron gun crossover d_0 divided by the products of the demagnification M of each lens in the electron optical system.

The probe-forming lens can also focus the final probe at various specimen working distances. Figure 2.12 illustrates this process. In both Figures 2.12a and 2.12b the condenser lens demagnifications S_0/S_i are the same. Also the same aperture size is used, so that α_a is the same in both cases. When the specimen is moved further below the final lens, the working distance S is increased. As shown in Figures 2.12a and 2.12b, the demagnification S'/S decreases as the working distance increases yielding

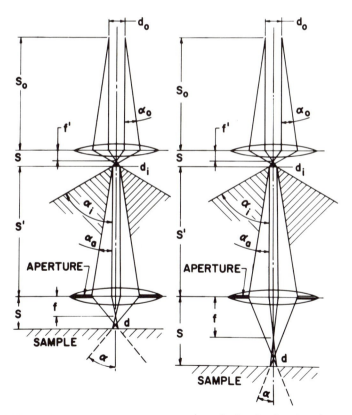

Figure 2.12. A schematic of the ray traces for two probe-forming lens focusing conditions: (a) small working distance, (b) large working distance.

a larger spot size on the specimen. However, the divergence angle α is decreased. To obtain this decrease in demagnification the final lens current is decreased, which in turn increases the focal length, f, of the lens. By measuring the final lens current, with the specimen in focus, a direct correlation between lens current and working distance can be obtained. The advantage of using a small working distance is that a small spot size is obtained which improves the resolution of the instrument. On the other hand a larger working distance yields a smaller α angle, which in turn increases the depth of field (see Chapter 4). With Figure 2.12 in mind, one can understand an alternate method for focusing whereby the working distance is set by controlling the current in the final lens and the specimen is moved vertically until it comes into focus on the viewing screen.

The condenser lens in combination with the aperture in the probe-forming lens determines the current in the final probe spot. As the demagnification of the electron image from the condenser lens increases, α_i increases (Figure 2.11). However, the amount of current which goes through the objective aperture is given by the ratio of the areas of the cones subtended by α_a and by α_i. Therefore the current which enters the final lens is given by the ratio $(\alpha_a/\alpha_i)^2$ times the current which is available in the intermediate image d_i. Considering both the condenser lens and the aperture in the probe-forming lens, as the strength of the condenser lens increases, M increases, α_i increases, and the amount of current available in the final focused spot decreases. Figure 2.13 illustrates how the focusing of the condenser lens controls the electron beam intensity in the electron probe (Birks, 1971). In one case (Figure 2.13a) the first lens is focused to allow most of the beam to pass through the aperture of the objective lens. As the cross-hatched region in Figure 2.13b shows, when the first lens is set for a shorter focal length, only a small portion of the beam passes through the second lens aperture. The intermediate image d_i obtained is larger in the first case (Figure 2.13a). Therefore if one attempts to minimize probe spot size d by minimizing d_i, one can only do this by losing current in the final electron probe. The operator must make a conscious decision as to whether minimum probe size or maximum beam current and signal is desired.

If no aberrations are inherent in the electron lens system then the minimum spot size d at the specimen can be calculated. The successive images of the crossover produced by the various reducing lenses are formed at constant beam voltage E_0 so that the brightness of the final spot is again equal to the brightness of β. Therefore the current density which is available in the final spot J_A is given by

$$J_A = \pi\beta\alpha^2 = J_c \frac{eE_0}{kT}\alpha^2 \tag{2.10}$$

Because of the infinite nature of the Gaussian distribution of the focused beam, the practical electron probe diameter d_k is defined as the diameter

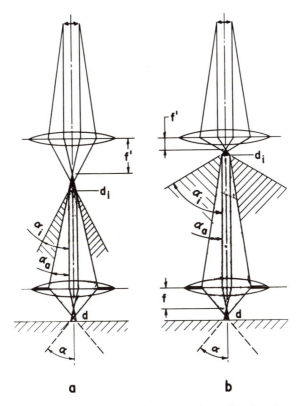

Figure 2.13. A schematic of the ray traces for two condenser lens focusing conditions: (a) weak condenser lens, (b) strong condenser lens. The lens plus the aperture in the probe-forming lens control the current in the final electron probe at the specimen (adapted from Birks, 1971).

out to a point where the current density has fallen to one-fifth of the maximum value, $J_A/5$ (Pease and Nixon, 1965). Note that the diameter at $J_A/5$ (which contains 85% of the total current) is larger than the diameter of a Gaussian probe which occurs at $J_A/1.65$ (which contains 68% of the total current). This diameter d_k is given by

$$d_k = \left(\frac{i}{B\alpha^2} \right)^{1/2} \tag{2.11}$$

where B is the factor defined as

$$B = 0.62 \frac{\pi}{4} \frac{eE_0}{kT} J_c \tag{2.12}$$

If there were no aberrations in the system it would only be necessary to increase α in order to increase the current at a constant probe diameter. However, because of the several aberrations present in the electron optical system, α must be kept small and the current available for a given probe diameter is limited.

2.2.3. Aberrations in the Electron Optical Column

2.2.3.1. Spherical Aberration

Spherical aberration arises because electrons moving in trajectories which are further away from the optical axis are focused more strongly than those near the axis. In other words the strength of the lens is greater for rays passing through the lens the larger the distance from the optical axis. This aberration is illustrated in Figure 2.14. The Z axis of the electron optical column is drawn horizontally in this figure. Electrons which diverge from point P and follow a path, for example, PA, close to the optical axis, will be focused to a point Q. Electrons that follow the path PB, which is the maximum divergence allowed by the aperture of the lens, will be focused more strongly. These rays are focused on the axis closer to the lens than point Q. As shown in Figure 2.14, these rays (path PB) are focused to a point Q' at the image plane rather than to point Q. This process causes an enlarged image δr of point P in the image plane, $\delta r = QQ'$. The minimum enlargement of point P occurs just in front of QQ' and is often referred to as the disk of least confusion. The diameter d_s of this disk can be written as

$$d_s = \tfrac{1}{2} C_s \alpha^3 \tag{2.13}$$

where α is the divergence angle at the image plane formed between BQ and the optical axis and C_s, the spherical aberration coefficient, is related to the beam voltage, E_0, and the focal length f of the lens. For most SEM instruments, the value of C_s is ~ 2 cm. The contribution of d_s to the final electron probe diameter can be made small by decreasing α. However to accomplish this, the objective aperture size must be decreased and therefore the current in the final spot will also be decreased.

2.2.3.2. Chromatic Aberration

A variation in the voltage E_0 and the corresponding velocity v of the electrons passing through the lens or a variation in the magnetic field H of the lens will change the point at which electrons emanating from a point P are focused. Figure 2.14 illustrates what may happen if the energies of the electrons in the diverging ray PB are different; ray paths of electrons of energies E_0 and $E_0 + \Delta E$ are observed. Some electrons will be focused to point Q and some to point Q' at the image plane. This process causes an enlarged image, $\delta r = QQ'$, of point P in the image plane. The diameter of the disk of least confusion which forms in front of QQ' d_c is usually written

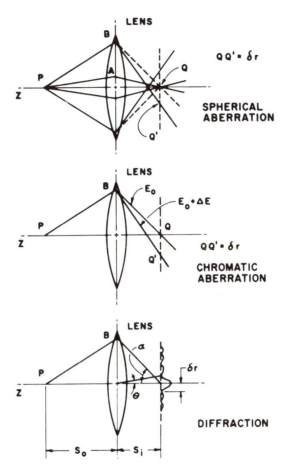

Figure 2.14. Schematic drawings showing spherical and chromatic aberration, as well as diffraction at a lens aperture. The drawings are adapted from Hall (1953) and Oatley (1972).

as

$$d_c = (\Delta E / E)C_c \alpha \qquad (2.14)$$

where α is the divergence angle at the image plane between BQ and the electron optical axis, $\Delta E / E_0$ is the fractional variation in the electron beam voltage, and C_c is the chromatic aberration coefficient. For most SEM instruments, the value of C_c is ~1 cm. The chromatic aberration coefficient is directly related to the focal length of the lens.

Variations in both E_0 and the magnetic field H (not illustrated in Figure 2.14) may occur from imperfect stabilization of the various power supplies. If the lens current or high voltage are stabilized to one part in 10^6 per minute the effect due to the variation in H and E_0 will be unimportant

(Oatley, 1972). Nevertheless, there is a variation ΔE in the energy of the electrons due to the spread of the initial energies of the electrons leaving the cathode. Typical values of this energy spread are 2 eV for the tungsten hairpin filament (Broers, 1973), 1 eV for LaB_6 depending on the bluntness of the tip of the cathode and the current drawn (Broers, 1974), and 0.2 to 0.5 eV for emitted electrons from field emission cathodes (Broers, 1973). The value of d_c can be minimized by decreasing the divergence angle α at the specimen as is the case for spherical aberration.

2.2.3.3. Diffraction

Even if the two aberrations previously discussed were insignificant, the image of a point P would still be of finite dimensions, owing to the wave nature of electrons and the aperture size of the final lens. This effect is illustrated in Figure 2.14. The intensity distribution of the point source at the image plane, caused by diffraction, is shown in this figure. The radius of the first minimum, δr, subtends an angle θ at the lens. As described by Hall (1953) the effect of diffraction also yields a disk of confusion of diameter d_d which is given by

$$d_d = 1.22\lambda / \alpha \qquad (2.15)$$

where λ is the wavelength of the electrons, given by $\lambda = 1.24(E_0)^{-1/2}$ with λ in nanometers, E_0 in electron volts, and α, the angle between the converging ray and the electron optical axis, in radians (Figure 2.14). For this aberration, the larger the value of α, the smaller will be the contribution of d_d.

2.2.3.4. Astigmatism

Although it has been tacitly assumed that magnetic lenses have perfect symmetry, this is not necessarily so. Machining errors and possible inhomogeneous magnetic fields within the iron or asymmetry in the windings, lead to loss of symmetry. If a lens system has elliptical rather than circular symmetry, for example, electrons diverging from a single point will focus to two separate line foci, at right angles to each other rather than to a point as the lens current is varied. The effect of astigmatism is to enlarge the effective size of the final electron probe diameter. For the final lens we can use a stigmator placed in the lens (Figure 2.10) to supply a weak correcting field in order to produce the desired symmetrical magnetic field. The stigmator usually has two major controls, one to correct for the magnitude of the asymmetry and one to correct for the direction of the asymmetry of the main field. Figure 2.15 shows the effect of astigmatism. When the probe-forming lens current is changed to produce first an underfocus and

Figure 2.15. Effect of astigmatism in the probe-forming lens: (a) underfocus, (b) overfocus, (c) image corrected for astigmatism. Marker = 200 nm.

then an overfocus condition, the image tends to stretch in two perpendicular directions (Figures 2.15a and 2.15b). One can usually correct for the astigmatism in the probe-forming lens by adjusting the stigmator magnitude and orientation alternately and by refocusing on the image at medium magnifications (5000 × to 10 000 ×). This cycle is repeated until the sharpest image is obtained. The stigmator can only correct for asymmetry in the magnetic field of the final lens. It cannot correct for a dirty aperture or incorrect alignment of the filament, etc.

2.2.3.5. Aberrations—Probe-Forming Lens

It is normally assumed that spherical and chromatic aberrations as well as diffraction and astigmatism are more significant in the final lens. This is justified since the images of the crossover produced by the preceding lenses have much larger diameters than the final spot size, and the effects of aberrations in these lenses are relatively small when compared to the size of the intermediate images. In most cases one can eliminate astigmatism in the probe-forming lens although the other three aberrations remain.

2.3. Electron Probe Diameter, d_p, vs. Electron Probe Current i

2.3.1. Calculation of d_{min} and i_{max}

Following Smith (1956) it is possible to determine the diameter d_p of an electron probe carrying a given current i. The current in the focused probe has approximately a Gaussian distribution. Therefore one must define the dimensions d_p of the probe. For practical purposes the probe diameter is defined as being the value within which some specified fraction, $\sim 85\%$, of the total current is contained. To obtain d_p it is usually assumed that all the significant aberrations are caused by the final lens. The aberrations considered are chromatic and spherical aberration as well as the effect of diffraction. The procedure that is followed is to regard the individual estimates of probe diameters, d_k, d_c, d_s, and d_d, as error functions and regard the effective spot size d_p equal to the square root of the sum of the squares of the separate diameters (quadrature), as described by

$$d_p = \left(d_k^2 + d_c^2 + d_s^2 + d_d^2 \right)^{1/2} \qquad (2.16)$$

From Equations (2.12)–(2.15) we obtain

$$d_p^2 = \left[\frac{i}{B} + (1.22\lambda)^2 \right] \frac{1}{\alpha^2} + \left(\frac{1}{2} C_s \right)^2 \alpha^6 + \left(\frac{\Delta E}{E} C_c \right)^2 \alpha^2 \qquad (2.17)$$

Pease and Nixon (1965) following Smith (1956) obtained the theoretical limits to probe current and probe diameter by considering only the first two terms of Equation (2.17), spherical aberration and diffraction. They differentiated d_p in Equation (2.17) with respect to the aperture angle α in order to obtain an optimum α. For the optimum α, the current in the beam will be a maximum and the beam size will be a minimum. Using this procedure values of α_{opt}, d_{min}, and i_{max} were obtained from Equation (2.17). These relations are

$$d_{min} = 1.29 C_s^{1/4} \lambda^{3/4} \left[7.92\left(\frac{iT}{J_c} \right) \times 10^9 + 1 \right]^{3/8} \qquad (2.18)$$

$$i_{max} = 1.26\left(\frac{J_c}{T} \right) \left[\frac{0.51 d^{8/3}}{C_s^{2/3} \lambda^2} - 1 \right] 10^{-10} \qquad (2.19)$$

$$\alpha_{opt} = (d/C_s)^{1/3} \qquad (2.20)$$

where i is given in amperes (A), J_c in A/cm^2, and T in degrees kelvin (K). To obtain d_{min} in nanometers, C_s and λ must be given in nanometers. The parameter outside the brackets in Equation (2.18) represents the limiting probe diameter of the microscope which occurs at zero beam current. It can be seen in Equation (2.19) that the incident beam current will vary with the 8/3 power of the probe diameter. Since secondary electron and x-ray emission vary directly with probe current, they fall off very rapidly as the probe diameter is reduced.

There are, however, a few ways in which i_{max} can be increased. As the voltage of the electron beam increases, λ decreases, the value of i_{max} increases, and d_{min} will decrease. However, to keep the x-ray emission volume small, as discussed in Chapter 3, the maximum voltage that can successfully be applied when x-ray analysis is desired is about 30 kV. The value of i_{max} can also be increased if C_s, the spherical aberration coefficient, can be reduced by decreasing the focal length of the objective lens. However, because of the need for an adequate working distance beneath the final lens, the focal length cannot be greatly reduced. Nevertheless, major changes in lens design could decrease C_s and hence increase current since i_{max} is proportional to $C_s^{-2/3}$ [Equation (2.19)]. A reduction in C_s by a factor of 10 could increase i_{max} by about a factor of 5 and decrease d_{min} by about a factor of 2. Such lenses as the "minilenses" pioneered by Mulvey (1974) are still at an early stage but appear to provide interesting possibilities for the future. At the present time, however, any large increases in i_{max} or corresponding decreases in d_{min} will come primarily from improvements in gun current density J_c as discussed previously.

Figure 2.16 illustrates the relationships between probe current and the size of the electron beam as given by Equations (2.18) and (2.19). A C_s value of 20 mm was taken from the measurements of the high-resolution instrument of Pease and Nixon (1965). Typical J_c values of 4.1 A/cm^2 for W at at 2820 K and 25 A/cm^2 for LaB$_6$ at 1900 K were chosen for the calculation. The relationship between probe current and electron beam size is given for three operating voltages, 10, 20, and 30 kV. The corresponding brightness β values according to Langmuir's equation [Equation (2.5)] are

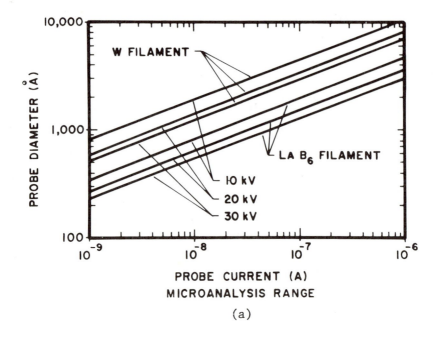

(a)

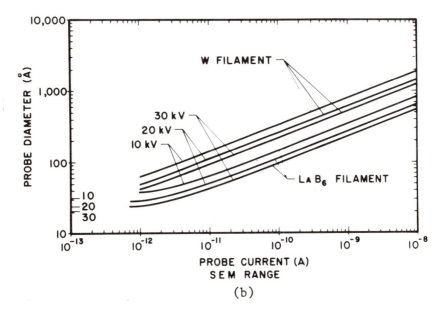

(b)

Figure 2.16. Relationship between probe current i_{max} and the size of the electron beam d_{min}. Calculations consider both the W hairpin filament and the LaB_6 gun operating at 15 and 30 kV.

5.3×10^4, 1.06×10^5, and 1.6×10^5 A/cm^2 sr for W at 10, 20, and 30 kV and 4.6×10^5, 9.1×10^5, and 1.37×10^6 A/cm^2 sr for LaB$_6$ at 10, 20, and 30 kV.

It can be observed from Figure 2.16a, for the EPMA microanalysis range, that for the particular C_s and J_c used, the maximum current available in a 1-μm electron beam using a conventional W filament is about 10^{-6} A at 10 kV and about 2×10^{-6} A at 30 kV. This amount of current is well above the minimum current (1 to 5×10^{-8} A) usually needed to perform satisfactory quantitative x-ray analyses with wavelength-dispersive spectrometers (WDS). X-ray analysis can, according to Figure 2.16a, be performed using a W filament with minimum electron beam sizes of the order of 0.2 μm (2000 Å). This spot size is well below the diameter of the region of x-ray emission from the sample ($\sim$1 μm, see Chapter 3). A small beam size of this order allows the operator the freedom to take electron scanning images of the analyzed areas as well without changing operating conditions. The LaB$_6$ gun provides additional advantages in the microanalysis range because it allows the analyst to do successful x-ray analysis with an electron beam below 0.1 μm in size. It should be pointed out that at normal SEM beam sizes of approximately 10 nm (100 Å) (Figure 2.16b), the beam current for W or LaB$_6$ filaments is below 10^{-10} A and is much too low for wavelength-dispersive x-ray analysis. This is, however, just the current range in which energy-dispersive x-ray analysis can be accomplished (see Chapter 5).

2.3.2. Measurement of Microscope Parameters (d_p, i, α)

All of the parameters characterizing the incident electron beam (i.e., the incident beam current i, the probe diameter d_p, and the convergence angle α) can be experimentally determined. While there is no need to monitor such quantities continuously, it is valuable to associate particular values of i, d_p, and α with specific operating conditions, both as a means of setting desired operating parameters and as a diagnostic device in the event of problems with the microscope.

The most straightforward quantity to measure is the incident current i, since this can be done with a "Faraday cup" which is simply a container completely closed except for a small entrance aperture (see Figure 2.17). An electron microscope aperture (3 mm diameter) with a hole 25–100 μm in diameter is convenient to use for this purpose. The container is made from a material (Ti or C) different from that used to fabricate the microscope stage. In this way any x-rays produced from the Faraday cup can be easily detected. The Faraday cup does not allow the backscattered and secondary electrons generated by the incident beam to escape. The current flowing to ground is therefore exactly equal to the incident beam current i, and it can

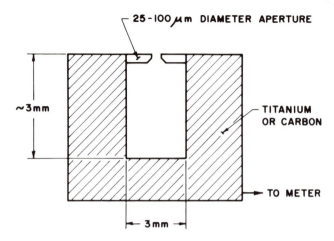

Figure 2.17. Schematic drawing of a cross section through a Faraday cup suitable for measuring incident current i.

conveniently be measured with a dc picoammeter or a calibrated specimen current amplifier. For cases where the highest accuracy is not required, a flat carbon block can be substituted for the Faraday cup. In this case the measured specimen current i_{sc} and incident beam current i are related as $i = i_{sc}/[1 - (\eta + \delta)]$ where η and δ are the backscatter and secondary electron yields (Chapter 3), respectively. For a carbon sample normal to the beam both η and δ are small so the error is only of the order of 10%.

The probe diameter, as previously defined, is measured by sweeping the beam across a sharp, electron-opaque edge and observing the change in signal as a function of the beam position. The profile has the form shown in Figure 2.18. Typically the diameter is taken as the distance between the 10% and 90% signal levels, although if the signal is very noisy the 25% and

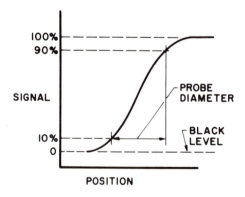

Figure 2.18. Schematic drawing of the electron signal emitted when a Gaussian electron beam is swept across a sharp edge of a specimen.

75% points may be easier to find. Suitable sharp edges are clean razor blades, cleavage edges in materials such as silicon, and fine-drawn wires. The edge must be clean, smooth, and nontransmitting to electrons. While these conditions are easy to satisfy for large probes (~ 1 μm diameter), it is very difficult to find, or manufacture, suitable edges for smaller probe diameters. Ideally the portion of the edge scanned should be over the entrance to a Faraday cup so that none of the scatter of the incident beam modifies the profile.

The probe convergence angle (total angle 2α) can be measured by using the same technique as described above for the probe diameter. The beam is focused on the edge and the probe diameter, D_1, is measured. Without changing the focus the test edge is then moved vertically a distance L using the Z control of the SEM stage, and the defocused diameter D_2 is measured. Then

$$\alpha = \frac{D_2 - D_1}{2L} \tag{2.21}$$

In most cases the answer is close to that predicted by using the diameter, D_A, of the aperture in the final lens and the distance S from this aperture to the specimen (see Figure 2.11), i.e.,

$$\alpha = D_A/2S \tag{2.22}$$

In most cases the working distance can be substituted for S. A more detailed discussion of these techniques can be found in Joy (1974).

2.3.3. High-Resolution Scanning Electron Microscopy

In the limit, when $0.51 d^{8/3}/C_s^{2/3}\lambda^2$ [Equation (2.19)] equals 1, $i_{\max} = 0$ and there is no current in the electron probe. Therefore one can calculate the minimum probe size, $d_{\min}$ [Equation (2.18)], as

$$d_{\min} = 1.29 C_s^{1/4}\lambda^{3/4} \tag{2.23}$$

This equation is similar to the formula for the limit of the resolving power of a conventional transmission electron microscope. At 10 kV and 30 kV the ultimate resolution is 3.2 and 2.1 nm, respectively, if $C_s = 2$ cm (Figure 2.16b). If the specimen is placed within the objective lens, as in the scanning transmission microscope (STEM), C_s is reduced from ~ 20 to ~ 3 mm (Joy and Maher, 1976) because of the shorter focal length, and the value of $d_{\min}$ will approach that of the standard transmission instrument. Since it is necessary to have the probe current i appreciably greater than zero to obtain useful secondary electron or x-ray signals, the value of $d_{\min}$ is greater than the theoretical resolution.

The minimum probe current generally accepted as being sufficient to form a satisfactory scanning picture using secondary electrons is about

10^{-12} A (see Chapter 4). Using Figure 2.16b it appears that a 5-nm electron beam can be attained using a tungsten filament at 30 kV and a 3-nm electron beam can be attained using a LaB_6 filament at 30 kV.

It is of interest to calculate the importance of the various aberrations d_c, d_s, and d_d on the final spot size of the electron probe. As an example, one can consider the various probe diameters, d, d_s, d_d, that, according to Equation (2.16), give a value of d_{min} of 5 nm. For an SEM operating at 30 kV, with a tungsten filament ($J_c = 4.1$ A/cm^2), a spherical aberration coefficient of 20 mm, and neglecting the effect of chromatic aberrations, $i_{max} = 1.64 \times 10^{-12}$ A and $\alpha_{opt} = 0.63 \times 10^{-2}$ rad according to Equations (2.19) and (2.20). From Equations (2.12), (2.13) and (2.15), the various contributions to the final 5-nm diameter are $d = 4.2$ nm, $d_s = 2.5$ nm, and $d_d = 1.4$ nm.

These calculations assume that chromatic aberration does not affect the final beam size ($d_c \simeq 0$). For high-resolution microscopy at low voltages and when tungsten hairpin filaments are used, the effect of chromatic aberration is not trivial. The effect of chromatic aberration can be calculated using Equation (2.14), $d_c = (\Delta E / E) C_c \alpha$, with a value of C_c, 0.8 cm (Pease and Nixon, 1965). For a thermionic cathode, the value of ΔE is 2–3 eV, as discussed previously. Using these values for the 5-nm beam at 30 keV and for $\alpha = 0.63 \times 10^{-2}$ rad previously discussed, the value of d_c is found to be about 4 nm. This is a significant contribution and according to Equation (2.16) will have the effect of increasing d_p from 5 nm to about 6.5 nm.

The effect of chromatic aberration for a tungsten hairpin filament is more important at lower voltages. For example, for a 5-nm beam, the value of d_c is 8 nm at 15 kV, using the same calculation scheme as discussed previously. The effect of chromatic aberration leads to an enlarged probe of about 9.5 nm. Since the energy spread ΔE of the emitted electrons in the LaB_6 gun is about the same as the tungsten filament (Broers, 1974), significant reductions in the effect of chromatic aberration cannot be expected. The brightness of the LaB_6 gun is, however, significantly higher, and smaller values of d_{min} are expected. Nevertheless, the effect of chromatic aberration is quite important in calculating the ultimate electron beam resolution using this electron gun. It is interesting to note that the energy spread in the field emission gun is 0.2–0.5 eV (Broers, 1973), much lower than the thermionic guns previously discussed.

In any attempt to achieve small probe sizes of $\leqslant 10$ nm in electron optical instruments, not only must the electron optics be designed to minimize C_s and C_c and maximize J_c, but the instrument must also be correctly aligned and vibration, ac stray magnetic field interference, and specimen contamination must be reduced to a minimum. Since filaments will warp with time and move from the alignment position, the filament must be occasionally recentered during operation. The final aperture, which defines the final value of α and the current, also requires constant care. It collects much of the beam current and can become easily contaminated. The apertures must be cleaned frequently and carefully positioned when replaced in the instrument. Stray ac magnetic fields from nearby apparatus and power supplies are troublesome to the operation of the instrument at high magnification ranges. These fields have frequencies of between 50 and 200 Hz and must be reduced in magnitude to a value on the order of 5–10 mG in the vicinity of the electron column. Attempts to minimize these effects have been described by Broers (1969). To reduce the effects of contamination as much as possible, an ion pump is used for high-vacuum pumping and the exposure of the system to oil is kept to a minimum. The specimen stage which is used at very high resolution is made in such a way that there is no mechanical contact between the specimen and the base of the chamber during analysis, to reduce vibration below a detectable level. Low-frequency mechanical vibrations (2–10 Hz) can cause the whole instrument to vibrate. The instrument must be isolated from the effect of these vibrations or high resolution will not be achieved. All of these effects can be eliminated by thorough careful engineering and therefore are not of great importance when considering ultimate resolution. However, they are of importance in the practical operation of the instrument.

3

Electron-Beam–Specimen Interactions

3.1. Introduction

The versatility of the scanning electron microscope for the study of solids is derived in large measure from the rich variety of interactions which the beam electrons undergo within the specimen. The interactions can be generally divided into two classes: (1) elastic events, which affect the trajectories of the beam electrons within the specimen without significantly altering the energy, and (2) inelastic events, which result in a transfer of energy to the solid, leading to the generation of secondary electrons, Auger electrons, characteristic and continuum x-rays, long-wavelength electromagnetic radiation in the visible, ultraviolet, and infrared regions, electron–hole pairs, lattice vibrations (phonons), and electron oscillations (plasmons). In principle, all of these interactions can be used to derive information about the nature of the specimen—shape, composition, crystal structure, electronic structure, internal electric or magnetic fields, etc. To obtain this information from the signals measured and the images recorded with the SEM, the microscopist needs a working knowledge of electron–specimen interactions, broadly qualitative and where possible, quantitative. This chapter does not provide an in-depth treatment of electron physics; rather, it attempts to provide an overview necessary for the analysis of SEM images and compositionally related signals. Additional detail will be given in later chapters devoted to qualitative and quantitative x-ray microanalysis.

3.2. Scattering

The electron–optical column which precedes the specimen has as its function the definition of the electron beam and control of the beam parameters: diameter, current, and divergence. Typical beams consist of electrons following paths which are nearly parallel, with a divergence of 10^{-2} rad (0.5°) or less and focused to a small diameter, typically from 5 nm to 1 μm. Since an SEM image is constructed from information derived at a matrix of beam locations, small beam diameters are obviously the first requirement for the production of scanning images with a high spatial resolution (see Chapter 2). Ideally, the diameter of the area sampled by the beam on the specimen should be equal to the beam diameter. This is not generally the case, however, because of the phenomenon of electron scattering. Scattering as a general term simply means an interaction between the beam electron and the specimen atoms and electrons which results in a change in the electron trajectory and/or energy. In discussing scattering, a key concept is that of the cross section or probability of an event. The cross section, denoted Q or σ, is defined in general as (Considine, 1976)

$$Q = N/n_t n_i \quad \text{cm}^2 \tag{3.1}$$

where N is the number of events per unit volume (cm^{-3}), n_t is the number of target sites per unit volume (cm^{-3}), and n_i is the number of incident particles per unit area (cm^{-2}). The cross section thus has dimensions of event/incident particle/(target particle/cm^2) or cm^2 and can be thought of as the effective size of an atom for a given interaction.

From the cross section for a process, the mean free path, or the average distance the electron travels between particular events, can be calculated. The mean free path, λ, is given by

$$\lambda = A/N_0 \rho Q \quad \text{cm} \tag{3.2}$$

where A is the atomic weight (g/mol), N_0 is Avogadro's number (6.02×10^{23} atom/mol), ρ is the density (g/cm^3), and Q is the cross section. To determine the mean free path for a particular event, λ_j, the cross section for that type of event, Q_j, is substituted in Equation (3.2). The mean free path for all possible events, λ_{tot}, is obtained by considering all possible scattering processes, and calculating the total mean free path according to the equation

$$1/\lambda_{tot} = 1/\lambda_a + 1/\lambda_b + 1/\lambda_c + \cdots \quad \text{cm}^{-1} \tag{3.3}$$

Note that the total mean free path must always be less than the smallest value among the mean free paths of the various possible processes.

3.2.1. Elastic Scattering

Electron scattering is divided into two categories, elastic and inelastic scattering, which are illustrated in Figure 3.1. When elastic scattering occurs, the direction component of the electron's velocity $\bar{v}$ is changed, but the magnitude $|\bar{v}|$ remains virtually constant, so that the kinetic energy,

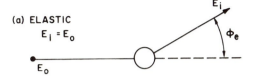

(a) ELASTIC

$E_i = E_0$

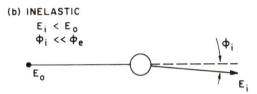

Figure 3.1. Schematic illustration of scattering processes which occur when an energetic electron of energy E_0 interacts with an atom. (a) Elastic scattering, instantaneous energy E_i after collision equals E_0; (b) Inelastic scattering, instantaneous energy E_i less than E_0.

(b) INELASTIC

$E_i < E_0$

$\phi_i \ll \phi_e$

$E = \frac{1}{2} m_e v^2$, where m_e is the electron mass, is unchanged. Less than 1 eV of energy is transferred from the beam electron to the specimen; this is negligible compared to the incident energy, which is typically 10 keV or more. The electron is deviated from its incident path by an angle ϕ_e, where the subscript e indicates "elastic." ϕ_e can range from 0° up to 180°, with a typical value of the order of 5°. Elastic scattering results from collisions of the energetic electrons with the nuclei of the atoms, partially screened by the bound electrons. The cross section for elastic scattering is described by the Rutherford model (Evans, 1955):

$$Q(> \phi_0) = 1.62 \times 10^{-20} \frac{Z^2}{E^2} \cot^2 \frac{\phi_0}{2} \quad \frac{\text{events}}{e^- (\text{atom}/\text{cm}^2)} \qquad (3.4)$$

where $Q(> \phi_0)$ is the probability of a scattering event exceeding a specified angle ϕ_0, Z is the atomic number of the scattering atom, and E is the electron energy (keV). As ϕ_0 approaches zero, the cross section increases toward infinity, as shown in Figure 3.2. Inspection of Equation (3.4) reveals a strong dependence on atomic number and beam energy, shown in Figures 3.2a and 3.2b, with the cross section increasing as the square of the atomic number and decreasing with the inverse square of beam energy. Considering elastic scattering events which result in scattering angles greater than 2°, the values of the mean free path between scattering events can be calculated from Equations (3.2) and (3.4), Table 3.1. Since the mean free path and the cross section are inversely related, it can be seen that the mean free path increases with decreasing atomic number and increasing electron energy. In transversing a given thickness of various materials, elastic scattering is more probable in high atomic number materials and at low beam energy.

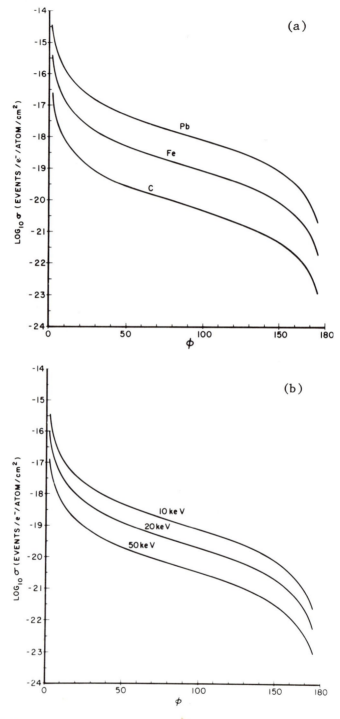

Figure 3.2. Plot of elastic scattering cross section [Equation (3.4)] (a) as a function of atomic number for $E_0 = 10$ keV, (b) as a function of beam energy for an iron target.

Table 3.1. Elastic Mean Free Path, in Nanometers
(Scattering Events Greater than 2°)

Element	10 keV	20 keV	30 keV	40 keV	50 keV
C	5.5	22	49	89	140
Al	1.8	7.4	17	29	46
Fe	0.3	1.3	2.9	5.2	8.2
Ag	0.15	0.6	1.3	2.3	3.6
Pb	0.08	0.34	0.76	1.4	2.1
U	0.05	0.19	0.42	0.75	1.2

3.2.2. Inelastic Scattering

The second general category of scattering is that of inelastic scattering. During an inelastic scattering event, energy is transferred to the target atoms and electrons, and the kinetic energy of the beam electron decreases. There are a number of possible inelastic scattering processes; we shall consider only the principal processes of interest in scanning electron microscopy and microanalysis (for a basic reference to the following terms, see Kittel, 1956). A brief description of these processes will be given here.

(a) Plasmon Excitation. The beam electron can excite waves in the "free electron gas" which exists between the ionic cores in a solid. This is a highly probable inelastic scattering process. In a metal such as aluminum, the excitation of a plasmon involves the transfer of about 15 eV to the solid.

(b) Excitation of Conduction Electrons Leading to Secondary Electron (Low-Energy) Emission. The interaction of the beam electron with the solid can lead to the ejection of loosely bound electrons of the conduction band. These ejected electrons are referred to as secondary electrons, and the majority receive an initial kinetic energy of 0–50 eV.

(c) Ionization of Inner Shells. A sufficiently energetic electron can interact with an atom and cause the ejection of a tightly bound inner-shell electron, leaving the atom in an ionized and highly energetic state. Subsequent decay of this excited state results in the emission of characteristic x-rays and Auger electrons.

(d) Bremsstrahlung or Continuum x-Rays. An energetic beam electron can undergo deceleration in the Coulombic field of an atom. The energy lost from the beam electron in this deceleration is converted into an x-ray photon, known as a bremsstrahlung ("braking radiation") x-ray. Since the energy loss in this deceleration process can take on any value, the bremsstrahlung x-rays form a continuous spectrum from zero energy up to the beam energy. Because the formation of such continuum x-rays is dependent on the direction of flight of the beam electron, the angular distribution of intensity of the continuum is anisotropic.

(e) Excitation of Phonons. A substantial portion of the energy deposited in the sample by the beam electron is transferred to the solid by the

excitation of lattice oscillations (phonons), i.e., heat. In the case of an electron beam incident on a bulk target, the region in which the electron deposits energy is in good thermal contact with the bulk of the sample which acts as an effective heat sink. Thus a significant temperature rise in the bombarded region is prevented. A temperature rise of 10°C or less is typically observed in bulk specimens for beam currents of the order of 1 nA. In thin specimens or at high beam currents (1 μA), significant heating may occur.

The cross section for several of these processes has been calculated by Shimizu *et al.* (1976) for an aluminum target as a function of energy, Figure 3.3. All of the cross sections are observed to decrease with increasing energy. Considering a range of atomic number, inelastic scattering is favored at low atomic numbers and elastic scattering at high atomic numbers.

Inelastic scattering occurs by a variety of discrete processes, with a variable amount of energy transferred to the solid depending on the strength of each interaction. Cross sections for the individual processes are difficult to obtain for all targets of interest. It is useful in many calculations to consider all inelastic processes grouped together to give a "continuous energy loss." Bethe (1933) derived a continuous energy loss relation which considers all energy loss processes. The energy loss per unit of distance traveled in the solid, dE/dx, is given by

$$\frac{dE}{dx} = -2\pi e^4 N_0 \frac{Z\rho}{AE_m} \ln(1.166E_m/J)$$

$$= -7.85 \times 10^4 (Z\rho/AE_m)\ln(1.666E_m/J) \quad \text{keV/cm} \quad (3.5)$$

where e is the electronic charge, N_0 is Avogadro's number, Z is the atomic number, A is the atomic weight (g/mol), ρ is the density (g/cm^3), E_m is the

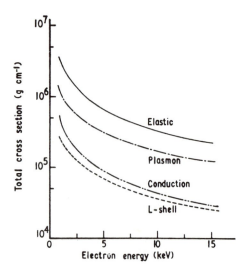

Figure 3.3. Plot of total cross section (dimensions $\sigma \cdot \rho$) for elastic scattering and several inelastic scattering processes (plasmon and conduction electron excitation and inner-shell ionization) for aluminum (Shimizu *et al.* 1976).

mean electron energy (keV) along the path, and J is the mean ionization potential (keV). The mean ionization potential is the average energy loss per interaction considering all possible energy-loss processes and has been given as (Berger and Seltzer, 1964)

$$J = (9.76Z + 58.5Z^{-0.19})10^{-3} \quad \text{keV} \tag{3.6}$$

The Bethe equation provides a convenient relation for determining the amount of energy lost by a beam electron while it travels in the specimen. Note that x is the distance along the trajectory, which, because of elastic scattering, deviates from a straight line. Thus, except for films with a thickness less than the mean free path for elastic scattering, a correction for the added path length due to elastic scattering must be made to calculate energy loss in thick films or bulk targets.

A concept related to the continuous energy loss approximation is that of the "stopping power," S, which is defined as

$$S = -\frac{1}{\rho}\frac{dE}{dx} \tag{3.7}$$

The behavior of the stopping power as a function of atomic number can be deduced from the following arguments. From Equation (3.6), the mean ionization potential J increases with increasing atomic number (Duncumb and Reed, 1968). In the definition of the stopping power, the density dependence of the Bethe equation is divided out, leaving the stopping power at a given energy proportional to $(Z/A)\ln[c/f(Z)]$, where c is a constant. Since the terms (Z/A) and $\ln[c/f(Z)]$ both decrease with increasing atomic number, the stopping power S decreases, being about 50% greater in aluminum than in gold at 20 keV (Cosslett and Thomas, 1966).

The processes of elastic and inelastic scattering operate concurrently. Elastic scattering causes the beam electrons to deviate from their original direction of travel, causing them to "diffuse" through the solid. Inelastic scattering progressively reduces the energy of the beam electron until it is captured by the solid, thus limiting the range of travel of the electron within the solid. The region over which the beam electrons interact with the solid, depositing energy and producing those forms of secondary radiation which we measure, is known as the interaction volume. An understanding of the size and shape of the interaction volume as a function of specimen and beam parameters is vital to proper interpretation of SEM images and microanalysis.

3.3. Interaction Volume

3.3.1. Experimental Evidence

The interaction of electrons in solids can be directly or indirectly visualized in a few special cases. Certain plastics, such as poly-

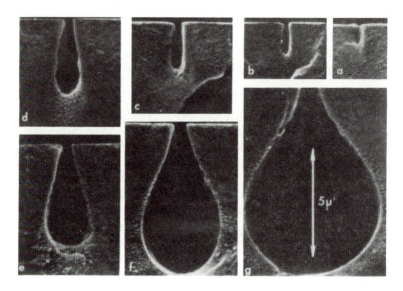

Figure 3.4. Direct visualization of the electron interaction volume in polymethylmethacrylate. In (a) through (g), the electron dose is the same, but the etching time is increased to reveal the decreasing energy deposition contours (Everhart *et al.* 1972).

methylmethacrylate (PMMA), undergo a chemical change during electron bombardment which renders the material sensitive to etching in a suitable solvent (Everhart *et al.*, 1972). Figure 3.4 shows the result of an experiment in which the interaction volume is indirectly revealed by such etching. The etching rate is controlled by the electron dose, e^-/cm^3, and hence etching for increasing time periods reveals contours of electron energy deposition. The etched structures allow us to visualize electron penetration in a low-atomic-number matrix. The interaction volume is found to have dimensions of several micrometers with the depth substantially greater than the width and to have a distinctive pear shape. The origin of this shape can be understood in terms of the characteristics of elastic and inelastic scattering. For a low-atomic-number matrix, such as a plastic, inelastic scattering is more probable, and hence the electrons tend to penetrate into the solid with relatively little lateral scattering initially in the "neck" region of the pear-shaped volume. The penetrating electrons lose energy, and as indicated by Equation (3.4), elastic scattering becomes more probable at lower energy. As a result of elastic scattering, the electrons deviate from their initial direction of travel and the lateral scattering contributes to the formation of the "bulbous" region of the pear-shaped interaction volume.

3.3.2. Monte Carlo Calculations

Electron bombardment of PMMA with subsequent etching provides experimental evidence on the interaction volume in a target of an average atomic number approximately equal to 6. In order to study the interaction volume in any target of interest, such as pure metals of high atomic number, the technique of Monte Carlo simulation of electron trajectories is especially useful (Berger, 1963; Shimizu and Murata, 1971; Heinrich *et al.*, 1976). In the Monte Carlo simulation technique, the detailed history of an electron trajectory is calculated in a stepwise manner. The length of the steps, which is a measure of the amount of detail in the simulation, is usually set equal to the mean free path between scattering acts, or a multiple thereof, as calculated from Equations (3.1)–(3.4). At each scattering step, a scattering angle is chosen which is appropriate to the type of event, elastic or inelastic. The choice of the type of scattering event and the value of the scattering angle are determined through the use of random numbers (hence, the name "Monte Carlo") which distribute the choices over the allowed range of values so as to produce a distribution of scattering events similar to the behavior of a real electron. Inelastic scattering can be treated as a discrete process (Shimizu *et al.*, 1976), or, more commonly, the Bethe relation, Equation (3.5), is employed to decrease the energy along the path. The electron trajectory is followed until the energy has been decreased through inelastic scattering to the energy of the electrons of the solid or to an arbitrary cutoff energy, usually chosen as an energy at which some process of interest can no longer be activated. An example of individual trajectories calculated by the Monte Carlo simulation is shown in Figure 3.5a. A single trajectory, while it can be accurately calculated, is not representative of the complete electron–solid interaction, and therefore a large number, typically 1000 to 10,000, of trajectories must be calculated to achieve statistical significance. When a large number of trajectories is plotted, the shape of the region of beam interaction in the solid, the interaction volume, can be visualized, Figure 3.5b. In the following sections, Monte Carlo calculations will be used extensively to illustrate the characteristics of the interaction volume as a function of beam and specimen parameters: beam energy, atomic number of the specimen, specimen thickness, and specimen tilt. Throughout this discussion, it must be emphasized that the numerical values listed to describe the interaction volume are only approximate. Examination of the plot of electron trajectories, Figure 3.5b, reveals that the boundaries of the interaction volume are not sharply defined. A change in the density of electron trajectories is observed, gradually approaching zero as the "limit" of the interaction volume is reached. Therefore, a single dimension used to describe the interaction volume can only be an approximation. Further, as shown by the experiment with PMMA, the deposition of energy within the interaction volume is not a constant. Contours of constant energy deposition per unit volume, as determined experimentally and calculated by Monte Carlo simulation for a low-atomic-number matrix, are shown in Figure 3.6 (Shimizu *et al.*, 1975). These contours reveal that the interaction volume has a dense core near the beam impact point, with the amount of deposited

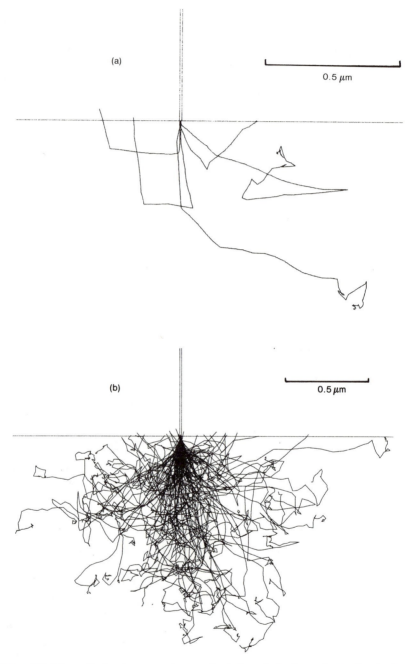

Figure 3.5. Monte Carlo electron trajectory simulation of the beam interaction in iron, $E_0 = 20$ keV. (a) Plot of five trajectories, showing random variations. (b) Plot of 100 trajectories, giving a visual impression of the interaction volume. Tilt = 0°

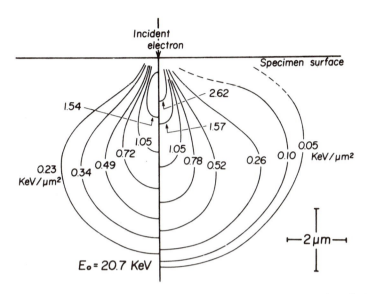

Figure 3.6. Energy deposition in a low-atomic-number solid (polymethylmethacrylate) as a function of position as calculated by Monte Carlo simulation and as measured experimentally by the etching technique (Everhart *et al.*, 1972).

energy decreasing approximately radially. A Monte Carlo plot of electron trajectories for a carbon target reveals the same pear-shaped region where a high density of trajectories is found. Note, however, that electron trajectories are also found near the surface outside the high-density region, but the density falls sharply away from the immediate vicinity of the beam.

3.3.2.1. Influence of Atomic Number

Monte Carlo calculations for targets of carbon ($Z = 6$), iron ($Z = 26$), silver ($Z = 47$), and uranium ($Z = 92$), reveal that the linear dimensions of the interaction volume decrease with increasing atomic number at a fixed beam energy, as shown in Figure 3.7. This is a direct consequence of the increase in the cross section for elastic scattering, since from Equation (3.4) $Q \propto Z^2$. In targets of high atomic number, the electrons undergo more elastic scattering per unit distance, and the mean scattering angle is greater, as compared to low-atomic-number targets. The electron trajectories in high-atomic-number materials thus tend to deviate out of the initial direction of travel and reduce the penetration into the solid. In low-atomic-number materials, the trajectories deviate less from the initial path into the solid, allowing for deeper penetration. The shape of the interaction volume

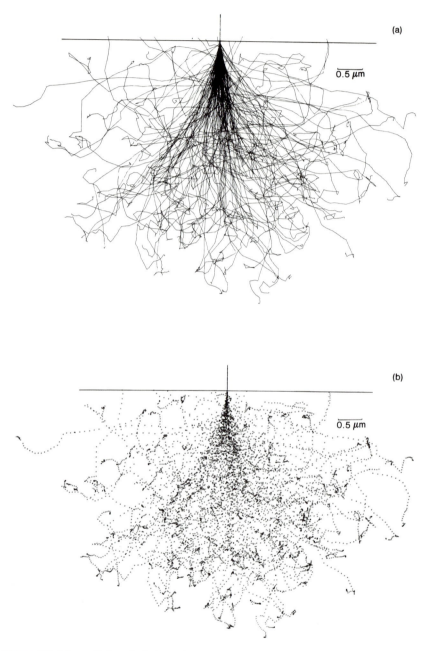

Figure 3.7. Monte Carlo calculations of the interaction volume in solids, $E_0 = 20$ keV. Sites of inner-shell ionization: (a), (b) carbon, K shell; (c), (d) iron, K shell; (e), (f) silver, L shell; (g), (h) uranium, M shell.

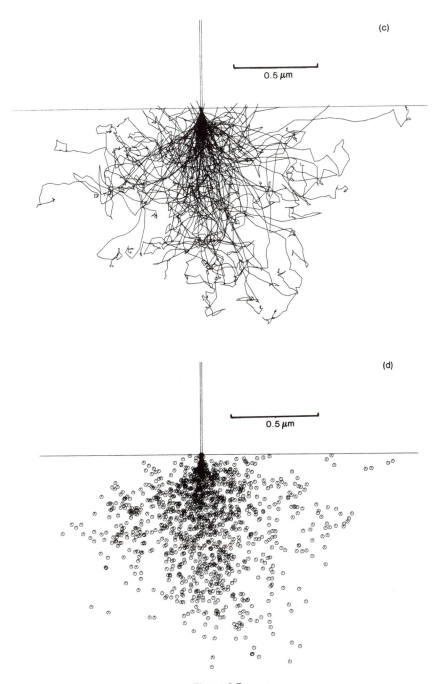

Figure 3.7. *cont.*

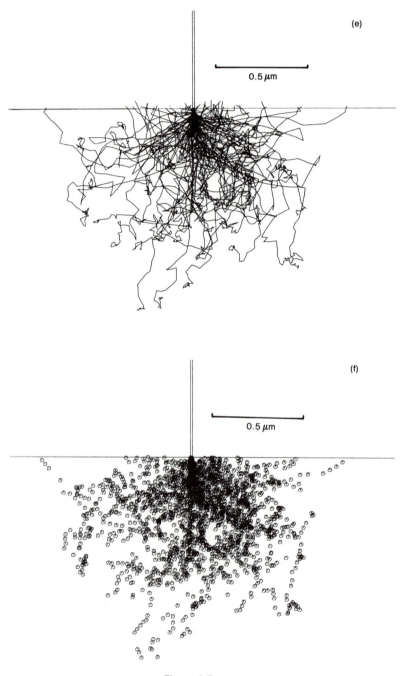

(e)

0.5 μm

(f)

0.5 μm

Figure 3.7. *cont.*

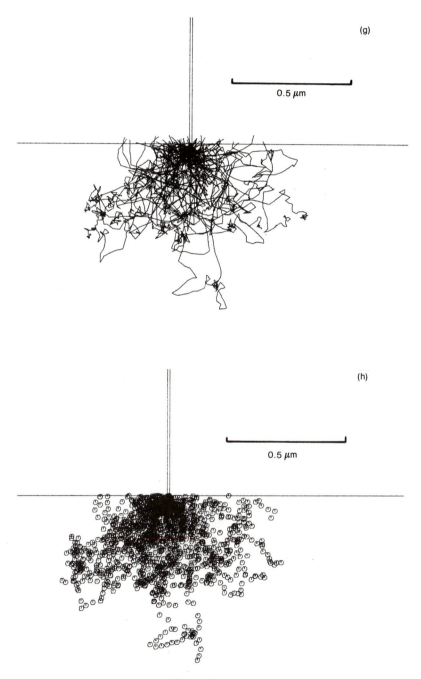

(g)

0.5 μm

(h)

0.5 μm

Figure 3.7. *cont.*

also changes significantly as a function of atomic number. The dense region of trajectories in Figure 3.7 changes from the pear shape of low atomic numbers, e.g., carbon, to a nearly spherical shape sectioned by the surface plane for high-atomic-number targets.

3.3.2.2. Effect of Beam Energy

The size of the interaction is a strong function of the energy with which the beam electrons interact with the sample. The interaction volume in iron is shown as a function of beam energy in Figure 3.8. The increase in size with beam energy can be understood from an examination of Equations (3.4) and (3.5). The cross section for elastic scattering has an inverse dependence on the square of the energy, $Q \propto 1/E^2$. Thus, as the energy increases, the electron trajectories near the surface become straighter and the electrons penetrate more deeply into the solid before the effects of multiple scattering cause some of the electrons to propagate back toward the surface. The rate of energy loss with distance traveled, as given by the Bethe relation, is inversely related to the energy, $dE/dx \propto 1/E$. At higher energy, the electrons can penetrate to greater depths since they retain a larger fraction of their initial energy after a given length of travel. Note that the shape of the interaction volume does not change form significantly with a change in beam energy. The lateral and depth dimensions scale in a similar manner with energy.

3.3.2.3. Effect of Tilt Angle

As the angle of tilt of a specimen is increased (i.e., the angle of beam incidence measured above the surface is decreased), the interaction volume becomes smaller, as shown in the Monte Carlo plots of Figure 3.9. This behavior can be rationalized in terms of the tendency of the electrons to undergo forward scattering in any individual scattering event. That is, the average angle of deviation from the direction of flight is small, of the order of 5° for a single event. At normal incidence, or 0° tilt, the tendency for forward scattering causes most of the electrons to propagate down into the specimen (Figure 3.10). In a tilted specimen, the tendency for forward scattering causes the electrons to propagate nearer to the surface, and the interaction volume thus has a reduced depth dimension. Note that the dimension parallel to the surface and perpendicular to the tilt axis increases as compared to the lateral dimensions at normal incidence, a consequence of forward scattering. The dimension parallel to the tilt axis is similar to that at normal incidence.

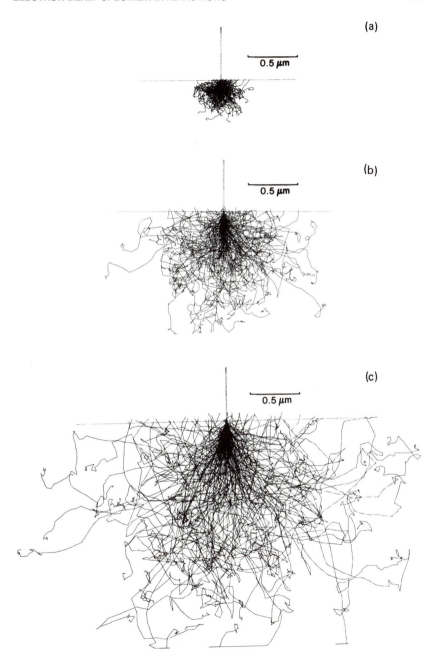

Figure 3.8. Monte Carlo calculations of the interaction volume in iron as a function of beam energy: (a) 10 keV, (b) 20 keV, (c) 30 keV.

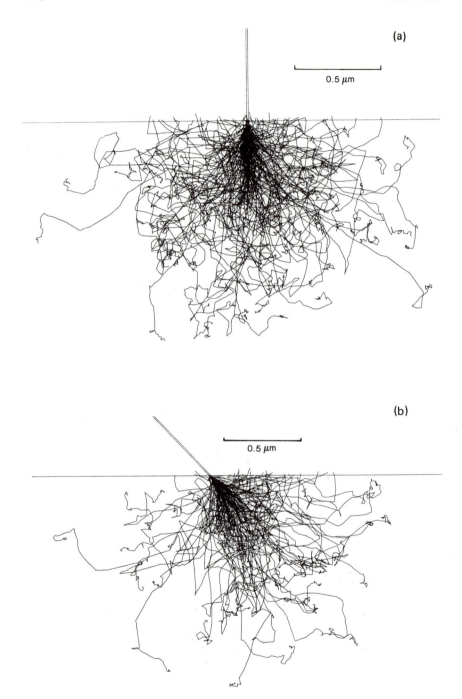

(a)

0.5 μm

(b)

0.5 μm

Figure 3.9. Monte Carlo calculations of the interaction volume in iron ($E_0 = 20$ keV) as a function of tilt: (a) 0° tilt; (b) 45°; (c) 60°.

(c)

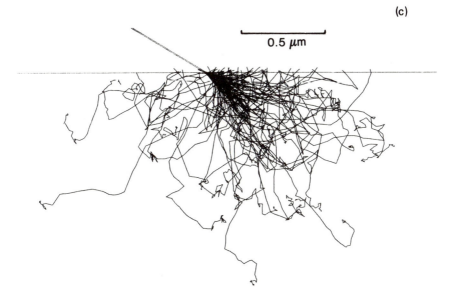

0.5 μm

Figure 3.9. *cont.*

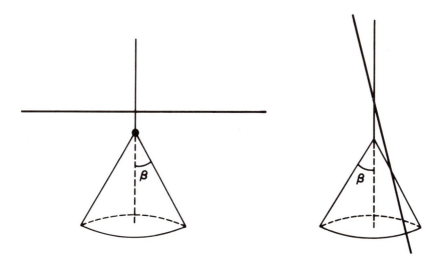

Figure 3.10. Schematic illustration of the origin of increased backscattering from tilted specimens. For a given average scattering angle β, at normal incidence, the electron tends to continue propagating into the solid. When the specimen is tilted, electrons can escape directly.

3.3.2.4. Measures of the Interaction Volume—The Electron Range

It is convenient to have a measure of the distance traveled by the electron in the solid, the so-called "electron range." A number of different definitions of the electron range exist in the literature; only those of principal interest for scanning electron microscopy and x-ray microanalysis will be considered here.

3.3.2.4.1. Bethe Range. If a suitable relation is available for the rate of energy loss with distance traveled, dE/dx, then a rigorous definition of the total distance traveled by an "average" electron is given by

$$R = \int_{E=E_0}^{E=0} \frac{1}{dE/dx}\, dE \tag{3.8}$$

By using the Bethe relation, Equation (3.5) for dE/dx, the so-called "Bethe range" is obtained. In an alternative expression, the density of the target is included to obtain the Bethe mass range, ρR_B, expressed in units of g/cm^2:

$$\rho R_B = \int_{E=E_0}^{E=0} \frac{1}{(dE/dx)(1/\rho)}\, dE = \int_{E=E_0}^{E=0} \frac{dE}{S} \tag{3.9}$$

where E_0 is the initial beam energy and S is the stopping power, Equation (3.7). The integrals in Equations (3.8) and (3.9) contain the expression $\int [E/\log(kE)]\, dE$ which must be integrated numerically. A convenient approximation to the integrated form of the equation is given by Henoc and Maurice (1976). Values of the Bethe range calculated from this equation are listed for several elements and beam energies in Table 3.2. The Bethe range expressed in units of micrometers is found to increase with increasing beam energy and decreasing atomic number. Conversely, the Bethe mass range, expressed in g/cm^2, is found to increase with increasing atomic number, because of the general increase in density with atomic number. The Bethe range is the average distance traveled by an electron along its trajectory. Because elastic scattering causes the direction of travel to change repeatedly along the path, the true trajectories deviate severely from a straight line drawn normal to the surface at the point of electron impact on the solid. The Bethe range is thus greater than the maximum dimension of the interaction volume measured from the surface.

3.3.2.4.2. Kanaya–Okayama Range. A number of workers have considered the effects of elastic and inelastic scattering (in the form of an energy-loss law) to derive an electron range which more closely approximates the depth dimension of the interaction volume. Kanaya and Okayama (1972) have derived such an expression for the maximum electron range:

$$R_{KO} = 0.0276 A E_0^{1.67} / (Z^{0.889}\rho) \quad \mu m \tag{3.10}$$

Table 3.2. Comparison of Various Electron Ranges (in Micrometers)

Target	5 keV	10 keV	20 keV	30 keV
Aluminum				
Bethe range	0.56	1.80	6.04	12.4
K–O range	0.41	1.32	4.2	8.3
Experimental:				
Maximum range	0.48	1.1	—	—
Practical range	0.33	0.85	—	—
Copper				
Bethe range	0.23	0.71	2.29	4.64
K–O range	0.15	0.46	1.47	2.89
Experimental:				
Maximum range	0.18	0.47	—	—
Practical range	0.11	0.34	—	—
Gold				
Bethe range	0.20	0.55	1.63	3.18
K–O range	0.085	0.27	0.86	1.70
Experimental:				
Maximum range	0.08	0.22	—	—
Practical range	0.05	0.15	—	—

where E_0 is given in keV, A in g/mol, ρ in g/cm^3, and Z is the atomic number of the target. Values of R_{KO} are also listed in Table 3.2. In Figure 3.8 this value of the range can be interpreted as the radius of a semicircle centered on the beam impact point which defines the envelope of trajectories.

    3.3.2.4.3. **Experimental Ranges.** Various workers have defined an electron range through experiments which measure the penetration of electrons through thin films, e.g., Cosslett and Thomas (1964a, b; 1965; 1966). Since the reader is likely to encounter such values of the electron range in the literature, it is useful to consider their definition and relation to other definitions of the range. Figure 3.11 shows the typical behavior of the transmission coefficient, defined as the fraction of the incident beam current which passes through a film, as a function of thickness for the case of copper with a 10 keV electron beam incident normal to the film (Cosslett, 1966). Two characteristics of such curves can be used to develop definitions of the electron range: (1) By extrapolating the linear portion of the curve to $\tau = 0$, the intercept on the depth axis is defined as the "extrapolated range" or "practical range." The "practical range" can be thought of as a measure of the dense region of trajectories in the interaction volume. (2) The experimental film thickness at which the transmitted beam is reduced to zero is defined as the maximum range. The maximum range can be thought of as a measure of the envelope containing all trajectories.

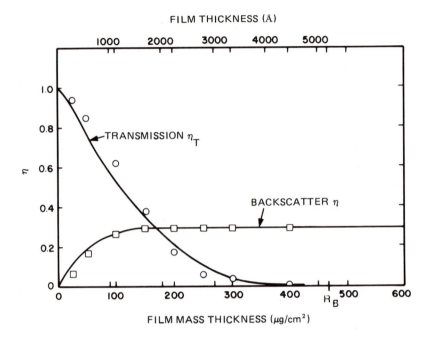

Figure 3.11. Transmission coefficient η_T and backscatter coefficient η as a function of film thickness for a copper target, $E_0 = 10$ keV. Solid lines represent experimental results (Cosslett, 1966); points represent individual Monte Carlo calculations (Newbury and Yakowitz, 1976).

3.3.2.4.4. Comparison of Range Values. Table 3.2 contains numerical values of the electron range as calculated by the equations listed above, and as experimentally measured according to the criteria for thin films. In general, the calculated Kanaya–Okayama range falls near the experimentally determined value of the maximum range. The Bethe range is found to deviate most significantly from the other measures of the range in those cases where the elastic scattering is maximized: high-atomic-number targets and/or low beam energy. The Bethe range deviates by more than 50% in the most extreme cases from the Kanaya–Okayama range.

When tilted samples are considered, the interaction volume becomes asymmetric about the tilt axis, and the depth dimension of the envelope containing a given fraction of the trajectories falls. As an approximation derived from Monte Carlo calculations such as those shown in Figure 3.9, the range $R(\theta)$ normal to the surface in a tilted specimen is given by

$$R(\theta) = R_{KO} \cos \theta \qquad (3.11)$$

where R_{KO} is the Kanaya–Okayama range given by equation (3.10).

3.4. Backscattered Electrons

It is found experimentally that a significant fraction of the beam electrons which strike a target subsequently escape. For example, if the beam current is measured in a Faraday cage (see Chapter 2) and the beam is then made to impinge on a positively biased copper target, only about 70% of the beam electrons expend all of their energy in the interaction volume and are absorbed by the target; the remaining 30% of the electrons are scattered out of the specimen. These reemergent beam electrons are collectively known as backscattered electrons. The backscattered electron coefficient, η, is defined as the number of backscattered electrons, n_{BS}, divided by the number of beam electrons incident on the target, n_B; or alternatively, electron currents can be considered:

$$\eta = \frac{n_{BS}}{n_B} = \frac{i_{BS}}{i_B} \tag{3.12}$$

Careful examination of the individual trajectories drawn with a Monte Carlo simulation reveals that the process of backscattering usually takes place as a result of a sequence of elastic scattering events in which the net change in direction is sufficient to carry the electron out of the specimen (Figure 3.12). It is possible for an electron incident normal to a specimen surface to scatter through an angle greater than 90° and thus escape the specimen after a single event. From Equation (3.4), the cross-section equation can be converted into a probability equation in the following way:

$$
\underset{\dfrac{\text{events}}{e^{-}\,(\text{atom}/\text{cm}^2)}}{Q}
\;\cdot\;
\underset{\dfrac{\text{atoms}}{\text{mol}}}{N_0}
\;\cdot\;
\underset{\dfrac{\text{mol}}{\text{g}}}{\dfrac{1}{A}}
\;\cdot\;
\underset{\dfrac{\text{g}}{\text{cm}^3}}{\rho}
\;\cdot\;
\underset{\text{cm}}{t}
\;=\;
\underset{\dfrac{\text{events}}{e^{-}}}{\text{probability}}
$$

$$\tag{3.13}$$

where t is a specified thickness. For the case of electrons incident on an iron layer 10 nm (100 Å) thick, the probability given by Equation (3.13) suggests that the average 20 keV electron will undergo 3.4 events of 3° or greater, but only one electron in 430 will undergo an event 90° or greater.

Backscattered electrons provide an extremely useful signal for imaging in scanning electron microscopy (Wells, 1977). For proper image interpretation, it is necessary to understand the properties of the backscattered electrons as a function of the parameters of the beam and the characteristics of the specimen (for a detailed review of properties, see Niedrig, 1978).

3.4.1. Atomic Number Dependence

The Monte Carlo trajectory plots, Figure 3.7, reveal that the fraction of trajectories which intersect the surface resulting in backscattering events

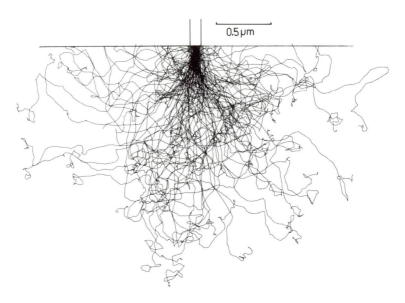

0.5 µm

Figure 3.12. Detailed single scattering Monte Carlo electron trajectory simulation for a copper target, $E_0 = 30$ keV, showing trajectories intersecting surface which results in backscattering.

increases with increasing atomic number. If the backscattering coefficient, η, is measured as a function of atomic number, Z, the relationship shown in Figure 3.13 is obtained (Heinrich, 1966a; Bishop, 1966). Generally, η is found to increase with increasing atomic number, with slight deviations in the first transition series (Heinrich, 1968). Reuter (1972) has obtained a fit to Heinrich's 20-keV (Heinrich, 1966a) data which is useful for calculations:

$$\eta = -0.0254 + 0.016Z - 1.86 \times 10^{-4}Z^2 + 8.3 \times 10^{-7}Z^3 \quad (3.14)$$

When the target is a homogeneous mixture of several elements, a simple rule of mixtures based on weight fractions of the components is found to apply (Heinrich, 1966a):

$$\eta_{\text{mix}} = \sum_i \eta_i C_i \quad (3.15)$$

where η_i is the backscattering coefficient for the pure element and C_i is the weight fraction.

Backscatter coefficients can be determined by the following procedure: (1) The specimen must be biased to a potential of $+50$ V relative to the grounded walls of the sample chamber to prevent the escape of secondary electrons (to be discussed subsequently) from the specimen. Ideally a biased

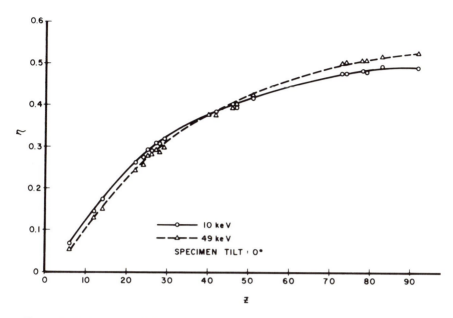

Figure 3.13. Variation of the backscatter coefficient as a function of atomic number at $E_0 = 10$ keV and $E_0 = 49$ keV. Data of Heinrich (1966a).

grid should surround the specimen to prevent collection of secondary electrons from the chamber walls. (2) The specimen acts as a current junction in a circuit, and the balance of currents is such that

$$i_B = i_{BS} + i_{SC} + i_{SE} \qquad (3.16)$$

where i_B is the beam current, which can be measured in a Faraday cup, i_{BS} is the backscattered electron current, i_{SC} is the specimen current (also known as the target current or absorbed current), and $i_{SE} = 0$. The specimen current i_{SC} can be measured by placing a high-sensitivity current-measuring device between the specimen and ground. The backscatter coefficient can then be calculated as

$$\eta = \frac{i_{BS}}{i_B} = \frac{i_B - i_{SC}}{i_B} \qquad (3.17)$$

3.4.2. Energy Dependence

The size of the interaction volume is a strong function of the beam energy (Figure 3.8). We might therefore expect the backscatter coefficient to vary significantly with energy as well. However, experimental measurements show that this is not the case. As shown in Figure 3.13, only a small variation of about 10% is found over the range 10–49 keV. The relative insensitivity of the backscatter coefficient to beam energy compared to the strong dependence of the interaction volume on beam energy can be understood from an examination of Equations (3.4) and (3.5). Although the increased mean free path for elastic scattering leads to greater penetration of the sample at high beam energy, the rate of energy loss decreases at high energy. The capacity of an electron to continue to travel in a solid and thus have a chance to reach the surface and escape is related to its energy. For a high beam energy, the electrons at any given depth retain more energy than they would have at a lower incident energy. Penetration effects are nearly perfectly offset by energy loss effects, so that the backscatter coefficient has only a slight dependence on energy.

3.4.3. Tilt Dependence

If the tilt angle, θ, of the specimen is increased, the size of the interaction volume decreases because the tendency of the electrons to undergo forward scattering causes them to propagate closer to the surface. The opportunity for backscattering events increases. A plot of the backscattering coefficient as a function of tilt angle, Figure 3.14, shows that η increases slowly with θ up to approximately 20°, and rapidly above 30°, with η tending toward unity at grazing incidence. The differences observed in η with atomic number, Z, at normal incidence are thus reduced at high

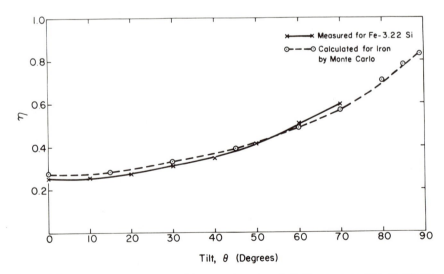

Figure 3.14. Backscatter coefficient of iron as a function of tilt. Beam energy 30 keV. Monte Carlo calculations and experimental measurements. Newbury *et al.* (1973).

tilt, since η for all elements tends toward the same value. For tilted samples, the backscatter coefficient is approximately described by (Arnal *et al.* 1969)

$$\eta(\theta) = \frac{1}{(1 + \cos\theta)^p} \tag{3.18}$$

where $p = 9/Z^{1/2}$ for pure elements.

3.4.4. Angular Distribution

The angular distribution of the trajectories of the backscattered electrons as they emerge from the surface is described by a cosine function when the beam is incident normally on a surface, Figure 3.15:

$$\eta(\phi) = \eta' \cos\phi \tag{3.19}$$

where ϕ is the angle between the surface normal and the direction of measurement and η' is the value of η along the surface normal. This cosine distribution is rotationally symmetric about the surface normal, so that the shape observed in Figure 3.15 would be found in any plane which contained the surface normal. From Figure 3.15 it can be seen that for a beam set normal to a surface, the maximum number of backscattered electrons travel back along the beam direction, with that number decreasing to zero as ϕ increases to 90°. The origin of the cosine distribution can be understood with the aid of Figure 3.16. Consider a beam electron penetrating to a depth P_0 before undergoing a large-angle collision or else multiple scattering to reverse its direction of flight. The path that this electron must travel

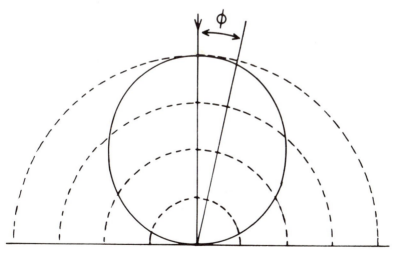

Figure 3.15. Angular distribution of backscattered electrons relative to surface normal showing a cosine law dependence.

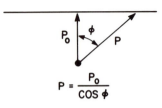

$$P = \frac{P_0}{\cos \phi}$$

$$N \propto \frac{1}{P} \Rightarrow N \propto \cos \phi$$

Figure 3.16. Origin of cosine angular distribution shown in Figure 3.15.

through the solid to reach the surface is related to the angle ϕ by

$$P = P_0/\cos \phi \qquad (3.20)$$

Because of inelastic scattering, the electron loses energy along this path. The probability of electron escape from the specimen decreases as its energy decreases, so as the path it must travel increases, escape becomes less probable. For any angle ϕ, the fraction of electrons which escape as

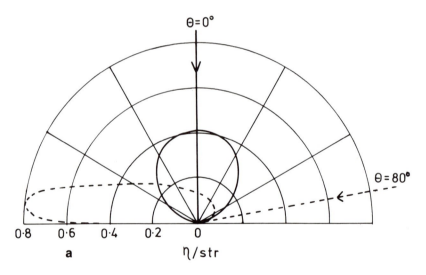

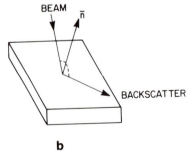

Figure 3.17. (a) Comparison of the angular distribution of backscattered electrons for sample tilts of $\theta = 0°$ and $\theta = 80°$. Note the distribution is strongly peaked in the forward direction at high tilt. (b) Schematic illustration of direction of maximum backscattering in a highly tilted specimen.

compared to those escaping along the shortest path P_0 for $\phi = 0$ is given by

$$N \propto 1/P \propto \cos \phi \qquad (3.21)$$

As the tilt angle, θ, is increased, the angular distribution of backscattered electrons changes, becoming asymmetrical about the tilt axis. At high tilt angles, this lobe becomes very highly developed in the forward scattering direction, Figure 3.17a, so that the largest fraction of backscattered electrons is observed traveling above the surface at about the same angle as the incident beam lies above the surface. The electrons tend to "skip" off the first few atom layers and exit the specimen after a few scattering events. Moreover, from a highly tilted specimen, the electrons tend to be emitted in a plane which is defined by the beam vector and the surface normal, Figure 3.17b.

3.4.5. Energy Distribution

As the beam electrons travel within the specimen, the various processes of inelastic scattering cause a transfer of energy to the atoms and electrons of the solid and a consequent reduction in the energy of the beam electrons. Typical rates of energy loss, as given by the Bethe expression, Equation (3.5), are of the order of 10 eV/nm and are dependent on the electron energy (Figure 3.18). The beam electrons which emerge from the surface of the specimen as backscattered electrons after having traveled some distance

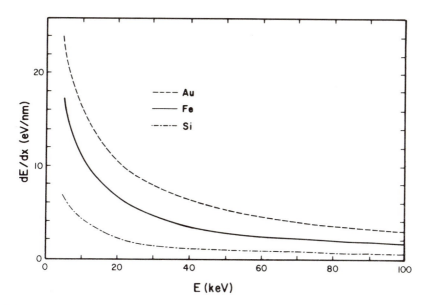

Figure 3.18. Plot of the rate of energy loss with distance traveled (eV/nm) for gold, iron, and silicon as a function of electron energy from the Bethe expression [Equation (3.5)].

within the solid escape with reduced energy. Plots of the energy distribution of backscattered electrons as a function of the fractional energy, $W = E/E_0$, are shown for several elements in Figure 3.19 (Bishop, 1966). For light elements, the distribution is broad while for heavy elements a distinct peak is observed. The position of this peak tends toward $W = 1$ and the peak height increases as the atomic number increases. If the energy distribution for a given element is measured at different take-off angles, ψ (where ψ is the angle of the detector above the surface) the energy distribution is observed to vary with angle. The distribution is more sharply peaked at higher take-off angles, Figure 3.20 (Bishop, 1966). This behavior can again be understood in terms of the length of path which the electron must travel to reach the surface. Those electrons observed at a low take-off angle are likely to have traveled a longer distance in the solid and thus lost more energy.

The energy distribution also varies with the tilt angle of the specimen.

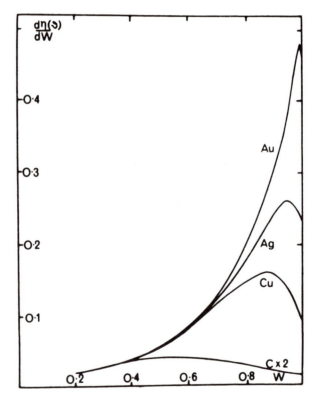

Figure 3.19. Energy distribution of backscattered electrons for several elements. The measurements were made at a take-off angle of 45° above the surface. $E_0 = 30$ keV (Bishop, 1966).

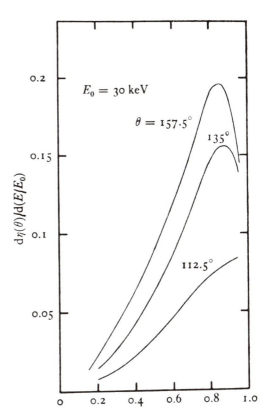

Figure 3.20. Energy distribution of backscattered electrons from copper at various take-off angles. The angles are measured relative to the beam and are equivalent to take-off angles of 22.5°, 45°, and 67.5° (Bishop, 1966).

Wells (1974a) has demonstrated that for highly tilted specimens those electrons which have lost the least energy are most strongly peaked in the forward scattering direction (Figure 3.21).

3.4.6. Spatial Distribution

Backscattered electrons can emerge from the surface of a specimen at a significant distance from the impact area of the beam electrons. This is a direct consequence of elastic scattering of the beam electrons in the solid and their subsequent spreading to form the interaction volume. Murata (1973) has calculated with Monte Carlo simulation techniques the number of backscattered electrons escaping as a function of position about an incident point beam (Figure 3.22a). For a beam normally incident on a flat sample, the distribution of backscattered electrons across a diameter of the emission area is symmetrical and peaked at the impact point. For high-atomic-number targets, the diameter of the distribution is smaller and the central peak is higher (Figure 3.23). When the sample is tilted (Figures

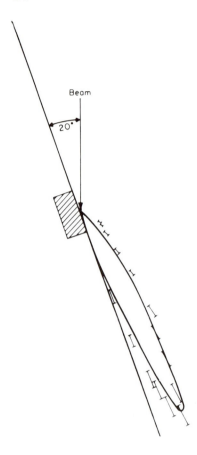

Beam

20°

Figure 3.21. Angular distribution of backscattered electrons ($E_0 = 20$ keV) which have lost less than 500 eV. I = Experimental values due to Wells (1974a); ———— = Monte Carlo calculations due to Newbury and Yakowitz (1976). Sample: 0.5 μm SiO$_2$ on Si.

3.22b and 3.23), the spatial distribution of the backscattered electrons becomes asymmetric about the tilt axis, with the peak displaced down the surface from the beam impact point.

The distributions of Figures 3.22 and 3.23 consider all electrons which are backscattered regardless of energy. If only those electrons which have lost less than 10% of their incident energy are considered, the spatial distribution is found to be smaller in diameter and to have a higher central peak (Figure 3.24). This result can be easily understood since those electrons which emerge at greater distances from the beam impact point must necessarily have traveled further in the specimen and undergone more inelastic scattering. From Figures 3.22, 3.23, and 3.24, it is clear that if the microscopist wishes to obtain information about the region of the sample near the beam impact area, it is advantageous to select those backscattered electrons which have not lost much energy.

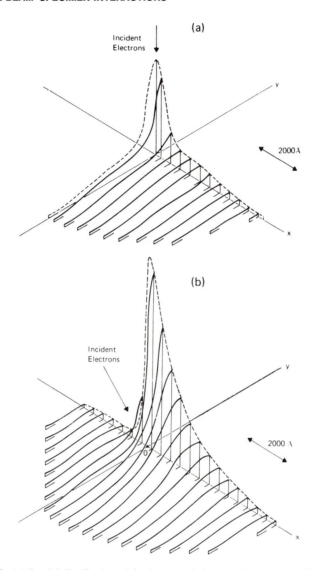

Figure 3.22. (a) Spatial distribution of backscattered electrons from copper ($E_0 = 20$ keV) with 0° tilt, $\eta = 0.300$. (b) Spatial distribution with 45° tilt, $\eta = 0.416$ (Murata, 1973).

3.4.7. Sampling Depth

As seen from the Monte Carlo simulations, the beam electrons generally penetrate some distance into the solid before undergoing a sufficient number of elastic scattering events to reverse the direction of travel and

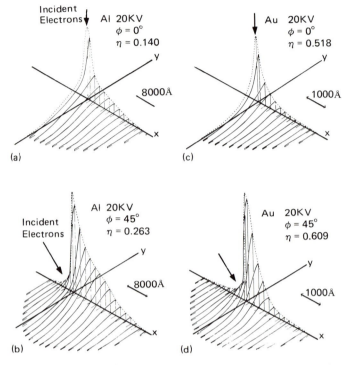

Figure 3.23. Spatial distribution of backscattered electrons from aluminum and gold ($E_0 = 20$ keV) for 0° tilt and 45° tilt (Murata, 1974).

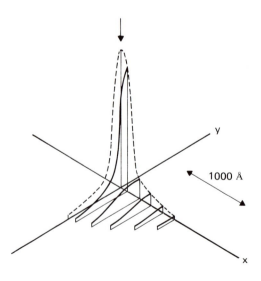

Figure 3.24. Spatial distribution of backscattered electrons which have lost less than 1 keV of energy ($E_0 = 20$ keV, $\phi = 0°$, $\Delta E < 1$ keV) (Murata, 1973).

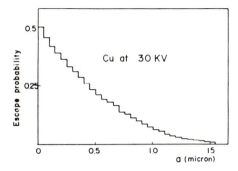

Figure 3.25. Escape probability of a backscattered electron which has penetrated to a given depth prior to backscattering (Shimizu and Murata, 1971).

lead to backscattering. The emerging backscattered electrons thus carry information about the nature of the specimen over a range of depth. The "depth of information" of backscattered electrons can be estimated from experiments with layered samples (Seiler, 1967) and from calculations (Shimizu and Murata, 1971). The depth of information is approximately 0.3 of the Kanaya–Okayama range. From the histogram of escape depths of backscattered electrons obtained by Monte Carlo calculations, Figure 3.25, it is clear that a single number cannot accurately define the depth of information, since a gradual distribution is observed to exist without a sharp cutoff. The implications of the sampling depth for image information depend greatly on the exact nature of the specimen.

3.5. Signals from Inelastic Scattering

The energy from the beam electrons deposited in the specimen is distributed among a number of secondary processes, several of which result in signals which are useful for microscopy and microanalysis. These signals include the formation of (1) secondary electrons, (2) characteristic x-rays, (3) bremsstrahlung or continuum x-rays, (4) cathodoluminescence radiation, e.g., emission of infrared, visible, and ultraviolet radiation. The fraction of the beam energy deposited in the sample which is associated with each of these processes is dependent on the nature of the sample.

3.5.1. Secondary Electrons

3.5.1.1. Definition and Origin

If the energy distribution of all electrons emitted from a sample is measured over the range 0 to E_0, a curve similar to that shown in Figure 3.26a is observed. The upper portion of the distribution, region I, is the

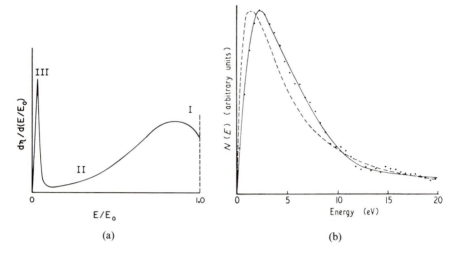

Figure 3.26. (a) Schematic illustration of the complete energy distribution of electrons emitted from a target, including backscattered electrons (regions I and II) and secondary electrons (region III). Note that the width of region III is exaggerated. (b) Energy distribution of secondary electrons as measured (points) and as calculated (dots), from Koshikawa and Shimizu (1974).

broad hump of electrons which have lost less than about 40% of their incident energy due to inelastic scattering. For most targets of intermediate and high atomic number, the majority of the backscattered electrons are found in this hump. A smaller fraction of beam electrons lose large amounts of energy, greater than 40%, before escaping the specimen, forming the tail of the distribution in region II. If region II were extrapolated to zero energy, the fraction of backscattered electrons having a particular energy would be expected to decrease smoothly to zero at zero energy. However, at very low energies, below approximately 50 eV, the number of electrons emitted from the sample is found to increase sharply to a level much greater than the extrapolated backscattered electron contribution at these energies. The increase in emitted electrons which forms region III in Figure 3.26a is due to the process of secondary electron emission (Bruining, 1954). Secondary electrons are defined as those electrons emitted from the sample with an energy less than 50 eV (an arbitrary cutoff). Although some backscattered beam electrons are included in this region, their inclusion in the definition of secondary electrons introduces a negligible error. The secondary electron coefficient, δ, is given by

$$\delta = \frac{n_{\rm SE}}{n_B} = \frac{i_{\rm SE}}{i_B} \qquad (3.22)$$

where $n_{\rm SE}$ is the number of secondary electrons emitted from a sample bombarded by n_B beam electrons, and i designates the equivalent currents.

Secondary electrons are produced as a result of interactions between energetic beam electrons and weakly bound conduction electrons (Streit-wolf, 1959). The interaction between the beam electron and the conduction band electron results in the transfer of only a few electron volts of energy to the conduction band electron. The resulting energy distribution of second-ary electrons, region III in Figure 3.26b, shows a peak at about 3–5 eV, with the distribution dropping off sharply as the energy increases.

3.5.1.2. Range and Sampling Depth

An important characteristic of secondary electrons is their shallow sampling depth, a direct consequence of the low kinetic energy with which they are generated. Secondary electrons are strongly attenuated during motion in a solid by energy loss due to inelastic scattering, which has a high probability for low-energy electrons. Moreover, to escape the solid, the secondary electrons must overcome the surface potential barrier (work function), which requires an energy of several electron volts. As the beam electron moves deeper into the sample and produces secondary electrons, the probability of escape of those secondary electrons decreases exponen-tially:

$$p \propto \exp(-z/\lambda) \qquad (3.23)$$

where p is the probability of escape, z is the depth below the surface where generation takes place, and λ is the mean free path of the secondary electrons. Seiler (1967) has determined that the maximum depth of emission is about 5λ, where λ is about 1 nm for metals and up to 10 nm for insulators. λ is dependent on the energy of the secondary electrons, so that a range of λ actually exists over the full energy range of secondary electrons. For purposes of estimation, the values given above suffice. The greater range in insulators is a direct consequence of the fact that inelastic scattering of secondary electrons takes place chiefly with conduction elec-trons which are abundant in metals and greatly reduced in insulators. The escape probability as a function of depth has been calculated with a Monte Carlo simulation by Koshikawa and Shimizu (1974). As shown in Figure 3.27, there is a sharp drop off in escape with depth. Compared to the information depth histogram for backscattered electrons, shown in Figure 3.25, the information depth for secondary electrons is about $1/100$ that for backscattered electrons.

The escape depth of secondary electrons represents only a small fraction of the primary electron range, about 1% for metals. Secondary electrons are generated throughout the interaction volume of the beam electrons in the specimen, but only those generated within the mean escape distance from the surface carry information which can be detected by the microscopist. Observable secondary electrons can be formed by the inci-

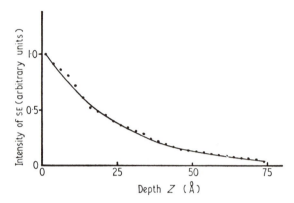

Figure 3.27. Probability of escape of secondary electrons generated at a depth Z below the sample surface (Koshikawa and Shimizu 1974).

dent beam electrons as they enter the specimen and by the backscattered electrons as they exit (Figure 3.28). Experimentalists have been able to distinguish the relative contributions of these two sources by measuring the secondary electron coefficients, δ_B and δ_{BS}, for each process. These coefficients are related in the following way to the total secondary-electron coefficient, which is detected (or a part thereof) in the scanning electron microscope:

$$\delta_{tot} = \delta_B + \delta_{BS}\eta \qquad (3.24)$$

where η is the backscatter coefficient. In general, the ratio δ_{BS}/δ_B is of the order of 3 or 4 (Seiler, 1967), that is, the generation of secondary electrons per high-energy electron is more efficient for the backscattered electrons

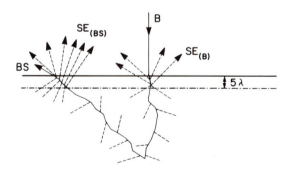

Figure 3.28. Schematic illustration of the two sources of secondary-electron generation in the sample. Incident beam electrons (B) generate secondary electrons (SE$_B$) upon entering the sample. Backscattered electrons (BS) generate secondary electrons (SE$_{BS}$) while exiting the sample. λ is the mean free path for the secondary electrons.

than for the incident beam electrons! This behavior arises because of two factors:

(1) The backscattered electrons are likely to approach the surface at a more shallow angle than the primary beam electrons which are incident normally. The backscattered electrons thus have more path length than the primary electrons in the layer corresponding to the mean escape depth of the secondary electrons, and are thus able to generate more secondary electrons in this critical layer.

(2) The backscattered electrons have a distribution in energy extending to much lower energy values than the monoenergetic beam electrons. The generation of secondary electrons is more efficient at low electron energies, which tends to make backscattered electrons more efficient at generating secondaries than the incident beam electrons.

Taking $\delta_{BS}/\delta_B = 3$ and using the values of η from Figure 3.13, the ratio of the numbers of secondary electrons generated from the two sources has a strong dependence on atomic number as shown in Table 3.3. Thus in images made with the secondary-electron signal, the secondary electrons generated by the beam electrons will be the dominant contribution to the signal for low-atomic-number matrices while secondary electrons generated by backscattered electrons will dominate in high-atomic-number targets.

It is also useful to consider the density of secondary electrons emitted per unit area from the two sources, incident beam electron and backscattered electron excitation. Everhart *et al.* (1959) have pointed out that the generation of secondary electrons by the primary beam occurs within $\lambda/2$ of the beam electron trajectory. For metals this generation distance is about 0.5 nm. Near the surface and within the escape range of 5λ the primary beam is essentially unscattered, so that the diameter of the area of secondary-electron escape from primary generation is the diameter of the incident beam enlarged by $2 \times \lambda/2 = \lambda$. The secondary electrons generated by the backscattered electrons are emitted across the entire escape area of the backscattered electrons, which may be a micrometer or more in diameter. In order to estimate this diameter, we shall consider the diameter containing 90% of the backscattered electron emergence as shown in Figure 3.22. Note that because the backscattered electron distribution is peaked in

Table 3.3. Ratio of Secondary Electrons Produced by Beam and Backscattered Electrons

Element	η	SE_{BS}/SE_B
C	0.06	0.18
Cu	0.30	0.9
Au	0.50	1.5

the center of this area, the density of backscatter-induced secondary-electron emission will be necessarily nonuniform.

The number of secondary electrons per unit area generated by the primary beam is much greater than that generated by the backscattered electrons. When the beam is scanned across the specimen, beam-produced secondary electrons respond to local surface features and carry information in the image while the secondary electrons generated remotely by the backscattered electrons act as a background noise. The implications of the presence of the two types of secondary electrons for image formation will be discussed in Chapter 4.

3.5.1.3. Influence of Beam and Specimen Parameters

3.5.1.3.1. Specimen Composition. Compared to the behavior of backscattered electrons, whose yield increases monotonically with the atomic number of the specimen, the secondary-electron coefficient is relatively insensitive to composition and fails to show any strong trend with atomic number, Figure 3.29, Wittry (1966). A typical value of δ is about 0.1 at an incident energy of 20 keV with certain elements such as gold having higher values of about 0.2. The slight periodic trends observed in Figure 3.29 are correlated to some extent with the number of outer-shell electrons, the atomic radius, and the density.

When compound targets are considered, as compared with the pure element values of Figure 3.29, a wider range of values of δ is observed.

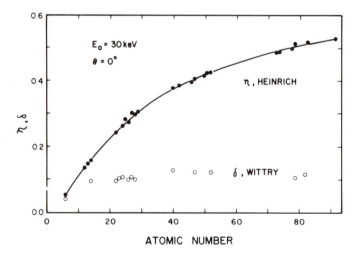

Figure 3.29. Comparison of backscattered electron coefficients and secondary-electron coefficients as a function of atomic number (Wittry, 1966; Heinrich, 1966).

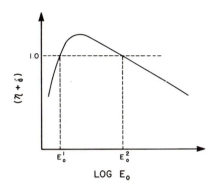

Figure 3.30. Schematic illustration of the total emitted electron coefficient ($\eta + \delta$) as a function of beam energy.

Under certain conditions, δ can approach or exceed unity for compounds such as Al_2O_3 or MgO. The dependence of δ on composition is a complicated function of the nature of the molecular bonds, the elements present, the crystal orientation, the conductivity, and the condition of the crystal surface, making it difficult to formulate a description of secondary-electron emission as a function of specimen composition. An example of pronounced variation in δ with composition is shown in Figure 4.28 of Chapter 4.

3.5.1.3.2. **Beam Energy.** The behavior of δ as a function of energy is shown schematically in Figure 3.30. Starting at zero beam energy, the secondary-electron coefficient rises with increasing energy, reaching unity at about 1 keV. A peak, slightly above unity for metals and as high as 5 for nonmetals is observed in the range 1–2 keV (Dawson, 1966). With a further increase in beam energy, δ decreases and passes through unity again in the range 2–3 keV, and continues to decrease with increasing beam energy to a value of about 0.1 for metals and 20 keV. Specific values determined by Reimer and Tollkamp (1980) show that the influence of beam energy varies with the atomic number of the target (Table 3.4).

3.5.1.3.3. **Specimen Tilt.** As the angle of specimen tilt, θ, is increased, δ is found experimentally to increase following a secant relationship (Kanter, 1961):

$$\delta(\theta) = \delta_0 \sec \theta \tag{3.25}$$

Table 3.4. Secondary-Electron Coefficients[a]

	5 keV	20 keV	50 keV
Aluminum	0.40	0.10	0.05
Gold	0.70	0.20	0.10

[a] From Reimer and Tollkamp (1980).

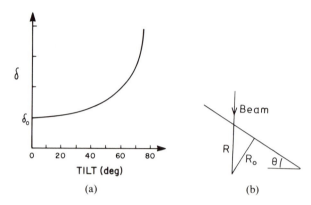

Figure 3.31. (a) Behavior of secondary-electron coefficient δ as a function of specimen tilt θ, following a secant law. (b) Origin of secant law behavior.

where δ_0 is the secondary-electron coefficient measured at normal incidence, Figure 3.31a. The origin of this secant relationship can be understood with the aid of Figure 3.31b. Consider that all secondary electrons created within a distance R_0 of the surface will escape. With the beam incident normal to the surface, the length of primary path, R, along which generated secondaries will escape is equal to R_0. As the specimen tilt, θ, is increased, the length of primary path within a distance R_0 of the surface increases as $R = R_0 \sec\theta$. Since R_0 is small, the primary beam does not change in energy significantly in traveling this distance and the rate of secondary-electron production due to the primary electron beam is essentially constant and proportional to R. Thus, since the path length increases with $\sec\theta$, the secondary-electron coefficient behaves similarly. Secondary electrons are also generated by the backscattered electrons. The backscattering coefficient increases with the tilt angle (Figure 3.15) and hence, the number of secondary electrons generated by the backscattered electrons also increases with tilt. Production of secondary electrons by both incident and backscattered electrons tends to increase with increasing tilt angle, leading to the approximate secant relationship, Equation (3.25).

3.5.1.3.4. Angular Distribution. In a fashion similar to the arguments used to explain the angular distribution of backscattered electrons, Figures 3.15 and 3.16, it can be shown that the secondary electrons follow a cosine distribution relative to the surface normal when the specimen is set perpendicular to the beam. For a tilted specimen the angular distributions of backscattered and secondary electrons differ. The angular distribution of backscattered electrons becomes asymmetric in the forward scattering direction. The angular distribution of secondary electrons remains a cosine distribution even for a tilted specimen. This behavior is a result of the

isotropic generation of secondary electrons by the primary b...
the directionality of emission is unaffected by tilting the specir...

3.5.2. X-Rays

3.5.2.1. X-Ray Production

During inelastic scattering of the beam electrons, x-rays can be formed by two distinctly different processes: (1) Deceleration of the beam electron in the Coulombic field of the atom core, which consists of the nucleus and tightly bound electrons, leads to formation of a continuous spectrum of x-ray energies from zero energy up to the value of the incident electron energy, as shown in Figure 3.32. This is referred to as the x-ray continuum or bremsstrahlung ("braking radiation"). (2) The interaction of a beam electron with an inner-shell electron can result in the ejection of the bound electron which leaves the atom in an excited state with a vacancy in the electron shell, Figure 3.33. During subsequent deexcitation, an electron transition occurs from an outer shell to fill this vacancy. The transition involves a change in energy, and the energy released from the atom can manifest itself either in the form of an x-ray or an ejected (Auger) electron. Since the energy of the emitted x-ray is related to the difference in energy between the sharply defined levels of the atom, it is referred to as a characteristic x-ray.

X-rays, as photons of electromagnetic radiation, have an associated wavelength, λ, which is related to the photon energy by

$$\lambda = hc/eE = 1.2398/E \quad \text{nm} \tag{3.26}$$

where h is Planck's constant, c is the speed of light, e is the electron charge,

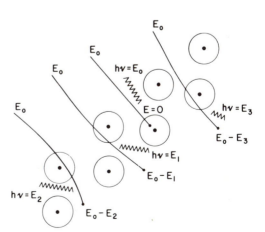

Figure 3.32. Schematic illustration of the origin of the x-ray continuum, resulting from deceleration of the beam electrons in the Coulombic field of the atoms.

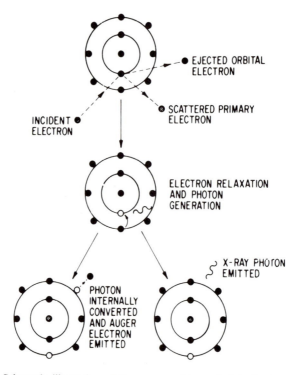

Figure 3.33. Schematic illustration of the process of inner-shell ionization and subsequent deexcitation by either Auger electron emission or x-ray photon emission.

E is the energy in keV, and λ is given in nm. Because x-ray spectrometry as employed in microanalysis involves the use of both wavelength-dispersive and energy-dispersive spectrometers, it is common to describe x-rays in both wavelength (nm or Å) or energy (keV) units. Both units will be employed in this book.

3.5.2.2. Continuum X-Rays

An electromagnetic spectrum which is generated within the sample as calculated by the Monte Carlo method is shown in Figure 3.34. The continuum extends from virtually zero energy (ultraviolet and visible light with an energy of a few electron volts) to x-rays having an energy equal to the energy of the incident electrons. The maximum energy corresponds to beam electrons which have lost all of their incident energy in one deceleration event. Since x-ray wavelength is inversely proportional to energy, the most energetic x-rays will have a minimum wavelength λ_{min}, called the short-wavelength limit λ_{SWL} or Duane–Hunt limit, which is related to E_0 through Equation (3.26).

The intensity of the x-ray continuum, I_{cm}, at any energy E or wavelength λ has been described by Kramers (1923) as

$$I_{cm} \sim i\overline{Z}\left[(\lambda/\lambda_{min}) - 1\right] \sim i\overline{Z}(E_0 - E)/E \qquad (3.27)$$

where i is the electron current and $\overline{Z}$ is the average atomic number of the target. Subsequent refinements of Kramers' equation will be described in Chapter 8. From Equation (3.27) and the calculated spectrum of Figure 3.34, it can be seen that the intensity of the x-ray continuum decreases as the continuum energy increases. The intensity of the continuum is a function of both the atomic number of the target and the beam energy. As the beam energy E_0 increases, the maximum continuum energy increases and λ_{min} decreases. The intensity at a given continuum energy increases because of the fact that with a higher energy the beam electrons have a greater statistical probability to undergo a given energy loss. The amount of continuum radiation increases with increasing atomic number because of the increased Coulombic field intensity in the core (nucleus plus inner-shell electrons) of heavy atoms compared to low-atomic-number atoms. The continuum intensity varies directly with the number of beam electrons, and therefore directly with the beam current, i_B.

The level of the continuum radiation plays an important role in determining the minimum detectable level for a particular element since the continuum forms a background against which characteristic signals must be measured. The continuum is therefore usually regarded as a nuisance to the analyst. However, it should be noted that, from Equation (3.27), the continuum carries information about the average atomic number (and hence, composition) of the specimen. Thus regions of different $\overline{Z}$ in a

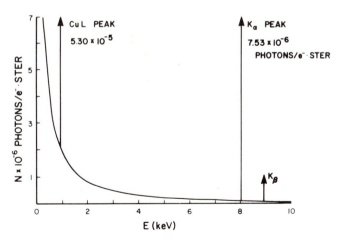

Figure 3.34. Calculated x-ray spectrum generated in a copper target by a 20-keV beam showing the continuum background and the Cu $K\alpha$, Cu $K\beta$, and Cu $L\alpha$ characteristic lines.

specimen will emit different amounts of continuum intensity at all energies. This fact can prove useful in analyzing a specimen and forms the basis for certain correction schemes for quantitative analysis, particularly of biological specimens and particles (see Chapter 7).

3.5.2.3. Characteristic X-Rays

Characteristic x-rays are produced during deexcitation of an atom following the ejection of an inner-shell electron caused by the interaction of an energetic beam electron. A sufficiently energetic beam electron may eject an inner-shell electron, K, L, or M, leaving the atom in an ionized or excited state, Figure 3.33. The atom relaxes to its original ground state (lowest energy) in about 10^{-12} s after ionization. In the process of relaxation, transitions of electrons occur from one shell to another, and in one possible result of such transitions, the excess energy which the excited atom contains can be released in the form of a photon of electromagnetic radiation. The energy of the photon is equal to the difference in energy between the shells involved in the transition, and for inner-shell transitions, the energy is such that the photon lies in the x-ray range of the electromagnetic spectrum.

A detailed treatment of the properties of characteristic x-rays is beyond the scope of this book, and the interested reader is referred to the literature (e.g., Bertin, 1975). Certain basic concepts, such as atomic energy levels, critical ionization energy, families of x-ray emission energies, and x-ray intensities, which are fundamental to x-ray microanalysis will be covered.

3.5.2.3.1. *Atomic Energy Levels.* The electrons of an atom occupy energy levels illustrated schematically in Figure 3.35, with each electron energy uniquely described by a set of quantum numbers (n, l, j, m).

(1) The principal quantum number, n, denotes a shell in which all electrons have nearly the same energy: $n = 1$ corresponds to the shell designated K in x-ray terminology, $n = 2$ (L shell), $n = 3$ (M shell), $n = 4$ (N shell), etc.

(2) The orbital quantum number, l, characterizes the orbital angular momentum of an electron in a shell. l is restricted to values related to n: 0 to $n - 1$.

(3) While orbiting, the electron is also spinning. The spin quantum number, s, describes that part of the total angular momentum due to the electron spinning on its own axis and is restricted to the values $\pm 1/2$. Because magnetic coupling between the spin and orbital angular momenta occurs, the quantum number describing the total angular momentum, j, takes on values $j = l + s$.

(4) Under the influence of a magnetic field, the angular momentum

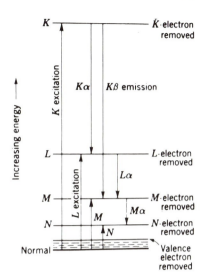

Figure 3.35. Energy level diagram for an atom, illustrating the excitation of the *K*, *L*, *M*, and *N* shells, and the formation of *Kα*, *Kβ*, *Lα*, and *Mα* x-rays.

takes on specific directions characterized by the magnetic quantum number m_j. The values of m_j are given by $m_j \leqslant |j|$, e.g., for $j = 5/2$, $m_j = \pm 5/2$, $\pm 3/2$, $\pm 1/2$.

The arrangement of electrons in an atom is controlled by the Pauli exclusion principle, which places the restriction that no two electrons can have the exact same set of quantum numbers. Thus, each electron has a particular set of quantum numbers (n, l, j, m) which describes it. The shells and subshells with their maximum electron occupation and with the corresponding x-ray notation are listed in Table 3.5.

3.5.2.3.2. **Critical Ionization Energy.** Ionization occurs when an electron is removed from a shell and ejected from the atom. Because the energy of each shell and subshell is sharply defined, the minimum energy necessary to remove an electron from a shell has a specific value, the so-called critical ionization energy (or x-ray absorption energy). Each shell and subshell requires a different critical ionization energy. As an example we shall consider the critical ionization energies for the *K*, *L*, and *M* shells and subshells of platinum ($Z = 78$); see Table 3.6. The critical ionization energy is an important parameter in calculating x-ray intensities. Extensive tabulations of values of the critical excitation energy have been prepared by Bearden (1964).

3.5.2.3.3. **Energy of Characteristic X-Rays.** The deexcitation of an atom following ionization takes place by a mechanism involving transitions of electrons from one shell or subshell to another. The transitions may be radiative, that is, accompanied by the emission of a photon of electromagnetic radiation, or nonradiative, e.g., the Auger electron emission process.

Table 3.5. Shells and Subshells of Atoms

X-ray notation	Quantum numbers				Maximum electron population
	n	l	j	m	
K	1	0	1/2	$\pm 1/2$	2
L_I	2	0	1/2	$\pm 1/2$	2
L_{II}	2	1	1/2	$\pm 1/2$	2
L_{III}	2	1	3/2	$\pm 3/2 \pm 1/2$	4
M_I	3	0	1/2	$\pm 1/2$	2
M_{II}	3	1	1/2	$\pm 1/2$	2
M_{III}	3	1	3/2	$\pm 3/2 \pm 1/2$	4
M_{IV}	3	2	3/2	$\pm 3/2 \pm 1/2$	4
M_V	3	2	5/2	$\pm 5/2 \pm 3/2 \pm 1/2$	6
N_I	4	0	1/2	$\pm 1/2$	2
N_{II}	4	1	1/2	$\pm 1/2$	2
N_{III}	4	1	3/2	$\pm 3/2 \pm 1/2$	4
N_{IV}	4	2	3/2	$\pm 3/2 \pm 1/2$	4
N_V	4	2	5/2	$\pm 5/2 \pm 3/2 \pm 1/2$	6
N_{VI}	4	3	5/2	$\pm 5/2 \pm 3/2 \pm 1/2$	6
N_{VII}	4	3	7/2	$\pm 7/2 \pm 5/2 \pm 3/2 \pm 1/2$	8

The fraction of the total number of ionizations which lead to the emission of x-rays is called the fluorescence yield, ω. For a given shell, the fluorescence yield increases as the atomic number of the atom of interest increases, Figure 3.36.

The x-rays emitted during a radiative transition are called characteristic x-rays because their specific energies (and wavelengths) are characteris-

Table 3.6. Critical Ionization Energy for Platinum

Shell	Critical ionization energy (keV)
K	78.39
L_I	13.88
L_{II}	13.27
L_{III}	11.56
M_I	3.296
M_{II}	3.026
M_{III}	2.645
M_{IV}	2.202
M_V	2.122

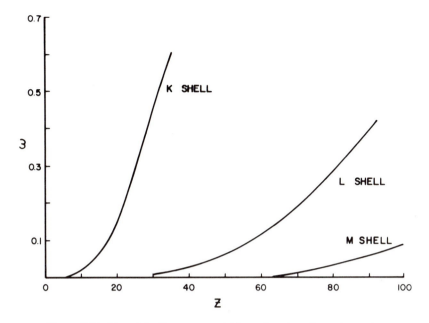

Figure 3.36. Plot of the fluorescence yield as a function of atomic number.

tic of the particular element which is excited. The energy levels of the shells vary in a discrete fashion with atomic number so that the difference in energy between shells changes significantly even with adjacent atomic numbers. This fact was discovered by Moseley (1913, 1914) and expressed in the form

$$\lambda = B/(Z - C)^2 \tag{3.28}$$

where B and C are constants which differ for each family and λ is the characteristic x-ray wavelength. Moseley's relation, which forms the basis for qualitative analysis, is plotted for three different x-ray lines ($K\alpha$, $L\alpha$, and $M\alpha$) in Figure 3.37.

When a vacancy is created in a shell by ionization, the transition to fill that vacancy can often occur from more than one outer shell. Thus, in the simplified diagram of Figure 3.35, following ionization of a K shell, a transition to fill the vacancy can occur from either the L shell or the M shell. Since these shells are at different energies, the x-rays created are at different energies and are designated differently. Thus the $K\alpha$ x-ray is formed from a transition from the L shell and $K\beta$ from an M-shell transition. Since these shells are divided into subshells of different energy, the $K\alpha$ x-rays are further subdivided into $K\alpha_1$ (transition from an L_{III} subshell) and $K\alpha_2$ (transition from the L_{II} subshell) and $K\beta$ is split into $K\beta_1$ (from M_{III}), $K\beta_2$ (from $N_{II,III}$), $K\beta_3$ (M_{II}), $K\beta_4$ (M_V), and $K\beta_5$ ($M_{IV,V}$).

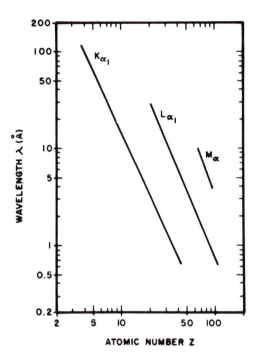

Figure 3.37. Moseley's relation between λ and Z for the $K\alpha_1$, $L\alpha_1$, and $M\alpha_1$ characteristic x-ray lines.

Since the characteristic x-rays involve transitions between shells whereas ionization involves moving an electron from a shell to an effective infinity outside the atom, the energy of the characteristic x-ray is always less than the critical ionization energy for the shell from which the original electron was removed. Thus, $E_{K\alpha} = E_K - E_L$ and $E_{K\beta} = E_K - K_M$, whereas $E_C = E_K - 0$.

3.5.2.3.4. **Families of X-Ray Lines.** The detailed diagram of Figure 3.38 demonstrates the family of x-ray lines which exists for each element. The complexity of the family is a function of the atomic number of the element. Thus for carbon, with two K-shell electrons, and four L-shell electrons, only $K\alpha$ x-rays can be created. Although the L-shell electrons in carbon can be ejected in a collision, there are no M-shell electrons to fill the vacancy. Sodium ($Z = 11$) has one M-shell electron, so that both $K\alpha$ and $K\beta$ x-rays can be emitted. For a heavy element, such as lead, with its complicated shell structure, the family of x-ray lines is far more extensive. Examples of x-ray spectra for the range 1–20 keV, as measured with an energy-dispersive x-ray spectrometer, are shown in Chapter 6 for titanium $K\alpha$, $K\beta$ (Figure 6.2), copper $K\alpha$, $K\beta$, $L\alpha$ (Figure 6.8), and terbium L series, M series (Figure 6.9). These spectra demonstrate the increase in complexity of the spectrum with atomic number. Note that in these figures many lines are not resolved, e.g., $K\alpha_1$–$K\alpha_2$, because of the poor resolution of the energy-dispersive spectrometer (see Chapter 5).

When a beam electron has sufficient energy to excite a particular characteristic x-ray energy, all other characteristic x-rays of lower energy for the element will be simultaneously generated. This occurs because of (1) direct ionization of those lower-energy shells by beam electrons and (2) x-ray formation resulting from the propagation of a vacancy created in an inner shell to outer shells as a result of electron transitions. Thus, if K x-rays of a heavy element are observed L and M x-rays will also be present.

3.5.2.3.5. **Weights of Lines.** Although many possible transitions can occur to fill a vacancy created in a shell, giving rise to x-rays of different energy, e.g., $K\alpha$ and $K\beta$, or up to 25 different L lines, the probability for each type of transition varies considerably. The "weights of lines" refers to the relative probabilities for the formation of lines within a family, i.e., arising from ionization in a given shell. Note that weights are stated within

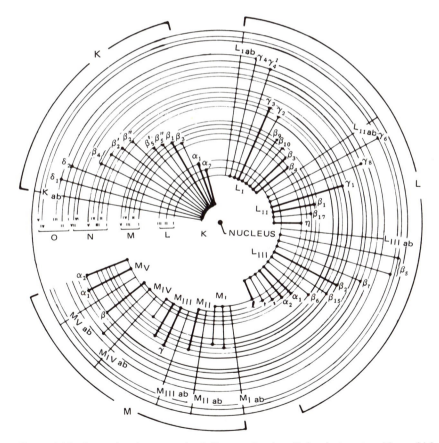

Figure 3.38. Comprehensive energy level diagram showing all the electron transitions which give rise to K, L, and M x-rays from Woldseth (1973).

Table 3.7. Weights of Lines (Approximate)[a]

$K\alpha(1)$, $K\beta(0.1)$
$L\alpha_{1,2}(1)$, $L\beta_1(0.7)$, $L\beta_2(0.2)$
$L\gamma_1(0.08)$, $L\gamma_3(0.03)$, $Ll(0.04)$
$L\eta(0.01)$
$M\alpha(1)$, $M\beta(0.6)$, $M\zeta(0.06)$
$M\gamma(0.05)$, transition $M_{II}N_{IV}(0.01)$

[a]The weights are given in parentheses.

a family, such as the L family; the values do not imply relative weights between families, such as the K versus L. The weights of lines in a family vary in a complex fashion with atomic number. K family weights are well known, but L and M family weights are less so. Table 3.7 gives, to a first approximation, the weights of lines of significant intensity; these weights are a useful guide in interpreting spectra as observed with an energy-dispersive x-ray spectrometer.

As can be seen from an examination of the weights of lines in Table 3.7, there is a relatively small set of intense x-ray lines for each element. Although many transitions are possible, only these "principal lines" are typically measured for analytical purposes. The notation used in x-ray spectroscopy denotes the relative weights of lines within a family. The alpha lines are more intense than the beta, e.g., $K\alpha > K\beta$. Within each subfamily, e.g., α, β, γ, etc., number subscripts denote the progression of intensity, with the smaller number being the most intense, e.g., $K\alpha_1 > K\alpha_2$. It is important to be aware of the possible existence of the minor members of the family of lines listed in Table 3.7 in a spectrum of an unknown, since these low-intensity lines might be misinterpreted as belonging to a trace element in a specimen.

3.5.2.3.6. Intensity of Characteristic X-Rays

(a) Cross Section for Inner-Shell Ionization. Numerous cross sections for inner-shell ionization can be found in the literature; these have been reviewed by Powell (1976). The basic form of the cross section is that derived by Bethe (1930):

$$Q = 6.51 \times 10^{-20} \frac{n_s b_s}{U E_c^2} \ln C_s U \quad \text{ionizations}/e^-/(\text{atom}/\text{cm}^2) \quad (3.29)$$

where n_s is the number of electrons in a shell or subshell (e.g., $n_s = 2$ for a K shell), b_s and C_s are constants for a particular shell, E_c is the ionization energy for the shell (keV), and U is the overvoltage:

$$U = E / E_c \quad (3.30)$$

where E is the beam electron energy. The cross section is plotted as a function of U in Figure 3.39. Powell recommends values of $b_k = 0.9$ and

$C_k = 0.65$ for the K shell and the overvoltage range $4 < U < 25$. Note that the overvoltage used in practical electron probe microanalysis can be as low as $U = 1.5$, which is outside the range over which the ionization cross section has been well characterized. Moreover, from Figure 3.39 it can be seen that the cross section is changing most rapidly in this low overvoltage range.

(b) *Thin Foil.* X-ray production from a thin foil (that is, a foil of thickness such that the beam electron undergoes only one scattering act) can be predicted from the following arguments.

To convert the cross section given in equation (3.29) which has dimensions of ionizations/electron/(atom/cm^2), to x-ray photons/electron, n_x, the following expression is used:

$$
\underset{\substack{\dfrac{\text{photons}}{e^-}}}{n_x} \quad = \quad \underset{\substack{\dfrac{\text{ionizations}}{\text{electron(atom/cm}^2)}}}{Q} \quad \cdot \quad \underset{\substack{\dfrac{\text{photons}}{\text{ionization}}}}{\omega} \quad \cdot \quad \underset{\substack{\dfrac{\text{atoms}}{\text{mol}}}}{N_0} \cdot \underset{\substack{\dfrac{\text{mol}}{\text{g}}}}{\dfrac{1}{A}} \cdot \underset{\substack{\dfrac{\text{g}}{\text{cm}^3}}}{\rho} \cdot \underset{\substack{\text{cm}}}{t}
$$

$$(3.31)$$

where ω is the fluorescence yield, N_0 is Avogadro's number, A is the atomic weight, ρ is the density, and t is the thickness.

(c) *Bulk Target.* For a bulk target, equation (3.31), with $t = dx$, must be integrated over the electron path, taking account of the loss of energy by the beam electron along this path and the finite range. The characteristic yield from the bulk, I_c, is given by

$$
I_c = \int_0^{R_B} \frac{Q_x \omega N_0 \rho}{A} \, dx = \frac{\omega N_0 \rho}{A} \int_{E_0}^{E_c} \frac{Q}{dE/dx} \, dE \tag{3.32}
$$

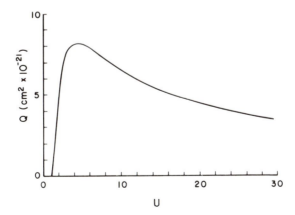

Figure 3.39. Plot of the cross section for inner-shell ionization as a function of overvoltage, U.

A discussion of this expression will be given in Chapter 7 which deals with quantitative x-ray analysis. Equation (3.32) is derived for the case of electrons which lose all of their energy within the target. In reality, backscattering occurs and a fraction of the incident electrons are lost from the target and carry away some portion of the incident energy. A correction factor for backscattering, R (Duncumb and Reed, 1968), has been calculated to take account of this loss of intensity and to give the actual generated intensity:

$$R = \frac{I_{c_{\text{gen}}}}{I_c} \qquad (3.33)$$

where $I_{c_{\text{gen}}}$ is the intensity generated in the sample considering backscatter loss.

(d) Bulk Target (Experimental). A number of workers have reported experimental measurements of $I_{c_{\text{gen}}}$ (Green, 1963; Lifshin et al, 1980). The experimental expressions have the general form

$$I_{c_{\text{gen}}} = a\left(\frac{E_0 - E_c}{E_c}\right)^n = a(U - 1)^n \qquad (3.34)$$

The results for total K x-ray generation ($K\alpha_{1,2} + K\beta$) obtained by Lifshin et al. (1980) on several elements are given in Table 3.8.

(e) Peak-to-Background. Consider the characteristic intensity to follow an experimentally determined relation of the form

$$I_c \propto i_B\left(\frac{E_0 - E_c}{E_c}\right)^n \qquad (3.35)$$

where i is the beam current. The bremsstrahlung continuum background follows a relation of the form

$$I_{\text{cm}} \propto i_B Z(E_0 - E)/E \qquad (3.36)$$

The peak-to-background, P/B, or ratio of intensity of characteristic to

Table 3.8. Experimental Measurements of X-Ray Yield[a]

Element	Z	n	Absolute efficiency (x-ray photons/electron/sr)
Mg	12	1.42	0.114×10^{-4}
Si	14	1.35	0.316×10^{-4}
Ti	22	1.51	0.631×10^{-4}
Cr	25	1.52	0.741×10^{-4}
Ni	27	1.47	0.933×10^{-4}

[a] From Lifshin et al. (1980).

continuum radiation within the same energy interval is

$$\frac{P}{B} = \frac{I_c}{I_{cm}} = \left(\frac{E_0 - E_c}{E_c}\right)^n \frac{1}{Z(E_0 - E)/E} \tag{3.37}$$

Assuming that the continuum energy of interest $E \cong E_c$

$$\frac{P}{B} = \frac{1}{Z}\left(\frac{E_0 - E_c}{E_c}\right)^{n-1} \tag{3.38}$$

The peak-to-background ratio thus increases as the difference $(E_0 - E_c)$ increases. Since, in principle, the sensitivity increases as P/B increases, it would seem advantageous to make $(E_0 - E_c)$ as large as possible. However, as we shall see, as E_0 increases, the depth of x-ray production increases, and the x-rays must therefore undergo additional absorption in order to escape from the sample, which decreases the available signal, limiting the sensitivity.

3.5.2.4. Depth of X-Ray Production

Characteristic x-rays are generated over a substantial fraction of the interaction volume formed by the electrons scattering in the solid, as shown in Figure 3.8. In order to predict the depth of x-ray production or "x-ray generation range" and the x-ray source size (x-ray spatial resolution), the electron penetration must be known. From the discussion on electron penetration, electron range equations have the general form [e.g., the Kanaya–Okayama range, Equation (3.10)]

$$\rho R = KE_0^n \tag{3.39}$$

where K depends on material parameters and n varies from 1.2 to 1.7. The electron mass range considers all electron energies from E_0 down to $E \cong 0$. The x-ray generation range for the production of either characteristic or continuum x-rays is smaller than the electron range. Characteristic x-rays can only be produced within the envelope containing electron energies above the critical ionization energy, E_c, for the line in question. Similarly continuum x-rays can only be formed by electrons of an energy E which is greater than or equal to the particular continuum energy of interest. To account for this energy limit, the mass range for the generation of characteristic or continuum x-rays of a particular energy is given by an equation of the form

$$\rho R_x = K(E_0^n - E_c^n) \tag{3.40}$$

where E_c is the critical ionization energy for characteristic x-rays and E_c can be taken equal to the continuum energy of interest for the continuum case.

From the Kanaya–Okayama electron range [Equation (3.10)], K in the equation has the value

$$K = 0.0276A / Z^{0.889} \qquad (3.41)$$

and thus depends on the target atomic number and weight. Other authors (Reed, 1966; Andersen and Hasler, 1966; and Heinrich, 1981) have set K equal to a constant independent of atomic number and weight. As an example, the x-ray generation range of Andersen and Hasler has the form

$$\rho R = 0.0064\left(E_0^{1.68} - E_c^{1.68}\right) \qquad (3.42)$$

R is given in micrometers. There is a substantial disagreement among the various equations for the x-ray generation range, as shown in the plot of Figure 3.40. The Andersen–Hasler range is a measure of the dense region of x-ray generation, whereas the Kanaya–Okayama range gives the full trajectory envelope. Both ranges will be employed in this book.

In Figure 3.40, the experimental electron range of Kanaya and Okayama (1972) is also plotted and, as expected, the x-ray generation range is smaller than the total electron range. The difference between the electron and x-ray production ranges, in this case for Cu $K\alpha$, is quite large since the critical ionization energy for the copper K shell is 8.980 keV. Electrons of that energy can still travel substantial distances in the target.

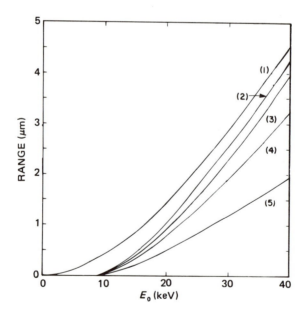

Figure 3.40. X-ray generation range for Cu $K\alpha$ radiation in copper as a function of beam energy. (1) Kanaya–Okayama electron range; (2) Kanaya–Okayama x-ray range; (3) Castaing x-ray range; (4) Andersen–Hasler x-ray range; (5) Reed x-ray range.

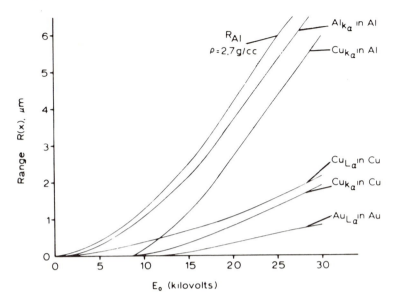

Figure 3.41. X-ray generation range for the Al $K\alpha$, Cu $K\alpha$, Cu $L\alpha$, and Au $L\alpha$ lines generated within aluminum, copper, and gold as a function of beam energy.

Figure 3.41 shows the depth of production for the principal analytical lines of elements representing low (aluminum, 2.70 g/cm³), intermediate (copper, 8.93 g/cm³), and high density (gold, 19.3 g/cm³) targets. The electron range for aluminum is also shown for comparison purposes. As the atomic number and density of the target increase, the depth of production for the principal line decreases. The depth of production is also a function of the critical ionization energy of the line. Thus, Cu $L\alpha$ has a greater depth of production than Cu $K\alpha$ at a given beam energy, and the Au $M\alpha$ line has a greater depth of production than Au $L\alpha$. This latter fact is illustrated for the case of a hypothetical alloy of silicon and germanium at a low concentration in a matrix of iron; see Figure 3.42. It can readily be observed in Figure 3.42 that the depth of x-ray production varies significantly with the element being analyzed.

The discussion of concepts involved in defining the depth of x-ray production is summarized in Figure 3.43. This figure shows a comparison of the depth of x-ray production at the same beam energy for aluminum $K\alpha$ and copper $K\alpha$ in a matrix of density ~3 g/cm³ (e.g., aluminum) and in a matrix of density ~10 g/cm³ (e.g., copper, nickel). In the low-density matrix, both aluminum and copper are produced at greater depths than in the high-density matrix. In addition, the shape of the interaction volume differs considerably, with the low-density matrix giving a pear-shaped volume and the high density giving a less distinct neck.

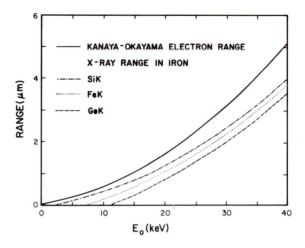

Figure 3.42. X-ray generation range (Andersen–Hasler) of Si $K\alpha$, Fe $K\alpha$, and Ge $K\alpha$ in an iron matrix as a function of beam energy.

The x-ray generation range is critical in estimating the "sampling volume" for analysis. In general, this sampling volume will be smaller than the interaction volume because of the necessity for the electrons to exceed the critical excitation energy for the element in question. As shown in Figure 3.43, the sampling volume will differ for different elements in a

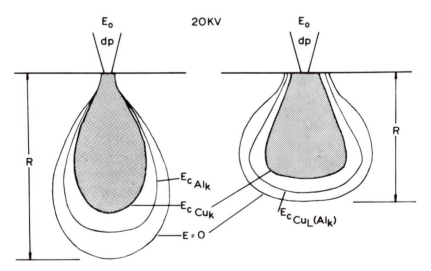

Figure 3.43. Comparison of x-ray production regions from specimens with densities of 3 g/cm^3 (left) and 10 g/cm^3 (right).

target. The sampling volume is also reduced in size relative to the interaction volume because of the phenomenon of x-ray absorption, to be discussed in Section 3.5.2.5. However, the sampling volume can actually be larger than the interaction volume in those cases in which there is significant fluorescence, to be discussed in Section 3.5.2.6.

In all of the discussion on x-ray generation ranges, it must be recognized that the density of x-ray production per unit volume is not constant through the interaction volume. The density of production is related to the number and length of electron trajectories per unit volume and the average overvoltage. The production of x-rays as a function of depth into the sample is known as the $\phi(\rho Z)$ function. In Figure 3.44, the sample is considered to be divided into layers of equal thickness. As the incident beam penetrates these layers, the length of trajectory in each successive

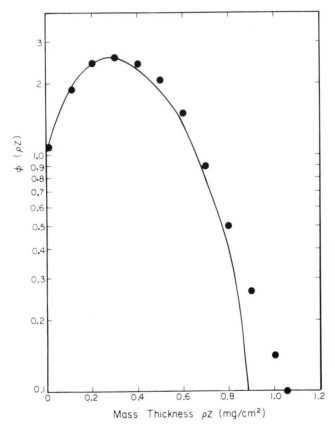

Figure 3.44. Depth distribution of the production of x-rays in aluminum. Experiment, Castaing and Henoc (1966); Monte Carlo calculation, Newbury and Yakowitz (1976). ● = Castaing–Henoc experimental; ——— = Monte Carlo calculation.

layer increases because (1) elastic scattering deviates the electron out of the normal, requiring a longer path to cross the layer and (2) backscattering results in electrons crossing the layer in the opposite direction. The x-ray production thus goes through a peak at a certain depth, and then decreases because energy loss and backscattering reduce the number of electrons available at increasing depth. Moreover, from the cross section in Figure 3.39, electrons at an overvoltage of about 4 make the most significant contributions to x-ray production. The density of x-rays is thus seen to vary considerably with depth. For analytical purposes, the $\phi(\rho Z)$ function is the more exact description of x-ray depth of production.

3.5.2.5. X-Ray Absorption

As x-rays of a given energy E and incident intensity I_0 pass through a slab of thickness x and density ρ, the intensity emitted, I, is attenuated according to the equation

$$I/I_0 = \exp\left[-(\mu/\rho)(\rho x)\right] \tag{3.43}$$

where (μ/ρ) is the mass absorption coefficient of the sample for x-rays. Some typical values of mass absorption coefficients for Ni $K\alpha$ radiation (7.470 keV) in several elements are listed in Table 3.9. Extensive compilations are available in the literature (Heinrich, 1966b). A selection of mass absorption coefficients for $K\alpha$ and $L\alpha$ x-ray energies is tabulated in Chapter 14.

The principal x-ray absorption process in the energy range of interest for microanalysis (1–20 keV) is the photoelectric effect. In this case, the energy of the x-ray photon is totally transferred to a bound inner-shell electron in an atom, ejecting the electron (designated a photoelectron) and annihilating the photon. X-rays can also be scattered inelastically, resulting in a change in energy, through the Compton effect, in which the x-ray interacts with a free electron. For the energy range of interest, the cross section or probability for the Compton effect is so low relative to the

Table 3.9. Energy and Mass Absorption Coefficient Data for Ni $K\alpha$ in Several Elements

Z	Matrix element	Energy (keV)			(μ/ρ) Ni $K\alpha$ (cm^2/g)
		$K\alpha$	$K\beta$	E_{edge}	
25	Mn	5.895	6.492	6.537	344
26	Fe	6.400	7.059	7.111	380
27	Co	6.925	7.649	7.709	53
28	Ni	7.472	8.265	8.331	59
29	Cu	8.041	8.907	8.980	65.5

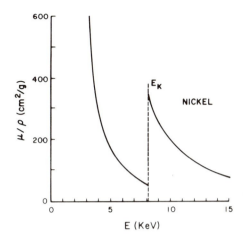

Figure 3.45. Plot of the mass absorption coefficient as a function of energy in nickel.

photoelectric process that it can be safely ignored. X-ray absorption, then, can be treated exclusively through the photoelectric process. For an individual photon, absorption is an "all or nothing" process, i.e., either the photon escapes the solid with its energy unchanged or it is totally absorbed. This fact is especially important to the analyst who depends on characteristic x-rays of specific energy to identify the elemental constituents of the sample.

Mass absorption coefficients, μ/ρ, generally decrease in a smooth fashion with increasing x-ray energy except in the energy region immediately above the energy of an "absorption edge" which corresponds to the energy necessary to eject an electron from a shell; see Figure 3.45 and Table 3.9. X-rays with an energy slightly greater than the critical ionization energy can efficiently couple their energy to eject a bound electron and are strongly absorbed. The x-ray mass absorption coefficient is thus found to increase abruptly at an edge, and then continue to decrease smoothly well above the edge. In the region of a few hundred electron volts above the edge, the variation is irregular. The extended x-ray absorption fine structure (EXAFS), e.g., Figure 3.46, arises from scattering processes in the neighborhood of the absorbing atom. The details of EXAFS can be useful for characterization of the electronic structure of matter.

Absorption edges can be directly observed in x-ray spectra. The x-ray continuum generated during electron bombardment represents a flux of x-rays of all energies through the sample. At the position of an x-ray edge, the abrupt increase in the mass absorption coefficient causes a sharp change in the intensity of the x-ray continuum which is emitted. Figure 3.47 shows the spectrum in the range 0–10 keV for nickel. Ni $K\alpha$ and Ni $K\beta$ peaks are also observed, as well as the discontinuity in the x-ray continuum

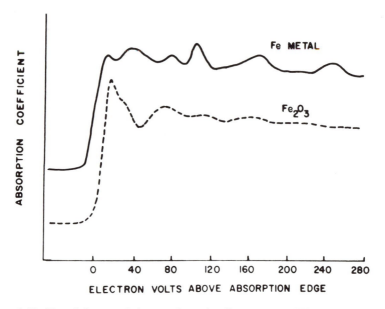

Figure 3.46. Plot of the extended x-ray absorption fine structure (EXAFS) near an x-ray absorption edge. Fe L edge, Nagel (1968).

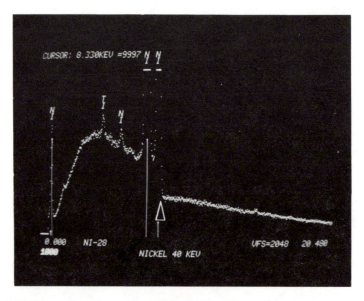

Figure 3.47. Energy-dispersive x-ray spectrum of nickel ($E_0 = 40$ keV) showing sharp step in the continuum background due to the Ni K absorption edge.

at the Ni K edge. The absorption edge at 8.331 keV is indicated by the arrow.

For a sample containing a mixture of elements, the mass absorption coefficient is found by taking the summation of the mass absorption coefficient at the x-ray energy of interest for each element multiplied by its weight fraction:

$$\left(\frac{\mu}{\rho} \right)^i_{\text{spec}} = \sum_j \left(\frac{\mu}{\rho} \right)^i_j C_j \tag{3.44}$$

where $(\mu/\rho)^i_j$ is the mass absorption coefficient for radiation from element i in element j and C_j is the weight fraction of the element.

3.5.2.6. X-Ray-induced Fluorescence

Since the photoelectric process operates in the absorption of x-rays, the atom of the sample is left in an excited, ionized state following absorption of an x-ray. The atom will subsequently undergo the same deexcitation mechanisms discussed for electron-induced ionization. Thus, because of x-ray absorption, characteristic x-ray emission can result. This phenomenon is known as x-ray-induced fluorescence or secondary radiation (as distinguished from primary radiation caused by direct electron ionization). Since secondary radiation can be induced by either characteristic or continuum x-rays, two phenomena are distinguished.

  3.5.2.6.1. Characteristic Fluorescence. If the energy of the characteristic radiation from element A exceeds the absorption energy for element B in a sample containing A and B, then characteristic fluorescence of B by A will occur. To examine this situation, consider a sample containing manganese, iron, cobalt, and nickel (data, Table 3.9). The absorption energy for manganese is lower than the $K\alpha$ energies for cobalt and nickel, therefore characteristic fluorescence will occur from these radiations. The $K\beta$ energies of iron, cobalt, and nickel all exceed the critical excitation energy for the manganese K shell and so these radiations can all induce fluorescence of manganese. The arguments can be repeated for each element in the specimen as shown in Table 3.10. When characteristic

Table 3.10. Fluorescence in a Sample Containing
Mn, Fe, Co, and Ni

Element	Radiation causing fluorescence
Mn	(Fe $K\beta$; Co $K\alpha$, $K\beta$; Ni $K\alpha$, $K\beta$)
Fe	(Co $K\beta$; Ni $K\alpha$, $K\beta$)
Co	(Ni $K\beta$)
Ni	None

fluorescence occurs, the primary radiation is strongly absorbed, which is indicated by the high coefficient of mass absorption for the primary radiation in the element being fluoresced. Thus the mass absorption coefficient of manganese for Ni $K\alpha$ which fluoresces manganese is about seven times greater than the absorption coefficient of cobalt for Ni $K\alpha$, which does not fluoresce cobalt (Table 3.9). The efficiency of x-ray-induced fluorescence is greatest for x-ray energies just above the absorption edge energy of interest. For example, characteristic fluorescence of iron ($E_c =$ 7.111 keV) is more efficiently excited by Ni $K\alpha$ (7.472 keV) than by Cu $K\alpha$ (8.041 keV). The efficiency of characteristic-induced fluorescence can be estimated from the mass absorption coefficient. In the example for iron, the mass absorption coefficient of iron for Ni $K\alpha$ is 380 cm^2/g and for Cu $K\alpha$ is 311 cm^2/g, thus indicating the greater fluorescence effect of Ni $K\alpha$.

3.5.2.6.2. **Continuum Fluorescence.** The continuum radiation spectrum provides x-rays at all energies up to the beam energy. Since the most efficient production of fluorescent radiation occurs from x-rays at an energy just above the absorption energy, there will always be continuum-induced fluorescence. The calculation of the intensity of continuum fluorescence involves considering the contribution of that portion of the continuum spectrum from the absorption energy of interest E_c to the beam energy, E_0. Henoc (1968) has discussed the phenomenon and calculation in detail.

The ratio of continuum-induced fluorescence ("indirect") to the total x-ray production from electron and continuum-induced fluorescence is designated S. Experimental values compiled by Green (1962) for the K and L_{III} edges versus atomic number are plotted in Figure 3.48 (solid lines) along with calculations by Green and Cosslett (1961) (dashed lines). The contribution of continuum fluorescence for light elements is small, but rises to 10% for copper K radiation and as high as 35% for gold L radiation.

3.5.2.6.3. **Range of Fluorescence.** The range of electron-induced x-ray production is constrained to lie within the interaction volume of the electrons in the target. The range of fluorescence-induced radiation comes from a much larger volume because of the large range of x-rays in a solid as compared to electrons. Considering the case of iron in a nickel matrix, the Ni $K\alpha$ radiation (7.472 keV) can fluoresce iron K radiation ($E_K = 7.111$ keV). The range of Ni $K\alpha$ radiation in an Ni–10% Fe matrix can be calculated from Equations (3.43) and (3.44). Considering the source of Ni $K\alpha$ in the specimen as the electron interaction volume, Figure 3.49, the Ni $K\alpha$ propagates with uniform intensity in all directions from this source. Ni $K\alpha$-induced fluorescence of Fe K radiation occurs over the volume with radius indicated in Figure 3.49. The relative volumes of the regions of 50%, 75%, 90%, and 99% of Fe K fluorescence by Ni $K\alpha$ are compared in Figure

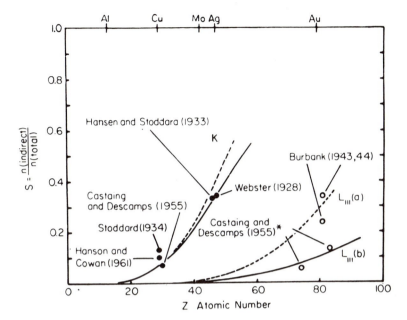

Figure 3.48. Variation of the ratio of indirect to total x-ray production with atomic number for the $K\alpha$ and $L\alpha$ lines (from Green, 1963).

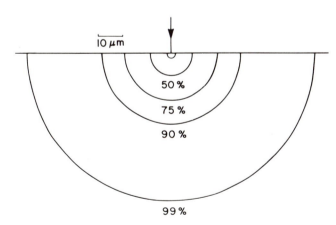

Figure 3.49. Range of secondary fluorescence of Fe $K\alpha$ by Ni $K\alpha$ in an alloy Ni–10% Fe. $E = 20$ keV. Innermost circle shows the extent of direct production of Fe $K\alpha$ by beam electrons.

3.49 with the volume of electron excitation. Note the enormous difference in the ranges of electron-induced and x-ray-induced radiation.

3.5.3. Auger Electrons

When an ionized atom undergoes deexcitation, the electron transitions can result in the emission of an x-ray as previously discussed, or in the ejection of an electron (the Auger effect). The ejected Auger electron has an energy which is characteristic of the atom because the electron transitions occur between sharply defined energy levels. In Figure 3.33, the particular Auger deexcitation path involves the filling of a K-shell vacancy by an L-shell electron with subsequent ejection of another L-shell electron.

The yield of Auger electrons is given by $1 - \omega$, where ω is the fluorescence yield for x-rays. Thus, the yield of Auger electrons will be greatest for light elements where the fluorescent yield is low. A spectrum of the electrons emitted from a pure element target under electron bombardment is shown in Figure 3.50; the number of electrons of a particular energy $N(E)$ is plotted against energy, E. The Auger electron peaks appear as a small perturbation on a high background. In order to enhance the peaks against the background, Auger electron spectra are frequently pre-

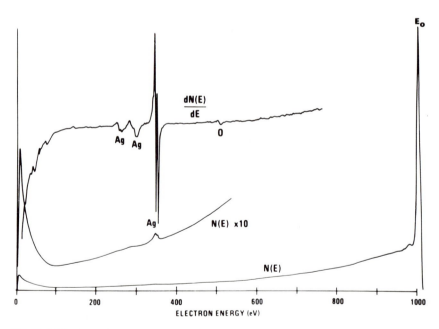

Figure 3.50. Auger electron spectra of silver with an incident beam energy of 1 keV. Derivative and integral spectra are compared (courtesy N. C. MacDonald).

sented in the differentiated mode $dN(E)/dE$ vs. E (Figure 3.50). The background consists of beam electrons which have lost varying amounts of energy before being backscattered from the specimen, forming a continuum of energy values, as well as Auger electrons which have lost energy while propagating through the sample.

The principal difference between the properties of characteristic x-rays and characteristic electrons (Auger electrons) in measuring the composition of a solid has to do with the depth of sampling of the two techniques. Both x-rays and Auger electrons are produced as a result of inner-shell ionizations caused by beam electrons, and therefore the Monte Carlo plots showing the sites of ionizations (e.g., Figures 3.7, 3.8, 3.9) give the same distribution of generation for x-rays and Auger electrons in the interaction volume. The subsequent propagation of the x-rays and Auger electrons through the sample to reach the surface occurs under radically different conditions. Inelastic scattering has a very low probability for characteristic x-rays, and so those x-rays that are not totally absorbed by the photoelectric process reach the surface unchanged. The characteristic x-ray signal is thus a measure averaged over whole interaction volume. Electrons, on the other hand, have a high probability of inelastic scattering and energy loss. For Auger electrons in the energy range 50 eV to 2 keV, the inelastic mean free path is of the order of 0.1–2 nm. Only Auger electrons emitted from atoms lying within a thin surface layer about 1 nm thick can escape the specimen carrying their characteristic energy and thus be useful for analysis. Auger electrons formed deeper in the interaction volume may escape, but they will have lost an indeterminate amount of energy and be unrecognizable in the background. Auger electron spectroscopy can provide a form of surface microanalysis, with a depth of sampling of the order of 1 nm, regardless of the depth dimensions of the interaction volume. The lateral spatial resolution of Auger spectroscopy is determined in part by the diameter of the region through which the backscattered electrons are emitted, since the backscattered electrons can create Auger electrons.

3.5.4. Cathodoluminescence

When certain materials, such as insulators and semiconductors, are bombarded by energetic electrons, long-wavelength photons are emitted in the ultraviolet and visible regions of the spectrum. This phenomenon, known as cathodoluminescence, can be understood in terms of the electronic band structure of the solid (Figure 3.51). Such materials are characterized by a filled valence band in which all possible electron states are occupied and an empty conduction band; the valence and conduction bands are separated by a band gap of forbidden states with an energy separation of E_{gap}. When an energetic beam electron scatters inelastically

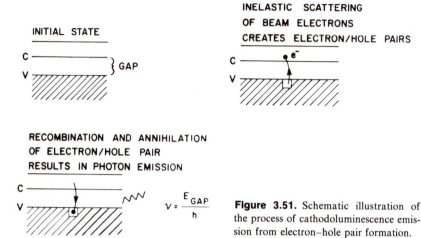

Figure 3.51. Schematic illustration of the process of cathodoluminescence emission from electron–hole pair formation.

in such a solid, electrons from the filled valence band can be promoted to the conduction band, leaving "holes," i.e., the absence of an electron, in the conduction band, which creates an electron–hole pair. In cadmium sulfide, the gap energy is 2.4 eV, while in silicon the gap energy is 1.1 eV. If no bias voltage exists on the sample to sweep the electron–hole pair apart, the electron and hole may recombine. The excess energy, equal to the gap energy, is released in the form of a photon. Since the band gap is well defined, the radiation is sharply peaked at specific energies and is characteristic of the composition. Spectra of cathodoluminescence radiation emitted from GaAlAs are shown in Figure 3.52. The decay of the electron–hole pair can be modified by the presence of impurity atoms or physical crystal defects such as dislocations, leading to shifts in the energy and intensity of the radiation, as well as the time constant of the decay.

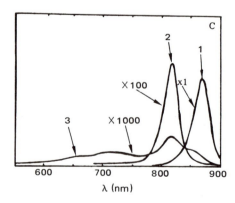

Figure 3.52. Spectra of cathodoluminescence emission from semiconductor GaAlAs: (1) bulk material; (2) thin material; (3) thin material in presence of lattice defect (Petroff *et al.*, 1978). Temperature = 25°C; $E_0 = 150$ keV.

3.6. Summary

The concepts of the interaction volume of the beam electrons and the sampling volume of the measured radiation are important both in the interpretation of images and in quantitative x-ray microanalysis. These concepts are summarized in Figure 3.53, which schematically shows the principal interactions and products for a target of intermediate atomic number (copper–10% cobalt) and a typical beam energy used for microscopy and microanalysis ($E_0 = 20$ keV). The points to note in Figure 3.53 include the following:

(1) The interaction volume for the beam is approximated as a hemisphere with a radius equal to the Kanaya–Okayama range.

(2) The sampling volume for the backscattered electrons is shown as a disk with a depth of 0.3 of the Kanaya–Okayama range. Note that the diameter of this disk is determined by the lateral extent of the interaction volume.

(3) The sampling volume of the secondary electrons is a disk with a thickness of 10 nm and a diameter equal to that of the backscattered electron disk.

(4) The x-ray generation ranges for Cu $K\alpha$ and Cu $L\alpha$ are calculated from the Kanaya–Okayama x-ray generation range equation. Since the

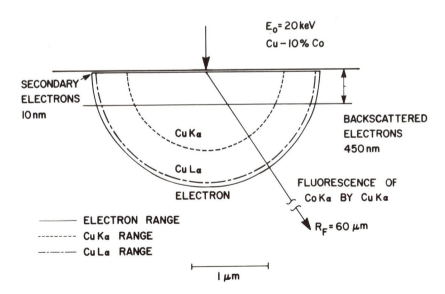

Figure 3.53. Schematic illustration of the electron beam interaction in a copper–10% cobalt alloy, showing (1) electron interaction volume, (2) backscattered electron sampling depth, (3) secondary-electron sampling depth, (4) x-ray generation range for Cu $K\alpha$ and Cu $L\alpha$, (5) fluorescence range for Co $K\alpha$ by Cu $K\alpha$.

self-absorption of copper for its own Cu $K\alpha$ x-rays is small, the sampling volume will be nearly the same size as the x-ray generation range. Cu L radiation is much more strongly absorbed. Even though the x-ray generation range is larger for Cu L than for Cu K, the sampling volume will be much smaller. The x-ray generation range for Co $K\alpha$ is slightly larger than that for Cu $K\alpha$ since the critical excitation energy is lower.

(5) The range of characteristic fluorescence of Co $K\alpha$ ($E_c = 7.71$ keV) by Cu $K\alpha$ ($E = 8.04$ keV) for 99% of the total fluorescence generation is about 60 μm, which is a factor of 50 larger than the range of direct electron excitation. Figure 3.53 can only be described as approximate since only the limits of the various processes are shown. The volume density of each process is not constant with position. Electron signals decrease exponentially with distance from the surface. X-ray signals increase to a peak value at about 0.3 of the electron range [see the $\phi(\rho Z)$ distribution in Figure 3.44] before undergoing an exponential decrease with further increases in depth.

Image Formation in the Scanning Electron Microscope

4.1. Introduction

One of the most surprising aspects of scanning electron microscopy is the apparent ease with which images of rough objects can be interpreted by newcomers to the field, or even by laymen unfamiliar with the instrument. However, there is frequently more than meets the eye in SEM images, even images of simple objects, and to gain the maximum amount of information, it is necessary to develop skills in image interpretation. Moreover, to ensure that the image has been properly constructed and recorded in the first place, a good working knowledge of the entire image formation process is necessary. In this chapter, we will consider the major aspects of the SEM imaging process: (1) the basic scanning action used for the construction of an image; (2) the origin of the commonly encountered contrast mechanisms which arise from the electron–specimen interaction; (3) the characteristics of detectors for the various signals and their influence on the image; (4) signal quality and its effect on image quality; and (5) signal processing for the final display.

4.2. The Basic SEM Imaging Process

4.2.1. Scanning Action

From Chapter 2, "Electron Optics," the functions of the electron gun, lenses, and aperture system can be briefly summarized. The electron optics of the SEM provides the microscopist with the capability of forming a beam defined by three parameters: beam current, i (in the range from 1 pA to 1 μA), beam diameter, d (5 nm to 1 μm), and beam divergence α (10^{-4}–10^{-2} sr). These parameters are not independent. The relation between them is given by the brightness equation, Equation (2.4) in Chapter 2, which may be thought of as the first of two important relations in scanning electron microscopy for determining and controlling performance. The second important relation, the threshold current/contrast equation, will be given in this chapter.

The electron beam, defined by the parameters d, i, and α, enters the specimen chamber and strikes the specimen at a single location. Within the interaction volume, both elastic and inelastic scattering occur as described in Chapter 3, producing detectable signals from backscattered electrons, secondary electrons, absorbed electrons, characteristic and continuum x-rays, and cathodoluminescent radiation. By measuring the magnitude of these signals with suitable detectors, a determination of certain properties of the specimen, e.g., local topography, composition, etc., can be made at the single location of the electron beam impact. In order to study more than a single location, the beam must be moved from place to place by means of a scanning system, illustrated in Figure 4.1. Scanning is usually accomplished by driving electromagnetic coils arranged in sets consisting of

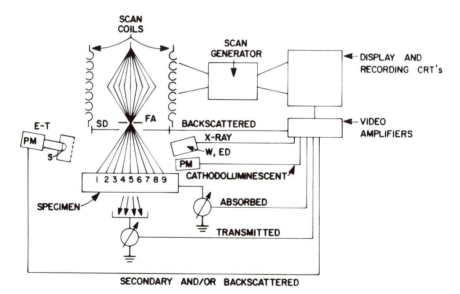

Figure 4.1. Schematic illustration of scanning system of the scanning electron microscope. Abbreviations: FA, final aperture; SD, solid state electron detector; ET, Everhart–Thornley detector; S, scintillator; PM, photomultiplier; W, wavelength-dispersive x-ray spectrometer; ED, energy-dispersive x-ray spectrometer; CRT, cathode ray tube. Numbers 1–9 indicate successive beam positions during a scanning sequence.

two pairs, one pair each for deflection in the X and Y directions. A typical double-deflection scanning system, as shown in Figure 4.1, has two sets of scan coils, located in the bore of the final (objective) lens, which drive the beam first off-axis and then back across the optic axis, with the second crossing of the optic axis taking place in the final aperture. This system has the advantage that by placing the scan coils within the lens, the region below is kept open and the specimen can be placed close to the lens. As discussed in Chapter 2, by minimizing the working distance, the spherical aberration coefficient is minimized. By locating the beam-defining aperture at the second crossover, low magnifications (large scan angles) can be obtained without cutting off the field of view on the aperture (Oatley, 1972). The beam is thus moved by the process of "scanning action" through a sequence of positions on the specimen (e.g., 1, 2, 3, . . . in Figure 4.1) as a function of time to sample the specimen properties at a controlled succession of points. In an analog scanning system, the beam is moved continuously along a line (the line scan), for example, in the X direction. After completion of the line scan, the position of the line is shifted slightly in the Y direction (the frame scan), and the process is repeated to produce a grid pattern. In a digital scan system, the beam is sent to a particular address in an X–Y grid. In this case, only discrete beam locations are selected, compared to the continuous motion of the analog system; however, the overall effect is the same. An added benefit of the digital system is that the numerical address of the beam location is accurately and reproducibly known, and therefore, the information on the electron interaction can be encoded in the form of an X, Y address with I_j values for the intensity of each signal j measured.

4.2.2. Image Construction (Mapping)

The information flow from the scanning electron microscope consists of the scan location in X–Y space and a corresponding set of intensities from each detector used to measure the electron–specimen interaction. This information can be conveniently displayed to the eye in two principal ways:

(a) Line Scans. In making a line scan, the beam is scanned along a single line on the specimen, e.g., in the X or Y direction. The same signal, which is derived from the scan generator, is used to drive the horizontal scan of a cathode ray tube (CRT). The resulting synchronous line scan on the specimen and CRT produces a one-to-one correspondence between a series of points in the "specimen space" and on the CRT or "display space." The intensity measured by one of the detectors, e.g., a secondary-electron detector, can be used to adjust the Y deflection of the CRT, which produces a trace such as that illustrated in Figure 4.2. In such a line scan

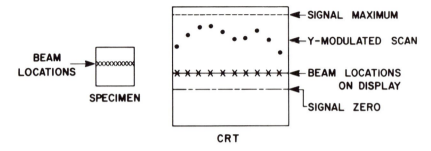

Figure 4.2. The principle of information display by line scanning, showing the locus of scan locations on the specimen and corresponding relative positions on the display CRT. The actual points on the CRT are plotted along the Y-axis depending on the strength of the signal.

displayed on a CRT, the horizontal position is related to distance along a particular line on the specimen and the vertical position to the signal strength. For example, if the intensity of a characteristic x-ray signal is used, the Y deflection of the CRT may be a measure of the amount of a particular element present. Line scans are extremely useful for diagnostic work, where the signal profile across a feature is needed. Line scans can be used to display small changes in signal which can be easily detected in Y modulation but which would be difficult to discern in a conventional intensity-modulated area image. In recording a line scan, it is necessary to obtain (1) the scan locus (i.e., the location in the area raster at which the line scan is taken), (2) the Y-modulated scan, (3) the signal zero level, and (4) the signal maximum.

(b) Area (Image) Scanning. In forming the SEM image with which we are familiar, the beam is scanned on the specimen in an X–Y grid pattern while the CRT is scanned in the same X–Y pattern, as illustrated in Figure 4.3. Again, a one-to-one correspondence is established between the set of beam locations on the specimen and the points on the CRT. To display the electron interaction information, the signal intensity S derived from the detector is used to adjust the brightness of the spot on the CRT ("intensity or Z modulation"), as shown in Figure 4.4. Thus, the creation of an SEM image consists of constructing a map on the CRT. Unlike an optical or transmission electron microscope, no true image actually exists in the SEM. In a true image, there are actual ray paths connecting points in the image as displayed on a screen or recorded on film with the corresponding points on the specimen. In the SEM, image formation is produced by the mapping operation which transforms information from specimen space to CRT space. That such an abstract creation can be readily interpreted is a considerable surprise. The information contained in the image conveys the proper shape of the object because the synchronous specimen–CRT scans are contrived to maintain the geometric relationship of any arbitrarily

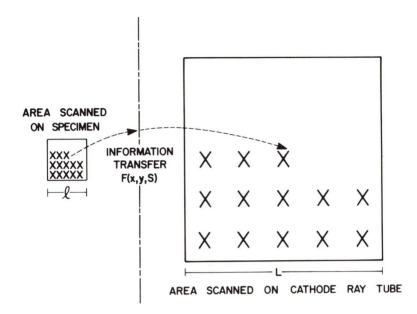

Figure 4.3. The principle of information display by area or image scanning. A correspondence is established between a set of locations on the specimen and on the CRT. Magnification = L/ l.

chosen set of points on the specimen and on the CRT. The nature of the intensity variations in the image which produce shading will be discussed subsequently. Thus in Figure 4.5, a triangle on the specimen remains a triangle of the same shape on the CRT. (Note that imperfections in the scan can lead to image distortions.) The relative size of the objects differs, as a result of the magnification.

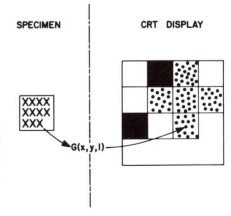

Figure 4.4. The principle of intensity or Z modulation used to display the magnitude of the signal produced by electron–specimen interaction at the locations scanned in Figure 4.3. Black represents low intensity; stippled, intermediate intensity; white, high intensity.

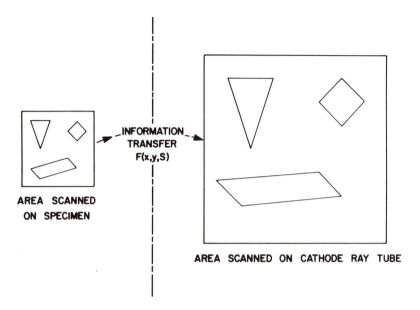

Figure 4.5. Shape correspondence between the scanned field on the specimen and the display on the CRT. In a perfect scan system, the shapes are transferred without distortion.

4.2.3. Magnification

Magnification in the SEM image is accomplished by adjusting the scale of the map on the CRT. If information along a length of line l in the specimen space is mapped along a length L in CRT space (Figure 4.3), the linear magnification M is

$$M = L/l \qquad (4.1)$$

There are a number of important points about SEM magnification:

(1) Since the CRT scan length, L, is fixed, e.g., 10 cm, an increase in magnification is achieved by decreasing the length, l, of the side of the area scanned on the specimen. It is interesting to note the size of the area sampled on the specimen as a function of magnification; see Table 4.1. The practical importance of Table 4.1 becomes clear when we consider the problem of surveying a specimen to determine its significant features. Clearly, taking only one or two images at a high magnification is not a satisfactory procedure since the total size of the area sampled represents a tiny fraction of a typical sample. For survey work, a combination of both low-magnification and high-magnification imaging must be used, and an adequate number of areas must be studied in order to gain a valid description of the specimen.

(2) Magnification in the SEM depends only on the excitation of the

Table 4.1. Area Sampled as a Function of
Magnification

Magnification	Area on sample[a]
$10 \times$	$(1 \text{ cm})^2$
$100 \times$	$(1 \text{ mm})^2$
$1\,000 \times$	$(100 \ \mu\text{m})^2$
$10\,000 \times$	$(10 \ \mu\text{m})^2$
$100\,000 \times$	$(1 \ \mu\text{m})^2$

[a] Assuming a 10×10 cm CRT.

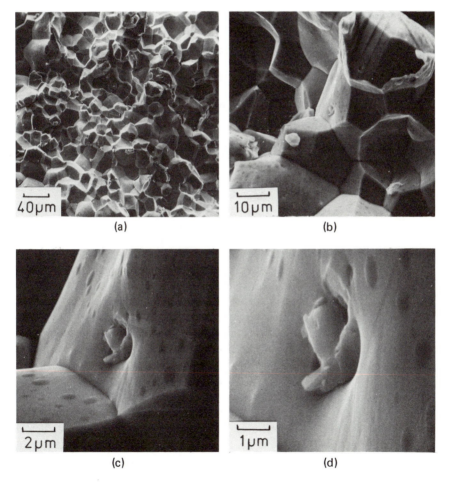

(a)   (b)

(c)   (d)

Figure 4.6. (a)–(d) Magnification series of an iron fracture surface at constant working distance and lens excitation, illustrating rapid surveying capability of the SEM and the lack of image rotation. Beam energy: 20 keV.

scan coils and not the excitation of the objective lens, which determines the focus of the beam. Thus, once the image is focused at high magnification, lower magnifications can be obtained with no further lens adjustments, which is very useful for rapid surveying of a specimen, as shown for a fracture surface in Figure 4.6.

(3) The image does not rotate as the magnification is changed, as can be seen in Figure 4.6, since the objective lens excitation is constant. As discussed in Chapter 2, a relative rotation of the image will occur if the working distance is changed, i.e., the pole piece to specimen distance, because the objective lens excitation must be then changed to maintain focus. Image rotation with a change in working distance is illustrated in Figure 4.7.

4.2.4. Picture Element (Picture Point)

An important SEM concept related to scanning action and magnification is that of the "picture element" or "picture point" size. The picture element is the region on the specimen to which the beam is addressed and from which information is transferred to a single spot on the CRT. On the high-resolution CRT used for photography, the minimum CRT spot size diameter is usually 0.1 mm (100 μm). The corresponding picture element diameter on the specimen depends on the magnification, by the following formula:

$$\text{picture element diameter} = \frac{100\ \mu\text{m}}{\text{magnification}} \tag{4.2}$$

Values of the picture element size as a function of magnification are tabulated in Table 4.2. The concept of picture element size is fundamental to an understanding of focusing, "hollow magnification," and depth of focus. An image is in true focus when the area sampled by the incident beam (allowing for the influence of the interaction volume which may have the major effect in defining the area sampled for small beam diameters) is smaller than the picture element size. Let us consider a situation in which a beam of energy 20 keV and 50 nm (500 Å) diameter is incident on gold. Most of the imaging signal is derived from backscattered electrons with energy such that $E/E_0 > 0.7$, so that the diameter at the surface of the area sampled is of the order of 100 nm (1000 Å) in diameter. From Table 4.2, images of a flat object (all points at the same working distance) below 1000 × will be in perfect focus because the picture elements are less than 100 nm in size. Above 1000 ×, the area sampled will begin to overlap more

---

Figure 4.7. (a)–(b) Images of a fracture surface at constant mangification but different working distances and lens excitation illustrating image rotation as lens strength is changed. (a) 15-mm working distance; (b) 45-mm working distance. Beam energy 20 keV.

(a)

(b)

Table 4.2. Picture Element Size as a Function of Magnification

Magnification	Picture element size
10 ×	10 μm
100 ×	1 μm
1 000 ×	0.1 μm (1000 Å)
10 000 ×	0.01 μm (100 Å)
100 000 ×	1 nm (10 Å)

than one picture element. At some point, which depends to a certain extent on the visual acuity of the individual observer, the overlap of picture elements will be perceived as a "blurring" at sharply changing features such as edges. This effect is illustrated in Figure 4.8. For the average observer, the overlap must extend to at least two picture elements to be perceived as blurring. The concept of hollow magnification arises because of the fact

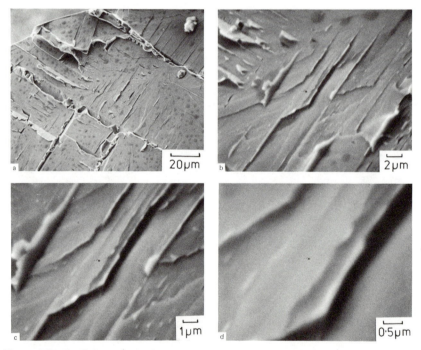

Figure 4.8. (a)–(d) Images of a fracture surface illustrating the effects of hollow magnification. The sequence (a)–(b)–(c) contains fine scale information whereas in (d) the edges are blurred. No new information is contained in (d) which cannot be seen in (c) at a lower magnification. Beam energy 15 keV.

that beyond a certain magnification, we do not gain any additional information in an image due to the overlapping of picture elements, e.g., Figures 4.8c and 4.8d.

The previous argument does not imply that we always wish to operate with the smallest possible beam. As discussed in Chapter 2, the beam current falls approximately as the square of the beam size. As will be demonstrated later in this chapter, the image quality and information content is related to the total signal which can be collected. For optimum image quality, we wish to have the maximum current in the probe compatible with the requirement that the beam diameter or sampling area is small enough relative to the picture element for adequate focusing. At low magnification where the picture element is large, e.g., 1 μm diameter at 100 $\times$, the beam size can be increased substantially to increase the total signal available without any significant degradation in apparent focus.

4.2.5. Depth of Field

The picture element concept can also be employed to understand more fully the concept of depth of field. As explained in Chapter 2, the angular divergence of the rays which constitute the focused beam causes the beam to broaden above and below the point of optimum focus. If we consider a rough specimen such that there are features at various working distances, the beam size which strikes these features will be different depending on the working distance (Figure 4.9). To calculate the depth of field, we must know at what distance above and below the optimum focus the beam has broadened to the extent that it overlaps a sufficient number of picture elements to produce noticeable defocusing. The geometrical argument illustrated in Figure 4.9 indicates that, to a first approximation, the vertical distance $D/2$ required to broaden the beam of a minimum radius r_0 to a radius r is

$$D/2 \cong r/\alpha \qquad (4.3)$$

where α is the beam divergence, as defined by the semicone angle. If we consider that defocusing will become objectionable when the beam size overlaps two picture elements (of a size 0.1 mm on the CRT) then $r = 0.1$ mm/magnification. The depth of field, D, is the vertical distance from $-D/2$ to $+D/2 = D$, so that

$$D = \frac{2r}{\alpha} = \frac{0.2 \text{ mm}}{\alpha M} \qquad (4.4)$$

Note that to increase the depth of field at a fixed probe size, the operator can either decrease the divergence or decrease the magnification. To observe a given feature, a certain magnification and probe size will be needed, so that the divergence is the only adjustable parameter. The

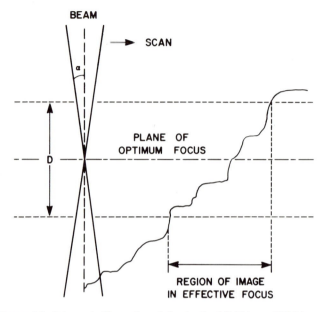

Figure 4.9. Schematic illustration of the depth of field in an SEM image.

divergence is adjusted by selection of the final aperture radius R and working distance WD:

$$\alpha = R/\text{WD} \qquad (4.5)$$

Typical final aperture sizes, specified by the diameter, available in the SEM are 100, 200, and 600 μm. The usual working distance is 10 mm, with a possible increase to 50 mm or more in some instruments. The depth of field calculated by Equation (4.4) for several combinations of possible operating parameters is listed in Table 4.3. Examples of the appearance of a rough

Table 4.3. Depth of Field at 10-mm Working Distance

Magnification	100-μm aperture ($\alpha = 5 \times 10^{-3}$ rad)	200-μm aperture ($\alpha = 10^{-2}$ rad)	600-μm aperture ($\alpha = 3 \times 10^{-2}$ rad)
10 $\times$	4 mm	2 mm	670 μm
50 $\times$	800 μm	400 μm	133 μm
100 $\times$	400 μm	200 μm	67 μm
500 $\times$	80 μm	40 μm	13 μm
1 000 $\times$	40 μm	20 μm	6.7 μm
10 000 $\times$	4 μm	2 μm	0.67 μm
100 000 $\times$	0.4 μm	0.2 μm	0.067μm

object, a fracture surface, under different conditions of the depth of field are given in Figure 4.10. Both the working distance and the aperture size are varied independently in Figure 4.10.

From the previous discussions, we can identify two distinctly different major operating modes for the SEM. (1) Depth of field mode: If we wish to study rough specimens with extensive topography, the depth of field should be maximized by choosing the smallest aperture available and the longest working distance. (2) High-resolution mode: If we wish to operate at high magnification and high resolution, the working distance should be minimized consistent with adequate signal collection and the aperture size selected according to Equation (2.20) of Chapter 2 to optimize the competing effects of the various lens aberrations. This choice of aperture and working distance will not yield the best value of depth of field. Considering the values in Table 4.3, even with a relatively large aperture and short

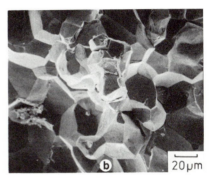

Figure 4.10. Appearance of a fracture surface with different depths of field obtained by varying the aperture size and the working distance. (a) 15-mm working distance, 600-μm aperture—small depth of field, most of field of view is out of focus; (b) 15-mm working distance, 100-μm aperture—intermediate case, some regions not in focus; (c) 45-mm working distance, 100-μm aperture—maximum depth of field, entire field in focus. Beam energy 20 keV.

working distance, the depth of field is quite substantial. Compared to the light optical microscope at the same magnification, the depth of field in the SEM is typically 10–100 times greater.

4.2.6. Image Distortions

4.2.6.1. Projection Distortions

The SEM image consists of a two-dimensional reconstruction of information derived from the specimen. Because the depth of focus of the instrument is considerable, possibly as large as the width of the scanned area, information from a three-dimensional volume of space is effectively projected onto a two-dimensional plane. In the conventional scanning system, shown in Figure 4.1, the scan plane is set at right angles to the optic axis of the instrument. Since the scanned beams diverge from a point in the final aperture, the image construction is a gnomonic projection. In a gnomonic projection, there are inherent distortions in the plane perpendicular to the optic axis. Since distance is related to the tangent of the scan angle, ϕ, an angular scan movement produces a smaller change in distance near the center of the field than at the edges. Hence the magnification varies across the field. At a nominal magnification of 10 and a working distance of 10 mm, this results in a 20% distortion near the edge of field relative to the center. At high magnification (greater than $100 \times$) the projection distortion becomes negligible because the scan angle is small and $\tan \phi \cong \phi$.

4.2.6.2. Tilt Correction

Only objects which lie in planes perpendicular to the optic axis will be reconstructed with minimum distortion. Planes which are tilted relative to the normal scan plane are foreshortened in the reconstruction, as shown in Figure 4.11a. The effective magnification varies depending on the angle of a feature relative to the normal scan plane. Only for the normal scan plane is the nominal magnification (that set on the instrument) equal to the magnification which is actually obtained. Thus, in an image of a faceted, rough object such as the fracture surface of Figure 4.6, the effective magnification varies from place to place in the image. For a highly tilted surface the magnification perpendicular to the tilt axis is less than the magnification parallel to tilt axis by a factor equal to $\cos \theta$, where θ is the tilt angle. Thus, distance measurement in such an image can only be meaningful if we know the local tilt of a surface.

To improve the collection of backscattered and secondary electrons, the specimen is commonly tilted to an angle of 30°–45° toward the

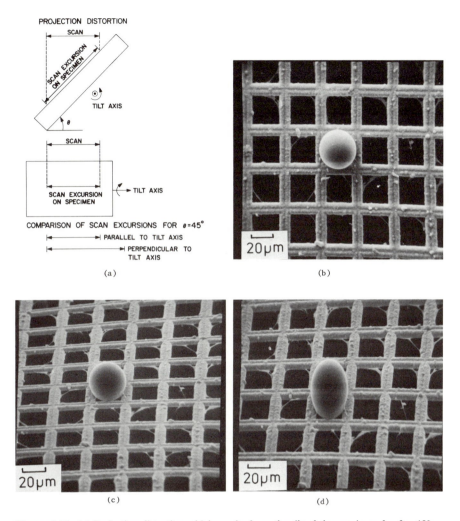

Figure 4.11. (a) Projection distortion which results from the tilt of the specimen for $\theta = 45°$. The scan has the correct length parallel to the tilt axis, but is lengthened perpendicular to the tilt axis, (b) Latex sphere on a copper grid, specimen normal to beam, (c) Specimen tilted to $\theta = 55°$, (d) Tilt correction applied. Note that the foreshortening of the grid squares in (c) is corrected in (d), but the image of the sphere is greatly distorted in the process. Beam energy 20 keV.

scintillator–photomultiplier detector. If the object is planar, the result of tilting the specimen is to introduce the projection distortion noted above. This distortion can be readily observed in the image of a grid, where the squares observed at 0° tilt (Figure 4.11b) become rectangles at 50° tilt (Figure 4.11c). A feature commonly available on the SEM is "tilt correction," illustrated schematically in Figure 4.11a. The "tilt correction" seeks to correct the image distortion by reducing the length of a scan perpendicular to the tilt axis by the factor $\cos\theta$ so that the magnification is identical in two orthogonal directions. This correction restores the grid openings to their proper shape (Figure 4.11d).

Note, however, the effect of tilt correction on the latex sphere which is also observed in Figure 4.11b. At 0° tilt, Figure 4.11b, the projection of the sphere yields the expected circle and the grid openings are squares. In the tilted, uncorrected image, Figure 4.11c, the sphere is still projected as a circle since the intersection of any plane with a sphere is a circle, while the square grid openings are distorted. When tilt correction is applied, Figure 4.11d, the spheres are projected as ellipses, while the grid openings are again seen as squares. This example illustrates the fact that tilt correction can only be applied to planar objects for which the tilt is accurately known and for which the tilt angle is everywhere the same in the field of view. If tilt correction is applied to an image such as Figure 4.11c, unnecessary distortions are introduced to the final image. Such distortions will not be obvious if the object is irregular in nature, but they will exist, and they will distort the three-dimensional reconstruction achieved by stereo pairs.

An additional feature which is found on some SEMs is that of

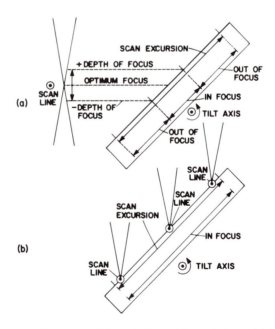

Figure 4.12. Schematic illustration of the technique of dynamic focus. (a) Normal situation for a highly tilted sample, with part of the field of view out of focus despite the depth of field, (b) Dynamic focusing with the lens adjusted as a function of scan position.

"dynamic focus," which is not to be confused with "tilt correction." In dynamic focusing, Figures 4.12a and 4.12b, the strength of the final lens is varied with the position of the scan to compensate for variations in probe size with working distance. When a highly tilted, planar object is viewed, the lens is strengthened when the scan is at the top of the field of view and weakened as the scan proceeds down the object, thus keeping the beam in optimum focus at all times. A highly tilted object will remain in focus even when its vertical displacement exceeds the depth of field, as shown in Figures 4.13a and 4.13b. Note, however, that dynamic focusing depends on the existence of a simple (and known) relationship between the scan position and the working distance. Dynamic focusing cannot be applied to rough, irregular objects; it can only be applied to flat, planar objects.

4.2.6.3. Scan Distortions

In the ideal case, the grid of picture points constructed by the scanning system would be free of distortions; that is, the distance between any adjacent pair of picture points is the same anywhere in the raster. However, a variety of distortions can exist in the scan. If meaningful measurements are to be made, it is necessary to correct these distortions, or at least be aware of them. The distortion introduced by gnomonic projection was considered above. If only irregular objects with no symmetrical shapes are examined, it will not be possible to recognize scan distortions. To detect distortions, a symmetrical object such as a sphere or grid is needed. The intersection of the normal scan plane having equal orthogonal magnification values with a sphere is a circle. Deviations from a circle reveal the existence of scan distortion, as shown in Figures 4.14a and 4.14b, where the distortion increases near the edge of the scan field. The simplest distortion is unequal scan strength in the orthogonal X, Y directions leading to unequal magnification. More complicated distortions arise from the nature of the projection and scanning system defects (barrel and pincushion distortion).

4.2.6.4. Moire Effects

Although the SEM image appears to be continuous to the human eye, it is a grid of picture points, with a periodicity. We are effectively looking at the specimen through a grating. Normally, this grating introduces no artifacts, but when the specimen itself has a periodic structure, the superposition of two gratings can lead to the formation of moire fringes, which are the interference patterns between two gratings of similar period. These patterns occasionally appear in SEM images. In Figure 4.15, a series of images of a copper grid at various magnifications shows the development of moire fringes as the magnification is reduced. At low magnification, the

(a)

(b)

Figure 4.13. (a) Image of a grid corresponding to the situation in Figure 4.12(a). (b) Image with dynamic focusing.

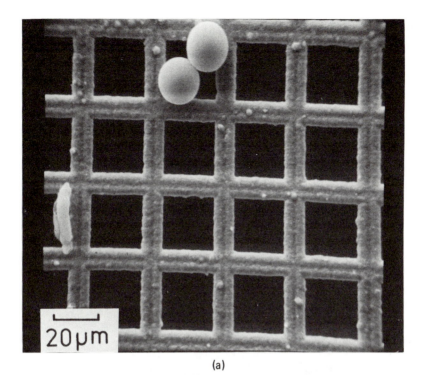

(a)

(b)

Figure 4.14. Illustration of edge distortion in an SEM image revealed by the use of latex spheres. Note pronounced distortion of the sphere to an elliptical shape as the sphere is translated from the center to the edge of the field. Beam energy 20 keV.

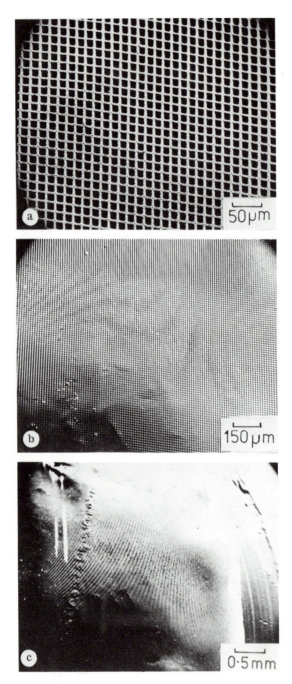

Figure 4.15. Development of moire fringes in the image of a grid as the magnification is decreased and the scanning raster periodicity approaches the grid periodicity. The dark lines are moire fringes and are artifacts; they do not exist on the specimen. (a) Regular grid pattern, high magnification; (b), (c) at lower magnifications moire fringes appear as wavy lines due to misalignment between specimen grid and scanning grid.

periodicity of the grid approaches that of the scanning grid, giving rise to the moire fringes. The dark fringes are thus an artifact of the imaging process and not a feature of the specimen.

4.3. Stereomicroscopy

The large depth of field, which typically can be equal to the scanned width or more, and the good resolution that is obtainable in the SEM make possible the effective use of stereo techniques in the examination of rough surfaces. The stereo effect is obtained by viewing two images of the same area taken with some angular difference between them. The perception of depth arises from the two slightly differing images presented to the brain by our eyes. The greater the difference in the angles the greater is the apparent depth, up to a limit where the brain can no longer "fuse" the two images together.

Two methods can be employed to produce stereo pairs on an SEM. The simplest is to translate the sample between exposures; the other is to tilt the specimen between the exposures. Both methods produce the slightly differing views needed to give stereoscopic vision. The translation or shift method is similar to that used by aerial photogrammetrists in that the specimen is moved across the field of view so that any given feature is imaged in different parts of the scan raster. The displacement between micrographs is obtained by using the stage shift controls. The area of overlap, which is common to both micrographs, can be viewed stereoscopically. In order to give a stereo image containing a sufficient depth effect, the displacement needed is fairly large, typically half the screen width at $20 \times$ magnification. This technique is not therefore suitable for high-magnification use as there would be no overlap region. The tilting method has two advantages. Firstly, it can be applied at all magnifications available on the microscope; and secondly, the whole image frame area is now usable. Depending on the magnification, a tilt difference between the two exposures of from 5° to 10° is usually optimum (Lane, 1970). This change in angle can be accomplished either by mechanically tilting the specimen in a goniometer stage, or by keeping the specimen fixed and tilting the incident beam. Because beam tilting can be accomplished very rapidly, real-time TV stereo imaging can be performed in this way (Dinnis, 1971, 1972, 1973).

The images forming the stereo pair must contain as much overlap area as possible, and this can be ensured by marking the position of some prominent feature on the first image on the visual CRT screen with a wax pencil and then aligning the same feature, after tilting, to the identical position. In general, tilting will also change the vertical position of the sample, and the second image must be restored to a correct focus. Focusing must not be done by altering the excitation of the final lens of the SEM because this will change the magnification of the second image relative to the first and cause an image rotation as well. Instead, the Z motion of the stage should be used to bring the specimen back into focus.

The stereo pair is examined with a suitable viewing optical system ("stereo viewer"), which by an arrangement of lenses or mirrors presents one micrograph image to each eye. The integrated image from the two photographs gives an appearance of depth through the brain's interpretation of the parallax effects observed. An example of a stereo pair obtained from an irregular surface is shown in Figure 4.16. When this figure is examined with a stereo viewer, the portions of the specimen that are elevated and those that are depressed can be easily discerned. Such information is invaluable for determining the elevation character of an object, i.e., whether it is above or below adjacent features. Because of the great depth of field of the SEM, such interpretation is difficult from a single image since the eye and brain lack the necessary "out of focus" detail which allows this interpretation in the optical case (Howell, 1975). It must be realized, however, that the depth effect is an illusion since the third-dimension plane normal to the plane of the photographs is formed by the observer's visual process. This may readily be illustrated by rotating the image pair through 180° such that the left micrograph is interchanged with the right. In this case, regions that originally appeared elevated are now depressed, and vice versa. A standardized convention for orienting the micrographs is thus necessary if correct interpretations are to be made. The following routine, due to Howell (1975), is suitable for all cases. The lower, less tilted, micrograph is placed for viewing by the left eye; the more tilted micrograph is placed for viewing by the right eye, with the tilt axis running parallel to the interocular plane (i.e., normal to the line between the left- and right-hand prints). When the stereo pair is "fused," surfaces which sloped towards the electron collector in the SEM slope from the right down to the left. On many instruments the tilt axis of the stage lies parallel to the

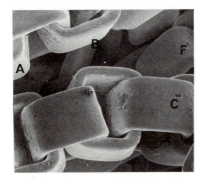 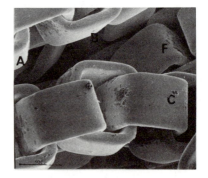

Figure 4.16. Stereo pair of a fine-link gold chain, oriented with the usual convention such that the tilt axis is vertical and the left image is taken at a lower tilt angle. The cross marks the optic axis.

bottom edge of the CRT screen, and the micrographs must therefore be rotated through 90° counterclockwise to orient them correctly. This then has the disadvantage that the image appears to be illuminated from the side rather than from the top. However, most modern SEMs have a tilt axis parallel to the specimen–detector line, so as to allow the effective illumination to be from the top of the stereo pair. Even when this desirable tilt axis is not provided, the same effect can be achieved by using rotation rather than tilt (Lane, 1970). The stage rotation R, tilt θ, and the desired parallax angle P are related by

$$\sin R = \frac{\sin P}{\sin \theta} \tag{4.6}$$

As an example, for a tilt $\theta = 34°$, and a required parallax of $P = 6°$, the rotation R must be 11°. With a stage giving clockwise motion for an increasing reading on the dial of the rotation control, the left member of the stereo pair is the image taken at the higher rotation reading. Also, the left-hand micrograph must be rotated clockwise relative to the right-hand micrograph in order to match their respective fields of view. This difficulty can be avoided if electrical image rotation is used to compensate for the mechanical stage rotation.

Quantitation of the topography of features in SEM micrographs can be carried out by measurements with stereo pairs. This can be done even if the operator is not personally able to perceive the stereo effect (as is quite often the case). The only measurements that can be made from the micrograph are the X and Y coordinates of the feature of interest. These coordinates are referred to the optic axis of the instrument, i.e., the place where the central ray of the scan raster meets the photo plane. This point is most easily found by turning the SEM up to maximum magnification and allowing the beam to produce a spot of contamination on the sample. In general the optic axis will not be at the geometrical center of the visual or record CRT, but the position once found will be constant unless the scan coil assembly is disturbed. The "+" on Figure 4.16 indicates the true position of the optic axis for the microscope that recorded these particular images.

The only parameters required are the X, Y coordinates of a feature, and the tilt angle difference α between the two halves of the stereo pair. The magnification M must also be known accurately if absolute values are needed. At normal working magnifications in an SEM it can usually be assumed that the scan is effectively moving parallel to the optic axis; very simple formulas can then be used for quantitation. With reference to a fixed point, such as the optic axis in the two photographs, the three-dimensional coordinates X, Y, Z of the chosen feature are given by

$$MZ = (P/2)\sin \alpha / 2 \tag{4.7}$$

$$MX = x_L - P/2 = x_R + P/2 \tag{4.8}$$

$$MY = y_L = y_R \tag{4.9}$$

where the parallax $P = (x_L - x_R)$, the subscripts L, R referring to the measured coordinates in the left and right micrographs, respectively. The optic axis point is thus $(0, 0, 0)$.

As an example consider the feature labeled F in Figure 4.16. In this case we have by direct measurement from the original micrographs, oriented as in the convention described above, $x_L = +5.45$ cm, $x_R = +4.30$ cm, $y_L = y_R = +2.50$ cm, $\alpha = 10.0°$, $M = 51$. So $P = (5.45 - 4.30) = 1.15$ cm, $\sin \alpha/2 = 0.087156$. Thus $Z = P/2M \sin \alpha/2 = +0.258$ cm, $X = (x_L - P/2)/M = 0.0956$ cm, and $Y = y_L/M = 0.0490$ cm.

The feature is thus over 0.25 cm above the reference at the optic axis. For other reference planes, or for cases where the parallel projection approximation does not apply, more lengthy calculations are required, and the reader is referred to the references (Wells, 1960; Lane, 1969; Boyde, 1973, 1974a, b) for further details.

4.4. Detectors

In order to form an image in the SEM, an appropriate detector must be employed to convert the radiation of interest leaving the specimen into an electrical signal for presentation to the amplification chain which provides the intensity modulation of the display–photographic CRT. From Chapter 3, we have a variety of signals available for possible selection: secondary electrons, backscattered electrons, x-rays, cathodoluminescence radiation, beam electrons absorbed by the specimen ("specimen current"), and currents induced in certain types of semiconductor specimens. In this chapter, we shall consider detectors for the electron signals and cathodoluminescence. X-ray detectors will be treated in Chapter 5.

Three important parameters in any detector system are (1) the angle relative to the specimen surface at which the detector accepts the signal of interest (the take-off angle or signal emergence angle), (2) the range of solid angle over which the detector accepts a signal (the solid angle of collection $\Omega = A/r^2$, where A is the area of the detector and r is the distance from the beam impact on the sample to the detector), and (3) the conversion efficiency, or percentage of the signal which strikes the detector and produces a response.

4.4.1. Electron Detectors

The electrons which escape the specimen fall into two classes with widely differing properties: (1) secondary electrons, which are emitted with an average energy of 3–5 eV, and (2) backscattered beam electrons which escape the specimen with a distribution in energy covering the range $0 \leqslant E \leqslant E_0$, where E_0 is the incident beam energy. The backscattered electron energy distribution is peaked at $0.8-0.9E_0$ for medium- and high-atomic-number materials.

4.4.1.1. Scintillator–Photomultiplier Detector System

The detector most commonly used in scanning electron microscopy is the scintillator–photomultiplier system, developed into its current configuration by Everhart and Thornley (1960). This detector, illustrated in Figure 4.17, operates in the following manner. An energetic electron strikes the scintillator material, which may be a doped plastic or glass target, or a compound such as CaF_2 doped with europium (for a review of scintillators, see Pawley, 1974). The electron produces photons which are conducted by a light pipe (a solid plastic or glass rod with total internal reflection) to a photomultiplier. Since the signal is now in the form of light, the signal can pass through a quartz window into a photomultiplier which is permanently isolated from the vacuum of the SEM. The photons strike the first electrode causing it to emit electrons, which then cascade through the remaining electrode stages eventually producing an output pulse of electrons with a gain of 10^5-10^6. This gain is obtained with very little noise degradation and a wide frequency bandwidth. In order to make use of the low-energy secondary electron signal, the scintillator is covered with a thin (10–50 nm) layer of aluminum and biased to approximately $+10kV$, which serves to accelerate the low-energy electrons. Note that for typical beam energies, e.g., 20 keV, most of the backscattered electrons can excite the scintillator without the aid of the high-voltage bias. In order to prevent the 10-kV bias from displacing the incident beam or introducing astigmatism, the biased scintillator is surrounded by a Faraday cage near ground potential. The Faraday cage has a mesh opening to permit the entrance of electrons. To improve the collection of secondary electrons, a positive potential of as much as $+300$ V can be placed on the cage. This voltage will not cause

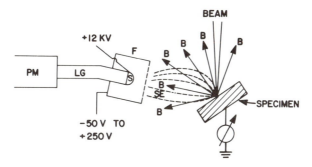

Figure 4.17. Schematic diagram of Everhart–Thornley scintillator–photomultiplier electron detector. B, backscattered electron; SE, secondary electron; F, Faraday cage; S, scintillator; LG, light guide; PM, photomultiplier.

significant degradation of the incident beam. To reject secondary electrons, the potential on the Faraday cage can be set to -50 V or else the high voltage on the scintillator can be removed.

The take-off angle and solid angle of collection for a scintillator–photomultiplier detector are illustrated in Figure 4.18a for a flat sample normal to the beam. Those high-energy backscattered electrons which leave the specimen with motion directly toward the face of the scintillator (that is, along a "line of sight") are collected; all other backscattered electrons are lost. The effect of the bias of the Faraday cage on backscattered electrons is negligible, regardless of whether the bias is positive or negative. Considering the cosine angular distribution over which the backscattered electrons are emitted at normal beam incidence, the portion of the distribution which is collected at different take-off angles is illustrated in Figure 4.18b. Because of the nature of the cosine distribution, at low take-off angles only a small fraction of the backscattered electrons are collected.

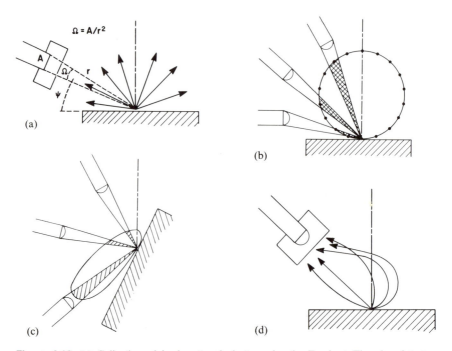

Figure 4.18. (a) Collection of backscattered electrons by the Everhart–Thornley detector showing the solid angle of acceptance, Ω, and the take-off angle, Ψ. (b) Relative collection of backscattered electrons emitted in a cosine distribution by ET detectors placed at various take-off angles. (c) Relative collection of backscattered electrons emitted from a highly tilted surface by ET detectors placed at various take-off angles. (d) Schematic illustration of deflection of trajectories of secondary electrons of various energies by positive potential on Faraday cage.

For a typical SEM scintillator–photomultiplier detector, the take-off angle for backscattered electrons is on the order of 30°, and the solid angle of collection is about 0.05 sr (1 cm diameter scintillator, 4 cm from specimen). When the specimen is highly tilted, the angular distribution of backscattered electrons becomes skewed in the forward direction. The solid angle of collection convolved with this angular distribution is shown in Figure 4.18c. For a tilted specimen, a detector placed at a low take-off angle in the forward scattering direction has the highest collection efficiency. The response of the scintillator as a function of beam energy is dependent on the exact material used but generally increases with increasing beam energy.

It is more difficult to define the detector take-off angle and the solid angle of collection for secondary electrons. The positive bias on the Faraday cage causes a substantial deviation of the secondary electron trajectories toward the detector, shown in Figure 4.18d, which increases the solid angle of collection quite substantially over that determined by the simple geometry used for backscattered electrons, illustrated in Figure 4.18a. Secondary electrons can often be collected even if no direct line of sight exists between the point of emission and the scintillator. From a flat surface, the collection efficiency can approach 100%. Only by applying a negative bias to the Faraday cage can the secondary electrons be rejected from the scintillator.

In summary, then, the conventional scintillator–photomultiplier or Everhart–Thornley electron detector has the following features:

(1) The electron signal is greatly amplified with little noise introduced and with a wide bandwidth compatible with TV-frequency scanning.

(2) Both secondary electrons and backscattered electrons can be detected.

(3) The geometric collection efficiency for backscattered electrons is low, about 1%–10%, while for secondary electrons it is high, often 50% or more.

(4) Secondary electrons can be selectively removed from the imaging signal by applying a negative potential to the Faraday cage.

(5) As long as a line of sight exists between the point at which the electron beam strikes the specimen and the scintillator, a backscattered electron signal component will exist in the image. In practice, a more complicated behavior may be encountered with the Everhart–Thornley detector. The high-energy backscattered electrons emitted from the sample which do not strike the scintillator may nevertheless contribute to the signal. These backscattered electrons may strike the walls of the sample chamber, creating secondary electrons which are efficiently collected by the positive potential on the Faraday cage. This increases the collection efficiency for the backscattered electron signal over the direct line of sight argument of Figure 4.18a.

4.4.1.2. Scintillator Backscatter Detectors

There are usually 2 to 5 times more backscattered electrons emitted from the sample than secondary electrons. Backscattered electrons also carry much useful information on specimen composition, topography, crystallinity, etc. Several detectors based on the scintillator have been developed to make greater use of the backscattered electron signal (Wells, 1977; Robinson, 1980).

(a) Large-Angle Scintillator Detector. Wells (1974a) and Robinson (1975) have developed large scintillator detectors placed in close proximity to the specimen to maximize the solid angle of collection, which in some arrangements can approach 2π sr. An example of such a detector is shown in Figure 4.19a, where the scintillator is placed above the specimen. This set-up provides a high take-off angle and high collection efficiency for a sample set normal to the beam. Note that if the specimen is tilted with the detector in this position, the collection efficiency drops sharply since the backscattered electrons are emitted in the forward direction away from the detector.

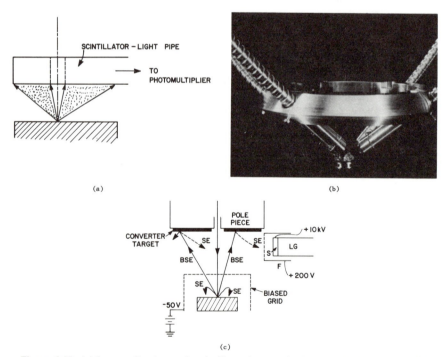

Figure 4.19. (a) Large collection angle scintillator detector for backscattered electrons (Wells, 1974; Robinson, 1975). (b) Multiple scintillator detector array (Jackman, 1980). (c) Conversion-type backscatter detector (Moll *et al.*, 1978; Reimer and Volbert, 1979).

(b) Multiple Scintillator Arrays. An alternative approach is to replace the single large scintillator of Figure 4.19a with an array of scintillator detectors, Figure 4.19b, each coupled by its own light pipe to a photomultiplier (Jackman, 1980). Through the use of optical switches, the signals from these detectors can be viewed individually to give asymmetric detection, summed to give the equivalent of a large symmetric detector, or subtracted. The utility of some of these mixed signal modes will be subsequently discussed.

(c) Conversion Detectors. In the backscattered electron conversion detector (Moll *et al.* 1978; Reimer and Volbert, 1979) the phenomenon of secondary electron generation by the backscattered electrons striking the chamber walls is utilized. The signal can be increased by placing a material with a high secondary electron coefficient, such as MgO, in proximity to the specimen. In practice, a convertor target is attached to the polepiece, as shown in Figure 4.19c. Secondaries generated at this target are collected by the positive bias on the conventional Everhart–Thornley detector. To exclude secondaries produced directly at the specimen, a biased grid is placed above. Since secondary-electron production increases with decreasing incident electron energy, the conversion detector produces the opposite energy response to a scintillator which is struck directly by the backscattered electrons.

4.4.1.3. Solid State Detectors

The solid state electron detector uses the electron–hole pair production induced in a semiconductor by energetic electrons. The electronic structure of the semiconductor consists of an empty conduction band separated by a band gap from a filled valence band. When the energetic electrons scatter inelastically, electrons are promoted to the conduction band, where they are free to move, leaving behind holes in the valence band, which can also move under applied potential, Figure 4.20 (Kimoto and Hashimoto, 1966; Gedcke *et al.* 1978). Left to themselves, the free electron and hole will eventually recombine. Under an applied potential, the electron and hole can be swept apart. This potential can be supplied by an external circuit or by the self-bias generated by a p–n junction. For silicon, approximately 3.6 eV is expended per electron–hole pair. For a 10-keV electron striking the detector, a current of up to 2800 electrons will flow from the detector. This signal must then be amplified with a current amplifier to produce a suitable signal for eventual video display.

A number of important features of solid state detectors can be noted:

(a) The solid state detector has the form of a flat, thin (several millimeters) wafer which can be obtained in a variety of sizes, from small squares to large annular detectors in the form of a ring with a hole to

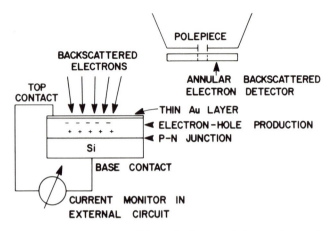

Figure 4.20. Schematic diagram of a solid state backscattered electron detector based upon a *p–n* junction; typical location of an annular-type backscattered electron detector on the bottom of the final lens.

permit passage of the electron beam. In the simplest case, an inexpensive solid state detector can be obtained from certain solar cells (with protective casings and overlayers removed) (Wells, 1978).

(b) The solid state detector can be placed close to the specimen, giving a high geometric efficiency. Multiple detector arrays can be easily employed, allowing for signal mixing.

(c) The detector is sensitive to high-energy backscattered electrons only and not to secondary electrons. The detector is also affected by x-rays, but these are relatively rare compared to the number of electron events. The solid state detector can of course detect secondary electrons providing they are first accelerated to a sufficient energy by suitable potentials as in the scintillator detector system (Crewe, *et al.*, 1970).

(d) The total efficiency depends on the energy of the electron detected. The response function usually increases linearly above a threshold energy level (caused by the necessity of the electron having sufficient energy to penetrate the outer electrode and inactive silicon layer), as shown in Figure 4.21. Figure 4.21 shows that the solid state detector gives a greater response (detector gain) from the desirable high-energy backscattered electrons (Chapter 3). The existence of a threshold serves to totally eliminate low-energy backscattered electrons (< 5 keV), which are least desirable since they originate farthest from the beam impact point. The actual response of a solid state detector can be conveniently measured by placing it directly under the beam and varying the beam energy at constant current.

(e) The solid state detector acts to boost the signal by about three orders of magnitude prior to current amplification. A current amplifier is required, preferably of the operational amplifier type, which can also be used to amplify the specimen current signal (Fiori *et al.*, 1974).

(f) Because of the capacitance of the silicon chip, the solid state detector usually has a relatively narrow bandwidth, which prevents its use at fast (TV) scanning rates. Advanced detectors have been developed which have a low enough capacitance to operate with wide bandwidth (Gedcke *et al.*, 1978).

4.4.1.4. Specimen Current (The Specimen as Detector)

The specimen is a junction with currents flowing into and out of it. The possible currents involved are illustrated in Figure 4.22. The beam

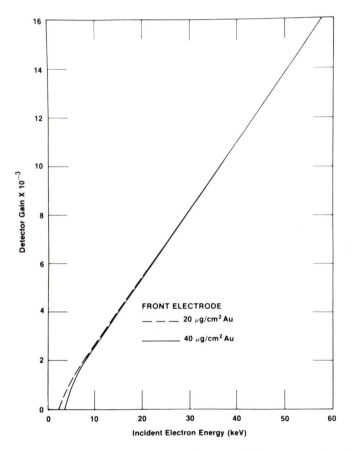

Figure 4.21. Response of an advanced solid state detector with electron energy (from Gedcke *et al.*, 1978).

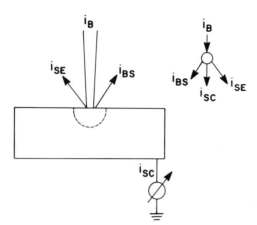

Figure 4.22 Illustration of currents which flow in and out of the specimen: i_B, beam current; i_{BS}, backscattered electron current; i_{SE}, secondary-electron current; i_{SC}, specimen current. The junction equivalent of the specimen is also shown.

electrons represent current flow into the junction (specimen) while the backscattered electrons and secondary electrons are currents flowing out of the junction. Note that in this discussion of current flow, all electrons are equivalent, regardless of their energy, since currents involve the passage of charges per unit time. Thus, a 15-keV backscattered electron and a 4-eV secondary electron both represent one charge exiting the specimen.

Thevinin's theorem states that the summation of all currents entering and leaving a junction must equal zero, or else charge accumulation will occur. In the case of the specimen in an SEM, the magnitude of the currents is as follows: considering a target such as copper, a beam with an energy of 20 keV and a current i_B will produce a backscattered electron current, i_{BS}, of about $0.3i_B$ and a secondary electron current, i_{SE}, of $0.1i_B$. The total current leaving the specimen as a result of backscattering and secondary-electron emission is thus only $0.4i_B$. Unless another current path is provided, the specimen will charge since $i_{out} < i_{in}$. A current path can be provided with a connection, usually a wire or conducting paint to the common electrical ground of the instrument (Figure 4.22). A current, the specimen current i_{SC} (also known as absorbed current or target current), will flow through this wire, with a value equal to

$$i_{SC} = i_B - i_{BS} - i_{SE} \qquad (4.10)$$

This equation assumes the specimen does not collect secondaries formed at the chamber walls by the backscattered electrons. A current balance is obtained:

$$i_{in} = i_B = i_{out} = i_{BS} + i_{SE} + i_{SC} \qquad (4.11)$$

The specimen current signal can be made sensitive to backscattering effects only by biasing the specimen positively by about 50 V to efficiently recollect secondary electrons and prevent their escape. Equation (4.11) then

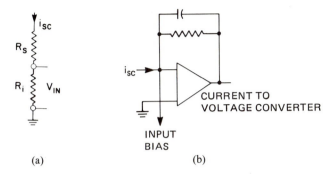

Figure 4.23. (a) Schematic illustration of high-resistance specimen current amplifier: R_S, specimen resistance; R_i, resistance of input resistor; i_{SC}, specimen current; V_{in} is the voltage developed across the input resistor. (b) Schematic diagram of a current-to-voltage converter type of specimen current amplifier.

becomes

$$i_B = i_{BS} + i_{SC} \tag{4.12}$$

In order to make use of the specimen current signal, the specimen current must be routed through a current amplifier on its way to ground. Two types of current amplifiers are typically encountered, as shown in Figure 4.23. (1) In Figure 4.23a, a large resistance is placed in series with the specimen current in order to produce a suitable voltage input to an amplifier. In order to produce a voltage drop of 1 V across the resistor, a resistance of 10^9–10^{10} Ω or more must be employed for beam currents in the 10^{-9}–10^{-11}-A range. This resistance is difficult to maintain between the specimen and ground in view of other possible paths in the specimen-stage area. (2) In a more modern form of specimen current amplifier, Figure 4.23b, the specimen current is directed to the virtual earth of an operational amplifier, alleviating the problem of providing a large resistance between the specimen and ground. This type of amplifier can be designed to operate with specimen currents as small as 10^{-11} A, while retaining an adequate bandwidth to pass high-frequency components of an image (Fiori *et al.*, 1974).

4.4.2. Cathodoluminescence Detectors

The phenomenon of cathodoluminescence, the emission of electromagnetic radiation in the ultraviolet, visible, or infrared wavelengths during electron bombardment, is useful for the characterization of minerals, semiconductors, and biological specimens. The radiation is detected with a photomultiplier and may be dispersed through an optical spectrometer prior to the photomultiplier to allow narrow-band spectral measurements.

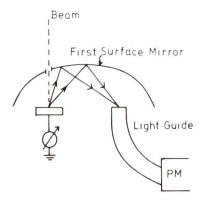

Figure 4.24. Ellipsoidal mirror collector for cathodoluminescent radiation. The specimen is placed at one focus and the light is reflected to the other focus, where it is transmitted to a photomultiplier.

The critical element in the design of a detection system is the coupling with the specimen. In the simplest case, this coupling consists of a lens and a light pipe to collect light over a reasonable solid angle of emission. In high-sensitivity detection schemes, an ellipsoidal mirror is placed over the specimen, with the specimen at one focus, as shown in Figure 4.24 (Horl and Mugschl, 1972; van Essen, 1974). An entry hole allows the beam access to the specimen. Light emitted by the specimen is reflected by the mirror to the second focus, where a light pipe is placed to transmit the light to the photomultiplier. Collection of nearly 100% of the emitted light is possible with such a system.

The ellipsoidal mirror which surrounds the specimen also precludes the efficient collection of any emitted electron signal. In this situation, the specimen current signal is useful for electron imaging, since it is unaffected by the presence of the mirror. The signal depends only on the exit of the backscattered and secondary electrons from the sample.

4.5. The Roles of the Specimen and Detector in Contrast Formation

4.5.1. Contrast

Contrast can be defined according to the following equation:

$$C = (S_{max} - S_{min})/S_{max} \qquad (4.13)$$

where S_{max} and S_{min} represent the signals detected at any two points in the scan raster; by this definition C is positive and $0 \leqslant C \leqslant 1$. Contrast represents the information in the signal which can be related to the properties of the specimen which we wish to determine. In discussing contrast, we must consider the specimen and the detector of interest as a closed system. The contrast which we can observe must be initially created by events within the specimen (e.g., scattering from different kinds of atoms) or in its immediate vicinity (e.g., by electric or magnetic fields just above its surface). Contrast can be subsequently modified by the characteristics of the particular detector used. However, the signal leaving the detector contains all the information available for the particular set of imaging conditions employed. Subsequent amplification and signal processing, described in following sections, can only serve to control the way in which the information is displayed. The information in the signal cannot be increased after the detector. In this section we will consider the two basic contrast mechanisms, atomic number (or compositional) contrast and topographic contrast, which will be encountered in the general examination of specimens, whether of biological or physical origin.

4.5.2. Atomic Number (Compositional) Contrast (Backscattered Electron Signal)

Let us consider the simplest possible specimen (Figure 4.25a), a solid block of an amorphous, pure element, which is effectively infinite in thickness relative to the electron range for the beam energy employed. The area scanned on the sample is much smaller than the lateral area of the sample, so that the beam (and the interaction volume) never approaches the edge of the specimen. Under such conditions, the signals emitted from all beam locations are identical, except for statistical fluctuations (to be discussed subsequently under Image Quality). Equation (4.13) shows there is no contrast observable. Now consider a slightly more complicated sample (Figure 4.25b), which consists of two distinct regions of different pure elements 1 and 2 with $Z_2 > Z_1$ separated by a sharp interface. The area scanned on the specimen crosses this interface, so that beam locations on element A and element B are present in the image. From the nature of beam–specimen interactions described in Chapter 3, the electron current leaving the specimen at locations 1 and 2 can differ for two reasons: (1) the number of backscattered electrons is a strong and predictable function of the atomic number, and (2) the number of secondary electrons is also dependent on atomic number, but to a smaller and less predictable degree. We shall assume that the secondary emission is similar in both elements. Considering only the backscattered electrons, contrast will exist in the signal measured between locations 1 and 2. If we make use of a negative-biased scintillator–photomultiplier detector or a solid state detector, a signal will be derived which is proportional to the number of backscattered electrons:

$$S_{\text{detector}} \propto \eta \qquad (4.14)$$

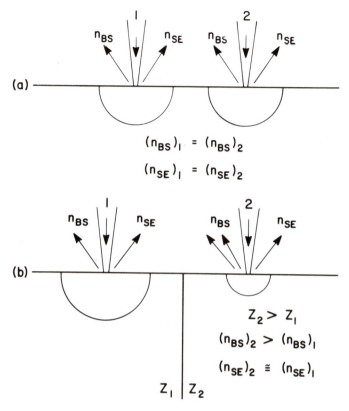

Figure 4.25. Schematic illustration of electron currents emitted from a sample which is (a) the same composition at two beam locations and (b) different in composition. n_{BS}, number of backscattered electrons; n_{SE}, number of secondary electrons.

The contrast, then, will be

$$C = \frac{S_2 - S_1}{S_2} = \frac{\eta_2 - \eta_1}{\eta_2} \tag{4.15}$$

Since the backscattering coefficient varies in a generally smooth, monotonic fashion with atomic number, the backscattered electron signal can be used to derive useful information about the relative difference in average atomic number of regions of a specimen. This contrast mechanism is known as "atomic number contrast" (also, "compositional contrast" or "Z contrast"). An example of this contrast mechanism is shown in Figure 4.26. In Figure 4.26a, the bright regions have a higher atomic number than the dark regions.

Equation (4.15) introduces the concept of a contrast calculation, that is, relating a property of the specimen, in this case composition as it affects

 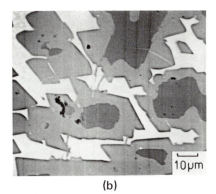

10μm

(a)

10μm

(b)

Figure 4.26. (a) Backscattered electron image derived from a reverse-biased Everhart–Thornley detector. (b) Direct specimen current image of the same region as shown in Figure 4.26a. (Note: the faint vertical lines arise from interfering stray signals.) Specimen: Raney nickel alloy (aluminum–nickel); beam energy 20 keV.

backscattering, to the contrast information which is potentially available in the image.

Considering the properties of backscattering discussed in Chapter 3, the following characteristics of atomic number contrast can be deduced:

(1) Because of the monotonic increase of η with Z (or $\bar{Z}$ in the case of compound targets), regions of high average atomic number will appear bright relative to regions of low atomic number in a specimen.

(2) The contrast calculated from Equation (4.15) and Reuter (1971) fit to η vs. Z, [see Equation (3.14), Chapter 3] for several combinations of elements is listed in Table 4.4. For example, elements separated by only one unit in atomic number, produce low contrast, e.g., Al and Si yield a contrast of 0.067 (6.7%). For elements with a great difference in atomic number, the contrast is much stronger, e.g., Al and Au yield a contrast of 0.69 (69%).

Table 4.4. Atomic Number Contrast

Z_A	Z_B	η_A	η_B	C
13(Al)	14(Si)	0.153	0.164	0.067
13(Al)	26(Fe)	0.153	0.279	0.451
13(Al)	79(Au)	0.153	0.487	0.686
	Adjacent pairs of elements			
13(Al)	14(Si)	0.153	0.164	0.067
26(Fe)	27(Co)	0.279	0.287	0.028
41(Nb)	42(Mo)	0.375	0.379	0.013
57(La)	58(Ce)	0.436	0.439	0.0068
78(Pt)	79(Au)	0.485	0.487	0.0041

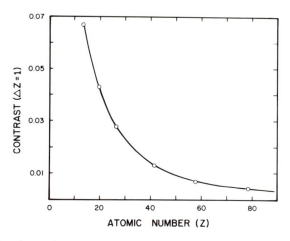

Figure 4.27. Atomic number contrast from adjacent pairs of elements ($\Delta Z = 1$). Contrast from backscattered electrons only.

(3) Considering adjacent pairs of elements in the Periodic Table, as the atomic number increases, the atomic number contrast decreases as a result of the decrease in the slope of the η vs. Z curve with increasing Z. For unit differences in Z, the contrast is 6.7% for Al–Si while for Au–Pt the contrast has fallen to 0.41%. The contrast between adjacent pairs of elements as calculated above is plotted in Figure 4.27 and several examples are listed in Table 4.4. In Figure 4.27, the possible contribution of secondary electrons has been ignored. Thus, the calculated values of contrast are appropriate to a detector sensitive to backscattered electrons only, such as a solid state detector or negatively biased Everhart–Thornley detector. If secondary electrons are considered as well, the contrast calculation becomes more complicated:

$$C = \frac{S_2 - S_1}{S_2} = \frac{(e_{BS}\eta_2 + e_{SE}\delta_2) - (e_{BS}\eta_1 + e_{SE}\delta_1)}{(e_{BS}\eta_2 + e_{SE}\delta_2)} \qquad (4.16)$$

where e_{BS} is the collection efficiency for backscattered electrons and e_{SE} is the collection efficiency for secondary electrons. In general, $e_{SE} > e_{BS}$ for an Everhart–Thornley detector for the reasons discussed previously. Thus, the actual magnitude (and sense) of the atomic number contrast can be significantly affected by the secondary electrons if δ_1 and δ_2 differ. The secondary-electron coefficient is not a strong function of atomic number for beam energies above 10 keV. However, below 5 keV, the increase in the secondary-electron coefficient can strongly affect the observed atomic number contrast. The lack of reliable data on secondary-electron coefficients, especially at low beam energy, makes interpretation of atomic number contrast difficult in this energy range.

(4) The directionality of the emission of backscattered electrons can affect atomic number contrast through the placement of the detector relative to the specimen. At normal beam incidence, the backscattered electrons follow a cosine distribution, so that a detector placed at a high take-off angle will intercept a larger fraction of backscattered electrons than that collected at a low take-off angle. Thus, an annular solid state detector placed above the specimen at a high take-off angle will be more effective for imaging atomic number contrast than a detector at a low take-off angle.

(5) The energy spectrum of backscattered electrons becomes progressively more peaked at high values of E/E_0 as the atomic number increases (Chapter 3). Thus a detector which is particularly sensitive to high-energy electrons will tend to enhance the atomic number contrast of high-Z elements relative to low-Z elements. The solid-state detector has such an energy-dependent response (Figure 4.21).

(6) The special properties (energy response, collection efficiency, etc.) of electron detectors such as the solid state detector or Everhart–Thornley detector can substantially change the atomic number contrast from the value predicted by the contrast calculation. The only signal which provides contrast similar to that calculated by Equation (4.15) is the specimen current signal. This is particularly true if the specimen is positively biased to recollect secondary electrons so that backscattering alone affects the specimen current signal. The contrast in the specimen current signal is insensitive to trajectory effects because the only requirement for affecting the magnitude of the specimen current is that an electron leave the specimen; the direction in which it leaves can have no effect.

The magnitude of the contrast in the specimen current signal can be related to the contrast in the total backscattered electron signal by the following argument. From Equation (4.12) for the biased specimen case we obtain

$$i_B = i_{BS} + i_{SC} \tag{4.17}$$

Considering the difference in these signals, Δ, between any two points in the image we have

$$\Delta i_B = 0 = \Delta i_{BS} + \Delta i_{SC} \tag{4.18}$$

since the beam current is constant during scanning. Rearranging Equation (4.18) gives

$$\Delta i_{BS} = -\Delta i_{SC} \tag{4.19}$$

That is, the sense of contrast between two points in an image is reversed when a backscattered electron and a specimen current image are compared, as shown in Figures 4.26a and 4.26b. Since the specimen current decreases as the atomic number increases, the dark regions in Figure 4.26b have the highest atomic number. Dividing both sides by the beam current, i_B, we obtain

$$\frac{\Delta i_{BS}}{i_B} = \frac{-\Delta i_{SC}}{i_B} \tag{4.20}$$

By definition $i_{BS} = \eta i_B$ and $i_{SC} = (1 - \eta)i_B$

$$\eta \frac{\Delta i_{BS}}{i_{BS}} = \frac{-(1 - \eta)\Delta i_{SC}}{i_{SC}} \tag{4.21}$$

$$\frac{\Delta i_{SC}}{i_{SC}} = -\left(\frac{\eta}{1 - \eta}\right)\frac{\Delta i_{BS}}{i_{BS}} \tag{4.22}$$

Thus, the values of $C = \Delta i_{BS}/i_{BS}$ in Table 4.4 must be modified by the factor $-[\eta/(1 - \eta)]$ to give the contrast in the specimen current signal.

4.5.3. Compositional Contrast (Secondary-Electron Signal)

Generally, the secondary electron coefficient does not vary strongly with atomic number when pure elements are measured, so that compositional contrast is not usually observed with this signal. Some interesting exceptions have been presented in the literature (Lifshin and DeVries, 1972; Sawyer and Page, 1978) in which strong compositional contrast is observed in the secondary electron signal. An example of such contrast, from the work of Sawyer and Page, is shown in Figure 4.28. The secondary plus backscattered electron image of the reaction-bonded silicon carbide shows strong contrast between the interior and exteriors of silicon carbide grains. The backscattered electron image of the same region shows no contrast between these areas; only normal atomic number contrast between the SiC grains and the intergranular silicon is observed. Sawyer and Page (1978) propose that the contrast in the secondary-electron signal arises from differences in the secondary-electron coefficient controlled by the impurity content. Since the silicon carbide is a semiconductor, the presence of trace levels of impurities can modify the electron acceptor levels in the electronic band structure, which controls the secondary-electron emission. The overall change in chemical composition due to the presence of these impurities is so small that there is not enough of a difference in the average atomic number to produce perceptible atomic number contrast in the backscattered electron image.

Compositional contrast in the secondary-electron signal is very sensitive to the condition of the sample surface. An evaporated carbon layer

$\longrightarrow$

Figure 4.28. (a) Compositional contrast observed in the secondary plus backscattered electron image of reaction bonded silicon carbide. (b) Backscattered electron image of the same region. Both images obtained with an Everhart–Thornley detector. Beam energy 20 keV. (Courtesy Trevor Page, University of Cambridge.)

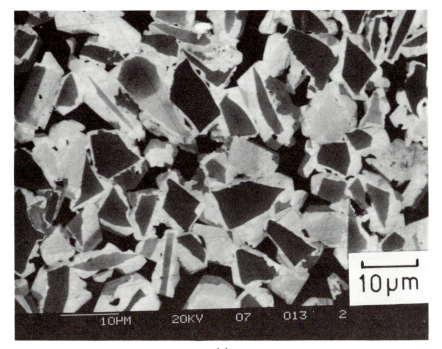

(a)

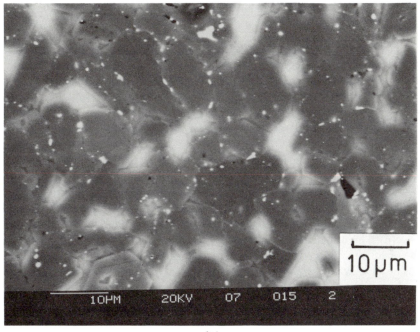

(b)

applied to the specimen or a contamination layer produced during electron bombardment can completely suppress the contrast seen in Figure 4.28a.

4.5.4. Contrast Components

The previous discussion demonstrates that the contrast which eventually resides in the signal leaving the detector is a complicated function of the characteristics of the specimen, the signal-carrying electrons, and the detector. We can identify various components of the contrast:

(a) Number Component. The number component refers to contrast arising as a result of different numbers of electrons leaving the specimen at different beam positions as a result of differences in specimen properties. Atomic number contrast in the specimen current signal is an example of pure number contrast.

(b) Trajectory Component. The trajectory component refers to contrast effects resulting from the path the electrons travel after leaving the specimen.

(c) Energy Component. The energy component of the contrast arises from the property that the contrast may be carried by a specific portion of the backscattered electron energy distribution.

4.5.5. Topographic Contrast

4.5.5.1. Origin

Topographic contrast arises because backscattering and secondary-electron generation depend on the angle of incidence between the beam and the sample. The angle of incidence will vary because of the roughness (topography) of the sample, leading to contrast formation which is related to the physical shape of the sample. Topographic contrast is the most frequently encountered contrast mechanism in general-purpose scanning electron microscopy.

From the information presented in Chapter 3, the following effects can be expected to contribute to the formation of topographic contrast:

(i) The total backscatter coefficient increases as a monotonic function of the specimen tilt. This effect produces a number component in the topographic contrast in the backscattered electron signal.

(ii) The angular distribution of backscatter electrons is dependent on the specimen tilt. At normal incidence, $\theta = 0°$, the distribution follows a cosine law, while at $\theta > 0°$, the distribution becomes progressively more peaked in the forward scattering direction and the maximum is contained in a plane defined by the beam and surface normal vectors. The directionality of backscattering from tilted surfaces provides a trajectory component in the topographic contrast in the backscattered electron signal.

(iii) The total secondary-electron coefficient varies with the angle of tilt of the surface, approximately as $\delta = \delta_0 \sec \theta$. Tilted surfaces thus produce more secondary electrons than surfaces normal to the beam, which introduces a number component in the topographic contrast in the secondary-electron signal. The angular distribution of secondary electrons does not vary significantly with tilt.

The contrast which is actually observed in the signal is a complex convolution of the beam–specimen interaction properties and the detector properties. We will consider the contrast as it appears with each type of detector, considering a representative of the general class of rough objects, a faceted fracture surface, such as that shown in Figure 4.10.

4.5.5.2. Topographic Contrast with the Everhart–Thornley (ET) Detector

(a) Backscattered Electrons. We shall first consider the appearance of the backscattered electron portion of topographic contrast with the ET detector, i.e., the Faraday cage is biased negatively to exclude secondary electrons, Figure 4.29a. Under this condition, three properties of the ET detector principally affect the contrast observed. (i) The detector views the sample anisotropically, since it is placed on one side; (ii) the solid angle of collection for backscattered electrons is small, so that only electrons scattered in the exact direction of the detector will be counted; (iii) the detector direction is at a high angle to the beam, usually greater than 50° (i.e., the take-off angle is low). The appearance of the sample under this condition is shown in Figure 4.29b and is characterized by having regions producing very high signals as well as regions producing essentially no signal whatsoever. As a result, the contrast in the final image is harsh, consisting mainly of white areas and black areas, with relatively few regions at intermediate gray levels. This appearance can be explained as follows. Although all surfaces struck by the primary beam produce backscattered electrons, only those surfaces which face the scintillator will direct at least some portion of their backscattered electrons toward it, producing a usable signal. Any surface tilted away from the detector will send very few backscattered electrons toward the detector, resulting in a low signal, and hence little information, other than the fact that the surface must be tilted away from the detector.

Since we human observers are used to interpreting images according to the environment in which we function, namely, light interacting with matter, it is useful to consider the light optical analogy to SEM images where possible (Oatley *et al.* 1965). The light optical analogy to Figure 4.29b is illustrated in Figure 4.29c. The image is equivalent to what we might see if a rough object were illuminated only by a strongly directional light source, e.g., a flashlight, placed on one side of the specimen at a low angle, i.e., at oblique incidence. The observer would look down on the

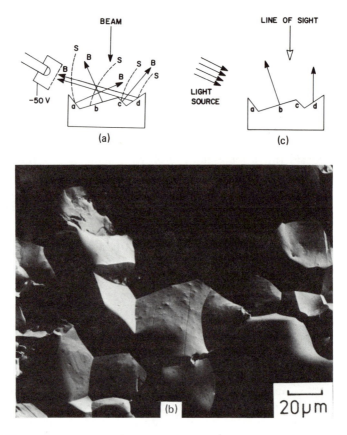

Figure 4.29. (a) Electron collection from randomly oriented surfaces. Everhart–Thornley detector biased negatively. Solid trajectories: backscattered electrons; broken trajectories: secondary electrons. (b) Image of iron fracture surface with this detector configuration. (c) Equivalent light illumination to obtain sample appearance in (b). Beam energy 15 keV.

specimen from above. Those surfaces of the specimen facing the light source appear bright, while those facing away from the light source are nearly completely black.

(b) Secondary Plus Backscattered Electrons. Let us now consider the case in which the Faraday cage is biased positively so that secondary electrons are collected as well, Figure 4.30a. Note that the backscattered electrons are collected in exactly the same manner as the previous situation, Figure 4.29a. We expect that secondary electrons will be emitted from all surfaces struck by the beam, with the number increasing sharply with the effective local tilt angle. The contrast between two surfaces at different tilts can be found by differentiating the secant law for secondary-electron emission. Thus, $d\delta = \delta \sec \theta \tan \theta \, d\theta$ and $C = d\delta / \delta = \tan \theta \, d\theta$, where θ is the

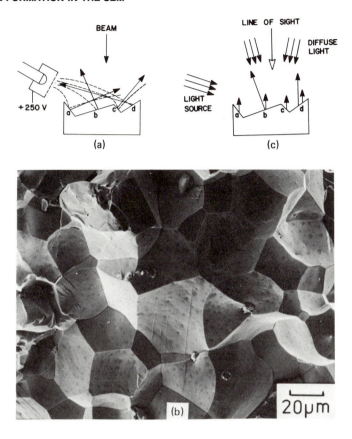

Figure 4.30. (a) Electron collection from randomly oriented surfaces. Everhart–Thornley detector biased positively. (b) Image of iron fracture surface with this detector configuration; same field as Figure 4.29b. (c) Equivalent light illumination to obtain sample appearance in (b). Beam energy 15 keV.

average angle of tilt of the two surfaces and $d\theta$ is the difference in tilt angle. The most important influence on the appearance of the image is the high collection efficiency of the detector for secondary electrons. Secondary electrons are collected in some degree from all surfaces of the specimen struck by the beam, including those surfaces which are inclined away from the detector and which did not produce a useful backscattered electron signal. The image obtained, Figure 4.29b, is more evenly illuminated, with signals obtained from all surfaces. Note that the image is properly considered as a secondary plus backscattered electron image, a fact which is often overlooked. It is important in interpreting topographic images to consider both contributions to the signal. In comparing Figures 4.29b (backscattered electrons only) and 4.30b (backscattered plus secondary electrons), note that those facets which were bright in the backscattered electron image remain bright in the secondary plus backscattered image. The principal difference with the addition of secondary electrons is the collection of some signal from all surfaces struck by the beam, which relieves the harsh contrast in the pure backscattered electron image and provides information from all parts of the field of view. It is the existence of the backscattered electron signal which provides most of the image information on the sense of the topography, that is, whether a feature is a protrusion or depression.

The light optical analogy for the secondary plus backscattered case, Figure 4.30c, consists of the oblique, direct light source from Figure 4.29c, augmented with a general diffuse light source which surrounds the specimen, so that the observer located above the sample sees all surfaces at least partially lighted.

4.5.5.3. Light Optical Analogy

A proper understanding of the light optical analogy is necessary to establish the correct point of view in analyzing an SEM image of rough objects prepared with a positively biased Everhart–Thornley detector. To establish the equivalence between the SEM situation and the human observer, we must match the components with similar characteristics. The eye is highly directional as is the electron beam and hence we imagine the observer looking down through the final aperture. The ET detector has both a highly directional characteristic (for backscattered electrons) and a diffuse characteristic (for secondary electrons). In our ordinary experience, we are used to directional top lighting (the sun) with omnidirectional diffuse lighting from atmospheric scattering. We thus imagine the specimen illuminated by a light source consisting of a highly directional light beam placed at the position of the ET detector as well as by a general diffuse light source throughout the sample chamber. Thus, to interpret an image, we imagine that we can place our eye above the sample and illuminate it from the position of the detector. Bright regions must face the light source. Because we are used to dealing with top illumination, it is most natural to interpret the sense of topography when the image is oriented so that the ET detector position is located at the top of the image. If an observer unknowingly views an image with the apparent illumination coming from a direction other than the top, it is possible that the sense of the topography may become inverted. The convention of top illumination must therefore always be used in presenting SEM micrographs.

Although this method of interpretation of topographic contrast is successful in most cases, occasionally images are encountered where the sense of the topography is not obvious, perhaps because of peculiarities of electron collection. In such cases, tilting and rotating of the sample or the

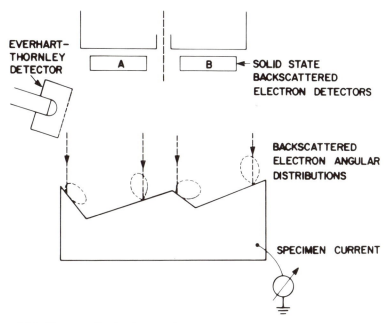

Figure 4.31. Electron collection from randomly tilted surfaces with a large annular solid state detector or pair of solid state detectors.

preparation of stereo pairs may be needed to properly assess the sense of the topography.

4.5.5.4. Topographic Contrast with Other Detectors

(a) Solid State Detector. The appearance of a rough specimen in an image prepared with a solid state detector varies significantly with the take-off angle of the detector. With a large solid angle detector placed near the beam at a high take-off angle (Figure 4.31a), a signal is obtained from all surfaces, with the brightness decreasing as the tilt of a surface increases, as shown in Figure 4.32a. With a large detector, or a pair of detectors acting in a sum mode, trajectory effects on the contrast are somewhat reduced, since electrons emitted from surfaces tilted in opposite directions relative to the beam are still likely to be detected with a large solid angle of collection. However, the contrast sense reverses, and so the topography appears "inside out" (compare Figures 4.29b and 4.32a). With split detectors operating in a difference mode the detector is much more sensitive to trajectory effects which restores the sense of the topography, providing the difference signal is arranged to produce apparent top illumination (Figure 4.32b).

(b) Specimen Current. The direct specimen current image of a rough surface, shown in Figure 4.33a, is markedly different from the emissive

Figure 4.32. (a) Image obtained in the sum mode with a pair of detectors; virtually the same image would be obtained with an annular detector. (Image field same as Figure 4.30b.) (b) Image obtained in the difference mode. Beam energy 15 keV.

mode images obtained with the ET detector. Because of the reversal of contrast, as given by Equation (4.22), the sense of the topography appears to be reversed from what the observer is familiar with since those surfaces more normal to the beam appear brightest. To aid in understanding the topography of a specimen, the contrast is often inverted in the subsequent amplification chain, producing the inverted specimen current image, Figure 4.33b, which has the same contrast sense as an emissive mode image. The

Figure 4.33. (a) Direct specimen current image of the same field of view as Figure 4.30b. (b) Inverted specimen current image corresponding to (a). Beam energy 15 keV.

inverted specimen current image of a rough surface is characterized by a lack of harsh contrast, as compared to a backscattered electron image obtained with the ET detector. Contrast in the specimen current signal develops from number contrast effects only; specimen current images are insensitive to trajectory effects or energy effects. Thus, shadowing is eliminated. The inverted specimen current image from a sample biased to retain secondaries is equivalent to what we would see with a solid state detector which filled the complete hemisphere above the sample so that all back-scattered electrons were collected.

The specimen current signal has several noteworthy advantages: (1) since the contrast is independent of what happens to electrons after they leave the sample, the specimen current signal can always be utilized for imaging, even in those cases where a conventional emissive mode detector is at a disadvantage due to poor collection geometry. (2) Since both backscattering and secondary emission increase monotonically with tilt, the specimen current image can be used in a quantitative sense to assess the tilt of a surface relative to the beam. Thus, in Figure 4.33b, all surfaces producing a similar signal level are at a similar tilt relative to the beam. (3) The specimen current signal can be used to separate number contrast effects from trajectory and/or energy effects.

4.6. Image Quality

Given that a specimen in the SEM is capable of producing contrast we shall now consider what criteria must be satisfied to yield a final image on the cathode ray tube which conveys this contrast information to the observer. These criteria can be separated into two classes: (1) the relationship of the contrast potentially available to the quality of the signal and (2) the techniques of signal processing which must be applied to actually render the contrast information in the signal visible to the eye.

4.6.1. Signal Quality and Contrast Information

We are all familiar with the everyday problem of attempting to tune in a distant television station on a home receiver. If the station's signal is weak, we find the visibility of detail in the picture is obscured by the presence of noise, that is, random fluctuations in the brightness of the image points which are superimposed on the true signal changes we wish to see, the contrast in the image. The presence of randomness or noise as a limitation to the information available in an image is a common theme in all imaging processes.

If a line scan is made across a region of a sample, the signal coming from a detector can be displayed on an oscilloscope, with the scan position along the horizontal axis and the signal plotted in the Y direction, Figure 4.34. We can identify the signals at any two points of interest, e.g., S_A and S_B and calculate the contrast from Equation (4.13). If the same line scan is repeated, the traces on the oscilloscope will not be found to superimpose exactly. If the signal from a single beam location is repeatedly sampled for the picture point time t, the nominally identical signal counts will be found

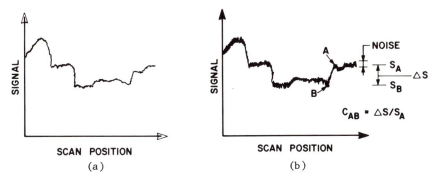

Figure 4.34. (a) Illustration of a single line scan across a field of view. (b) Illustration of multiple scans along the same line in the sample; A and B represent two arbitrarily chosen points of interest.

to vary. The scanning electron microscope imaging process is basically the counting of discrete events, e.g., secondary electrons, arriving with a random distribution in time during a sampling period. Measuring the signal, S, involves counting a number of events, n, at the detector. Because of the random distribution of the events in time, subsequent counts of the same point will vary about the mean value $\bar{n}$ by an amount $\bar{n}^{1/2}$. The signal quality can be expressed by the signal-to-noise ratio S/N:

$$S/N = \bar{n}/\bar{n}^{1/2} = \bar{n}^{1/2} \tag{4.23}$$

As the mean of the counts increases, the S/N ratio improves. The S/N ratio of the SEM image can be estimated from a line scan displayed on an oscilloscope, shown schematically in Figure 4.34b. The noise is estimated from the thickness of the trace, and the signal at the point of interest can be measured directly.

Rose (1948) made an extensive study of the ability of observers to detect contrast between two points in a scanned TV image in the presence of noise. He found that for the average observer to discern the difference between two points, the change in the signal due to the contrast, ΔS, had to exceed the noise N, by a factor of 5:

$$\Delta S > 5N \qquad \text{(Rose criterion)} \tag{4.24}$$

This visibility criterion can be used to develop the relation between the threshold contrast, that is the minimum level of contrast potentially available in the signal, and the beam current. Considering the noise in terms of the number of signal events,

$$\Delta S > 5\bar{n}^{1/2} \tag{4.25}$$

Equation (4.25) can be converted to a contrast equation by dividing through by the signal, S:

$$\frac{\Delta S}{S} = C > \frac{5\bar{n}^{1/2}}{S} = \frac{5\bar{n}^{1/2}}{\bar{n}} \tag{4.26}$$

or

$$C > 5/\bar{n}^{1/2}$$
$$\bar{n} > (5/C)^2 \tag{4.27}$$

Equation (4.27) indicates that in order to observe a given level of contrast, C, a mean number of signal carriers, $\bar{n}$, given by $(5/C)^2$, must be collected per picture point. Considering electrons as signal carriers, the number of electrons which must be collected in the picture point time, τ, can be converted into a signal current, i_S:

$$i_S = \bar{n}e/\tau \tag{4.28}$$

where e is the electron charge (1.6×10^{-19} coul.). Substituting Equation (4.27) for $\bar{n}$ we obtain

$$i_S > 25e/C^2\tau \tag{4.29}$$

The beam current i_B differs from the signal current i_S, by the efficiency of signal collection, ϵ, which depends on the details of the beam–specimen interaction and the detector characteristics:

$$i_S = i_B \times \epsilon \tag{4.30}$$

Combining Equations (4.29) and (4.30),

$$i_B > \frac{25(1.6 \times 10^{-19} \text{ coul.})}{\epsilon C^2\tau} \tag{4.31}$$

The picture point time, τ, can be replaced by the time to scan a full frame, t_f, with the relation $\tau = t_f/n_{PP}$, where n_{PP} is the number of picture elements in the frame. For a high-quality image, $n_{PP} = 10^6$ (1000×1000 matrix). The threshold current can therefore be expressed as

$$i_B > \frac{4 \times 10^{-12}}{\epsilon C^2 t_f} \quad \frac{\text{coul.}}{\text{s}} \quad \text{or} \quad \text{A} \quad \text{(threshold equation)} \tag{4.32}$$

The threshold equation allows one to calculate the minimum beam current which must be employed to detect a specified level of contrast, C, between two points in an image for a specified frame time and collection efficiency (Oatley *et al.*, 1965). Alternatively, if we specify that a given beam current must be used, the threshold equation allows us to calculate the lowest value of contrast which we can possibly image. Objects in the field of view which do not produce this threshold contrast cannot be distinguished from random background fluctuations.

A useful way to understand the interrelations of the parameters in the threshold equation is the graphical plot shown in Figure 4.35. This plot has

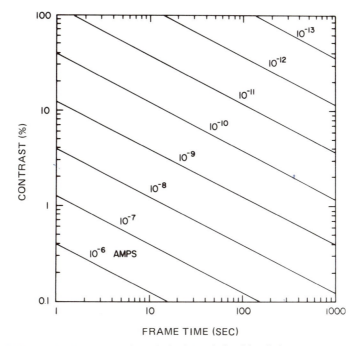

Figure 4.35. Graphical representation of the interrelationship of the parameters contrast, frame time, and beam current, in the threshold equation.

been derived considering that the collection efficiency, ϵ, is 0.25; that is, one signal-carrying electron (backscattered or secondary) is collected for each four beam electrons which strike the sample. This collection efficiency is a reasonable assumption for a target with high backscattering and secondary-electron coefficients, such as gold, when the electrons are detected with a ET detector or a large solid state detector. From the plot, we can immediately see that in order to image a contrast level of 0.10 (10%) with a frame time of 100 s, a beam current in excess of 10^{-11} A must be supplied. However, to observe a contrast of 0.005 (0.5%) a beam current of 10^{-8} A must be used. Alternatively, if a beam current of 10^{-10} A is supplied, all objects producing contrast less than 0.05 (5%) against background will be lost in the noise.

4.6.2. Strategy in SEM Imaging

The threshold equation [Equation (4.32)] and the brightness equation [Equation (2.4) in Chapter 2] form the basis for developing strategy in obtaining optimal SEM images. The threshold equation relates key properties of the specimen/beam/detector system, i.e., the contrast and collection efficiency, to parameters of the microscope, i.e., beam current and frame time. The brightness equation relates this beam current to the minimum probe size and divergence which can be obtained.

To illustrate the strategy in planning an SEM imaging experiment, let us consider the following situation. We wish to image a flat, polished sample consisting of alternate layers of aluminum and iron of various thicknesses viewed edge-on. We want to know the minimum beam current required to photographically image this structure, and the finest detail we could hope to observe. From the earlier discussion on atomic number contrast, we know that the contrast in the backscattered electron signal between aluminum and iron is about 0.45 (45%), Table 4.4. Assuming a collection efficiency of 0.25, the threshold beam current to image a contrast level of 0.45 is 7.9×10^{-13} A. If we assume that the finest detail which can be imaged is limited by the probe size, the brightness equation can be used to calculate the electron probe size which contains this current. For a brightness of 5×10^4 A/cm^2/sr and a divergence of 10^{-2} rad (200 μm aperture, 10 mm working distance), the minimum probe size d is 2.5 nm (25 Å). Such a probe size would in fact be considerably enlarged by lens aberrations and the sampling volume of the signal would also substantially increase the effective size of the probe. Now consider the same structure with layers of platinum and gold. For such a structure, the contrast from Table 4.4 is only 0.0041 (0.41%). The threshold beam current now increases to 9.5×10^{-9} A. The minimum probe size containing this current increases to 280 nm (2800 Å). Thus, in order to image low levels of contrast, we are forced to substantially increase the beam current in order to satisfy the threshold equation, and the penalty which must be paid is the necessity of working with a larger beam size, and the consequent loss of fine detail in the image.

4.6.3. Resolution Limitations

The ultimate resolution obtainable in an SEM image can be limited by any of a number of factors: (1) the instrument's electron–optical performance; (2) the contrast produced by the specimen/detector system; and (3) the sampling volume of the signal within the specimen. We will consider examples of each case.

4.6.3.1. Electron Optical Limitations

Let us consider a sample which produces a contrast of 1.0 (100%), e.g., particles of a heavy metal such as gold on a light substrate such as boron, or the edge of an object placed across a Faraday cup. We will assume a signal collection efficiency of 0.25 (25%). The threshold beam current for a 100-s photographic frame would be 1.6×10^{-13} A from Equation (4.32). Considering a beam brightness of 5×10^4 A/cm^2/sr (typical tungsten hairpin source at 20 keV) and a divergence of 5×10^{-3} rad (100 μm aperture, 1 cm working distance), the minimum probe size given by the brightness equation would be 2.3 nm. The lens aberrations discussed in Chapter 2 would significantly degrade this probe:

Diffraction: $d_d = 1.22 \, \lambda / \alpha$. For a 20-keV beam and $\alpha = 5 \times 10^{-3}$,

$$d_d = 2.1 \text{ nm}.$$

Chromatic aberration: $d_c = (\Delta E / E_0) C_c \alpha$. For $\Delta E = 2$ eV and

$$C_c = 0.8 \text{ cm}, d_c = 4 \text{ nm}.$$

Spherical aberration: $d_s = \frac{1}{2} C_s \alpha^3$. For $C_s = 2$ cm, $d_s = 1.3$ nm.

Adding these contributions in quadrature the actual probe size which could be obtained is 5.2 nm (52 Å). Thus, even with a specimen which produces maximum contrast and a detector giving reasonable collection efficiency, the smallest probe size and best resolution which could be obtained is about 5 nm. This resolution could be improved only through the use of better lenses (reduced C_C, C_S) and higher brightness electron sources. Examination of the edge resolution actually demonstrated by commercial instruments reveals values typically in the range 3–10 nm.

4.6.3.2. Specimen Contrast Limitations

A far more dramatic effect on the spatial resolution is found if we consider the contrast actually generated by typical specimens. Table 4.5 contains values of the minimum probe size predicted by the threshold and brightness equations for contrast values in the range 1.0 to 0.001. To calculate d_{min} the following parameters were used: $\epsilon = .25$, $t_f = 100$, $\beta = 5 \times 10^4$, and $\alpha = 5 \times 10^{-3}$. Many specimens of practical interest produce contrast in the range 0.01–0.10. For such specimens, the limit of spatial

Table 4.5. Minimum Probe Size for Various Contrast Values

Contrast	d_{min}[a]	Contrast	d_{min}
1.0	2.3 nm	0.025	91 nm
0.5	4.6 nm	0.01	230 nm
0.25	9.1 nm	0.005	460 nm
0.10	23 nm	0.0025	911 nm
0.05	46 nm	0.001	2.3 μm

[a] Aberrations not considered.

detail is likely to be in the range 230–23 nm (2300–230 A). It is this lack of specimen contrast which frequently limits SEM performance. Thus, although we may be able to discern fine spatial details on certain high-contrast specimens or at the edges of a specimen where contrast is high, the spatial resolution may be significantly poorer for typical specimens. The microscopist may belive the microscope to be at fault when the spatial resolution is poor, but more often it is the specimen which limits performance.

4.6.3.3. Sampling Volume Limitations

Frequently, situations can arise in which the beam current necessary to observe a given level of contrast can be obtained in a small beam diameter, but nevertheless, the limiting spatial resolution observed on the sample is substantially larger than the beam size. The reason for this loss of resolution usually lies in the sampling volume of the signal used to form the image. As discussed in Chapter 3, both backscattered electrons and secondary electrons are generated in an interaction volume whose effective size in the sample is substantially larger than a finely focused beam.

Consider an interphase boundary between two materials with a large difference in average atomic number. Although the boundary is atomically sharp, the backscatter coefficient will change gradually across the boundary as a result of the finite size of the interaction volume, as shown in Figure 4.36. Thus, an image of the boundary formed with backscattered electrons will not show a transition which is as sharp as the real boundary. That is,

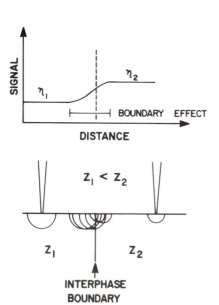

Figure 4.36. Schematic illustration of the influence of the beam interaction volume on the measured signal as the beam is scanned across an interphase boundary. Note that the signal does not change sharply at the boundary.

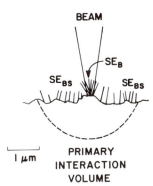

Figure 4.37. Schematic illustration of the two sources of secondary-electron production by the incident beam electrons (SE_B) and production by the backscattered electrons as they exit the sample (SE_{BS}).

our ability to determine the position of the boundary is limited by the nature of the electron-beam–specimen interaction.

A more frequently encountered resolution limitation which arises as a result of electron beam spreading is the lack of detail in images prepared at high magnifications with secondary electron signals. In Table 3.3, Chapter 3, it is shown that the secondary electrons created by the backscattering electrons can form a substantial fraction of the total secondary electron signal collected, reaching 50% or more for high-atomic-number elements. These secondary electrons, which are generated at points remote from the beam entrance area, are unlikely to be significantly affected by fine details in the specimen which are on the same scale as the beam, Figure 4.37. The secondary electrons generated by the primary electrons within the beam impact area do respond to fine-scale detail, but the usable signal resides on a high background caused by the remote secondary electrons. From the point of view of the threshold equation, the efficiency of signal collection, ϵ, is reduced since not all of the signal actually collected is carrying the required information. From the point of view of the specimen, the signal used for imaging is not confined to the beam impact area.

When we examine high-magnification images prepared with a conventional ET detector, Figure 4.38a, the only fine-scale details which can be observed are often the edges of structures. No fine details can be observed away from the edges, which may give the impression that the interior is nearly "atomically smooth." This apparent smoothness is misleading. The situation illustrated schematically in Figure 4.38b actually occurs. At an edge, the interaction and sampling volumes are small because of beam penetration effects. Moreover, because of the increased surface-to-volume ratio, secondary electrons can be created (1) by the beam electrons upon entering the specimen, (2) by backscattering electrons escaping the specimen through the entrance surface and the sides, and (3) by the electrons transmitted through the specimen. The excellent collection of secondary

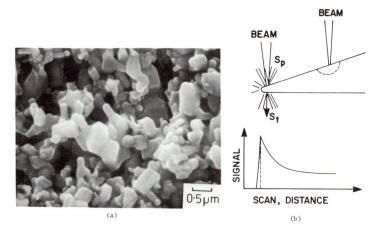

0·5μm

(a)

(b)

Figure 4.38. (a) High-magnification image of copy paper showing sharp edges but no interior structure. Beam energy equals 25 keV. (b) Schematic illustration of the signal generation situation which takes place in producing an image such as (a).

electrons by the positively biased ET detector results in a large secondary electron signal when the beam is at or near the edge compared to locations in the interior of a structure. The leading edges are thus observed in high contrast compared to the interior. From the threshold equation, this allows us to work with a small beam current and therefore a small beam. The high spatial resolution obtained for thin edges does not ensure that similar fine-scale details can be observed in the interior of a sample. The contrast for fine details in the interior against the general background is greatly reduced owing to the enlarged interaction and sampling volumes. The contrast of such details will be below the threshold contrast for the value of beam current which can be successfully used to observe the leading edges. This argument demonstrates that there is always some threshold level of contrast below which objects cannot be distinguished. For high spatial resolution beam conditions, this threshold contrast will be quite substantial and numerous objects may be lost. Thus, the absence of detail in a focused image is unfortunately no guarantee that the object is smooth.

In order to overcome these limitations of the signal and obtain high-resolution images which are representative of the true nature of the surface, it is necessary to devise means to discriminate against signal carriers which are created outside the beam impact area. Looking at the available signals, it quickly becomes clear that it will be very difficult to achieve this discrimination with secondary electrons. There is no observable difference between the physical properties of the secondary electrons created by the incoming beam electrons and the exiting backscattered electrons, and thus no way to separate the two signal components.

4.6.3.4. "Low-Loss" Backscattered Electron Imaging

The backscattered electron signal does offer some interesting possibilities for improving the spatial resolution. Wells (1974a, b) has extensively studied the properties of backscattered electrons with regard to improved spatial resolution and has developed an effective method which makes use of "low-loss electrons." Wells' method is based on the observation that the farther an electron travels in the specimen from the primary impact point, the greater will be its loss of energy. The backscattered electrons which have lost only about 1% of their incident energy, the so-called "low-loss electrons," can only have traveled a few nanometers before being scattered out of the specimen. These low-loss electrons are believed to escape the specimen mainly as a result of a single high-angle elastic scattering event. In order to maximize the generation of low-loss electrons and to direct their trajectories in a small solid emission angle, the specimen is highly tilted, which produces an angular distribution sharply peaked in the forward direction. The electron detector is placed in the forward scattering direction to maximize collection of the desirable portion of the signal. An adjustable biased grid system is used to reject all electrons below a certain energy KE_0, where K is typically set at 0.95–0.99. The high-energy electrons with $E/E_0 > K$ are then accelerated after the grid by a high-voltage bias and detected with a scintillator–photomultiplier system. Images from this detector system, coupled with a high brightness electron gun, show some of the finest structural detail from solid objects ever obtained in the SEM. Examples of low-loss electron images of an etched SiO_2 surface and of a bacteriophage are shown in Figure 4.39 (Broers, 1974a, b). Close examination of low-loss images in comparison with conventional ET detector images of the same field of view provides convincing evidence that many fine-scale details are simply lost in the conventional images. In addition to being confined to a small lateral range, low-loss electrons are also necessarily generated near the surface of the specimen. Thus, Wells (1974a) has demonstrated the possibility of imaging thin surface films (3–5 nm thick) of aluminum oxide which remained over voids in aluminum metal damaged by electromigration. In a conventional image, the subsurface voids could be observed but not the film.

The characteristics of the solid state detector for backscattered electrons are also attractive for high-resolution imaging. Since this detector produces a signal proportional to the electron energy and may produce no response below a cutoff energy, it provides some emphasis of the high energy fraction of the signal which is desirable for high-resolution imaging. The small gain of the solid state detector (10^3) compared to the photomultiplier (10^6) leads to limitations on the scanning speed which can be employed in high-resolution imaging as well as difficulty in operating at extremely low current.

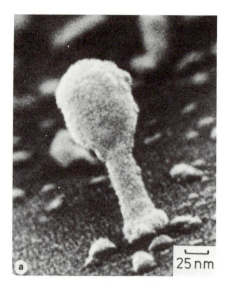

Figure 4.39 Low-loss backscattered electron images. (a) Low-loss scanning electron micrograph of T4 coliphage. (Broers *et al.*, 1975; courtesy A. Broers, IBM.) (b) Low-loss scanning electron micrograph of etched SiO₂ step on silicon wafer, showing fine scale surface structure. Sample coated with 20 nm of gold–palladium. Beam energy 40 keV. (Broers, 1974; courtesy A. Broers, IBM.)

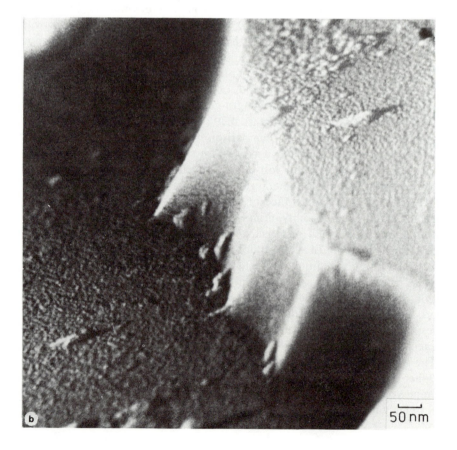

4.7. Signal Processing for the Display of Contrast Information

4.7.1. The Visibility Problem

The proper sequence of events in the operation of an SEM includes the following steps: (1) consideration of the contrast mechanisms which a specimen might exhibit; (2) evaluation of the effects of the detector characteristics and placement on that contrast; (3) selection of the adjustable instrument parameters, including beam current and beam energy, to ensure that the desired contrast actually exists in the signal; (4) recognition of the limitations imposed on the beam parameters by the brightness equation and lens aberration effects. Providing a satisfactory specimen is used and the above steps are followed, contrast information of interest may exist in the signal produced by the detector. This signal is of too small a magnitude to be used directly for modulation of the brightness of the final display or photographic cathode ray tubes. The detector signal must be linearly amplified to a suitable level by one or more amplifiers, which form an integral part of an SEM. Even after such amplification, however, the contrast in the signal may not be visible in the CRT display viewed by the microscopist. The reason for this lack of visibility lies in the basic limitations of the human eye in discerning slight changes in intensity on a brightness-modulated CRT. Although the signal can be divided into many discrete intensity levels, the average observer can only distinguish about 12 distinct shades from pure black to pure white (gray scale) on a typical CRT. This "gray scale" corresponds to the full range of input voltage from the amplification system, Figure 4.40. In such a display, the minimum contrast level in the input signal which the eye can perceive is about 0.05 (5%).

If linear amplification is used to amplify the detector signal to a level suitable for display, the natural contrast which leaves the specimen–detector system will be equal to the image contrast in the final display. That is, if a linear amplification factor K is applied to the signal S to produce the display intensity I:

$$I = KS \qquad (4.33)$$

$$C_{\text{natural}} = \frac{S_{\max} - S_{\min}}{S_{\max}} = \frac{KS_{\max} - KS_{\min}}{KS_{\min}} = \frac{I_{\max} - I_{\min}}{I_{\max}} = C_{\text{image}} \qquad (4.34)$$

If the natural contrast is high, as it typically is in the case of topographic contrast, then linear amplification will produce a satisfactory final image in which the natural contrast spans the gray scale range and is easily perceived, as shown in Figure 4.41. Values of the natural contrast which fall below 5% will not be visible to the observer in such an image, and natural contrast in the range 5%–10% will be perceived with difficulty. As a

Figure 4.40. Distinct levels of intensity which can be discerned by the eye in a CRT display (gray levels).

general observation, images are most easily understood when the contrast information spans as much of the dynamic range of the display or photographic medium as possible. A line scan in a satisfactory situation would resemble that shown in Figure 4.42, with all gray levels excited. It is the function of signal processing techniques to manipulate the natural contrast in the signal into suitable image contrast for the observer.

In the discussion which follows, we will always distinguish between the contrast leaving the specimen–detector system, the "natural contrast," and the contrast on the visual CRT or photographic record, the "image contrast." This distinction is important because the contrast which must be used in the threshold equation for determination of the required beam current is the natural contrast. We shall see that signal processing provides the capability for extensive modification of the image contrast to make it more visible to the eye. Such manipulations of the image contrast cannot, however, produce any increase of information which is not resident in the signal leaving the detector.

The video signal of the SEM is in an ideal form for the application of

Figure 4.41. Example of linear amplification applied to a sample which produces high natural contrast by the topographic contrast mechanism (specimen: pollen; courtesy J. Geller, JEOL). Beam energy, 15 keV.

signal processing methods, since it is time-resolved, effectively allowing us to process a single picture element at a time. The processing can, in fact, be applied in either an analog form, i.e., with various specialty amplifiers placed in series with the linear amplifier and the final display, or in digital form. In digital signal processing, the scanned image is converted into a

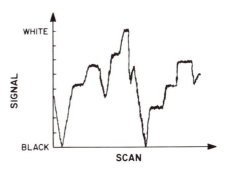

Figure 4.42. Schematic illustration of the signal characteristics which would produce a "good" image to an observer: the signal spans the full dynamic range of the display medium.

digital form in which each picture element is represented as an $X-Y$ address with a third value for the signal intensity at that point: (X, Y, I). The digital representation can be processed in a computer to give a new, modified form (X', Y', I'), where any or all of the parameters can be altered. This modified form can then be reconstituted through a digital-to-analog converter to provide a signal which can be displayed on an SEM CRT. Digital SEM image processing is still in its infancy. The vast majority of signal processing on existing SEMs is carried out by analog means, and we shall concentrate on these methods here. The reader interested in digital techniques is referred to the recent literature on the subject (Jones and Smith, 1980).

4.7.2. Signal Processing Techniques

In the examination of a wide variety of specimens, the natural contrast available for direct display by linear amplification will occasionally be found to be too weak, too strong, or certain features of interest will be dominated by others, decreasing their visibility. A variety of signal processing methods has been developed to cope with each of these limitations. In this section we will consider the signal processing techniques commonly available on SEMs, including (1) contrast reversal, (2) differential amplification; (3) nonlinear amplification; (4) signal differentiation, (5) signal mixing, (6) Y modulation and (7) intensity contouring.

4.7.2.1. Contrast Reversal

Many SEMs are equipped with the facility of contrast reversal. Reversal is achieved by subtracting the signal from a fixed level corresponding to the maximum value allowed:

$$S_{\text{out}} = S_{\text{max}} - S_{\text{in}} \qquad (4.35)$$

Contrast reversal is useful in cases where the nature of the detector signal is such that the sense of the contrast is reversed from what the observer expects, as is the case for specimen current. Contrast reversal of a topographic image observed in specimen current is illustrated in Figures 4.43a and 4.43b. Note that contrast reversal of an emissive mode image produces an undesirable contrast sense, which can mislead the observer with regard to the proper sense of topography, as shown in Figures 4.43a and 4.43b.

$\longrightarrow$

Figure 4.43. Contrast reversal leading to a false sense of topography. (a) Normal image obtained from an Everhart–Thornley detector; (b) same field of view after contrast reversal. Beam energy 15 keV.

(a)

(b)

4.7.2.2. Differential Amplification

One of the most commonly encountered contrast problems in SEM imaging is the case of having a sample which produces a low natural contrast. Atomic number contrast and topographic contrast can often fall below 0.10 (10%). In addition, special contrast mechanisms of interest in materials science studies, including electron channeling contrast, types I and II magnetic contrast, and voltage contrast, may produce contrast levels in the range 0.001 (0.1%) to 0.05 (5%). A method which is very effective for the enhancement of low contrast is the technique of differential amplification (also known on SEMs as contrast expansion, black level, dark level, or dc suppression). In studying the application of signal processing techniques, it is very useful to observe the changes in a signal plotted as a line scan on an oscilloscope with the scan position on the horizontal sweep and the signal intensity on the vertical. In fact, a waveform monitor, an oscilloscope which continuously displays the Y-modulated line scan, is a nearly indispensable tool for effective microscopy.

If we examine the image for a low-contrast situation following linear amplification, for example, an aluminum–silicon couple [atomic number contrast 0.067 (6.7%)], the change in signal as the beam passes from the aluminum to the silicon is barely enough to cause a one-step change in the gray level recorded on the film. The image, shown in Figure 4.44a thus appears to be lacking in detail, with only poor definition of the Al and Si regions. The signal can be thought of as having a spatial frequency spectrum which consists of a high-frequency component of the image detail in which we are interested superimposed on a dc level which corresponds to

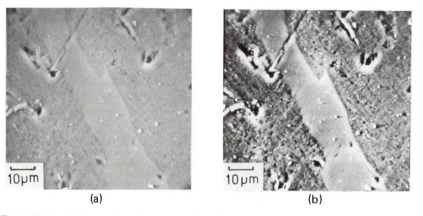

(a)   (b)

Figure 4.44. (a) Image of an aluminum–silicon eutectic (natural contrast approximation 7%) with linear amplification only. (b) Same image as (a) but with differential amplification applied. Beam energy 20 keV.

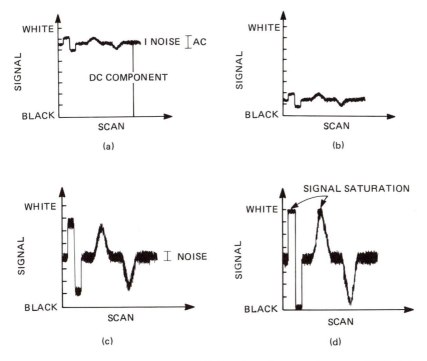

Figure 4.45. (a) Line trace of a specimen which produces low natural contrast. (b) First stage of differential amplification: a fixed dc level is subtracted. (c) Second stage of differential amplification in which the differential signal is linearly amplified, expanding the range of gray levels over which the contrast information is displayed. (d) Excessive application of differential amplification in which signal saturation occurs.

the average signal (Figure 4.45a). In signal processing by differential amplification, a fixed, dc level is first subtracted from the signal, producing the line trace shown in Figure 4.45b. The remaining signal, the differential, is then linearly amplified, producing the line trace in Figure 4.45c. The high-frequency image components, which formerly occupied only one or two gray levels, now span a substantial portion of the dynamic range of the film, which yields the image in Figure 4.44b with the aluminum and silicon regions much more easily discerned.

The portion of the dynamic range of the display or film which is used depends on the gain of the linear amplifier. The gain can be increased until regions of the image are black and white, which is an undesirable situation. When the signal is at either end of the dynamic range, there will be loss of information due to the saturation of the display. That is, the display is unable to respond to higher signals when it has already reached white or lower signals than black. The line trace for this situation is illustrated in Figure 4.45d. Saturation at the white end of the dynamic range can also lead to the phenomenon of "blooming," in which the beam on the CRT is so intense that it excites neighboring picture elements around the element to which it has been addressed. In addition, a speckled, "grainy" texture may be observed in the image. This speckle is the noise in the signal which is also a high-frequency component of the image spectrum and which is consequently amplified along with the desired information. The noise becomes evident when it begins to occupy more than one gray level above and below the average signal. The observation of noise in the image after differential amplification provides graphic proof of the inability of the signal processing techniques to recover information not already present in the signal. The noise can only be reduced by increasing the dwell time of the beam on each picture element in order to accumulate more signal carriers and thus more information in the signal leaving the detector. Differential amplification cannot make a contrast level visible unless the threshold current for that contrast is exceeded. However, when the threshold current is exceeded, differential amplification circuits typically found on SEMs can provide a satisfactory display of natural contrast levels as low as 0.001 (0.1%).

4.7.2.3. Nonlinear Amplification (Gamma)

A second type of contrast problem which arises frequently is the situation in which the specimen produces contrast spanning the whole dynamic range, but the features of interest only occupy a limited range of gray levels located near the black or white end of the dynamic range resulting in poor visibility, as shown in the line trace of Figure 4.46a.

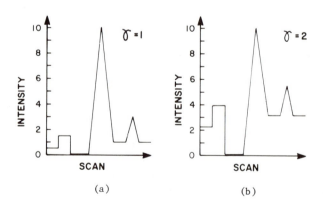

Figure 4.46. (a) Schematic illustration of a line trace for which the signal spans the dynamic range, but where the information of interest is in a narrow gray scale range. (b) Same signal trace but with $\gamma = 2$ applied.

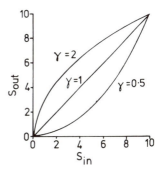

Figure 4.47. Signal response curve for nonlinear amplification (gamma). For $\gamma = 2$, the contrast at the dark end of the scale is expanded while the bright end is compressed; for $\gamma = 1/2$, the reverse is true.

Nonlinear amplification (gamma processing) can provide selective contrast expansion at either the black or white end of the dynamic range while preventing saturation or clipping of the display. This is accomplished by the application of a nonlinear function to the input signal:

$$S_{out} = S_{in}^{1/\gamma} \qquad (4.36)$$

where the exponent, γ, typically has the integer values 2, 3, 4 or fractional values $1/2$, $1/3$, $1/4$. Following the nonlinear amplifier, the signal is amplified linearly to make use of the full dynamic range, which yields the signal response functions for $\gamma = 2$ and $\gamma = 0.5$ shown in Figure 4.47. For $\gamma = 2$, a small range of input signals is distributed over a greater range of output gray levels than in the linear case, which results in the signal modification shown in Figure 4.46b. The signals at the white end of the range are compressed into fewer output gray levels. Note, however, that the signal is not allowed to saturate except at the maximum input signal, as in the linear case. For $\gamma < 1$, the expansion is obtained at the white end at the expense of the black end. The application of nonlinear amplification to a real imaging situation is illustrated in Figure 4.48, in which details in the holes are difficult to discern in the linear image (Figure 4.48a) but which can be observed with improved visibility in the $\gamma = 2$ image (Figure 4.48b). If this contrast expansion were achieved with differential amplification (black level), many areas of the image would "wash out" due to saturation.

4.7.2.4. Signal Differentiation

It is useful to consider the features in an image in terms of spatial frequencies. Thus, the edges of objects across which the signal changes rapidly, such as the facets on the fracture surface in Figure 4.49a, are high-frequency components. Across the interior of a facet, the signal changes slowly, so such features are low-spatial-frequency components. The edges of objects are generally of the most interest to us in defining the position, size, and shape. Signal differentiation is a signal processing operation which

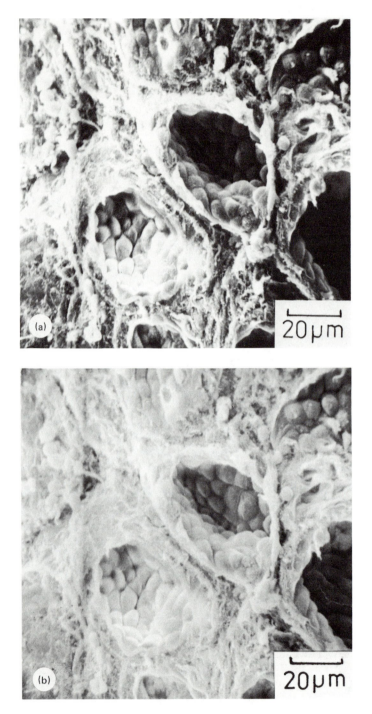

Figure 4.48. Application of gamma processing to improve visibility of detail in a hole. Sample, mouse thyroid: (a) linear image; (b) gamma processing with $\gamma = 2$. Beam energy 20 keV.

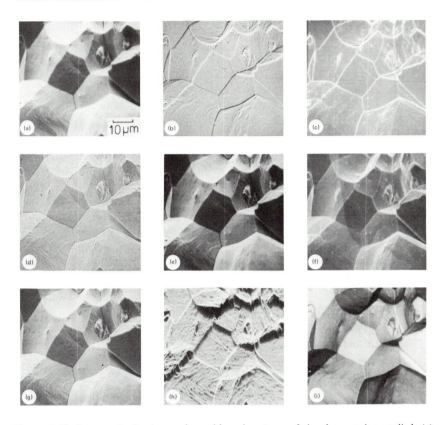

Figure 4.49. Images of a fracture surface with various types of signal processing applied: (a) direct image; (b) first time derivative; (c) absolute value of first time derivative; (d) second time derivative; (e) 50% direct image plus 50% first time derivative; (f) 50% direct image plus 50% absolute first time derivative; (g) 50% direct image plus 50% second time derivative; (h) Y-modulation image; (i) reversed contrast. Beam energy 20 keV.

emphasizes the high-frequency component of the image and deemphasizes the low-frequency components:

$$S_{out} = \frac{dS_{in}}{dt} \tag{4.37}$$

where t is time. In this case, the output signal has a high absolute value when the input signal is changing rapidly, as at an edge. Note that the sign of S_{out} depends on the sense of change of dS_{in}. In the usual arrangement, S_{out} from low spatial frequency regions is set at the middle of the gray scale, and positive-going signal changes produce signals near the white end of the dynamic range. Negative-going signal changes are placed near the black end of the range.

The pure derivative image of a rough object has an unfamiliar appearance, Figure 4.49b. While the edges of the facets of the fracture surface are easily seen, the sense of the topography is difficult to assess. It is, in fact, the low spatial frequency shading which provides the information to the observer about the sense of topography.

A second failure of the first time derivative arises from the fundamental anisotropy of the derivative processing operation. The magnitude of the derivative depends on the rate of change of the signal as the scan crosses a feature. In a conventional scanning pattern, a line is rapidly scanned in a time of typically 1–10 ms. The vertical scan takes place at the frame scan rate, typically 1–100 s, which is three orders of magnitude slower than the horizontal line scan. Thus, the rate of change of signal across a feature will depend on its orientation relative to the line scan. If the object is oriented perpendicular to the line scan, a high value of the derivative is obtained, and the object is enhanced relative to the background, as shown in the image of a grid in Figure 4.50. If the object is parallel to the line scan, a low derivative value is obtained, and the visibility decreases in the image as shown in Figure 4.50b, where the scan is parallel to the grid bars.

The effect of this anisotropy can be overcome somewhat through the use of orthogonal scanning, in which the field is scanned twice at right angles and the two images are summed (Fiori *et al.*, 1974). Furthermore, additional derivative functions can be employed, such as the absolute value of the first derivative, ds/dt, in which negative-going signals are reversed, and the second time derivative d^2S/dt^2 (Fiori *et al.*, 1974). Examples of the waveforms which result from these processes are shown schematically in Figure 4.51. The effects of the derivative processing and orthogonal scan-

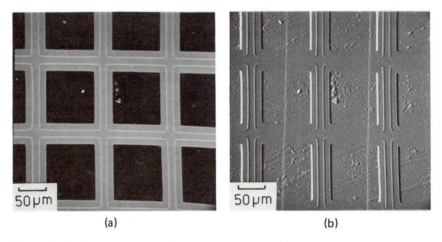

(a)   (b)

Figure 4.50. Images of a grid illustrating anisotropy of time derivative processing. (a) Inverted specimen current image. (b) Pure first derivative image with horizontal scan lines. Note loss of horizontal bars of grid and strong enhancement of vertical bars. Note simultaneous display of fine detail on substrate and grid bars in derivative image, while in the original image only the grid can be seen. Beam energy 20 keV.

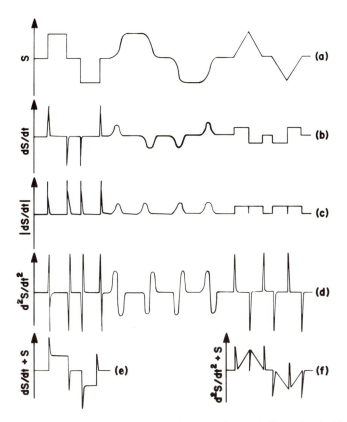

Figure 4.51. Waveforms of derivative processing operations: (a) direct signal; (b) first time derivative; (c) absolute value of first derivative; (d) second derivative; (e) signal mixing, direct plus first derivative; (f) signal mixing, direct plus second derivative.

ning on the image of the edge of a hole are shown in Figure 4.52. In the unidirectionally scanned, first derivative image, Figure 4.52c, the edge of the hole is bright on one side and darker than the background on the other. At the point where the scan line is tangent to the hole, the image of the edge is lost against the background. The orthogonally scanned, first derivative image, Figure 4.52d, has only one quadrant dark. The line of information loss is still present, but it is rotated 45° to that of Figure 4.52c. The absolute value of the first derivative, unidirectionally scanned, shows a line of information loss, Figure 4.52e, whereas the orthogonally scanned, absolute first derivative shows almost uniform edge enhancement without any line of information loss, Figure 4.52f. When the second time derivative is considered, unidirectional scanning again gives a line of information loss, Figure 4.52g, while orthogonal scanning yields a uniform circle edge, Figure 4.52h. The orthogonally scanned, second derivative transformation

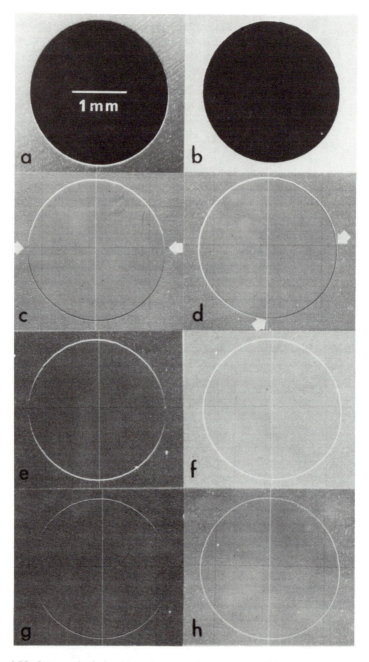

Figure 4.52. Images of a hole with various time derivatives applied: (a) direct emissive mode image; (b) inverted specimen current image used for processing; (c) derivative image, vertical scan lines bottom to top; (d) orthogonally scanned first time derivative; (e) absolute value of first derivative; (f) orthogonally scanned first derivative; (g) second time derivative; (h) orthogonally scanned second time derivative.

can be demonstrated mathematically to be an isotropic transformation with no preferential orientation enhancement (Fiori *et al.*, 1974). In an analog scan system, implementing a coincident orthogonal scan is difficult. Higher-order derivative processing is best done in the digital mode. Smith *et al.* (1977) have demonstrated the utility of image gradient processing based upon derivatives in the digital domain for image enhancement. In addition to image effects, the calculated image gradient has mathematical properties which can be used for characterizing the texture of a surface.

Another drawback of the derivative processing techniques arises from their sensitivity to noise in the signal. Derivative operators are essentially high pass filters designed to exclude low-frequency background signals. Noise, however, resides in the high-frequency end of the image spectrum and is thus enhanced by derivative processing. Derivative processing is only effective in situations of high signal-to-noise ratio.

The final drawback to derivative processing occurs because of the possibility of improper interpretation of the derivative image (Fiori *et al.*, 1974). Since edges are enhanced by the derivative, image "crispening" or sharpening of edges is obtained (Goldmark and Hollywood, 1951). Since we normally recognize objects by their edges, such crispening is useful. However, above a certain magnification, the crispening may lead to a false sense of image resolution. Although derivative processing can apparently sharpen the image of an edge, it does not necessarily identify the precise position of the edge. The derivative merely highlights the position of most rapid signal change, which is a function of the beam profile, specimen profile, and electron beam interaction volume. Thus, the use of derivative processing at high magnifications is to be treated with caution, especially if measurements are to be made with the images.

4.7.2.5. Signal Mixing

One answer to the dilemma posed by the flatness of the pure derivative image is to mix the signals (Heinrich *et al.*, 1970). The final signal for display can be made up as a summation of signals from various points in the signal processing chain as shown schematically in Figures 4.51e and 4.51f. Thus, the direct signal and the differentiated signal can be combined:

$$S_{out} = aS_{in} + b\frac{dS_{in}}{dt} \qquad (4.38)$$

where a and b are adjustable in the range $0 \leqslant a, b \leqslant 1$. An example of such a mixed signal image is shown in Figure 4.49e. The edges are enhanced while a portion of the direct signal is used to retain the general shading so that the sense of the topography is not lost.

Signal mixing can be used successfully with other combinations of signals, such as those derived from different detectors. Frequently x-ray information ("dot maps") is combined with a structural imaging signal to show qualitatively the location of constituents in a structure.

4.7.2.6. *Y*-Modulation

In the *Y*-modulation form of image processing, a series of successive line scans are superimposed to form a nearly continuous image. The *Y*-modulation transformation represents the most radical departure from the basic concepts of scanning illustrated in Figures 4.3–4.5. In forming the *Y*-modulation image, a conventional raster scan is made on the specimen. However, on the CRT the horizontal position of the corresponding point is determined by the horizontal location of the scan along a line and the vertical location is determined by the strength of the signal measured at that point (Figure 4.53). The intensity of all points plotted on the CRT is maintained at a constant value. The resulting image can be quite distorted as a result of the lack of a one-to-one correspondence between the specimen grid and the CRT grid. As a result, *Y*-modulation images should not be used for spatial measurements. The value of the *Y*-modulation image is its tendency to enhance fine-scale structure which can improve the visibility of surface texture, e.g., Figures 4.49a and 4.49h. Small relative changes in the signal which result from this fine structure may not be visible in a conventional intensity-modulated image unless extensive differential amplification is employed (Oron and Tamir, 1974).

4.7.2.7. Intensity Contouring (Gray Levels)

Normally, the signal used for intensity modulation of the final CRT image is continuous in nature, i.e., it can take on any value between defined limits. The nature of the CRT display is such that only a limited number of distinct changes in intensity, or gray levels, can be discerned, of the order of 12. If the signal-to-noise ratio is poor, the number of valid gray levels into which the signal can be divided may be even less than 12. The random fluctuations in the signal cause unavoidable changes in the gray level

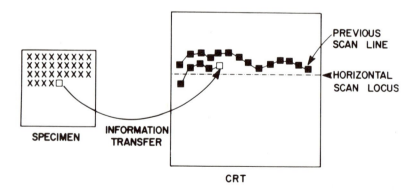

Figure 4.53. Schematic illustration of the generation of a *Y*-modulation image.

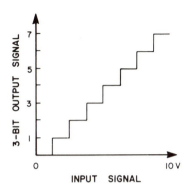

Figure 4.54. Staircase response function obtained by conversion of analog signal into a digital form with discrete levels only.

displayed, with the result that the noise is manifested as graininess in the image. This graininess can be controlled to some extent if the number of discrete levels is limited and matched to the signal. The analog signal is first converted into a digital signal with an analog-to-digital converter (ADC) with a resolution of 4 bits ($2^4 = 16$ distinct levels). Once the signal is in the digital domain, the number of allowed levels can be defined. The digital signal is converted back to the analog domain for CRT display by means of a DAC, with the result that the analog signal now contains only discrete values (Figure 4.54). An illustration of an image processed in this way is shown in Figure 4.55.

The discrete gray level image can be used to produce an intensity contouring effect which can be useful in identifying features of a common type, such as certain particles in a field of mixed objects, providing the objects of interest produce a characteristic signal level. Discrete gray level images are usually required as input (or, in fact, are generated from analog input) to digital image processing systems used in quantitative metallography or particle characterization.

4.7.3. Combinations of Detectors

The choice of the signal detector(s) can also have a major effect on the appearance of the image. As demonstrated earlier, the appearance of the fracture surface imaged with an Everhart–Thornley detector varied depending on whether the secondary electrons were included in the signal or not (Figures 4.29, 4.30). The same field of view imaged with specimen current or a large solid state backscattered electron detector differed substantially from the ET detector image (Figures 4.32, 4.33).

The special characteristics of certain signal detectors can sometimes be used to separate the individual components of contrast when competing contrast mechanisms operate. For example, if a sample is rough and has regions of different composition, it is possible to have topographical and atomic number contrast in the same image. As described in the discussion

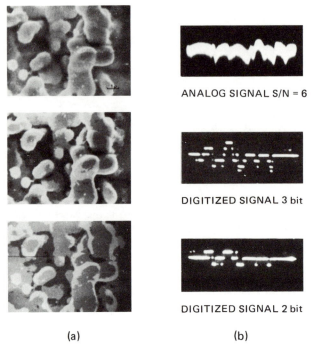

ANALOG SIGNAL S/N = 6

DIGITIZED SIGNAL 3 bit

DIGITIZED SIGNAL 2 bit

(a)   (b)

Figure 4.55. Example of an image with discrete gray levels obtained by analog-to-digital signal conversion: (a) direct image; (b) discrete gray level image.

on the Everhart–Thornley detector, topographic contrast has a strong trajectory component and a somewhat weaker number component, while atomic number contrast has mainly a number component and a weaker trajectory component. Thus, by choosing a detector which is sensitive to the number components alone, the atomic number contrast can be viewed separately from the topographic contrast. Two detector schemes can achieve this separation: (1) specimen current and (2) a pair of backscattered electron detectors placed at high take-off angles.

(1) Specimen Current Detection. The specimen current signal depends only on the escape of electrons from the specimen and is insensitive to the trajectories which they follow. Figures 4.56a and 4.56b compare the images of a specimen of a lightly etched two-phase lead–tin alloy which has regions of different atomic number as well as surface topography. The topography is apparent in the Everhart–Thornley detector image (Figure 4.56a), but is virtually absent in the specimen current image inverted to give the usual sense of atomic number contrast (Figure 4.56b).

(2) Paired Solid State Detectors. In making use of paired solid state detectors, Figure 4.31, we can utilize the sum of the signals derived from

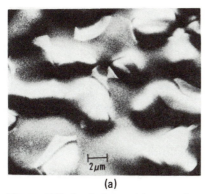 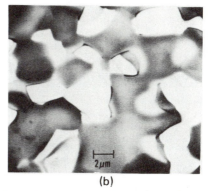

| (a) | (b) |

Figure 4.56. Lead–tin eutectic superplastic alloy, polished and lightly etched. (a) Image obtained with an Everhart–Thornley detector, positively biased. (b) Image obtained with specimen current signal, reversed. Beam energy 20 keV.

two (or more) separate detectors placed symmetrically about the beam (Kimoto and Hashimoto, 1966). Thus, signals emitted from surfaces at different orientations are likely to be collected at one of the detector locations, which reduces sensitivity to trajectory effects. The sum image contains mainly atomic number contrast, Figure 4.57a. The difference of the signals from the detectors can also be used to form an image, Figure 4.57b. In the difference image, trajectory effects are emphasized since the signals reaching the detectors from a given location are greatly different. Topography is strongly emphasized in the difference image.

4.7.4. Beam Energy Effects

The choice of the beam energy used to irradiate the specimen can also have a great effect on the appearance of the image, especially if secondary electrons are included in the imaging signal. The beam energy can influence the image through (1) the depth of penetration of the beam electrons which affects the sampling depth of the backscattered electrons; (2) changes in the secondary-electron coefficient, particularly at beam energies less than 5 keV; and (3) charging effects and voltage contrast of surface contaminant layers. A detailed study of the origin of these effects is beyond the scope of the present discussion. The example below is presented to alert the microscopist to the existence of these effects.

Figures 4.58a–4.58d show the appearance of a scratched, carbon-contaminated platinum surface at various beam energies, with the imaging signal derived from an Everhart–Thornley detector. The images show a remarkable change in contrast as the beam energy is lowered. Maximum contrast is observed at a beam energy of 5 keV. At 1 keV little contrast is

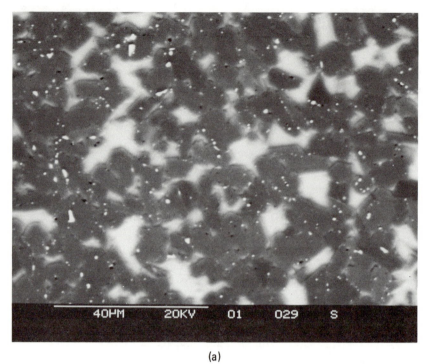

(a)

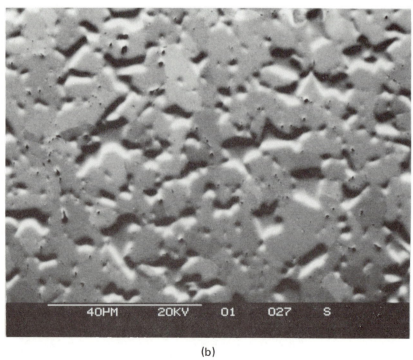

(b)

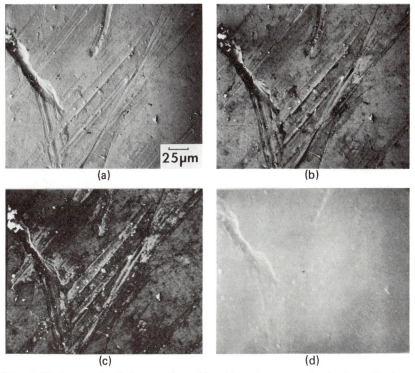

Figure 4.58. Images of a platinum surface with a thin carbon contamination layer, Everhart–Thornley detector, secondary plus backscattered electron signals: (a) 20 keV; (b) 10 keV; (c) 5 keV; (d) 1 keV.

observed. The beam electrons have such a short range that penetration of the surface contamination is not achieved. Above 5 keV, the secondary-electron coefficient decreases and the primary beam penetration becomes so great that the interaction with the surface layer is greatly reduced. At an energy of 20 keV, all impression of a surface layer is lost.

4.7.5. Summary

The appearance of an image can be greatly changed by applying the various modes of signal processing which are available. Figure 4.49 con-

---

Figure 4.57. Separation of contrast components with a pair of solid state backscatter detectors. Specimen: reaction bounded silicon carbide (a) summation image, showing atomic number contrast between silicon (bright) and silicon carbide (dark); (b) difference image, showing reduction in atomic number contrast and a great enhancement in the surface topography. Beam energy 20 keV.

3

y of signal processing techniques applied to the same field of
fracture surface. The apparent topography of the image
nding on the processing, with the pure derivative images
parently flat object compared to the direct signal, which gives
the impression of surfaces tilted at a variety of angles. The derivative
images, however, yield an enhanced sense of the edges of the facets and of
the texture of the surfaces. The absolute value of the first derivative
provides a strong outlining effect which is useful in delineating the extent of
a facet. The mixed signal images are very effective for combining the depth
information of the direct image with edge and surface texture enhancement
of the derivative image.

The microscopist usually does not have the luxury of recording a given
field of view with a large variety of signal processing techniques. Moreover,
the utility of the various techniques is somewhat dependent on the nature
of the specimen being examined. However, it is a good idea for the
microscopist to record at least some selected samples in all available signal
processing modes in order to investigate the imaging flexibility of his SEM.
The art of obtaining the most useful SEM images requires the user to
extend his instrument beyond the narrow range of "optimum" operation
conditions usually described in the instruction manual. Even the more
modest SEMs are capable of a surprising variety of user-selected operating
modes.

X-Ray Spectral Measurement: WDS and EDS

5.1. Introduction

Chemical analysis in the scanning electron microscope and electron microprobe is performed by measuring the energy and intensity distribution of the x-ray signal generated by a focused electron beam. The subject of x-ray production has already been introduced in the chapter on electron-beam–specimen interactions (Chapter 3), which describes the mechanisms for both characteristic and continuum x-ray production. This chapter is concerned with the methods for detecting and measuring these signals as well as converting them into a useful form for qualitative and quantitative analysis.

5.2. Wavelength-Dispersive Spectrometer

5.2.1. Basic Design

Until 1968, when solid state detectors were first interfaced to microanalyzers, the wavelength-dispersive spectrometer (WDS) was used almost exclusively for x-ray spectral characterization. The basic components of the WDS are illustrated in Figure 5.1. A small portion of the x-ray signal generated from the specimen passes out of the electron optical chamber and impinges on an analyzing crystal. If Bragg's law is satisfied,

$$n\lambda = 2d\sin\theta \qquad (5.1)$$

(where n is an integer, $1, 2, 3, \ldots$; λ is the x-ray wavelength; d is the interplanar spacing of the crystal; and θ is the angle of incidence of the

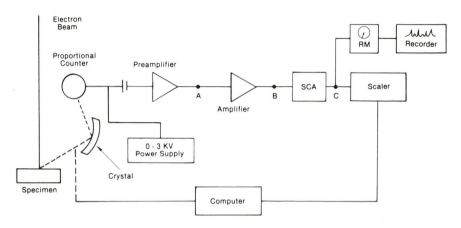

Figure 5.1. Schematic representation of a wavelength-dispersive spectrometer.

x-ray on the crystal), the x-rays will be diffracted and detected by a proportional counter. The signal from the detector is amplified, converted to a standard pulse size by a single-channel analyzer (SCA), and then either counted with a scaler or displayed as a rate meter (RM) output on a strip chart recorder. A typical qualitative analysis therefore involves obtaining a strip chart recording of x-ray intensity as a function of crystal angle, converting peak positions to wavelengths through Bragg's law, and then using the Moseley relationship [equation (3.28), Chapter 3] to relate the detected wavelengths to the presence of specific elements. In practice, crystal spectrometer readings are either proportional to wavelength or calibrated directly in wavelength. Standard tables can then be used for elemental identification.

Bragg's law can be easily derived with the aid of Figure 5.2. A beam of coherent (in-phase) x-rays is assumed to be specularly reflected from parallel crystal planes spaced d units apart. Of the two x-ray beams shown in Figure 5.2, the lower one will travel an additional path length $ABC = 2d\sin\theta$ prior to leaving the crystal. If this distance equals an integral number of wavelengths n, then the reflected beams will combine in phase and an intensity maximum will be detected by the proportional counter. If measured with a high-quality analyzing crystal the intensity distribution resulting from the diffraction phenomenon can be relatively narrow. For example, the measured full width, half-maximum (FWHM) for Mn $K\alpha$ is about 10 eV, compared to the natural value of 2 eV. X-rays having a wavelength which will not satisfy the Bragg equation are absorbed by the crystal or pass through it into the crystal holder.

The x-ray signal in focused electron beam instruments is fairly weak and can be thought of as originating from a point source; therefore to

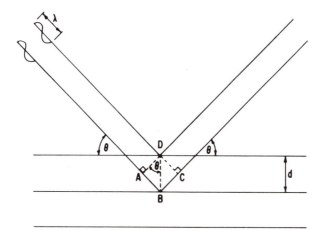

Figure 5.2. Diffraction according to Bragg's law.

maximize the signal at the detector, curved-crystal, fully focusing x-ray spectrometers are used in preference to flat-crystal spectrometers of the type normally associated with tube-excited x-ray emission analysis. In a fully focusing spectrometer of the Johansson type, illustrated in Figure 5.3, the x-ray point source, the specimen, the analyzing crystal, and the detector are all constrained to move on the same circle with radius R, called the

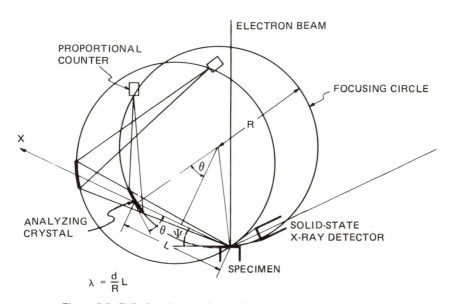

Figure 5.3. Fully focusing wavelength-dispersive spectrometer geometry.

focusing circle. Furthermore, the crystal planes are bent to a radius of curvature of $2R$ and the surface of the crystal itself is ground to a radius of curvature of R. As a result of this geometry, all x-rays originating from the point source will have the same incident angle θ on the crystal and will be brought to a focus at the same point on the detector, thereby maximizing the overall collection efficiency of the spectrometer without sacrificing good wavelength resolution. Clearly, if a flat crystal were used, the angle of incidence of the x-ray beam would vary across the length of the crystal, giving rise to broad and possibly overlapped peaks with reduced maximum intensity and peak-to-background ratios. Although Soller slits could be used to obtain a more parallel beam striking the crystal, they would have the adverse effect of reducing the peak intensity.

Figure 5.3 also illustrates the additional geometric requirement of a constant x-ray take-off angle Ψ imposed by having a small, fixed x-ray port for the x-ray signal to leave the electron optical chamber. The fully focusing requirement is maintained by moving the analyzing crystal along a straight line away from the sample while rotating the crystal, and moving the detector through a fairly complex path so as to make the focusing circle rotate about the point source. An interesting consequence of this arrangement is that the crystal-to-source distance L is directly proportional to the wavelength. This can be shown with the aid of Figure 5.3: First we write

$$L/2 = R \sin\theta \quad \text{or} \quad L/2R = \sin\theta \tag{5.2a}$$

Combining this with Bragg's law, Equation (5.1) gives

$$n\lambda = 2d \sin\theta = 2dL/2R \tag{5.2b}$$

or for first-order reflections

$$\lambda = (d/R)L \tag{5.2c}$$

with higher-order reflections occurring simply at multiples of the L value for the first-order reflections. Most spectrometers read out directly in L. Actually, the majority of fully focusing spectrometers use crystals that are only bent to a radius of curvature of $2R$ but not ground to be completely tangent to the focusing circle, since grinding a crystal will tend to degrade its resolution by causing defects and a broader mosaic spread. This compromise, known as Johann optics, results in some defocusing of the image at the detector, but does not seriously impair resolution. Another type of Johann optics spectrometer involves maintaining a constant crystal-to-source distance and bending the crystal so that λ varies with R as given by Equation (5.2). Although the mechanics of this type of spectrometer are somewhat simpler than the linear spectrometer, only a few crystals, such as mica and LiF, can tolerate repeated flexing without being seriously damaged, and for this reason bending-crystal spectrometers are no longer commonly used in microanalysis. A further variation of the use of Johann optics is incorporated in the "semifocusing spectrometer," which also uses a fixed source-to-crystal distance, but in this arrangement several crystals bent to various fixed radii of curvature are mounted on a carousel so that each can be switched into position rather than using a single bending crystal. However, since the focusing condition is strictly obeyed for only one wavelength per crystal, a certain degree of defocusing and consequent loss of resolution and peak intensity will be experienced at wavelengths other than the optimum value. The advantage of this approach is that placement of the x-ray source on the focusing circle is less critical, so that when x-ray images are obtained by scanning the electron beam over the surface of the sample, they are less susceptible to defocusing effects since the entire image is, in fact, defocused.

In instruments with fully focusing crystal spectrometers, there is a finite volume of the sample which can be considered to be "in focus." The shape and dimensions of this region depend on a number of factors, such as the design of the spectrometer, the choice and size of crystal, the wavelength being measured, and the criterion for focus (i.e., 99% or 95% of the maximum intensity). The criterion for focus is important when comparing capabilities required for quantitative analysis with those for x-ray mapping. For quantitative applications the volume of the sample which can be considered to be "in focus" to the x-ray spectrometer can be approximated by a flattened ellipsoid with dimensions on the order of 1 mm by 100 μm by 2 μm. For qualitative analysis and x-ray mapping these values can be doubled or even tripled. This region is oriented such that the long axis is perpendicular to the focusing circle, the intermediate axis coincident with the take-off direction, and the short axis perpendicular to both. In the typical spectrometer arrangement in an electron microprobe, the so-called "vertical spectrometers," the maximum sensitivity to spectrometer defocusing occurs along the Z direction of the sample (i.e., the electron optical axis).

Normally, in instruments with a coaxial light microscope, the light and x-ray optics are prealigned. Thus, by positioning the sample in optical focus with the stage controls, it should also be in x-ray focus. The most critical direction is of course along the short axis. In instruments with low take-off angles, this direction is most nearly parallel to the Z-axis motion, which makes it the most critical adjustment. At higher take-off angles, there is a $1/\cos\theta$ increase in the Z component due to the tilt of the volume of focus, which in turn, slightly decreases sensitivity to variations in height. An alternate approach is to rotate the plane of the focusing circle about the take-off direction (the so-called "horizontal spectrometer"). With this orientation the long axis is most nearly parallel to the Z direction, and vertical positioning is least critical. The tradeoff is that defocusing in the $X-Y$ plane becomes more likely. It should be noted that in an SEM equipped with a WDS system, the absence of an optical microscope with a shallow depth of focus to aid in locating the spectrometer focus can lead to severe problems in quantitative analysis. In this case, the excellent depth of focus of the SEM is a liability, since it is difficult to observe changes in working distance of a few micrometers, which is critical to the x-ray measurement.

Most electron microprobes and scanning electron microscopes can be equipped with more than one crystal spectrometer. Multiple spectrometers, each containing several crystals, are necessary not only for analyzing more than one element at a time, but also to include the variety of crystals required for optimizing performance in different wavelength ranges. Table 5.1 lists some of the most commonly used analyzing crystals, showing their comparative resolution, reflectivity, and $2d$ spacings. Since $\sin\theta$ cannot exceed unity, Bragg's law establishes an upper limit of $2d$ for the maximum wavelength diffracted by any given crystal. More practical limits are imposed by the spectrometer design itself, since it is obvious from Figure 5.3 that for $\sin\theta = 1$, i.e., $\theta = 90°$, the detector would have to be at the x-ray source point inside the electron optical column. A lower wavelength limit is imposed by Equation (5.2) since it becomes impractical to physically move the analyzing crystal too close to the specimen.

5.2.2. The X-Ray Detector

The most commonly used detector with microanalyzer crystal spectrometer systems is the gas proportional counter shown in Figure 5.4. It consists of a gas-filled tube with a thin wire, usually tungsten, held at a 1–3-kV potential, running down the center. When an x-ray photon enters the tube through a thin window on the side and is absorbed by an atom of the gas it causes a photoelectron to be ejected, which then loses its energy

Table 5.1. Crystals Used in Diffraction

Name	$2d(\mathring{A})$	Lowest atomic number diffracted	Resolution	Reflectivity
α-Quartz($10\bar{1}1$)	6.687	$K\alpha_1$ 15-P	High	High
		$L\alpha_1$ 40-Zr		
KAP($10\bar{1}0$)	26.632	$K\alpha_1$ 8-O	Medium	Medium
		$L\alpha_1$ 23-V		
LiF(200)	4.028	$K\alpha_1$ 19-K	High	High
		$L\alpha_1$ 49-In		
PbSt	100.4	$K\alpha_1$ 5-B	Medium	Medium
PET	8.742	$K\alpha_1$ 13-Al	Low	High
		$L\alpha_1$ 36-Kr		
RAP	26.121	$K\alpha_1$ 8-O	Medium	Medium
		$L\alpha_1$ 33-As		

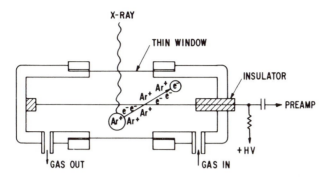

Figure 5.4. A gas flow proportional counter.

by ionizing other gas atoms. The electrons thus released are then attracted to the central wire, giving rise to a charge pulse. If the gas fill used is P10 (90% argon–10% methane), approximately 28 eV is absorbed per electron–ion pair created. For Cu $K\alpha$, which has an energy of slightly over 8 keV, about 300 electrons will be directly created by the absorption of a single photon. This would be an extremely small amount of charge to detect without a special cryogenic preamplifier system (low temperature, low noise). However, if the positive potential of the anode wire is high enough, secondary ionizations occur which can increase the total charge collected by several orders of magnitude. Figure 5.5 schematically shows the effect on the gas amplification factor of increasing the bias voltage applied to a tube. The initial increase corresponds to increasing primary charge collection until it is all collected (a gas amplification factor of 1) and then the curve levels off in the "ionization" region. Increasing the potential beyond this point initiates secondary ionization, the total charge collected increases

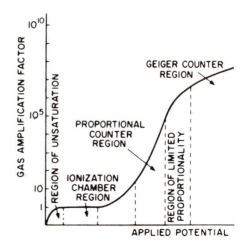

Figure 5.5. The effect of applied counter tube bias on the gas amplification factor.

drastically, and the counter tube enters what is termed the "proportional" region, because the collected charge remains proportional to the energy of the incident photon. Further increasing the voltage causes the tube to enter the Geiger region, where each photon causes a discharge giving rise to a pulse of a fixed size independent of its initial energy, thereby losing the information necessary for pulse height analysis. A further disadvantage is that the counter "dead time," which is the time required for the tube to recover sufficiently to accept the next pulse increases from a few to several hundred microseconds. Any increase in applied voltage beyond the Geiger region will result in permanent damage to the tube. In practice, operation in the lower part of the proportional region is preferred to minimize the effect of gain shifts with counting rate.

The proportional counter shown in Figure 5.4 is of the gas flow type, normally used for detecting soft x-rays ($\lambda > 3$ Å). A flowing gas, usually P10, is chosen because it is difficult to permanently seal the thin entrance windows necessary to reduce absorption losses. Hendee *et al.* (1956) have shown that the window transmission of Al $K\alpha$ is 1.2% through 34 μm (4.5-mil) Be foil, 55% through 7.5 μm (1-mil) Be foil, 30% through 1.5 μm (0.2-mil) Mylar and 84% through Formvar thin enough to give interference fringes. Since crystal spectrometers are normally kept under vacuum to eliminate absorption of the x-ray beam in the air, it is usually necessary to support ultrathin windows like Formvar and cellulose nitrate on fine wire screens in order to withstand a pressure differential of 1 atm; however, this causes an additional decrease in detector collection efficiency. Recently, unsupported stretched polypropylene films have come into use with considerable success. Sealed counters containing krypton or xenon are used for shorter x-ray wavelengths since, as shown in Figure 5.6, they have a higher

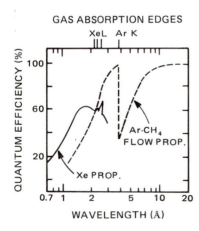

Figure 5.6. Representative collection efficiencies for proportional counters filled with Xe and Ar–CH$_4$ gases. The Ar–CH$_4$ counter has a 25-nm (250 Å) Formvar window.

quantum counter efficiency (the fraction of input pulses detected $\times 100\%$) than argon-filled detectors at 1 atm. The efficiency of argon-filled detectors for shorter x-ray wavelengths can, however, be increased by increasing the gas pressure to 2 or 3 atm.

5.2.3. Detector Electronics

The role of the detector electronics is to integrate the total charge produced by each x-ray photon and convert it into a single voltage pulse to be further processed for counting or display purposes. On the average the number of electrons per incident x-ray photon entering the detector is given by

$$n = (E/\epsilon)A \tag{5.3a}$$

where E is the incident x-ray energy, ϵ the average energy absorbed per electron–ion pair created (about 28 eV for argon), and A is the gas amplification factor (see Figure 5.5), determined by the potential applied to the counter tube. The charge collected by the preamplifier is given by

$$q = en = (E/\epsilon)eA \tag{5.3b}$$

where e is the charge of an electron $(1.6 \times 10^{-19}$ coul.). Because this quantity is very small, the preamplifier is normally located close to the counter tube to avoid stray electronic pickups. The output of the preamplifier is a voltage pulse V_P, where

$$V_P = \frac{q}{C} = \left(\frac{G_p eA}{\epsilon C} \right)E \tag{5.3c}$$

C is the effective capacitance of the preamplifier, and G_p is a gain factor, usually about unity. A typical pulse shape at this point (A in Figure 5.1) is given in Figure 5.7A, which shows a rapidly falling negative pulse with a tail a few microseconds long. The signal from the preamplifier is now suitable for transmission several feet over a coaxial cable to the main amplifier where it is usually inverted, further shaped, and amplified to give a Gaussian pulse of the type shown in Figure 5.7B. The magnitude of this pulse is given by

$$V_A = G_A V_P = \left(\frac{G_A G_p Ae}{\epsilon C} \right)E \tag{5.3d}$$

where G_A is the main amplifier gain. The value of V_A for a given x-ray energy is typically set within the range of 2–10 V by operator adjustment of either the G_A or the bias voltage applied to the detector. Since during an analysis all of the quantities shown in the parentheses in Equation (5.3) are held fixed it follows that

$$V_A = KE \tag{5.4}$$

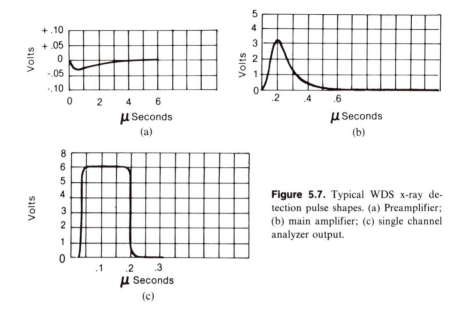

Figure 5.7. Typical WDS x-ray detection pulse shapes. (a) Preamplifier; (b) main amplifier; (c) single channel analyzer output.

where $K = G_A G_p Ae / \epsilon C$ is a constant. In other words, the average pulse size for a series of fixed-energy x-ray pulses, is directly proportional to the x-ray energy providing the counter tube voltage is set for the proportional range and the main amplifier is run in a linear region.

Pulse shapes of the type shown in Figure 5.7 are easily measured with most general-purpose laboratory oscilloscopes (having a 0.1 μs/cm sweep rate or better). Periodic monitoring of amplifier output pulses is highly recommended because it provides a convenient way to observe the manner in which typical pulses are processed by the detector electronics. Undesirable effects such as peak clipping, base line instabilities, noise and pulse overshoots, indicative of faulty electronics or improper control setting, become readily apparent and can often be easily corrected. The use of an oscilloscope to observe amplifier output pulses is, furthermore, the best method for correctly adjusting amplifier gain and counter tube bias. Information about the actual voltage distribution of pulses processed for a preselected time period can be readily obtained by means of either a single-channel analyzer (SCA) or multichannel analyzer (MCA). Basically the SCA serves two functions. As a discriminator it is used to select and transmit pulses within a predetermined voltage range for further processing. As an output driver the selected pulses trigger rectangular pulses of a fixed voltage and time duration compatible with scalar and rate meter input requirements. A typical pulse shape at this point (C in Figure 5.1) is given in Figure 5.7C. In this example the output pulse is 6 V high and lasts for 0.2 μs.

The process of pulse height analysis with an SCA is illustrated schematically in Figure 5.8. The operator sets either a combination of a base line voltage E_L and a window voltage ΔE by means of potentiometer controls in the SCA, or a base line voltage E_L and an upper window voltage E_U. In the example shown, only pulses from 5 to 7 V (pulse II, 6 V) are accepted. Pulses larger (pulse III, 8 V) or smaller (pulse I, 4 V) are rejected. In practice pulse voltage distribution curves with an SCA can be obtained by tuning the WDS to a specific characteristic line, selecting an SCA window value of a few tenths of a volt, and recording the pulse intensity as a function of the base line, E_L, setting. This can be done by either scanning the base line at a fixed rate from 0 to 10 V and recording the SCA output with a rate meter (see Figure 5.9) or manually stepping through a range of base line values, counting the pulses with a scaler and plotting the results.

A detailed description of the operating principles of the MCA will be deferred to the section on energy dispersive spectrometry. At this point, let it suffice to say that the MCA has the capability of determining the voltage of each pulse from the main amplifier and assigning it to one of a series of memory locations corresponding to preestablished voltage ranges. For example, if 100 channels of memory are used to span 10 V, then pulses from 0 to 0.1 V are assigned to memory location 1, 0.1 to 0.2 V to memory location 2, and so on. The contents of the memory locations are normally displayed as a pulse distribution histogram on a CRT either in real time during the data collection or at the completion of a data collection run as shown for chromium $K\alpha$ in Figure 5.10. The principal features to note are

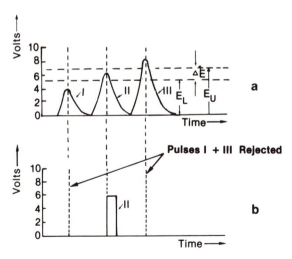

Figure 5.8. Schematic representation of pulse height analyzer behavior. (a) Main amplifier output; (b) single channel analyzer output. $E_L = 5$ V; $\Delta E = 2$ V; $E_U = 7$ V.

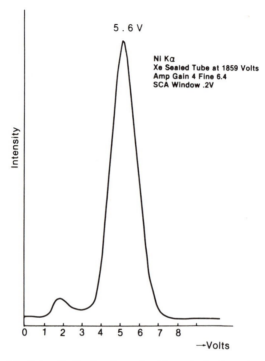

Figure 5.9. Pulse distribution determined by a single-channel analyzer.

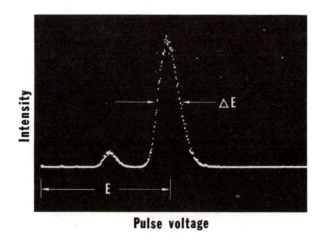

Figure 5.10. Multichannel analyzer display of Cr *Kα* pulse amplitude distribution from a flow proportional counter on a wavelength-dispersive spectrometer.

two relatively broad peaks. The larger of the two corresponds to the main Cr K pulse distribution. The smaller peak is known as an "escape peak." Its center occurs at a pulse voltage corresponding to an x-ray energy equal to the Cr $K\alpha$ energy minus the characteristic line energy of the element used for counter tube gas which in this case is argon (argon $K\alpha = 2.3$ keV) for the example shown in Figure 5.10.

The escape peak corresponds to a special type of proportional detector response during which either the incoming x-ray photon or the primary photoelectron ionizes a core rather than outer-shell electrons in the counter tube gas. This process is then followed by the emission of a characteristic x-ray photon which will have a high probability of escaping from the counter tube entirely. If it does, the remaining energy for electron–ion pair production will be correspondingly diminished. For these events the amplifier output equation (5.4) must be modified to the form

$$V_A(\text{escape peak}) = K(E - E_{CT}) \tag{5.5}$$

where E_{CT} is the characteristic line energy of the counter tube gas.

The natural spread of both peaks arises from the fact that each monoenergetic photon entering the detector does not give rise to the same number of electron–ion pairs. Several competing processes occur by which the initial photoelectron may dissipate its energy. The percentage resolution of a detector is defined as 100 times the width of the pulse distribution curve at half-maximum divided by the mean peak voltage. The resolution of a properly functioning tube is about 15%–20%. The pulse distribution should be approximately Gaussian and free of large asymmetric tails. It is desirable to periodically check this distribution since failure of the electronics or degradation of the counter tube can lead to a shift in the peak position, width, and/or symmetry, making any prior SCA settings improper.

There has always been some confusion about the precise role of the energy discrimination capability of the SCA. First of all, it cannot improve the energy selection of the spectrometer system for wavelengths close to that of the characteristic line being measured, for this has already been done by the crystal itself, which clearly has much higher energy resolution than the flow counter. The SCA can, however, eliminate both low- and high-energy noise as well as higher-order reflections ($n > 1$ in Bragg's law) arising from the diffraction of higher-energy characteristic or continuum x-rays with the same $n\lambda$ value since such events will occur at the same λ setting as the line of actual interest. Balancing the benefits versus the problems introduced in eliminating these effects often leads to the simple solution of restricting the discriminator settings to a high enough base line value to remove low-energy noise and adjusting the window or upper level

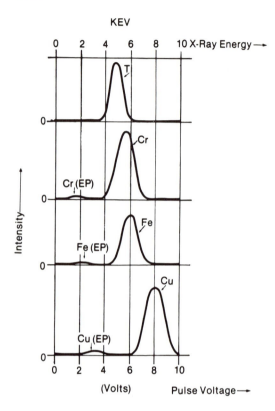

KEV

0 2 4 6 8 10 X-Ray Energy →

Intensity →

Ti

Cr

Cr (EP)

Fe

Fe (EP)

Cu

Cu (EP)

0 2 4 6 8 10

(Volts)   Pulse Voltage →

Figure 5.11. Four elemental pulse distributions determined by a multichannel analyzer.

to a value that allows the acceptance of all other pulses. One reason for this approach is dramatized with the aid of Figure 5.11, which shows a series of pulse distributions obtained with an MCA for crystal spectrometer settings corresponding to several elements readily accessible to a single LiF crystal during a typical scan. It can be seen in scanning from Ti to Cu, that in accordance with Equation (5.3) the main pulse distribution shifts from a mean value of about 4.5 to 8.0 V. Note that the constant K was intentionally set by means of the amplifier gain to give 1000 eV/V in order to make the voltage and energy scales numerically the same. It is evident that if the SCA window is set from 3 to 6 V for the main Ti pulse distribution, all of the pulses would be excluded when the spectrometer reached the copper setting. Therefore, the use of such a window is totally incompatible with qualitative line scanning and should only be used for fixed spectrometer setting quantitative analysis. It should further be pointed out that even when narrow-band SCA settings are used for fixed spectrometer positions, as in the case of a quantitative analysis, the analyst is not free from problems. It has been shown that under certain conditions the entire pulse distribution may shift to lower voltages with increasing count rate. The exact origin of this phenomenon is not understood, but it is probably related to either main amplifier base line stability or a decrease in the effective potential across the tube caused by positive ion buildup around the counter tube walls.

Typical crystal spectrometer scans of ratemeter output vs. wavelength for a nickel-base superalloy are illustrated in Figure 5.12a for LiF and 12b for RAP. Separation of $K\alpha_1$ and $K\alpha_2$ peaks (6 eV for vanadium) for the major elements in Figure 5.12a demonstrates the high-energy resolution that can be expected with a crystal spectrometer. Two other capabilities, namely, light element detection and peak shift measurements, are illustrated in Figure 5.13, which shows superimposed computer-controlled scans for boron $K\alpha$ in pure boron, cubic boron nitride, and hexagonal boron nitride. The peak shifts and satellite lines are due to shifts in the outer electron energy states associated with differences in the chemical bonding. Measurements of this type can also be used to fingerprint various cation oxidation states in metal oxides (Holliday, 1963). A more complete discussion of this effect is given in Chapter 8.

In addition to performing chemical analysis at a fixed point it is often desirable to analyze variations in the x-ray intensity of one or more selected elements across a line on a specimen or even over a two-dimensional field of view. In the line scan mode illustrated in Figure 5.14 the rate-meter output corresponding to a fixed spectrometer setting is used to modulate the vertical deflection of the CRT as the electron beam is scanned across the sample. Multiple exposures of the CRT are used to superimpose the x-ray and beam position information on a secondary-electron image so that the analyst can readily interpret the results. Data presented in this manner give semiquantitative information since a complete quantitative evaluation requires conversion of the intensity data to chemical composition by one of the mathematical methods described in Chapter 7. Furthermore, since deflection of the beam can lead to defocusing of the x-ray spectrometers, quantitative line scans can be performed by stepping the sample under a static beam.

Two-dimensional scanning known as x-ray mapping involves taking the output of the SCA and using it to modulate the brightness of the CRT during normal secondary-electron raster scanning. Each x-ray photon detected appears as a dot on the CRT with regions of high concentration characterized by a high dot density. It is in this respect that the appearance of an x-ray map differs from a secondary-electron image since the latter uses an analog rather than a digital signal to modulate the CRT intensity. Use of an analog signal capable of giving rise to a range of grey levels for each point in an image is a luxury only available when there is a high signal intensity. The x-ray intensity even for a single element phase is, however,

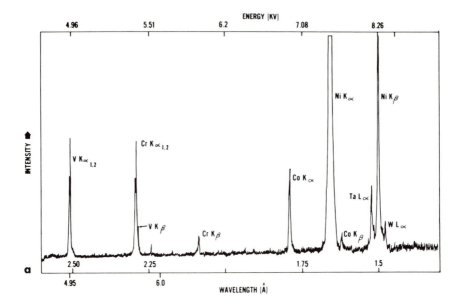

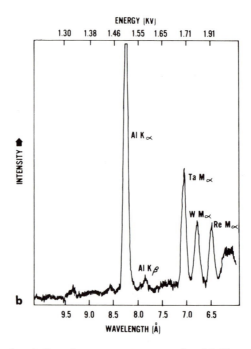

Figure 5.12. Wavelength-dispersive spectrometer scans of a nickel-base superalloy. (a) Scan using a LiF crystal; (b) scan using an RAP crystal.

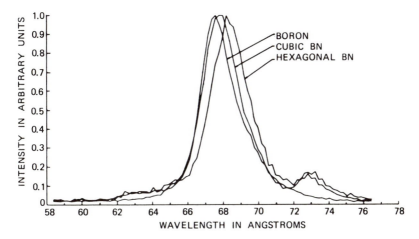

Figure 5.13. Boron $K\alpha$ scans obtained from pure boron, cubic boron nitride, and hexagonal boron nitride.

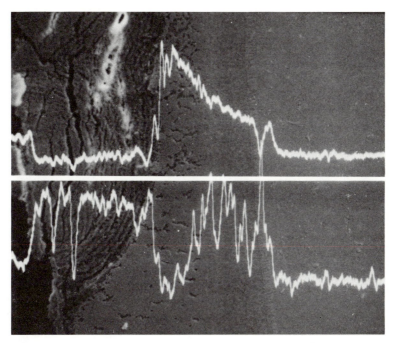

Figure 5.14. Co and Cr line scans across an oxidized high-temperature alloy. Straight line indicates the position of the scan on the secondary electron image. (Top portion represents cobalt, bottom chromium.)

several orders of magnitude less than the secondary-electron signal. Consequently, it is unusual to have more than one event recorded per picture element. For example, if a 1000-line picture is scanned in 100 s and a 1000 picture elements are defined along each line, a count rate of 10,000 counts/s would be required to average one photon per picture element.

Even though some improvement in image quality is possible by using fewer picture elements, longer counting times, and higher beam currents, it is still difficult, in view of the limitations imposed by x-ray generation statistics, to collect an adequate signal for a grey scale map. In other words, for a fixed data collection time, the analyst must always be aware of the tradeoff between the signal precision and spatial distribution information obtained by line scanning or area mapping. Another point to consider in any of the modes of analysis used is that, since the x-ray excited volume is usually considerably larger than the source of secondary electrons, any attempt to localize or quantify chemical information for submicrometer structural details in bulk specimens will generally be meaningless. A number of examples of the x-ray mapping technique can be found in Chapter 6.

5.3. Energy-Dispersive X-Ray Spectrometer

5.3.1. Operating Principles

Fitzgerald, Keil, and Heinrich (1968) published a very significant paper in which they first described the use of a solid state x-ray detector on an electron beam microanalyzer. Although their system was barely capable of resolving adjacent elements, it did demonstrate the feasibility of interfacing the two instruments. The next few years saw a period of rapid improvement in detector resolution from 500 eV to less than 150 eV thereby making the technique much more suitable for microanalysis requirements. Today the idea of using solid state detectors for midenergy x-ray spectroscopy (1–12 keV) is no longer a novelty for they can be found on a large percentage of scanning and transmission electron microscopes as well as electron microprobe analyzers.

The operating principles of the solid state detectors are illustrated in Figure 5.15. The x-ray signal from the sample passes through a thin beryllium window into a cooled, reverse-bias p-i-n (p-type, intrinsic, n-type) lithium-drifted silicon detector. Absorption of each individual x-ray photon leads to the ejection of a photoelectron which gives up most of its energy to the formation of electron–hole pairs. They in turn are swept away by the applied bias to form a charge pulse which is then converted to a voltage pulse by a charge-sensitive preamplifier. The signal is further amplified and shaped by a main amplifier and finally passed to a

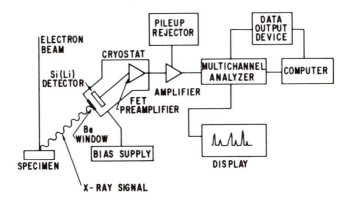

Figure 5.15. Schematic representation of an energy-dispersive spectrometer.

multichannel analyzer (MCA), where the pulses are sorted by voltage. The voltage distribution can be displayed on a cathode ray tube or an $X-Y$ recorder. The contents of the MCA memory in most recent instruments either reside directly in a computer or can be transmitted to a computer for further processing such as peak identification or quantification.

The typical physical appearance of a detector can be seen in more detail in Figure 5.16. The lithium-drifted silicon crystal is mounted on a cold finger connected to a liquid-nitrogen reservoir stored in the familiar appearing Dewar. The mounting chamber is light tight to stop non-x-ray photons from producing unwanted electron–hole pairs in the detector crystal. It is also sealed under vacuum both to prevent contamination and to more easily maintain the low temperature essential for reducing noise. Low temperature is also needed to limit the mobility of the lithium ions initially introduced in the silicon crystal to neutralize recombination centers, thereby making it possible to create a large intrinsic region of the desired size. Under no conditions should the bias be applied to a noncooled detector. Many systems, in fact, incorporate safety features to turn the bias off in the event of either the crystal warming up or a vacuum failure. Dewars presently available are capable of maintaining cryogenic conditions for several days without refilling. Note also that the crystal and supporting cold finger are well separated from the housing assembly to prevent condensation on the latter, and to provide a fair amount of electrical isolation. The feature of being able to mechanically move the detector crystal in and out relative to the specimen without breaking the vacuum is quite useful. As will be discussed later, the situation may arise where, for a fixed beam current, the x-ray signal has to be increased to obtain better counting statistics or decreased for better energy resolution. In many cases the desired count rate can be obtained by simply varying the crystal-to-sample distance, which changes the solid angle of detector acceptance.

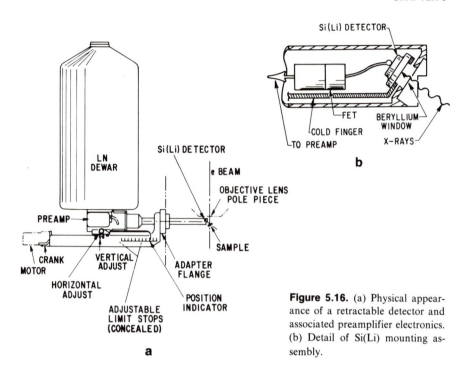

Figure 5.16. (a) Physical appearance of a retractable detector and associated preamplifier electronics. (b) Detail of Si(Li) mounting assembly.

The development of the energy-dispersive x-ray spectrometer has made "easy" x-ray spectrometry available to virtually all types of electron beam instruments. However, one must recognize that, because of the nature of the EDS technique, distortions are introduced into the ideal x-ray spectrum ("spectral artifacts") during the measurement process which must be dealt with in practical analytical spectrometry. In the discussion which follows, we will consider these artifacts at each stage of the detection and amplification process.

5.3.2. The Detection Process

The key to understanding how an energy-dispersive spectrometer (EDS) works is to recognize that the voltage pulses produced by the detector are on average proportional in size to the incoming x-ray photon energy. The basic detection process by which this proportional conversion of photon energy into an electrical signal is accomplished is illustrated in Figure 5.17. The quiescent Si(Li) crystal has a band structure (for a description of band structure see the discussion on cathodoluminescence in Chapter 3) in which the conduction band states are empty while the valence band states are filled. When an energetic photon is captured, electrons are promoted into the conduction band, leaving holes in the valence band. Under an applied bias, these electrons and holes are swept apart and collected on the electrodes on the faces of the crystal. The process of photon capture is photoelectric absorption. The incident x-ray photon is first absorbed by a silicon atom and an energetic electron is emitted. This photoelectron then creates electron–hole pairs as it travels in the detector silicon and scatters inelastically. The silicon atom is left in an energetic condition because of the energy required to eject the photoelectron. This energy is subsequently released in the form of either an Auger electron or a silicon x-ray. The Auger electron scatters inelastically and also creates electron–hole pairs. The silicon x-ray can be reabsorbed, which initiates the process again, or it can be scattered inelastically. Thus, a sequence of events takes place leading to the deposition of all of the energy of the original photon in the detector, unless radiation generated during the sequence, such as a silicon $K\alpha$ photon, escapes the detector (Figure 5.17). The detector is also sensitive to energetic electrons which enter; such electrons can directly form charge carriers.

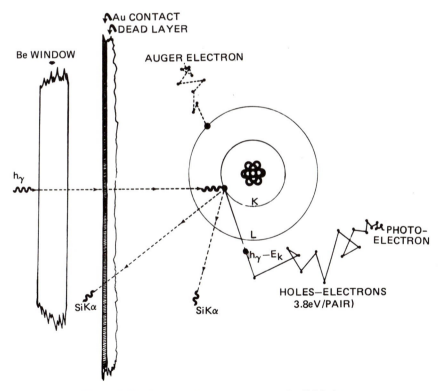

Figure 5.17. The x-ray detection process in the Si(Li) detector.

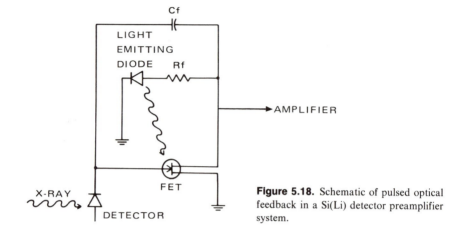

Figure 5.18. Schematic of pulsed optical feedback in a Si(Li) detector preamplifier system.

The ideal number of charges created per incident particle with energy E (eV) is given by

$$n = E/\epsilon \qquad (5.6)$$

where $\epsilon = 3.8$ eV for silicon. For example, if the detector captures one photon having an energy of 5 keV, then from Equation (5.6) the total number of electrons swept from the detector is approximately 1300, which represents a charge of 2×10^{-16} coul. This is an extraordinarily small charge. The subsequent circuitry must be capable of amplifying this signal by about 10^{10}. Because of the small charge collected, noise reduction is essential. Cooling of the detector crystal and first stage FET of the preamplifier has already been mentioned. An additional measure is the use of pulsed optical feedback (POF), as is illustrated in Figure 5.18. With this method the noise normally associated with resistive feedback in pre-amplifiers is eliminated by simply not using any feedback to drain the charge accumulated from the detector. This condition cannot exist indefinitely, so when the preamplifier output reaches a predetermined voltage level, a light-emitting diode is turned on, which causes a leakage current to flow in the FET, restoring it to its original operating point. Since considerable noise is generated when the POF is turned on, it is necessary to gate the main amplifier off during that period. POF preamplifiers are now in common use by most manufacturers, with the exception of ORTEC, which accomplishes the same effect with a proprietary technique called "dynamic charge restoration," which does not require special amplifier gating.

5.3.3. Artifacts of the Detection Process

Deviations from the ideal detector process result in the appearance of artifacts, principally peak broadening, peak distortion, silicon x-ray escape peaks, silicon and gold absorption edges, and the silicon internal flourescence peak.

5.3.3.1. Peak Broadening

The natural width of an x-ray peak is of the order of 2 eV measured at half the maximum of the peak intensity (FWHM). For example, for manganese $K\alpha$ radiation (5.898 keV), the FWHM is approximately 2.3 eV, which makes the natural width about 0.039% of the peak energy. The measured peak width from the Si(Li) spectrometer is degraded to a typical value of 150 eV for Mn $K\alpha$, or 2.5% of the peak energy. This degradation in width of the peak occurs because (1) there is a statistical distribution in the final number of charge carriers created by capturing photons of a single energy due to the discrete nature of the counting process and (2) an uncertainty is introduced by the thermal noise of the amplification process. The distribution of numbers of charge carriers for a single photon energy is reasonably well described by a Gaussian distribution, shown schematically in Figure 5.19. The FWHM of this distribution can be calculated from the two sources of noise by quadrature addition (see Chapter 2 for an explanation of quadrature) according to the equation

$$\text{FWHM} \propto \left(C^2 E + N^2 \right)^{1/2} \tag{5.7}$$

where C is the measure of the uncertainty in the formation of charge carriers and N is the FWHM of the electronic noise of the amplification process. C is given by the expression

$$C = 2.35 \left(F\epsilon \right)^{1/2} \tag{5.8}$$

where F is a constant known as the Fano factor. Figure 5.20 shows values calculated by Woldseth (1973) of the observed FWHM as a function of energy for different contributions of electronic noise. It can readily be seen that even if the noise contribution were totally eliminated the theoretical energy resolution limit would still be greater than 100 eV for Fe $K\alpha$ at 6.4 keV.

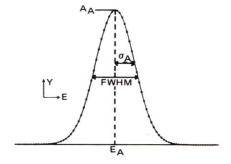

Figure 5.19. Theoretical Gaussian distribution, which describes an EDS peak. The standard deviation of the distribution is σ, the peak amplitude A_A, and the amplitude at any energy E is Y.

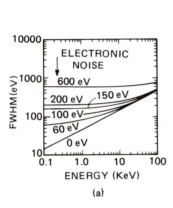 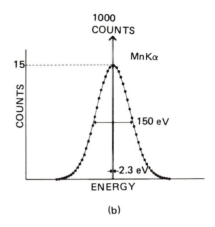

Figure 5.20. (a) Si(Li) energy resolution, including intrinsic and electronic noise effects as a function of energy (adapted from Woldseth, 1973). (b) Redistribution of peak counts for Mn $K\alpha$ with 150-eV resolution peak.

Since the noise term is a constant for a given detector, operated under a fixed set of conditions, Equation (5.7) can be restated in a form which is useful for comparing the width of a peak at an energy of interest to the width of a peak located at any other place in a spectrum:

$$\text{FWHM} = \left[2.5(E - E_{\text{ref}}) + \text{FWHM}_{\text{ref}}^2 \right]^{1/2} \qquad (5.9)$$

All units in this expression are in electron volts.

The immediate consequence of the peak broadening is a reduction in the height of the peak (counts per energy interval) as compared to the natural peak and an accompanying decrease in the peak-to-background ratio. Figure 5.20b shows the effect of detector broadening on the natural peak width of Mn $K\alpha$. In this example, a natural peak of width 2.3 eV and 1000 counts high is broadened into a peak of 150 eV width and 15 counts high. Since the energy resolution varies as a function of energy as described by Equation (5.9), one observes that, for a fixed number of counts, as the peak broadens with increasing energy, the amplitude must decrease, as shown in Figure 5.21.

The value of the FWHM is useful in estimating the extent of the overlap of peaks which have similar energies. An estimate of the extent of overlap is vital when considering peak interferences in qualitative analysis, where the identification of a low-intensity peak near a high-intensity peak may be difficult, and in quantitative analysis, where the removal of the interference is necessary for accurate determination of the composition. An example of the overlap of peaks is illustrated in the spectrum of KCl in Figure 5.22. With a detector of 170-eV resolution (Mn $K\alpha$), the potassium $K\alpha$ and $K\beta$ peaks are nearly resolved, while the chlorine $K\alpha$ and $K\beta$ peaks are not.

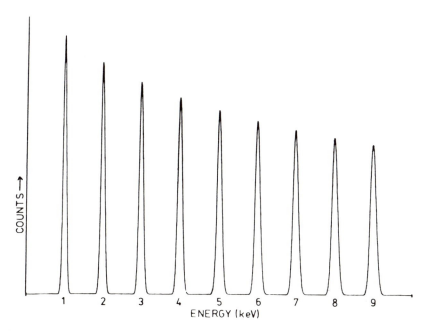

Figure 5.21. Resolution and peak intensity as a function of energy. Mn $K\alpha$ has a resolution of 150 eV.

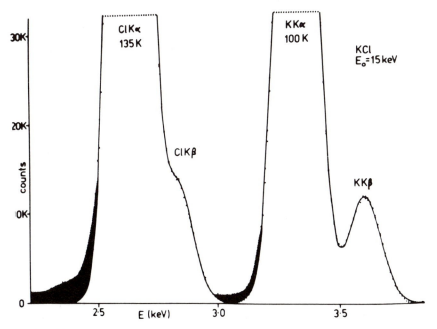

Figure 5.22. EDS spectrum of KCl illustrating peak overlap: K $K\alpha$, $K\beta$ are nearly resolved. The solid line is a Gaussian fit to the data points. The shaded area represents the deviation caused by incomplete charge collection.

5.3.3.2. Peak Distortion

Two different artifacts cause distortion, i.e., deviation from a Gaussian shape on the low-energy side of a peak: (1) The collection of charge carriers created in certain regions of the detector near the faces and sides is imperfect due to trapping and recombination of the electron–hole pairs, leading to a reduction in the value of n predicted by Equation (5.6) for the incident photon. The resulting distortion of the low-energy side of the peak is known as "incomplete charge collection" (Freund *et al.*, 1972; Elad *et al.*, 1973) and its effect is illustrated in Figure 5.22 for chlorine $K\alpha$ and potassium $K\alpha$ peaks. The deviation from a Gaussian distribution (shown as a solid line) is a function of energy. For example, the magnitude of the effect is significantly different for chlorine and potassium, which are separated by an atomic number difference of only 2. (2) The background shelf, Figure 5.23, is a phenomenon in which the presence of a peak increases the background at all energies below the peak value. Additional counts above the expected background result both from the incomplete charge collection phenomenon extending to low energy and from the escape from the detector of some of the continuum x-rays generated by the photoelectron as it scatters inelastically in the silicon. Any radiation lost

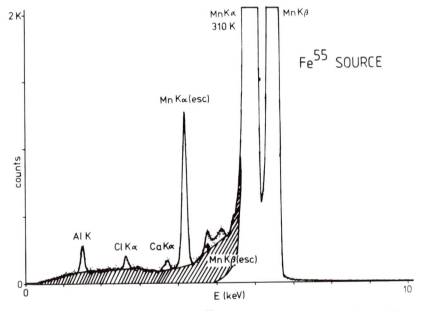

Figure 5.23. EDS spectrum derived from ^{55}Fe radioactive source (emits Mn K radiation). The background shelf is easily observable at energies below Mn $K\alpha$ down to the threshold at 300 eV. The Mn $K\alpha$ and $K\beta$ silicon escape peaks are noted. Extraneous characteristic peaks from the source holder are observed.

from the detector reduces the number of charge carriers created. Both the incomplete charge collection and the loss of continuum x-radiation lead to a transfer of counts from the peak to the entire energy region down to zero. Typically, the background shelf at one-half the energy of a peak has a relative intensity of about 0.1% of the parent peak. The total number of counts lost to the full peak due to this effect is approximately 1%.

5.3.3.3. Silicon X-Ray Escape Peaks

The generation of a photoelectron leaves the silicon atom in an ionized state. If the photoelectron is emitted from the K shell, the atom can subsequently undergo an electron transition to fill this K-shell vacancy, with subsequent emission of a silicon K x-ray or an Auger electron, Figure 5.17. The range of the Auger electron is only a fraction of a micrometer, and hence, it is highly probable that this electron will be reabsorbed in the detector and contribute its energy to the formation of charge carriers, yielding the correct value of energy deposited in the detector. The Si K x-ray, on the other hand, has a finite probability of escaping the detector (10% of the Si K initial intensity will remain after 30 μm of travel through silicon). When such an x-ray escape occurs from the detector, it robs the cascade being measured by the energy carried off in the x-ray, 1.740 keV

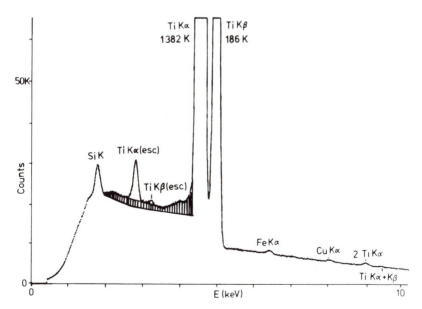

Figure 5.24. Electron-excited EDS spectrum of titanium. The Ti $K\alpha$ and $K\beta$, silicon x-ray escape peaks, and the 2 $K\alpha$ and ($K\alpha + K\beta$) sum peaks are noted. Extraneous peaks from the specimen chamber are also observed.

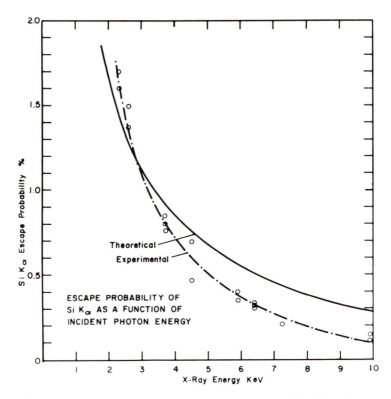

Figure 5.25. The escape probability of Si $K\alpha$ as a function of incident photon energy (adapted from Woldseth, 1973).

for Si $K\alpha$ and 1.832 keV for Si $K\beta$. Thus, an artifact peak, the "escape peak" (Reed and Ware, 1972) is formed at an energy equal to the energy of the parent line minus the energy of the silicon x-ray. In principle, both Si $K\alpha$ and Si $K\beta$ escape peaks are formed, but the probability for $K\beta$ formation is about 2% of the $K\alpha$, hence only one escape peak is observed per parent peak. Escape peaks are illustrated in Figures 5.23 and 5.24. In Figure 5.24 the parent peaks are Ti $K\alpha$ (4.51 keV) and Ti $K\beta$ (4.93 keV) and escape peaks are found at 2.77 keV (Ti $K\alpha$–Si $K\alpha$) and 3.19 keV (Ti $K\beta$–Si $K\alpha$). The magnitude of the escape peak relative to the parent peak varies from about 1.8% for phosphorus to 0.01% for zinc K x-rays, Figure 5.25. Silicon x-ray escape peaks cannot occur for radiation below the absorption energy of the silicon K shell (1.838 keV).

5.3.3.4. Absorption Edges

The typical Si(Li) spectrometer has a protective window of beryllium (approximately 7.6 μm thick), a front surface electrode of gold (approxi-

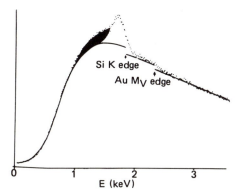

Figure 5.26. Electron-excited EDS spectrum from carbon. The silicon and gold absorption edges are illustrated. The solid line represents a theoretical fit to the continuum. $E_0 = 10$ keV.

mately 20 nm thick), and an inactive layer of silicon (20–200 nm thick). X-rays must pass through each of these layers to reach the active silicon and be detected. During passage of x-rays through the layers, absorption occurs. Absorption in the beryllium window eliminates nearly all x-rays below about 600 eV. Above 2 keV, virtually all x-rays are transmitted through the window. Between these limits, the absorption increases with decreasing energy such that at 1.5 keV about 70% of the x-rays are transmitted, while for an energy of 1 keV, the transmission is 45%. Absorption by the gold and silicon layers is much less significant because of the small mass thickness of these layers. However, a noticeable change in the x-ray continuum is observed at the absorption edge of silicon and to a lesser degree at gold, Figure 5.26. Just above the energy of the absorption edge, the mass absorption coefficient increases abruptly, resulting in a decrease in the measured continuum x-radiation. The height of the resulting step is an indication of the thickness of the layer. Note that the action of the broadening effect of the detection process causes the absorption edge, which in reality is a sharp change of absorption over a range of about 1 eV, to be smeared over a much broader range, typically 100 eV for the silicon absorption edge.

5.3.3.5. Silicon Internal Fluorescence Peak

The photoelectric absorption of x-rays by the silicon dead layer results in the emission of Si K x-rays from this layer into the active volume of the detector. These silicon x-rays, which do not originate in the sample, appear in the spectrum as a small silicon peak, the so-called silicon internal fluorescence peak. An example of this effect is shown in the spectrum of pure carbon, Figure 5.26, which also contains a significant silicon absorption edge. For many quantitative analysis situations, this fluorescence peak corresponds to an apparent concentration of 0.2 wt% or less apparent silicon in the specimen.

5.3.4. The Main Amplifier and Pulse Pileup Rejection

Achieving optimum energy resolution is dependent not only on the quality of the detector crystal, its environment, and associated preamplifier electronics but also on the operating characteristics of the main amplifier. In a Si(Li) detector system it is particularly critical because unlike the WDS all of the spectral dispersion is done electronically. Special circuity must be used to ensure maximum linearity, low noise, rapid overload recovery, and stable high count-rate performance. Most commercial amplifiers use pole zero cancellation networks to compensate for pulse overshoot when internal ac coupling is used, and dc restoration circuits to clamp the pulse base line to a stable reference voltage. Clearly any loss of linearity will lead to a breakdown in Equation (5.6) and the incorrect assignment of x-ray pulse energy.

Achieving the maximum energy resolution possible with an EDS system necessitates that the main amplifier have sufficient time to process each pulse so as to achieve a maximum signal-to-noise ratio. In practice this means that the operator must select a long time constant (T.C.), typically 6–10 μs. The resulting main amplifier output pulses are given in Figure 5.27, which compares pulse shapes for 1, 6, and 10 μs, respectively. What is important to note is that the time required for pulses processed with the 10-μs T.C. to return to the base line is more than 35μs while less than 5μs are required with the 1-μ setting. Therefore, use of the long T.C. required to achieve maximum resolution also increases the likelihood of a second pulse arriving at the main amplifier before it has completed processing the first. This point is also illustrated in Figure 5.27. It can be seen that pulse II arriving 20 μs after pulse I will correctly be evaluated as 4 V for the 1-μs T.C. but as 4.5 V for the 6-μ T.C. and 6.5-V for the 10-μs T.C. If in an actual experimental situation such pulses were accepted, those associated with the longer T.C.'s would be given an incorrect assignment in the

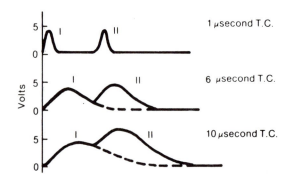

Figure 5.27. EDS main amplifier pulse shapes for different time constants (T.C.).

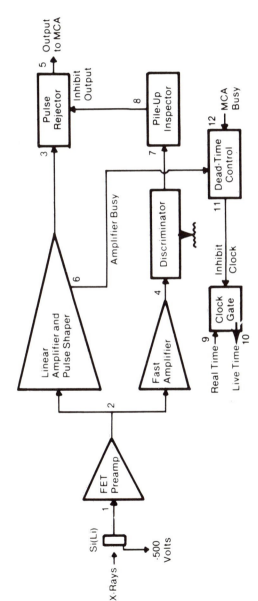

Figure 5.28. Schematic representation of the principal components used with pulse pileup rejection.

memory of the MCA and consequently would appear in the wrong channel of the CRT display. The electronic elimination of such events is accomplished by means of a pulse pileup rejection circuit of the type illustrated schematically in Figure 5.28, taken from the work of Fiori and Newbury (1978). Schematic waveforms at the designated output points are given in Figure 5.29. The collection of charge from the detector occurs very rapidly relative to other processes, typically on the order of 100 ns (point 1). Integration of this charge in the preamplifier produces a "staircase" pulse (point 2), which is seen both by the main amplifier and a parallel fast amplifier. Much less stringent pulse shaping and signal-to-noise criteria are placed on this "fast channel" so that it can be used to sense the presence of each pulse arriving at the main amplifier even if the main amplifier is busy processing a previous pulse. If the fast amplifier pulse exceeds a predetermined discriminator setting (point 4) then a signal (point 7) is sent to the pulse pileup inspector which can prevent the output of the main amplifier from going to the MCA (point 6). The decision can be made to block both pulses if the second one arrives before the first one reaches its peak value or to only block the second one if the first has peaked and been processed by the MCA, but the signal has not returned to the base line. Proper setting of the discriminator is crucial, for if it is too low, noise will be accepted as pulses, causing unnecessary pulse rejection; however, if it is too high, low-energy pulses might be missed. Pileup rejection is therefore more difficult for low-energy x-rays which may be difficult to separate from the background noise. Figure 5.30 shows a comparison between two iron spectra taken with and without pulse pileup rejection. The first feature to

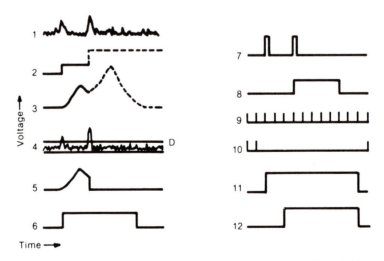

Figure 5.29. Waveforms corresponding to numbered points in Figure 5.28.

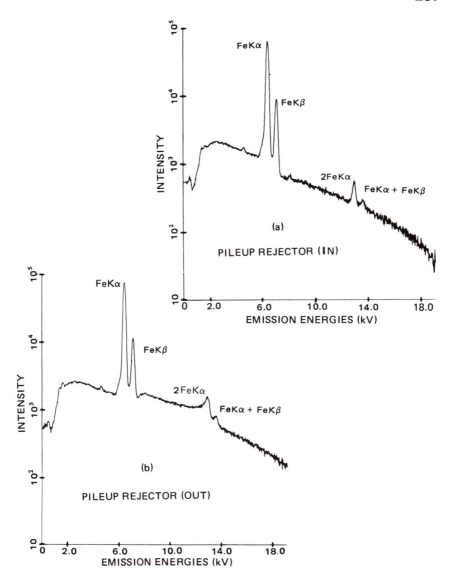

Figure 5.30. The use of pileup rejection to improve an iron spectrum (from Lifshin and Ciccarelli, 1973).

note is the presence of the 2 Fe $K\alpha$ and Fe $K\alpha$ + Fe $K\beta$ peak in both spectra. These peaks are due to the simultaneous arrival of two Fe $K\alpha$ photons or an Fe $K\alpha$ and an Fe $K\beta$ photon in the detector crystal creating electron–hole pairs in numbers corresponding to a single photon with the sum energy. They cannot be eliminated by the pulse pileup rejector, but the fraction of such peaks relative to the main characteristic peak can be reduced to a very low level by keeping the count rate down. In any case when a quantitative analysis is performed the number of $K\alpha + K\beta$ and twice the number of $2K\alpha$ pulses should be added to the $K\alpha$ peak to give the correct number of $K\alpha$ pulses.

Even in an optimally adjusted discriminator, there is an energy below which the desired pulses are so close to the noise that the discrimination becomes inadequate. This situation is illustrated schematically for magnesium and silicon in Figure 5.31. The magnesium pulses are poorly defined above the noise and therefore the proper setting of the discriminator for magnesium is difficult. Moreover, because of drift in the electronics which results from temperature changes, humidity, time, etc., the discriminator level may change. Thus, although adequate pulse coincidence discrimination may be achieved for silicon, Figure 5.32a, the discrimination is almost negligible for magnesium, Figure 5.32b, where a pileup continuum, double and triple energy peaks are observed.

Figure 5.28 also illustrates the significance of distinguishing between "real time" and "live time." The evenly spaced clock pulses at point 9 (Figure 5.29) correspond to actual elapsed time (real time). The value of this quantity may differ, however, from the live time, which is the actual period during which the system is not busy processing pulses. Note in Figure 5.29 that in the time interval shown, 14 real time pulses occur (point 9). During this period the dead-time control circuit inhibit pulse (point 11), activated by a combination of the amplifier busy (point 6) and MCA busy

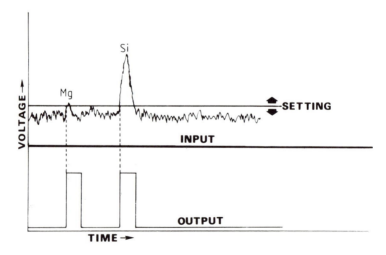

Figure 5.31. Effect of window setting on the discrimination of Mg and Si characteristic peaks.

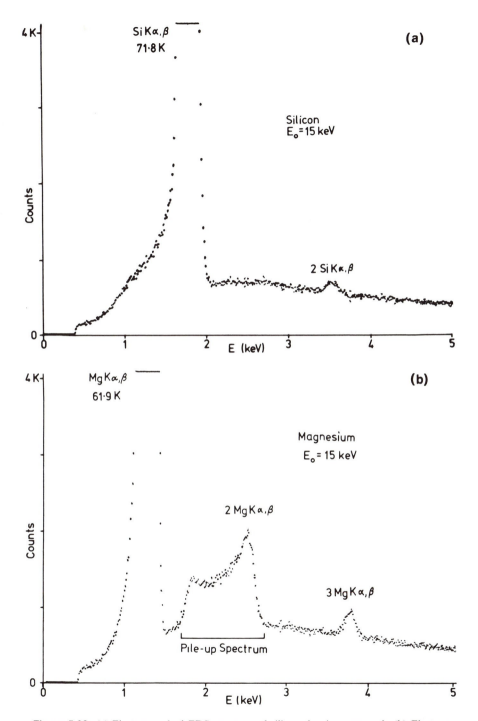

Figure 5.32. (a) Electron-excited EDS spectrum of silicon showing sum peak. (b) Electron-excited spectrum of magnesium showing pileup continuum and sum peaks.

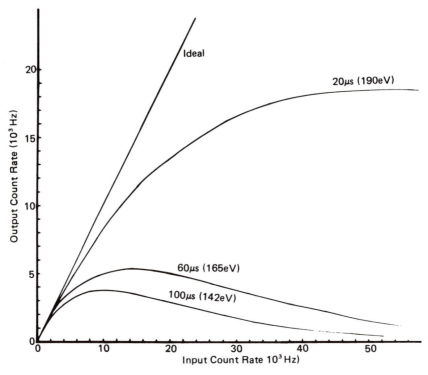

Figure 5.33. Relationship of output and input count rates for three different values of pulse width and resolution.

signals (as well as POF busy, not shown), has limited the number of live-time pulses (point 10) for the same real time interval to only three. The effect of this loss of pulses is illustrated in Figure 5.33, where it can be seen that only at low count rates (less than about 2000 counts per second) are the MCA input and main amplifier count rates equal. For reasons already described, as the amplifier input count rate increases, pulse pileup effects become more serious, particularly with the large amplifier time constants. In a qualitative analysis it may, therefore, be necessary to count for a longer period of time than anticipated on the basis of real time to achieve a desired level of precision based on counting statistics. In a quantitative analysis live time must be used in all cases since ratios of x-ray intensities, between samples and standards, taken under identical operating conditions serve as inputs to all of the quantitative correction models. Figure 5.33 also shows that increasing the amplifier input count rate by changing the probe current or moving the detector closer to the sample will initially result in a linear increase current in MCA input count rate followed by a nonlinear region in which the rate of increase in MCA input is less than the rate of increase of main amplifier input pulses. Eventually a point of diminishing returns is reached when increasing the main amplifier input count rate actually results in a decrease in the count rate seen by the MCA. Further increases beyond this point lead to essentially 100% dead time and consequently a total locking-up of the system. Figure 5.33 also shows that the onset of the various regions described is determined by a choice of operating curves based on acceptable resolution criteria. It is important to note that Figure 5.33 is not a universal curve and similar curves must be established for each system. Furthermore, the maximum input count rate is not simply that of the element of interest but rather that for the total of all pulses measured at any energy. Therefore, system count rate performance is determined by the major elements and not by the specific minor constituents which may be under investigation.

5.3.5. Artifacts from the Detector Environment

The user of an EDS system is usually responsible for the installation of the spectrometer and its associated electronics on the electron beam instrument. In order to obtain an optimum spectrum, the user may have to overcome a number of artifacts which result from interactions between the EDS system and its environment. These include microphonics, ground loops, accumulation of contaminants, and the entry of spurious radiation, including electrons, into the detector. A suggested approach to the setup of a detector to avoid the effects described below is found in the Appendix to this chapter.

5.3.5.1. Microphony

The Si(Li) spectrometer contains a detector and electronic circuitry of extraordinary sensitivity which can respond to radiation of energies other than x-rays. In particular, stray electromagnetic and acoustic radiation can affect the recorded x-ray spectrum. The coaxial cable through which the detector–preamplifier communicates with the main amplifier must be carefully routed to prevent it from becoming an antenna. The detector must be shielded against mechanical and acoustic vibration to which the detector acts as a sensitive microphone. The analyst can move a poorly routed cable to eliminate electromagnetic interference but is generally powerless to do anything about the mechanical isolation of the detector. Thus, it is important when evaluating a new detector prior to acceptance to check for microphonic and antenna effects. In Figure 5.34a, a spectrum obtained under nonmicrophonic conditions contains characteristic peaks and a continuum spectrum with a typical shape showing cutoff at low energies due to absorption in the beryllium window. In Figure 5.34b, the same spectrum

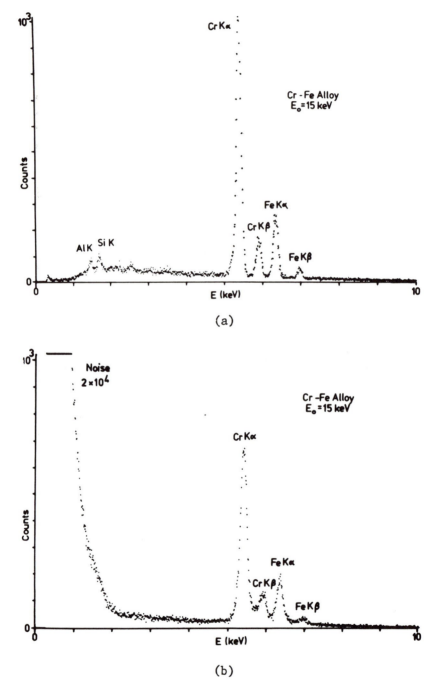

Figure 5.34. (a) Electron-excited spectrum of chromium–iron alloy. (b) Specimen obtained under same beam conditions as (a) but with acoustic interference.

was recorded with several sources of mechanical and acoustic vibrations in the vicinity of the detector—the operation of a wavelength dispersive spectrometer motor, conversation, etc. The detector responsed to these sources, producing an extremely high background in the region from 2 keV down to 0 keV. The characteristic peaks are broadened due to the noise in Figure 5.34b as compared to Figure 5.34a. While virtually every detector has some microphonic response to high-intensity noise, the detector should be isolated well enough to be insensitive to normal laboratory environment noise and vibration. The response in Figure 5.34b is quite unacceptable.

The main amplifier should also be carefully positioned. In general, it should be kept isolated from transformers and devices, such as computers or scalers, containing extensive logic circuits.

5.3.5.2. Ground Loops

One of the most insidious artifacts associated with the installation of a detector on an electron beam instrument is the occurrence of one or more "ground loops." We might normally assume that the metal components of the microscope–spectrometer system are all at ground potential with no current flowing between them. In fact, small differences in potential, of the order of millivolts to volts, can exist between the components. These potential differences can cause currents to flow which range from microamperes to many amperes. These extraneous currents are commonly referred to as "ground loops" or "ground currents" since they are flowing in components of the system which are nominally at ground potential, such as the chassis or outer shields of coaxial cables. Since alternating current ground loops have associated electromagnetic radiation, such currents flowing in coaxial cable shielding can modulate low-level signals passing through the center conductor. In EDS systems, the signals being processed are at extremely low levels, particularly in the region of the detector and preamplifier, hence, ground loops must be carefully avoided if signal integrity is to be maintained. The interference from a ground loop can manifest itself as degraded spectrometer resolution, peak shape distortion, background shape distortion, and/or dead-time correction malfunction. An example of the influence of a ground loop on a measured spectrum is shown in Figure 5.35. A normal Mn $K\alpha - K\beta$ spectrum, Figure 5.35a, can be degraded into an apparent multiple peak configuration, Figure 5.35b, in which each of the main peaks has a subsidiary peak. An intermediate condition can also be observed in which the main peaks are degraded in resolution, Figure 5.35c, without a distinct second peak observed. An explanation for this particular ground-loop-induced artifact is illustrated in Figure 5.36. If one were to examine the waveform of a ground loop signal passing through the slow channel of the processing chain, typically it would be found to be periodic in nature but not necessarily sinusoidal, with a

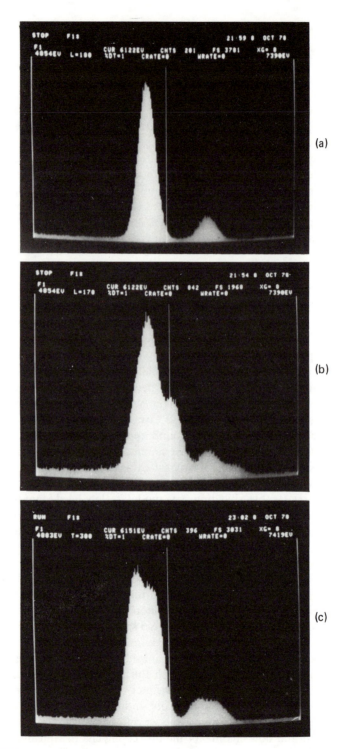

Figure 5.35. Effect of ground loops on an EDS spectrum of electron-excited manganese Mn $K\alpha$, $K\beta$. (a) Normal; (b), (c) affected by ground loops.

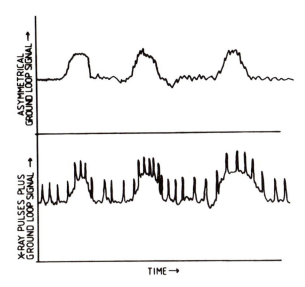

Figure 5.36. Schematic illustration of EDS signal pulse train in the presence of an asymmetrical ground loop signal.

large variety of possible forms such as that illustrated in Figure 5.36. When the random signal pulses corresponding to characteristic x-rays of a certain energy pass through the slow channel Figure 5.36, a certain fraction of them are superimposed on the high excursion caused by the ground loop. Consequently the pulses produce an apparently higher voltage output, resulting in an incorrect placement of the characteristic x-ray pulse in a higher channel in the multichannel analyzer.

It must be emphasized that the direct influence of the ground loop may not manifest itself in the spectrum as displayed in the multichannel analyzer, but there may still be deleterious effects on other important analytical functions, especially the dead-time correction. The EDS user should not presume that the dead-time correction circuitry must always be working correctly. After initial installation and periodically thereafter, a check should be made of the accuracy of the dead-time correction.

Ground loops are particularly insidious because of the many diverse ways in which they can affect the signal chain. Ground loops can enter the signal chain at any point between the detector and the multichannel analyzer. Moreover, a ground loop may occur intermittently. Because of this complexity, it is not possible to describe all of the manifestations of ground loops or a sufficiently general procedure to locate and cure them.

In dealing with ground loops, prevention is preferable to seeking a cure. Proper attention to eliminating possible ground loop paths during installation of the system will usually minimize difficulties. Figure 5.37 shows the major components of an EDS system and an electron beam

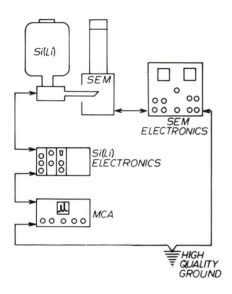

Figure 5.37. Schematic illustration of ideal ground paths in SEM/EDS system. The individual ground paths of the SEM and EDS should not be linked except at a high-quality ground.

instrument. The grounding path connecting the components of each system should be created in a logical fashion avoiding cross connections between microscope components and spectrometer components. An important ground loop path to avoid is that between the cryostat housing and the microscope. The resistance between the cryostat assembly (disconnected from its cabling) and the microscope column should typically exceed 5×10^6 Ω. Cross connections can be inadvertently introduced when the EDS main amplifier and/or mutlichannel analyzer units share common racks with microscope components such as scan generators, video amplifiers, power supplies, etc. Ideally, it is best to keep the two systems electrically isolated through the use of separate racks. The ground path should be established separately in each system terminating in a high-quality ground, Figure 5.38. Note that a high-quality ground is not typically available at the wall power plug. A high-quality ground might consist of a copper wire, 1 cm in diameter or greater, leading through the shortest distance possible to an external assembly consisting of three separate 3 m or longer copper rods driven vertically into the ground and reaching into the water table. The EDS user should note that modifications to establish a high-quality ground necessarily involve altering the electrical distribution network of the microscope–EDS system, and as a result, such modifications must be carried out under supervision of a qualified electrician.

5.3.5.3. Ice–Oil Accumulation

The accumulation of contamination in the EDS detector system during long-term operation can lead to degraded performance. Ice can accumulate

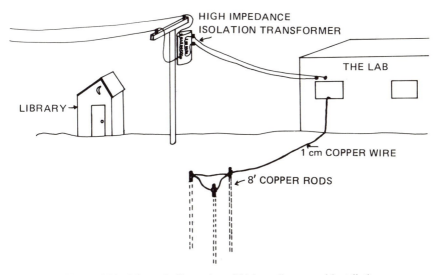

Figure 5.38. Schematic illustration of high-quality ground installation.

in two locations. (1) Moisture condensing in the liquid-nitrogen cryostat can form small fragments of ice which "dance" within the cryostat during boiling of the liquid nitrogen, Figure 5.39. This vibration can be transmitted to the detector and the sensitive field effect transistor. The vibration can degrade the resolution by as much as 30 eV. An accumulation of ice at the bottom of the cryostat can also reduce the thermal conduction between the liquid-nitrogen reservoir and the detector assembly, raising the detector temperature above the desired operating value. Some accumulation of ice in the cryostat is inevitable over a long period of time, but the rate of accumulation can be minimized if simple precautions are followed. The liquid nitrogen supply for the cryostat is typically placed in a transfer Dewar from a sealed main tank. It is important that the liquid nitrogen in the transfer Dewar not be exposed to humid atmosphere since ice crystals will quickly accumulate and then be inadvertently poured into the cryostat. When an accumulation of ice does develop in the cryostat, the EDS user

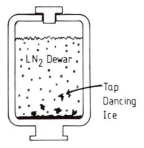

Figure 5.39. Accumulation of ice within the liquid-nitrogen Dewar of the EDS system.

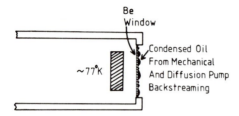

Figure 5.40. Oil accumulation from vacuum system on the window of the EDS system.

should contact the manufacturer for the proper procedure to remove the ice. (2) Ice and oil can accumulate on the detector surfaces if the vacuum integrity of the assembly is compromised by a pin-hole leak which might be located in the beryllium window. Also, in a windowless detector system, the detector will act as a cold finger to condense residual water and oil from the sample chamber. The consequence of having ice and/or oil on the detector is decreased surface resistance, which introduces resistor noise leading to degraded resolution. Correction of this condition requires servicing by the manufacturer and should not be attempted by the user.

The beryllium window is usually several degrees cooler than ambient temperature because of its proximity to the cooled detector chip. As a result, residual oil and water vapor in the vacuum of the specimen chamber can condense on the window, Figure 5.40, leading to increased x-ray absorption and loss of sensitivity at low x-ray energies. Oil removal from the window can be accomplished in the field, but only with extreme care. Again, the manufacturer should be contacted for details.

5.3.5.4. Sensitivity to Stray Radiation

(a) Origin. One of the features of the Si(Li) detector which is normally considered a great advantage is the relatively large solid angle of collection as compared to the focusing wavelength-dispersive spectrometer. The solid angle of collection is usually considered from the point of view of the electron-excited source in the specimen, with the apex of the cone placed at the beam impact point, Figure 5.41. To appreciate the complete collection situation, however, we must also consider the solid angle of collection from the point of view of the detector, Figure 5.41. It is obvious from Figure 5.41 that the true solid angle of collection is very large indeed, including not only the entire specimen but often a large portion of the sample stage and chamber walls. The difference in the collection angle between the points of view represented in Figure 5.41 would be immaterial if the excitation were really confined to the volume directly excited by the focused electron beam. Unfortunately, excitation can occur at a considerable distance from the region of impact of the focused beam. A schematic diagram of some typical

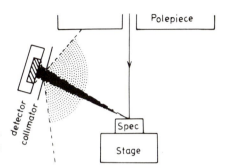

Figure 5.41. Solid angle of x-ray collection from the point of view of the specimen (dark shading) and the detector (light shading).

sources of this remote excitation is shown in Figure 5.42. Electron-induced remote sources include scattering from apertures, backscattering from the specimen, and rescattering from the pole piece. In this regard it should be noted that a significant fraction of the backscattered electrons from heavy elements retain 50% or more of the incident energy and are thus capable of exciting x-rays from the specimen environment (walls, stage, pole piece). Interaction with several surfaces is possible before an electron comes to rest. X-ray-induced remote sources originate principally from characteristic and continuum x-rays generated by those electrons which strike the upper surface of the final beam-defining aperture. These x-rays can propagate through the aperture and illuminate a large portion of the sample chamber as an x-ray fluorescence source. The significance of this source of remote

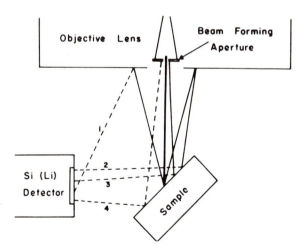

Figure 5.42. Possible sources of remote x-ray excitation in an electron-beam–EDS system. Solid lines, electron paths; dashed lines, x-ray paths. (1) Excitation of pole piece; (2) Remote excitation of sample by backscattered electron; (3) Remote excitation of sample by electron scattered by aperature; (4) Remote excitation of sample by x-rays produced at aperture.

excitation depends on (a) the material of the aperture, (b) its thickness, and (c) the beam energy. Thin-film "self-cleaning" apertures of molybdenum produce a strong source of remote excitation, since self-absorption of the Mo $K\alpha$ (critical excitation energy, 20 keV) radiation is low and the thin film (12.5 μm) transmits about 78% of the radiation. Considering that this final aperture may intercept 80% of the total beam, it is obvious that the x-ray fluorescence effect can be extremely deleterious to accurate analysis (Bolon and McConnell, 1976).

(b) **Recognition.** It is not always obvious that a remote source exists, but its effects can usually be recognized by employing the following procedure. A Faraday cup is fabricated from a block of metal (iron, brass, titanium, etc.) e.g., by drilling a blind hole of 3 mm diameter and a few millimeters in depth and press-fitting a microscope aperture (20 to 100 μm in diameter) into this hole, Figure 5.43. The aperture material should be different from that of the block. Spectra are then recorded with the focused beam alternately in the hole, on the aperture material, and on the block. The results of such a test of the system are shown in Figure 5.44. If no remote sources exist, then with the beam falling into the aperture no spectrum should be obtained. If this so-called "in-hole" spectrum contains x-ray signals of the Faraday cup materials, then the ratios of the intensities of the characteristic lines to the values obtained from the spectra of the aperture and block are an indication of the magnitude of the problem. For example, a strong signal for the aperture material in the in-hole spectrum indicates a source of radiation in the vicinity of about 1 mm of the beam, while a high signal for the block material indicates a more distant source. The extraneous radiation has been observed to be as high as 20% and is typically between 0.01% and 1% (Bolon and McConnell, 1976) of that obtained with the beam directly on the aperture. Examination of the peak-to-background ratio of characteristic peaks observed in the in-hole spectrum can also indicate the type of remote excitation—electron or x-ray. Excitation by x-rays produces a much lower continuum. If the peak-to-background ratio is higher in the in-hole spectrum than in the directly excited spectrum of the aperture or block containing the element, the remote excitation is most likely by x-rays.

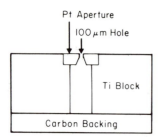

Figure 5.43. Schematic illustration of a cross section through a Faraday cup suitable for detecting remote sources of x-ray excitation in an electron beam instrument.

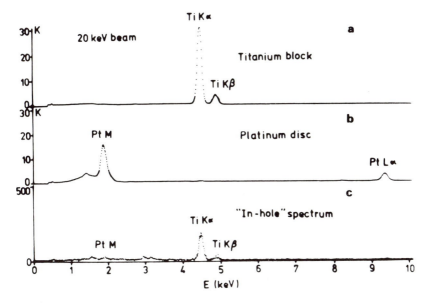

Figure 5.44. Spectra obtained in an electron probe microanalyzer from the components of a Faraday cup: (a) titanium block directly excited by electrons; (b) platinum aperture; (c) "in-hole" spectrum.

Observation of characteristic lines of the chamber wall or stage materials indicates excitation at great distances from the beam impact on the specimen. In this respect, it should be noted that a flat nontilted specimen and a specimen of rough surface may behave differently. A rough specimen is more likely to scatter electrons in all directions to the surroundings, whereas the backscattered electrons from a flat specimen normal to the beam follow a cosine distribution peaked about the normal, with little scattering in the horizontal plane.

(c) **Correction of Stray-Radiation Problem.** Eliminating stray radiation following its recognition is very much dependent on the particular instrumental configuration. Only general guidelines can be given here.

X-ray fluorescence problems originating in the final aperture can be minimized by (1) use of thick apertures and (2) choosing aperture materials of higher atomic number, such as platinum or tantalum. Typical beam energies (30 keV or lower) are insufficient to generate K x-rays, and the L lines of both materials have energies of less than 10 keV. For apertures of 12.5 μm thickness, the transmission of $L\alpha$ x-rays of the aperture material is less than 3% for Pt and 4% for Ta. Transmission of 20 keV continuum is 6% for Pt and 15% for Ta.

Electron scattering problems are more difficult to deal with. Apertures should be kept clean, because debris particles on the circumference can produce unwanted scattering. Electrons often scatter in the column and pass around apertures if passages exist. Aperture alignment should be optimized. Double apertures can be employed with some success, although their alignment is difficult.

Scattering from the specimen is difficult to control, especially if the sample is rough, such as a fracture surface. Adjacent stage, pole piece, and chamber wall surfaces can be coated with carbon-dag or beryllium sheets to prevent generation of characteristic x-rays in the energy region of interest by the scattered electrons. After all obvious sources of remote excitation have been minimized, a remnant "in-hole" spectrum may still exist. This "in-hole" spectrum can be subtracted from that of an unknown, but the procedure is risky, since the background spectrum may depend on scattering from the specimen and the surroundings of the specimen and the standard.

(d) Direct Entrance of Electrons into the Detector. The Si(Li) detector is capable of responding to an energetic electron which enters the active region of the detector. A pulse is developed, the height of which is a measure of the energy of the electron. When the beam electrons strike the sample, a significant fraction, about 30% for copper with a beam normally incident, are backscattered with a wide energy range. Many of these backscattered electrons retain a substantial fraction of the incident energy. It is inevitable that some of those electrons will be scattered in the direction of the detector. The beryllium window with a typical thickness of 7.6 μm (0.3 mil) is capable of stopping electrons with an energy below about 25 keV. Above energies of 25 keV, electrons will begin to penetrate the window and activate the detector, although with a loss of energy due to inelastic scattering in the beryllium. When higher-energy (> 25 keV) beams are employed, the electrons which enter the detector can have a substantial effect on the background. An example is shown in Figure 5.45 for a 40-keV beam incident on arsenic. The background below 20 keV is greatly distorted by the electron contribution added to the expected x-ray continuum shown by the solid curve. Above 25 keV, the background is mostly due to the normal x-ray continuum; virtually no electrons are able to penetrate the beryllium window and retain energies in this range. Note that the analyst usually examines the region between 0 and 10 keV or 0 and 20 keV. In the example of Figure 5.45, it may not be obvious that the background is anomalous unless the entire spectrum is examined.

Artifacts arising from scattered electrons entering the detector can be eliminated with magnetic shielding in front of the detector snout. In the windowless variety of Si(Li) spectrometer, such shielding is an absolute necessity. The artifact may become more pronounced for samples with a high atomic number or with surfaces highly tilted relative to the beam. These two conditions will produce the greatest number of high-energy

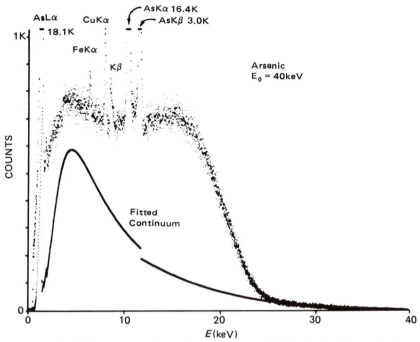

Figure 5.45. EDS spectrum of arsenic excited by a 40-keV electron beam showing an anomalous background below 25 keV due to direct entry of electrons scattered from the specimen into the detector. The x-ray continuum was fitted from the high-energy end of the spectrum.

backscattered electrons. Light element targets which are flat produce a relatively minor effect. Operation with an acceleration voltage below 20 kV will also minimize the effect.

5.3.6. The Multichannel Analyzer

The principal components of a multichannel analyzer are illustrated in Figure 5.46. They include an analog-to-digital converter (ADC), a memory unit, and various output devices. Each voltage pulse from the main amplifier is transformed into a digital signal by the ADC. The output of the ADC then becomes the address of a channel in the memory unit where an add-one operation is performed. In effect, the memory unit acts as a set of independent scalers counting pulses in preselected voltage increments. In the example shown channel 0 corresponds to pulses from 0 to 1 V, channel 1 to pulses from 1 to 2 V, and so on up to channel 7, which counts pulses from 7 to 8 V. In actual operation the memory is initially cleared, then the first pulse (2.5 V) is counted in channel 2, the second (4.3 V) in channel 4,

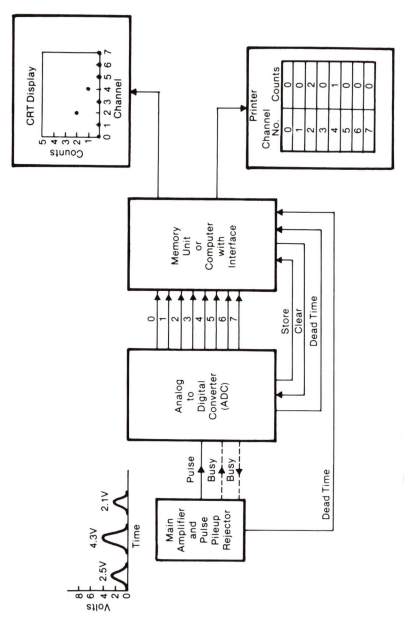

Figure 5.46. Schematic representation of a multichannel analyzer.

and the third (2.1 V) is also counted in channel 2. Upon completion of a preselected data collection time (clock or live) the contents of the MCA memory can be printed out directly, displayed on a CRT, or plotted on a strip chart recorder (not shown). In the CRT mode the electron beam of the display tube is scanned along the X axis in such a way as to generate a series of dots through the use of appropriate gating circuitry. The number of dots corresponds to the number of channels selected and the horizontal displacement of each one is made directly proportional to the channel number. Vertical deflection is then used to indicate the number of counts per channel. This is accomplished by converting the digital signal stored in memory back to a proportional analog voltage. The proportionality constant and, therefore, the counts "full scale" is set by the operator using a multiposition range switch on the instrument panel. Since many instruments use 16-bit memories, the maximum number of counts per channel is typically 65,535 although software or hardware options for greater storage capacity are often available. Analog horizontal and vertical scale displacement and expansion controls as well as a logarithmic display mode are also usually provided.

A block diagram of a typical ADC is presented in Figure 5.47 along with a corresponding waveform diagram given in Figure 5.48. Pulses from the main amplifier are first tracked by a voltage follower required to ensure compatibility with the next stage of electronics. The output (point A) is inspected by both upper- and lower-level discriminators set by the operator.

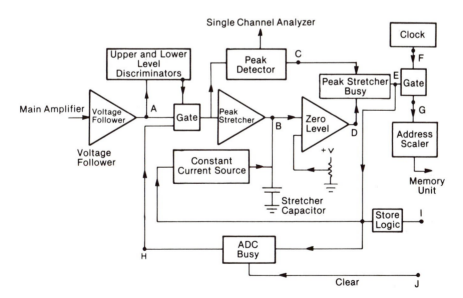

Figure 5.47. Block diagram of an analog-to-digital converter.

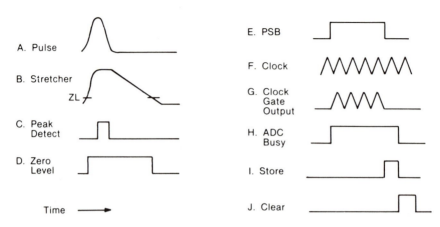

Figure 5.48. Waveforms corresponding to points designated on Figure 5.47.

Unacceptable pulses are blocked by a gate. Each accepted pulse, however, simultaneously triggers an SCA output pulse, is examined by peak detection circuitry, and also charges the peak stretcher capacitor to the maximum pulse voltage (point B). When the pulse voltage exceeds the analog zero level value ZL (also set by the operator) an appropriate logic signal (point D) combined with the peak detected pulse (point C) activates the pulse stretcher busy (PSB) and ADC busy circuits.

The PSB signal (point E) causes a gate to open which allows pulses from a clock to be counted by a scaler. In addition it causes a constant current source to linearly discharge the stretcher capacitor. When the zero level is again crossed the PSB signal is lifted, closing the gate to the clock (points F and G) thereby, leaving the address scaler containing a count of clock pulses directly proportional to the original pulse voltage. Lifting of the PSB signal, combined with other test conditions not shown in Figure 5.47, results in the output of a store pulse (point I) which causes the address to be transferred to the memory unit (see Figure 5.24). The ADC busy signal (point H) which has been blocking all incoming pulses is then reset by a clear pulse (point J) from the memory unit, making it possible to accept new pulses. The entire process can then be repeated until the predetermined data collection time is reached.

Examination of Figures 5.47 and 5.48 should now make it possible to understand a number of terms used in conjunction with MCAs. The conversion gain of an ADC refers to the total number of increments (addresses) used to characterize a measured pulse distribution. Switch-selectable values typically range from 32 to 8192 in multiples of 2. In effect the conversion gain determines the ADC resolution by controlling the discharge rate of the stretcher capacitor. This determines the number of clock pulses to be counted for a main amplifier pulse of a given size. In terms of Figure 5.48, waveform B, the slope and therefore the time required to reach the zero level (ZL) are determined by the switch setting. For a maximum acceptable main amplifier pulse of 8 V a conversion gain of 1024 means that at least 1024 pulses must be counted to achieve an ADC resolution of $(8/1024) = 0.0078$ V. Use of a 100-MH ADC implies that the clock frequency is 100 million cycles per second. For the example just given the time required to convert an 8-V pulse to a digital signal will, therefore, be 1024×10^{-8} s = 10 μs. Recalling Figure 5.27, it can be seen that for the long time constant (10 μs) associated with achieving maximum energy resolution, at least 20 μs are required for the pulse to return to the base line. During this time period the pileup rejector will block any incoming pulses and consequently the conversion of even the largest acceptable pulse will be completed before the main amplifier is ready to accept the next one. Thus, for this example the conversion process will not contribute to the overall system dead time. This will not be the case, however, when either a 25-MH ADC or a short time constant (1 μs) is used, and it is important for the operator to appreciate this point in selecting an ADC.

Memory group size refers to how large a section of the memory is actually used for spectral storage. In the simple example presented in Figure 5.46, eight lines, each corresponding to a single address, were shown connected, one for one, to an eight-channel memory. In reality separate lines are not needed to each channel since the address can be binary coded. Sixteen lines are sufficient to carry any address from 0 to 65,535 ($2^{16} - 1$). In the earliest MCAs coupled to EDS systems, hard-wired magnetic memory units were used which could be divided by the operator into specific groups. Consider, for example a system with a total of 2048 channels. Switch-selectable memory groups include

1/1 full memory	2048 channels
1/2, 2/2 first and second halves	1024 channels each
1/4, 2/4, 3/4, 4/4 four quarters	512 channels each

If the conversion gain is set equal to the memory size, either one, two, or four complete spectra can be stored in memory depending on the resolution required. The user is not really interested in channel numbers, however, but x-ray energies. According to Equation (5.6) the main amplifier pulse voltage is directly proportional to the x-ray photon energy being measured. Since the ADC responds linearly to the pulse voltage, the x-ray energy can be related to the channel number N_e by the following expression:

$$E = Z_0 + SN_e \qquad (5.10)$$

where S is a constant equal to the number of electron volts per channel and Z_0 is a constant equal to the zero offset to be described later. Note that MCA calibration requires setting Z_0 such that x-ray pulses of known energy from a standard occur at the positions given in Equation (5.10).

The number of electron volts per channel can be operator selectable with typical values ranging from 5 to 40. If, for example, Z_0 is set equal to zero, a value of $S = 20$ eV/channel is used, and 512 is selected for both the memory group size and the conversion gain, then the recorded spectra will cover x-ray energies of from 0 to 10.240 keV. Under these conditions copper K (8.048 keV) will give rise to a pulse distribution centered around channel 402. Almost all of the peak will be contained within a band defined by plus or minus one FWHM. Therefore, for a 160-eV detector measured at Cu $K\alpha$, the entire peak will contain about 16 channels, which is a reasonable number for most applications. Needless to say if one wishes to accurately measure detector resolution to better than 20 eV it is definitely desirable to use fewer eV/channel.

The analog zero offset controls the value of ZL in waveform B of Figure 5.48. It provides a way of either adding or subtracting a constant number of clock pulses to all main amplifier pulses being processed and therefore can be used to set Z_0 in Equation (5.10). From the point of view of the MCA CRT display, varying ZL produces a linear displacement of the spectrum on the viewing screen. In the example just mentioned, a ZL value of 2 V would result in a displacement of the spectrum by 25%. The display would cover a region of from 2.56 to 12.80 keV and the copper $K\alpha$ would have the same width but appear centered around channel $402 - 256 = 146$. In practice the analog zero adjust is usually restricted to setting $Z_0 = 0$ for calibration purposes. Shifts of the type just described are normally performed by using digital offset switches which add fixed numbers to the ADC scaler addresses. Their use makes it possible to use memory group sizes smaller than the conversion gain. High ADC resolution can then be combined with limited memory capacity for the examination of selected regions of a spectrum. This feature is particularly useful in accurately measuring detector resolution to a few electron volts by going, for example, to a combination of a conversion gain of 8192 with 512 channels of memory and then using digital offset to bracket the peak position.

The trend in advanced MCA development has been from completely hard-wired instruments to computer-based units, with the latest systems containing the best features of both. In the earliest MCAs the user literally had to count dots to determine channel numbers, which were then converted by hand to energy values. Elements were identified by comparing the calculated peak positions with tables of known x-ray line energies. The next generation included a dial-controllable feature known as a "bug" which made it possible to intensify the dot corresponding to the contents of any specific channel on the CRT. At the same time its position, in either channel number or energy, as well as the corresponding number of counts could be read directly from a numerical panel display. With the advent of low-cost character generators, scale information and labels could also be displayed directly on the CRT. Later came a series of special features including regions of interest (ROI) displays, line markers, and access to a variety of auxiliary spectral storage and retrieval devices. The ROI approach allows the user, often with the help of the bug, to define a series of energy bands in which a running count of the integrated intensity are recorded. In this manner integrated peak intensities can be monitored either as a prerequisite for quantitative analysis or pulses corresponding to specific elements can be coupled to SEM displays for line scans or area maps. Line markers are a series of vertical bars appearing at energies corresponding to the principal lines of any selected element. Introduced one element at a time and manipulated by the bug control or keyboard, they can be superimposed over an unknown spectrum to facilitate rapid qualitative analysis (see Figure 5.49).

Data readout with the earliest systems generally consisted of unlabeled $X-Y$ recordings, printed tables, or photographs of the CRT. Links to both large centralized computer systems and dedicated minicomputers were generally slow and awkward. Although there was some effort to interface the hard-wired systems to auxiliary tape units for spectral storage and

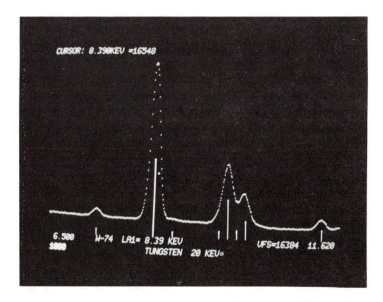

Figure 5.49. EDS CRT display for tungsten illustrating use of line markers.

retrieval this approach also proved limited. In the exclusively computer-based MCA, information from the ADC is transmitted directly to the central processor unit of a dedicated minicomputer which, through a combination of hardware and software, sees to it that pulse distribution information is directed to specified memory locations. The operator normally interacts with the system through an alphanumeric keyboard and various pushbutton switches. Desired actions can then be initiated under computer control. These include data acquisition, spectral storage and retrieval on auxiliary devices, and spectrum processing by a variety of techniques. Central to this type of system is the existence of an executive program which responds to specific mnemonic commands issued by the user. Unfortunately, unlike the more self-explanatory hard-wired controls of early MCAs, the software command structures of the computer-based system tend to be unique to each manufacturer's product. The advantages are, however, much more flexibility and greater speed in data interpretation.

With the computer-based MCA the operator normally has the option of changing parameters such as the memory group size, conversion gain, digital offset, the number of electron volts per channel, and a number of CRT display features under program control. Furthermore, since the memory of the computer is used as the MCA memory, applications programs written in high-level languages can readily access the stored data for quantitation. The storage and retrieval of both spectral data and processing or control programs is also a relatively straightforward matter since devices like tape drives and floppy disks are easily interfaced to almost all minicomputers. The major shortcoming of such systems has been that simultaneous data collection and processing on an internal time-shared mode tends to slow them down since pulse sorting and storage can be almost a full-time job. The solution to this problem has been to use a combination of minicomputers and microprocessors which handle specific tasks. For example, a dedicated microprocessor system with its own memory can be used for rapid pulse collection, sorting, and storage. Its memory can be coupled directly to a minicomputer which handles data processing. By using a common input bus, keyboard commands from the user can be interpreted by whichever unit is appropriate.

5.3.7. Summary of EDS Operation and Artifacts

The processes by which the characteristic lines and continuum distribution are modified in going from the generated spectrum in a Mn sample to the observed spectrum in the EDS are summarized in Figure 5.50. Point *a* and the associated spectrum correspond to what the distribution of characteristic lines and continuum might look like if the measurement could be made by an ideal detector located within the sample. Point *b*

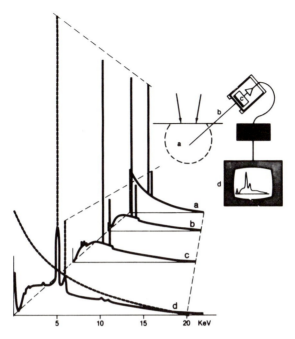

Figure 5.50. Schematic drawing of the relationship between the measured and generated spectral distributions of manganese (courtesy of R. Bolon). See text for details.

shows the spectral distribution as it would exist in the emitted x-rays. Note the principal change is due to absorption which sharply truncates the low-energy portion of the continuum, introduces a discontinuity associated with continuum absorption at the Mn K absorption edge, and also slightly diminishes the characteristic peak intensities. At point c the signal has passed through the Be window, Au surface contact, and silicon dead layers causing a further reduction in intensity at low energies and also introducing some fine structure caused by absorption and fluorescence effects. Finally at point d the signal has been processed by the Si(Li) diode and associated electronics. The processing results in a smearing effect which distributes most of the information corresponding to each sharply defined energy increment over a range of MCA channels encompassing plus or minus one FWHM. The effect is to noticeably broaden peaks and absorption edges well beyond their natural width, which leads to reduced peak-to-background ratios and the high probability of peak overlap. In addition, it is at this point that the electronic artifacts of escape peaks and peak-on-peak pileup make their appearance. Although the EDS does introduce a large number of modifications to the generated spectrum, the resultant signal is very useful and compares favorably with measurements from the WDS. A comparison of the two spectrometers is given in the next section.

5.4. Comparison of Wavelength-Dispersive Spectrometers with Energy-Dispersive Spectrometers

An ideal x-ray detector would be small, inexpensive, easy to operate, collect most of the x-rays emitted from a sample, have a resolution better than the natural x-ray line width being measured (a few electron volts), and be capable of collecting spectral data rapidly without losing information. Neither wavelength-dispersive spectrometers nor Si(Li) detectors individually have all of these characteristics, but when used together the two techniques do, in fact, complement each other. Table 5.2 summarizes a comparison of the major features of both modes of detection. An item-by-item analysis of Table 5.2 follows.

5.4.1. Geometrical Collection Efficiency

Geometrical collection efficiency refers to the solid angle of spectrometer acceptance $[(\Omega/4\pi)100\%]$. As was illustrated in Figure 5.3 the angle subtended in the plane of the focusing circle of a WDS does not change with λ. However, orthogonal divergence perpendicular to that plane will lead to a decreased collection efficiency with increasing λ for a given crystal. In the EDS case the higher collection efficiency is a result of the greater ease in positioning the detector close to the sample (often less than a centimeter). Furthermore, although the solid angle can be varied for retractable detectors (see Figure 5.16a) the nature of the detection process does not require that the detector be physically moved to encompass its entire energy range as is the case for a WDS.

5.4.2. Quantum Efficiency

Overall quantum efficiency is a measure of the percentage of the x-rays entering the spectrometer which are counted. At low beam currents EDS systems generally have higher count rate capability per unit beam current partially due to a higher geometric collection efficiency and partially due to higher inherent detector quantum efficiency. Figure 5.51, calculated by Woldseth (1973), shows that for a 3-mm-thick detector sealed with an 8-μm (0.3-mil) Be window close to 100% of the x-rays in the 2.5–15-keV energy range striking the detector will be collected. At higher energies a certain percentage of x-ray photons will be transmitted through the silicon crystal, while at low energies a certain percentage of photons will be absorbed in the Be window. Significant absorption of soft x-rays can also occur in the surface "dead layer" or gold contact layer on the detector crystal. Absorption and detector noise generally limit light element analysis to $Z \geqslant 9$, although windowless detectors capable of detecting carbon have

Table 5.2. Comparison between X-Ray Spectrometers

Operating characteristic	WDS Crystal diffraction	EDS Silicon energy dispersive
(1) Geometrical collection efficiency	Variable $< 0.2\%$	$< 2\%$
(2) Overall quantum efficiency	Variable $< 30\%$ Detects $Z \geqslant 4$	$\sim 100\%$ for 2–16 keV Detects $Z \geqslant 11$ (Be window) Detects $Z \geqslant 6$ (windowless)
(3) Resolution	Crystal dependent ~ 5 eV	Energy dependent (150 eV at 5.9 keV)
(4) Instantaneous acceptance range	$\sim$ The spectrometer resolution	The entire useful energy range
(5) Maximum count rate	$\sim 50,000$ count/s on an x-ray line	Resolution dependent < 2000 cps over full spectrum for best resolution
(6) Minimum useful probe size	~ 2000 Å	~ 50 Å
(7) Typical data collection time	Tens of minutes	Minutes
(8) Spectral artifacts	Rare	Major ones include: escape peaks, pulse pileup, electron beam scattering, peak overlap, and window absorption effects

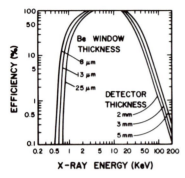

Figure 5.51. Calculated silicon detector efficiency as a function of x-ray energy (from Woldseth, 1973).

been reported. By comparison, the quantum counting efficiency of crystal spectrometers is generally less than 30%, partially due to transmission losses in the proportional counter tube (see Figure 5.6) and partially due to losses in the diffraction crystal. Large-d-spacing crystals and thin-window proportional counters make it possible to detect elements down to beryllium. Consideration of the overall quantum counting efficiency leads to the conclusion that the useful energy range for WDS systems is usually taken to be about 0.1–15 keV while for EDS it more typically spans 1.0 to 20 keV.

5.4.3. Resolution

Resolution for both WDS and EDS systems is usually measured as the FWHM. As already described in Equation (5.6) and illustrated in Figure 5.20a, EDS resolution is energy dependent in a predictable way, with values normally determined for Mn $K\alpha$ as measured with a [55]Fe source. Even at 150 eV it is still about 30 times poorer than that obtainable with a good quartz or lithium fluoride crystal. The principal effect of reduced resolution is to lower the peak-to-background ratio P/B at a given energy and hence the sensitivity or minimum detectability limit of a given element (see Chapter 7 for a discussion of trace element detection). The reduction in P/B occurs because it is necessary to sample a wider energy interval containing more background counts to obtain a major fraction of the peak counts P. Figure 5.52, taken from the work of Geller (1977), illustrates this point for the case of a 160-eV detector, used to obtain P and P/B for Fe $K\alpha$ and Si $K\alpha$ on pure iron and silicon, respectively. The term P^2/B, is often used for comparison of sensitivities for the detection of trace amounts of a particular element. The value of P^2/B goes through a maximum when the energy band chosen for peak integration is approximatly equal to the FWHM. Note that the peak-to-background ratio must be obtained by integrating both the peak and background counts in the ROI and taking

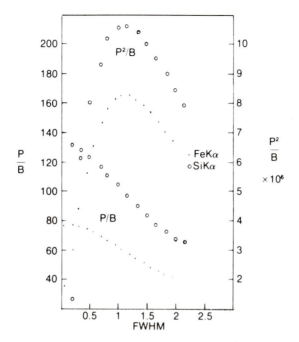

Figure 5.52. Integration limits and their effects on P/B and P^2/B for Fe and Si measured with an EDS (from Geller, 1977).

the ratio of the sums. This quantity is known as the integrated peak-to-background ratio. The situation is somewhat different in evaluating the performance of a WDS system since optimum performance is obtained by tuning the spectrometer to a diffraction angle corresponding to the maximum peak intensity and then detuning it to obtain corresponding background levels.

Integrating WDS peaks by scanning through them is basically an exercise in seeing the system's response to a varying of the spectrometer efficiency. Such scanning in the vicinity of a peak may have to be used, however, as a means of determining the position of a maximum for those quantitative measurements in which the spectrometer is detuned between sample and standard readings. This process does not add to the precision but is a means of avoiding the introduction of systemmatic errors.

The superior energy resolution of the crystal spectrometer results in significantly higher peak-to-background ratios and better spectral dispersion thereby minimizing the possibility of peak overlap. This can be readily seen by comparing spectra from the same superalloy standard obtained using a crystal spectrometer (Figures 5.12a and 5.12b) with that obtained from a Si(Li) detector (Figure 5.53). Figure 5.12a shows clearly distin-

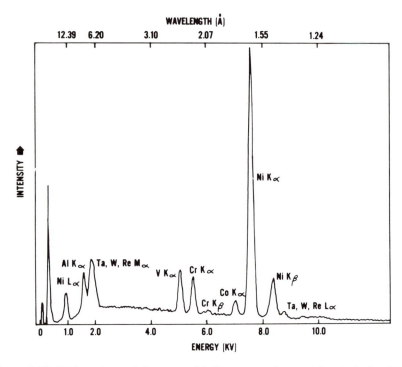

Figure 5.53. EDS spectrum of the same nickel-base superalloy used to obtain the WDS spectra in Figures 5.12a and 5.12b.

guished Ta $L\alpha$, Ni $K\beta$ and W $L\alpha$ lines while these same lines are hardly discernible in Figure 5.53. Similarly, the Ta, W, and Re $M\alpha$ lines are easily separated using an RAP crystal but remain unresolved with the Si(Li) detector. The inferior resolution of the solid state detector often makes it necessary to establish the presence of a series of spectral lines for a given element when the identification of a particular peak is ambiguous or the peak of a suspected line is obscured by another element. In such cases it is common practice to use the line markers or collect one or more pure elemental spectra and display them simultaneously with that of the un-known in order to make a direct comparison.

5.4.4. Spectral Acceptance Range

The term "instantaneous acceptance range" refers to that portion of the spectrum that can be measured at any instant of time. For the WDS only those pulses very close to the selected Bragg angle will be measured while all others are essentially ignored. The EDS, on the other hand, has a large acceptance range and will sequentially process all accepted pulses.

The use of the term "simultaneous detection" is to be avoided, however, since as described earlier, two photons entering the detector simultaneously will lead to a single and incorrect MCA assignment of the event at the sum energy.

5.4.5. Maximum Count Rate

As already shown in Figure 5.33 the maximum useful count rate for systems operating at optimum resolution is about 2000 to 3000 counts per second over the entire energy range of the excited x-rays $(0-E_0)$. If the composition of a minor component is sought this number may drop well below a hundred counts per second for the element of interest and therefore necessitate long counting times to achieve needed levels of precision. In the WDS case with the spectrometer tuned to a specific element count rates in excess of 50,000 cps are possible without loss of energy resolution.

5.4.6. Minimum Probe Size

The previous point leads directly to a discussion of the minimum useful probe size for x-ray analysis. As described in detail in Chapter 2 (see Figure 2.16), for each type of source and voltage there exists a maximum probe current which can be associated with any given probe size. For conventional tungsten sources the probe current varies with the eight-thirds power of the beam diameter with typical values at 20 kV ranging from 10^{-10} A for 20 nm (200 Å), 10^{-8} A for 100 nm (1000 Å) and 10^{-6} A for 1000 nm (10,000 Å). For an EDS system utilizing a 4-mm-diam detector placed 1 cm from a sample of pure nickel a count rate of about 10^4 cps is obtained for a 35° takeoff angle, a 20-nm (10^{-10} A) probe, and 100% quantum efficiency. According to Figure 5.33 the count rate of 10^4 would be too high to be consistent with maximum energy resolution so the operator would either have to retract the detector, decrease the EDS time constant, or decrease the probe current by going to a smaller spot. On the other hand the corresponding WDS count rate would probably be more like 100 cps, which would be too low for practical use. For bulk samples (more than a few micrometers thick) spatial resolution for chemical analysis does not improve for probes much less than 1 μm in diameter since the volume of x-ray production is determined by electron beam scattering and penetration rather than by the probe size. This point is dramatically demonstrated in Figure 5.54, which shows a series of Monte Carlo simulations of electron beam scattering and x-ray production for a 0.2-μm-diam probe with a hypothetical 1-μm TaC inclusion in a Ni–Cr matrix. It can readily be seen that the electron trajectories and consequently the region of

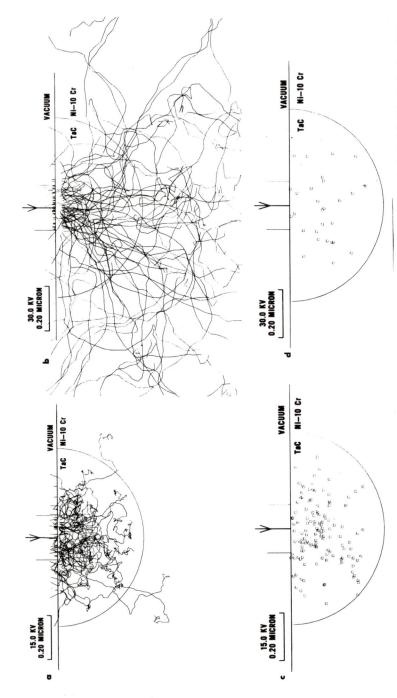

Figure 5.54. Monte Carlo simulated interaction of a 0.2-μm-diam beam with a hypothetical 1-μm-diam hemispherical TaC inclusion in a NiCr matrix. (a) Electron trajectories, 15 kV; (b) electron trajectories, 30 kV; (c) Ta Mα x-rays at 15 kV; (d) Ta Mα x-rays at 30 kV.

x-ray production, particularly at the high voltage, can easily exceed 1 μm or five times the beam diameter. Since probes of under a few hundred nanometers are of limited value in the examination of such samples a more complete analysis can be accomplished by boosting the probe current to 10 nA and using the WDS systems where the benefit of high count rate capability combined with high energy resolution can be realized. The situation is quite different, however, in the examination of thin foils and biological sections where spatial resolution equal to or even smaller than the foil thickness is possible. In this case the low x-ray yields necessitate having a detector with both the high geometrical collection and overall quantum efficiency of the EDS system. It is for this reason that they have been so successful when coupled to both scanning and analytical electron microscopes.

5.4.7. Speed of Analysis

From a practical point of view one of the best features of the EDS system is the speed with which data can be collected and interpreted. The continuous acceptance of a large energy range has distinct advantages for qualitative analysis, which partially offset some of the disadvantages previously described. When a WDS system is mechanically scanned it dwells at each resolvable wavelength for only a short part of the total scan duration. Therefore while looking at one element or even a section of the background it is throwing away information about all of the other elements. Unless the WDS is specifically programmed to go to peak positions, anywhere from only 1/100 to 1/1000 of the total data collection time may be associated with the measurement of each individual peak. In the EDS case a 100-s counting time coupled with a count rate of 2000 cps leads to a spectrum containing 200,000 counts. Even if half of those counts are background, most of the measurable constituents present in amount greater than a few tenths of a percent will probably be detected. Furthermore, use of the line markers and other interpretive aids can result in a qualitative analysis in a matter of minutes. In the WDS case several crystals covering different wavelength ranges would be employed with typical data collection and interpretation times being 10–30 min.

5.4.8. Spectral Artifacts

The WDS system is relatively free of spectral artifacts of the type which cause peak position shifts or incorrect elemental assignments. The EDS, on the other hand, is subject to a number of difficulties which can lead the unsuspecting analyst into trouble. Artifacts arise at each stage of the spectral measurement process. Detection artifacts include peak broadening, peak distortion, silicon escape peaks, absorption, and the silicon internal fluorescence peak. Pulse-processing artifacts include pulse pileup, sum peaks, and sensitivity to errors in dead-time correction. Additional artifacts arise from the EDS–microscope environment, including microphonics, ground loops, and oil and ice contamination of the detector components. Both WDS and EDS can be affected by stray radiation (x-rays and electrons) in the sample environment, but because of its much larger solid angle of collection, the EDS is much more readily affected by stray radiation. However, because of the large collection angle, the EDS is less sensitive to spectrometer defocusing effects, with sample position.

In summary, this comparison suggests that the strengths of the EDS and WDS systems compensate for each system's weaknesses. The two types of spectrometers are thus seen to be complementary rather than competitive. At the present stage of development of x-ray microanalysis, it is clear that the optimum spectrometry system for analysis with maximum capabilities is a combination of the EDS and WDS systems. As a result of the on-going revolution in the development of laboratory computers, automated systems which efficiently combine WDS and EDS spectrometers are readily available from several manufacturers.

Appendix: Initial Detector Setup and Testing

The wide variety of Si(Li) spectrometer–MCA systems precludes a specific description for proper set up and testing which is applicable to all instruments. The manufacturer provides specific instructions on the proper installation and adjustment of the instrument. It is usually left to the analyst in the field to make the actual installation. In this section, some general guidelines and suggestions to supplement the manufacturers' procedures are given in order to highlight the critical areas in the operation of these systems.

(1) Before the Si(Li) detector is installed on the instrument, it is very useful to test the system separately by activating the detector with a radioactive source, preferably ^{55}Fe. The ^{55}Fe source emits Mn $K\alpha$ and Mn $K\beta$ x-radiation with negligible continuum. The x-ray spectrum of this source should be recorded with the detector placed on a vibration-free surface such as plastic foam at least 5 cm from the source. The amplifier should be electrically isolated from other electronics. The manufacturer's instructions for setup for optimum resolution should be followed. The total spectrum count rate should be 1000 Hz or less. A total of 100,000, or more counts should be collected. It is best if this spectrum can be retained in digital form by storage on magnetic or paper tape. If digital recording is unavailable, the spectrum should be written on graph paper in several scale

expansions centered about Mn $K\alpha$. The merits of this source spectrum are the following:

(a) The source spectrum serves as a permanent reference to the as-received condition of the system. In the event of a question arising in the future about the quality of the system, a benchmark is available.

(b) The resolution in the as-delivered condition can be measured at the usual value, the FWHM for Mn $K\alpha$.

(c) To measure the degree of incomplete charge collection, the peak-to-tail ratio and the asymmetry of the Mn $K\alpha$ peak should be determined. The peak-to-tail ratio is typically measured by taking the ratio of the counts in the peak channel of Mn $K\alpha$ to the counts in the background channel at half the energy of Mn $K\alpha$. Because of the low background counts, an average over at least 10 channels should be used. For a typical detector, this ratio should be about $1000:1$. The asymmetry of the Mn $K\alpha$ peak is typically determined by measuring the full width of the peak at one-tenth the maximum amplitude (FWTM). The FWTM should be no worse than 1.9 FWHM.

(2) After installing the spectrometer on the SEM or EPMA, a spectrum should be obtained from a manganese target excited with a 15–20-keV electron beam at the same count rate and total Mn $K\alpha$ counts as the reference spectrum. The resolution measured on this spectrum should not be degraded by more than a few electron volts over the reference spectrum. If the resolution has deteriorated significantly and/or large numbers of counts are observed in the low-energy region, e.g., Figure 5.34b, there are several possible sources of trouble: microphony, lack of electrical ground isolation between circuit components (ground loops), and stray radiation from transformers, computers, etc. Remedies for these problems depend on the local situation. Ground loops can usually be eliminated by connecting all components of the system to a single high-quality ground and not interconnecting them.

(3) An electron-excited spectrum should be obtained from spectro-graphically pure carbon with a scanning beam to minimize possible con-tamination. This specimen should provide a continuum spectrum which is devoid of characteristic peaks. Such a spectrum from a typical detector, Figure 5.26, has the following characteristics:

(a) The intensity of the lowest-energy channels should nearly reach the base line, within about 3% of the intensity maximum in the continuum hump. If not, this failure may be indicative of a poorly set slow channel discriminator, microphony, stray radiation, or ground loops.

(b) A silicon K absorption edge and a silicon internal fluorescence peak are always observed, as shown in the spectrum obtained with a normal detector, Figure 5.26. In an abnormal detector, Figure 5.55, the silicon absorption edge and the silicon fluorescence peak are much more

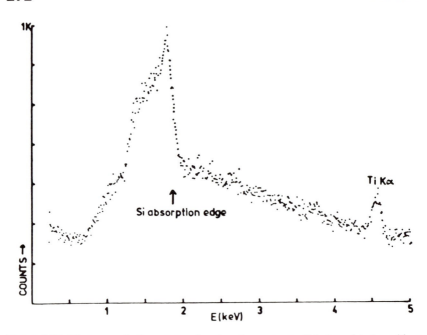

Figure 5.55. Electron-excited spectrum of carbon from an unsatisfactory detector with an extremely thick silicon dead layer producing an unusually high silicon absorption edge.

pronounced, and the magnitude of the absorption edge will be a function of the bias voltage on the silicon detector.

(4) At this point, the analyst should examine the entire carbon continuum spectrum up to the beam energy for other artifacts. Specifically, the presence of any characteristic peaks, e.g., Figure 5.55, is an indication of stray radiation and a clue to its source.

(5) Spectra should be obtained from elements which give characteristic lines throughout the energy range of interest to check the energy linearity of the system.

(6) The performance of the pulse pileup rejector should be examined as a function of x-ray energy and count rate. It can be expected that the pileup rejector should work well for radiation as energetic as silicon $K\alpha$ and greater. Below silicon $K\alpha$, the pileup rejector frequently becomes progressively less satisfactory. Spectra illustrating the pileup performance from an optimally adjusted system (maximum resolution and maximum allowed count rate), Figures 5.32a and 5.32b, show adequate pulse pileup rejection for silicon but an almost total failure for magnesium. A pileup continuum below the double energy peak and even a triple energy peak are observed in the magnesium spectrum. In the silicon spectrum, only the double energy peak, which corresponds to coincidence within the time resolution of the

pulse rejector, is observed. As a figure of merit, the area of the silicon $K\alpha$ double energy peak should be less than 1/200 of the parent peak at allowed system count rates.

(7) For accurate quantitative analysis, the performance of the dead-time correction circuit should be tested. The x-ray production in the specimen at any beam energy is proportional to the beam current striking the sample. This fact provides a way to vary the input count rate to the spectrometer system in a controlled fashion. A flat, pure element target such as iron should be scanned with a beam energy of 15–20 keV. The current should be set initially to give a total spectrum count rate of about 500 H. The beam current is measured in a Faraday cup. The integrated counts across the peak (peak plus background) are then plotted as a function of beam current for a fixed live time (e.g., 100s). The plot of counts versus current will be linear over that count rate where the dead-time correction mechanism is functioning properly.

6

Qualitative X-Ray Analysis

6.1. Introduction

The first stage in the analysis of an unknown is the identification of the elements present, i.e., the qualitative analysis. Qualitative x-ray analysis is often regarded as straightforward, meriting little attention. The reader will find far more references to quantitative analysis than to qualitative analysis, which has been relatively neglected in the literature, with a few exceptions (e.g., Fiori and Newbury, 1978). It is clear that the accuracy of the final quantitative analysis is meaningless if the elemental constituents of a sample have been misidentified. As a general observation, the identification of the major constituents of a sample can usually be done with a high degree of confidence, but, when minor or trace level elements are considered, errors can arise unless careful attention is paid to the problems of spectral interferences, artifacts, and the multiplicity of spectral lines observed for each element. Because of the differences in approach to qualitative EDS and WDS analysis, these techniques will be treated separately.

The terms "major," "minor," and "trace" as applied to the constituents of a sample in this discussion are not strictly defined and are therefore somewhat subjective. The following arbitrary working definitions will be used: major, 10 wt% or more; minor, 0.5–10 wt%; trace, less than 0.5 wt%.

In performing qualitative x-ray analysis, we have to make use of several different types of information. Foremost is the specific energy of the characteristic x-ray peaks for each element. This information is available in the form of tabulations, which may be in the convenient form of an "energy slide rule" (United Scientific, 1978; Ortec, 1977) or a graph, or, in the case of a sophisticated computer-based multichannel analyzer, the x-ray peak energy is given as a visual display marker ("KLM markers"). In employing these aids, the analyst should be aware of some of the potential shortcomings. The energy slide rules and even some computer "KLM" markers may not list all x-ray peaks observed in practice. Frequently, the

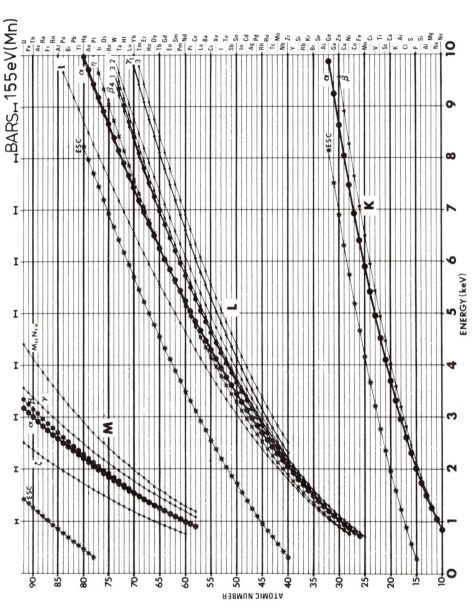

Figure 6.1. Plot of the energy of the x-ray emission lines observed in the range 0.75–10 keV by energy dispersive x-ray spectrometry (Fiori and Newbury, 1978).

Ll, *Lη*, *Mζ*, and $M_{II}N_{IV}$ lines are neglected, but these lines are frequently detected in practice. Without knowledge of their existence, errors may be made in ascribing these lines to other elements which are not actually present. A comprehensive table of x-ray lines, such as Bearden's compilation (Bearden, 1967), is the ultimate reference, but is often too detailed. For example, this compilation lists 25 *L*-family lines. Even when the integrated spectral count is as high as 5,000,000, only nine of these will usually be observed in high-quality EDS spectra, because of the low relative intensity of the other 16, or the overlap of other members of the family. Note, however, that the higher resolution of WDS systems may increase the number of lines which can be detected, requiring reference to a detailed x-ray compilation.

As an aid to qualitative EDS analysis, Fiori and Newbury (1978) published a graphical representation of all x-ray lines observed in high-quality (5,000,000 counts integrated intensity) EDS spectra in the 0.70–10-keV range, shown in Figure 6.1. This plot provides a convenient compilation of x-ray energies and also allows for a rapid evaluation of possible interferences. The spectral broadening effect for a 155-eV EDS spectrometer is also plotted, which allows for estimation of peak overlaps. Figure 6.1 is intended as an aid to qualitative analysis in conjunction with a table (or KLM markers) of x-ray energies. Accurate values (to 10 eV) of the x-ray energies are needed for proper peak identification.

The second important piece of information with which the analyst must be familiar is the concept of a family of x-ray peaks. When the beam energy exceeds the critical x-ray ionization energy for a shell or subshell so that it is ionized, all possible transitions involving that ionized shell will take place, producing a family of peaks, which will become more complicated as the electronic structure of the atom increases in complexity. With a beam energy of 15 keV or more, all possible lines of an element in the range 0.75–10 keV will be excited. The presence in the spectrum of all possible members of a family of lines increases the confidence which can be placed in the identification of that element. Since the family members must all exist, the absence of a particular line should immediately raise suspicion in the analyst's mind that a misidentification may have been made and other possible elements in that energy range should be considered.

6.2. EDS Qualitative Analysis

6.2.1. X-Ray Lines

The energy-dispersive x-ray spectrometer is an attractive tool for qualitative x-ray microanalysis. The fact that the total spectrum of interest, from 0.75 to 20 keV (or to the beam energy) can be acquired without scanning allows for a rapid evaluation of the specimen. Since the EDS detector has virtually constant efficiency (near 100%) in the range 3–10 keV, the relative peak heights observed for the families of x-ray lines are close to the values expected for the signal as it is emitted from the sample. On the negative side, the relatively poor energy resolution of the energy-dispersive spectrometer compared to the wavelength-dispersive spectrometer leads to frequent spectral interference problems as well as the inability to separate the members of the x-ray families which occur at low energy ($<$ 3 keV). Also, the existence of spectral artifacts, such as escape peaks or sum peaks increases the complexity of the spectrum, particularly when low relative intensity peaks are considered.

To aid in the identification of unknowns, it is useful to consider the appearance of the K, L, and M families in EDS spectra as a function of position in the energy range 0.7–10 keV. For x-ray lines located above approximately 3 keV, the energy separation of the members of a family of x-ray lines is large enough so that, despite the line broadening introduced by a typical 150-eV (Mn $K\alpha$) EDS system, it is possible to recognize more than one line. The appearance of typical K, L, and M family lines in the 3–10 keV range is shown in Figures 6.2–6.4.

The approximate weights of lines in a family are important information in identifying elements. The K family consists of two recognizable lines $K\alpha(1)$ and $K\beta(0.1)$. The ratio of $K\alpha$ to $K\beta$ is approximately 10 : 1 when these peaks are resolved and this ratio should be apparent in the identification of an element. Any substantial deviation from this ratio should be viewed with suspicion as originating from a misidentification or the presence of a second element. The L series is observed to consist of $L\alpha(1)$, $L\beta_1(0.7)$, $L\beta_2(0.2)$, $L\beta_3(0.08)$, $L\beta_4(0.05)$, $L\gamma_1(0.08)$, $L\gamma_3(0.03)$, $Ll(0.04)$, and $L\eta(0.01)$. The observable M series consists of $M\alpha(1)$, $M\beta(0.6)$, $M\gamma(0.05)$, $M\zeta(0.06)$, and $M_{II}N_{IV}(0.01)$. The values in parentheses give only approximate relative intensities since they vary with the element in question.

Below 3 keV the separation of the members of the K, L, or M families becomes so small that the peaks are not resolved with an EDS system. The appearance of these families below 2 keV is illustrated in Figures 6.5 (Si K, 1.74 keV), 6.6 (Y L, 1.92 keV), and 6.7 (Ta M, 1.71 keV). Note that the K peaks appear to be nearly Gaussian (because of the decrease in the relative height of the $K\beta$ peak to about 0.01 of the height of the $K\alpha$), while the L and M lines are asymmetric because of the presence of several unresolved peaks of significant weight near the main peak.

Since all x-ray lines for which the critical excitation energy is exceeded will be observed, all lines for each element should be located. Considering the 0.7–10-keV range, if a high-energy K line is observed [6.4 keV (iron) and above], then a low-energy L line will also be present for the element. Figure 6.8 shows this situation for copper K and L. Similarly, if a high-energy L line is observed [4.8 keV (cerium) or above], then a low-energy M

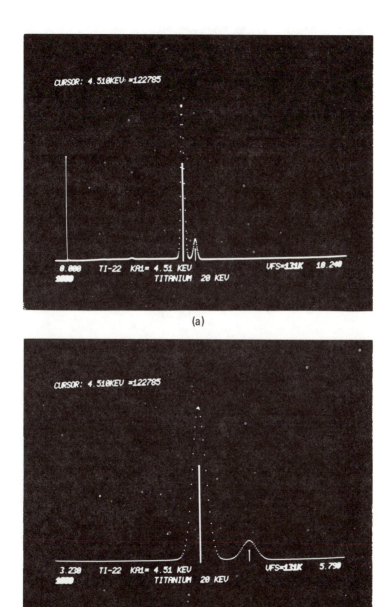

Figure 6.2. (a) Titanium, 20 keV (0–10.24 keV). (b) Expanded, Ti K series.

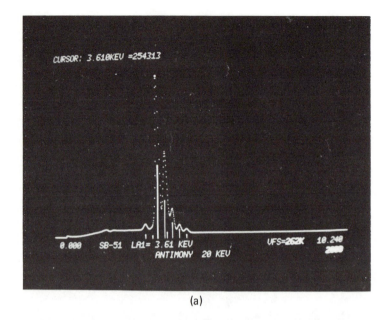

(a)

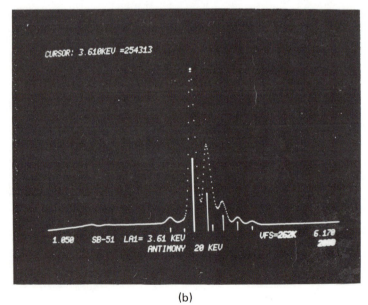

(b)

Figure 6.3. (a) Antimony, 20 keV (0–10.24 keV). (b) Expanded, Sb *L* series.

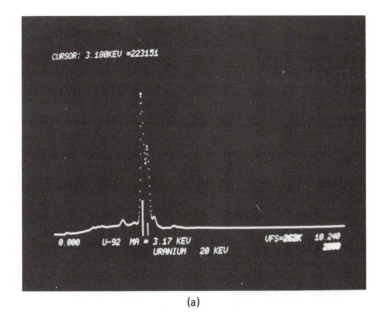

(a)

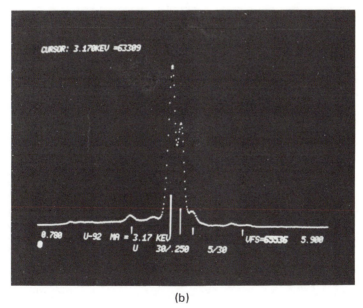

(b)

Figure 6.4. (a) Uranium, 20 keV (0–10.24 keV). (b) Expanded, U M series.

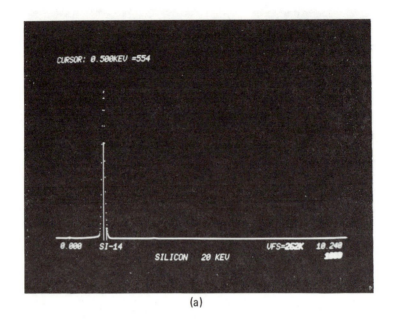

(a)

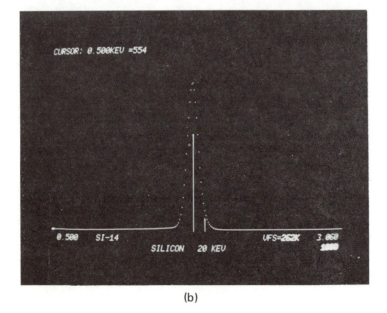

(b)

Figure 6.5. (a) Silicon, 20 keV (0–10.24 keV). (b) Expanded, Si K.

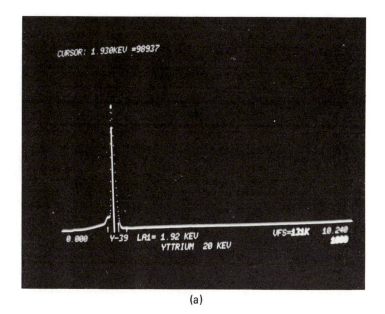

(a)

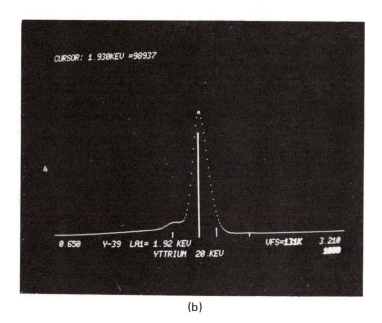

(b)

Figure 6.6. (a) Yttrium, 20 keV (0–10.24 keV). (b) Expanded, Y L.

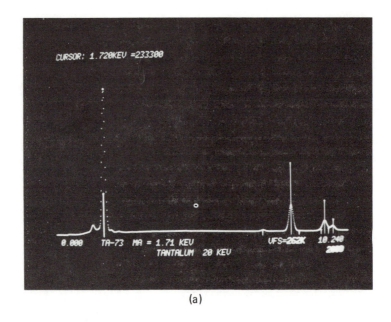

(a)

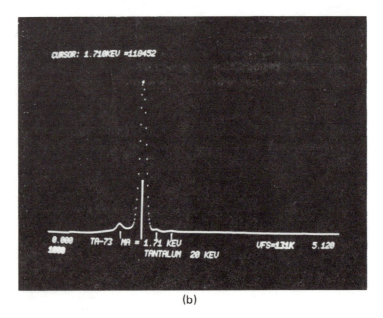

(b)

Figure 6.7. (a) Tantalum, 20 keV (0–10.24 keV). (b) Expanded, Ta *M*.

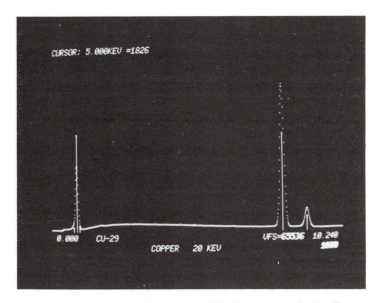

Figure 6.8. Copper, 20 keV (0–10.24 keV) showing *K* and *L* families.

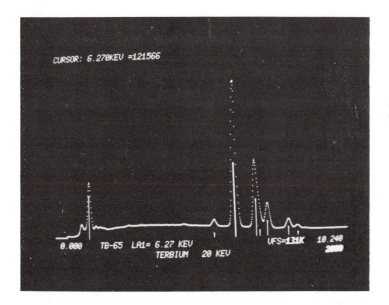

Figure 6.9. Terbium, 20 keV (0–10.24 keV) showing *L* and *M* families.

line will also be present. Figure 6.9 illustrates this situation for terbium L and M. Because of large differences in the generation and absorption of low- and high-energy x-rays it is not possible to make use of relative peak heights between K, L, or M families in qualitative analysis.

6.2.2. Guidelines for EDS Qualitative Analysis

From the previous discussion, we can construct a set of guidelines for qualitative analysis:

General Guidelines

(a) Only peaks which are statistically significant should be considered for identification. The minimum size of the peak P should be three times the standard deviation of the background at the peak position, i.e., $P > 3(\overline{N}_B)^{1/2}$. This peak height can be approximately estimated directly on the EDS display from the statistical scatter in the background on either side of the peak. The "thickness" of the background trace due to statistical fluctuations in the MCA channel counts is a measure of $(\overline{N}_B)^{1/2}$. The peak, then, should be at least three times this thickness. If it is difficult because of statistical fluctuations in the count to decide whether a peak exists above the continuum, then more counts should be accumulated in the spectrum to "develop" the peak. (For more information on detectability limits, see Chapter 8.)

(b) In order to satisfy (a) and obtain adquate counts in the spectrum, it is tempting to use a high count rate. However, EDS systems become increasingly susceptible to the introduction of artifacts, such as sum peaks, as the count rate increases. For an EDS system operating at its best energy resolution the maximum total spectrum input count rate should be kept below 3,000 counts per second. An alternative criterion is that the dead time should be kept below 30%.

(c) The EDS spectrometer should be calibrated so that the peak positions are found within 10 eV of the tabulated values. Note that, because of amplifier drift, the calibration should be checked frequently.

(d) To provide an adequate overvoltage to excite x-ray lines in the range 1–10 keV, a beam energy in the range 15–30 keV should be used. A beam energy of 20 keV is a good compromise between the need for adequate overvoltage and the need to minimize absorption in the sample which increases as the beam energy and depth of penetration increase.

(e) In carrying out accurate qualitative analysis, a conscientious "bookkeeping" method must be followed. When an element is identified, all x-ray lines in the possible families excited must be marked off, particularly low relative intensity members. In this way, one can avoid later misidentification of those low intensity family members as belonging to some other element at a minor or trace concentration. Artifacts such as escape peaks and sum peaks, mainly associated with the high intensity peaks, should be marked off as each element is identified. Peaks arising from stray excitation in the sample chamber ("system peaks") previously described in Chapter 5 should also be marked off.

Specific Guidelines

(a) Set the vertical gain of the MCA so that all peaks are contained in the display field. Begin at the high-energy end of the spectrum and work downwards in energy since the members of the K, L, and M families are more widely separated at high energies and likely to be resolved.

(b) Determine the energy of a large peak. If it corresponds closely to a $K\alpha$ line of an element, immediately look for a $K\beta$ line with about 10% of the $K\alpha$ peak height. $K\alpha$ and $K\beta$ lines of elements starting at sulfur (2.31 keV) will be resolved with a typical EDS spectrometer.

(c) If the $K\alpha$ and $K\beta$ pair does not fit the unknown, try the L series, noting the multiplicity of high-intensity L lines: [$L\alpha(1)$, $L\beta_1(0.7)$, $L\beta_2(0.2)$, $L\gamma_1(0.08)$, $L\gamma_3(0.03)$, $Ll(0.04)$, and $L\eta(0.01)$] which must be found to confirm the identification.

(d) M family lines can be observed for elements starting at cerium. While $M\alpha(1)$ and $M\beta(0.6)$ are poorly resolved, the lower-intensity $M\zeta \times$ (0.06), $M\gamma(0.05)$, and $M_{II}N_{IV}(0.01)$ are separated from the main peak and usually visible. Note that even for uranium, the M lines occur at an energy of 3.5 keV or less.

(e) At low x-ray energy (less than 3 keV), peak separation and the limited resolution of the EDS spectrometer will likely restrict element identification to only one peak. Note that low-energy L or M lines will be accompanied by high-energy K or L lines in the 4–20 keV range, which can aid identification.

(f) When an element is identified, all lines of all families (K, L or L, M) of that element should be marked off before proceeding. Next, escape peaks and sum peaks associated with the major peaks of the element should be located and marked. The magnitude of the escape peak is a constant fraction of the parent peak, ranging from about 1% for P $K\alpha$ to 0.01% for Zn $K\alpha$. The sum peak magnitude depends on the count rate.

(g) When all of the high-intensity peaks in the spectrum have been identified, and all family members and spectral artifacts have been located, the analyst is ready to proceed to the low-intensity peaks. Any low-intensity peaks which are unidentified after the above procedure should belong to elements at low concentrations. Note that for such minor or trace elements, only the α line in a family may be visible. The lines at lower relative intensity in a family will probably be lost in the statistical fluctuations of the background. As a result, the confidence with which minor or trace elements can be identified is necessarily poorer than for major elements. If the positive identification of minor and trace elements is important, a spectrum containing a greater number of counts will be needed. In many cases it may be necessary to resort to WDS qualitative analysis for greater confidence in identifying minor or trace elements.

(h) As a final step, by using Figure 6.1 the analyst should consider what peaks may be hidden by interferences. If it is important to know of the presence of those elements, it will be necessary to resort to WDS analysis.

6.2.3. Pathological Overlaps in EDS Qualitative Analysis

The limited energy resolution of the energy-dispersive spectrometer frequently causes the analyst to be confronted with serious peak overlap problems. In many cases, the overlaps are so severe that an analysis for an element of interest cannot be carried out with the EDS. Problems with overlaps fall into two general classes: (1) misidentification of peaks and (2) the impossibility of separating two overlapping peaks even if the analyst knows both are present. It is difficult to define a rigorous overlap criterion owing to considerations of statistics. In general, however, it will be nearly impossible to unravel two peaks separated by less than 50 eV no matter what peak stripping method is used. The analyst should check for the possibility of overlaps within 100 eV of a peak of interest. When the problem involves identifying and measuring a peak of a minor constituent in the neighborhood of a main peak of a major constituent, the problem is further exacerbated, and overlaps may be significant even with 200 eV separation in the case of major/minor constituents. When peaks are only partially resolved, the overlap will actually cause the peak channels for both peaks to shift by as much as 10–20 eV from the expected value. This phenomenon is encountered in the copper L family spectrum, Figure 6.8, where the unresolved Cu $L\alpha$ and Cu $L\beta$ peaks form a single peak whose peak channel is 10 eV higher than the Cu $L\alpha$ energy. Since copper K and L lines are often used as a set of calibration lines for EDS, the peak overlap effect in the L series can lead to a 10-eV calibration error at the low-energy end of the spectrum if the Cu $L\alpha$ energy is used directly. It is better to calibrate the system on a low-energy and a high-energy K line, e.g., pure Mg and pure Zn.

Because of the multiplicity of x-ray lines and the number of elements, it is not possible to list all significant interferences. Figure 6.1 can be conveniently used to assess possible interferences at any energy of interest by striking a vertical line and noting all lines within a given energy width on either side.

It is valuable to list elemental interferences which arise from peak overlaps in frequently encountered compositional systems. This is done

Table 6.1. Common Interferences in Biological X-Ray Microanalysis

Element in stain or fixative	Interfering x-ray line	Interferes with	Interfered x-ray line
U	M	K, Cu, Ti	$K\alpha$
		Cd, In, Sn, Sb, Ba	$L\alpha$
Os	M	Al, P, S, Cl	$K\alpha$
		Sr	$L\alpha$
Pb	M	S, Cl	$K\alpha$
		Mo	$L\alpha$
	L	As, Se	$K\alpha$
Ru	L	S, Cl, K	$K\alpha$
Ag	L	Cl, K	$K\alpha$
As	L	Na, Mg, Al	$K\alpha$
Cu (grid)	L	Na	$K\alpha$
Biological elements			
K	$K\beta$	Ca	$K\alpha$
Zn	$L\alpha$	Na	$K\alpha$

below for typically prepared biological samples and certain problems in materials science.

(1) Biological EDS Analysis. Many interferences which are encountered in biological x-ray microanalysis arise from the use of heavy metal fixatives and stains in sample preparation. The heavy metals have L and M family x-ray lines in the 1–5 keV range which can interfere with the K lines of low atomic materials, Na to Ca, which are important biologically. Table 6.1 lists the typical heavy metal stains and the elements for which significant peak overlap occurs.

(2) Materials Science EDS Analysis. Because of the large number of elements encountered in materials science analysis, the number of possible interferences is much greater than in biological analysis, and so the analyst

Table 6.2. Common Interferences in Materials Science X-Ray Microanalysis

Element	Interfering x-ray line	Interferes with	Interfered x-ray line
Ti	$K\beta$	V	$K\alpha$
V	$K\beta$	Cr	$K\alpha$
Cr	$K\beta$	Mn	$K\alpha$
Mn	$K\beta$	Fe	$K\alpha$
Fe	$K\beta$	Co	$K\alpha$
Pb	$M\alpha$	S	$K\alpha$
		Mo	$L\alpha$
Si	$K\alpha$	Ta	$M\alpha$
Ba	$L\alpha$	Ti	$K\alpha$

must constantly check his work to avoid errors. Particularly insidious is the interference in the first transition metal series, in which the $K\beta$ line of an element interferes with the $K\alpha$ line of the next higher atomic number, as indicated in Table 6.2. Quantitative EDS analytical systems can correct for these interferences. In qualitative analysis, however, when a minor element is interfered with by a major element x-ray line, it is often impossible to detect the presence of the minor element.

An example of a frequently encountered problem is the mutual interference of the S K, Mo L, and Pb M lines which lie within an energy range of only 50 eV. Spectral interference of this degree cannot be resolved by peak stripping methods to yield accurate qualitative analysis.

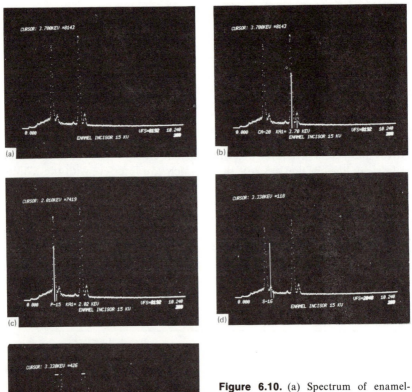

Figure 6.10. (a) Spectrum of enamel-forming region of rat incisor, 15 keV. (b) Identification of calcium. (c) Identification of phosphorus. (d) Identification of sulfur. (e) Identification of potassium, spectrum accumulated for additional time.

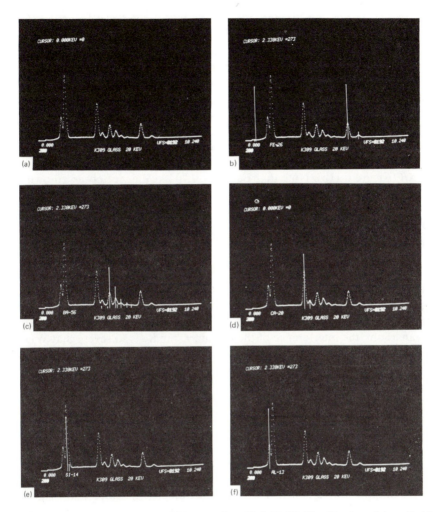

Figure 6.11. (a) Spectrum of multielement glass, 20 keV. (b) Identification of iron K. (c) Identification of barium L. (d) Identification of calcium K. (e) Identification of silicon K. (f) Identification of aluminum K.

6.2.4. Examples of EDS Qualitative Analysis

(1) Incisor of Rat. The spectrum shown in Figure 6.10a was obtained from the region of mineralized tissue formation in a rat incisor. Following the above procedure, we first find that the peaks at 3.69 and 4.01 keV correspond to Ca $K\alpha$ and Ca $K\beta$, Figure 6.10b. At lower energy, the peaks at 2.015 and 2.310 keV appear to also be a $K\alpha$–$K\beta$ pair, but while P $K\alpha$ is found to correspond to the 2.015-keV peak, P $K\beta$ is unresolved, Figure 6.10c. Further study reveals that the 2.310-keV peak is S $K\alpha$, Figure 6.10d.

There is an interesting discontinuity in the background just below Ca $K\alpha$ which may be a peak but which is close to the $3(\overline{N}_B)^{1/2}$ limit. Neither the escape peak of Ca $K\alpha$ nor the sum peak of P $K\alpha$ occurs in this energy region. Further accumulation of the spectrum, Figure 6.10e, reveals that the peak is K $K\alpha$, with K $K\beta$ lost under Ca $K\alpha$. The possible interferences of the main peaks (Ca $K\alpha$: Sb, Te; P $K\alpha$: Y, Zr) do not correspond to elements likely to exist in this system.

(2) Multielement Glass. The EDS spectrum of a complicated, multielement glass is shown in Figure 6.11a. Beginning at the high-energy end, the peaks at 6.40 and 7.06 keV are found to correspond to Fe $K\alpha$ and Fe $K\beta$ (Figure 6.11b). Note that Fe $L\alpha$ at 0.704 keV is not observed owing to high absorption. The next series of four peaks near 4.5 keV might at first be thought to be Ti $K\alpha$, $K\beta$ and V $K\alpha$, $K\beta$. However, the peak positions are not proper, nor are the relative peak heights. This series is the Ba L family Ba $L\alpha$(4.47), $L\beta_1$(4.83), $L\beta_2$(5.16), $L\gamma$(5.53), with others in the family unresolved, Figure 6.11c. The peaks at 3.69 and 4.01 keV are Ca $K\alpha$ and Ca $K\beta$, Figure 6.11d. At low energy, the peak at 1.74 keV is Si $K\alpha$ (Figure 6.11e) and at 1.49 keV the peak is Al $K\alpha$ (Figure 6.11f). Expansion of the vertical gain reveals no low-intensity peaks except those associated with artifacts. Note that in this example, if it were important to detect titanium or vanadium, the interference of the barium L family is so severe that WDS analysis would certainly be needed.

6.3. WDS Qualitative Analysis

6.3.1. Measurement of X-Ray Lines

For qualitative analysis with wavelength-dispersive spectrometers, one seeks to identify the elements in the sample by determining the angles at which Bragg's law is satisfied as the spectrometer is scanned through a range of angles. Peaks are observed at those angles, θ_B, at which the condition

$$n\lambda = 2d \sin \theta_B \qquad (6.1)$$

is satisfied. The strategy for qualitative WDS analysis is distinctly different from qualitative EDS analysis. The resolution of the WDS spectrometer is much better, typically < 10 eV compared to 150 eV for EDS, which leads to a peak-to-background ratio at least 10 times higher for WDS. As a result of the better resolution and peak-to-background, more members of the family of x-ray lines for a given element can be detected and must therefore be accounted for in a logical fashion to avoid subsequent misidentification of minor peaks. This is illustrated in Figure 6.12, where the EDS and WDS

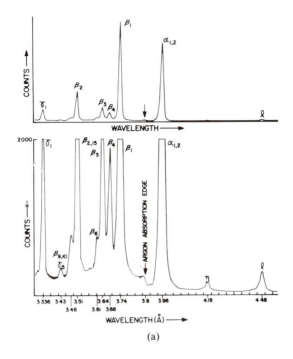

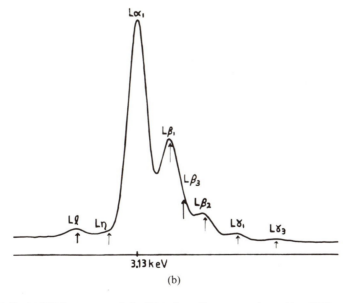

(b)

Figure 6.12. (a) WDS spectrum of the Cd L lines. Data were taken with a PET crystal. (b) EDS spectrum of the Cd L lines.

spectra for the cadmium L family are compared. Many of the peaks observed in the WDS spectrum are not detected in the EDS spectrum.

The analyst must recognize the consequences of all of the terms in Bragg's law, Equation (6.1). For a particular value of λ, as the term n varies from $n = 1, 2, 3 \ldots$, i.e., the order of the reflection changes, the value of θ_B changes. If the parent line ($n = 1$) is obtained on a given crystal, there may be several other values of θ at which peaks corresponding to this same λ will be found for $n = 2$, 3, etc. If other crystals, i.e., other d spacings are considered, additional reflections may be detected, e.g., corresponding to $n = 4$, 5, etc. Thus, the complete collection of lines for an element which must be considered not only includes all of the family of x-ray lines but also the higher-order reflections associated with each of the parent lines.

A second consequence is that for a given setting of θ on a particular crystal with spacing d, x-rays of different energies which satisfy the same value of the product $n\lambda$ will all be diffracted. Thus, if diffraction occurs for $n = 1$ and $\lambda = \lambda_1$, diffraction will also occur for $n = 2$ and $\lambda = (\lambda_1/2)$ and so on. As an example, consider sulfur in an iron–cobalt alloy. The S $K\alpha$ ($n = 1$) line occurs at 5.372 Å while Co $K\alpha$ ($n = 1$) occurs at 1.789 Å. The third-order Co $K\alpha$ line falls at 3×1.789 Å $= 5.367$ Å. These lines are so close in wavelength that they would not be adequately separated by an ordinary WDS spectrometer. While such interferences are unusual, the analyst must be aware of their possible occurrence. Although the interference cannot be resolved spectroscopically, the S $K\alpha$ and Co $K\alpha$ x-rays are of different energies and can be separated electronically. This method takes advantage of the fact that the x-rays of different energies create voltage pulses of different magnitude. In the sulfur–cobalt example, the voltage pulses produced by the Co $K\alpha$ are about three times larger than those of the S $K\alpha$. By using the pulse height discriminator previously described (Chapter 5) a voltage window can be set around the S $K\alpha$ pulse distribution which excludes the Co $K\alpha$ pulse distribution. It should be noted that once a discriminator window has been established, the spectrometer should not be tuned to another peak without removal of the discriminator window or resetting to the appropriate value for another peak.

From these considerations, it is obvious that WDS qualitative analysis is not as straightforward as EDS qualitative analysis. Nevertheless, WDS qualitative analysis offers a number of valuable capabilities which are not possessed by EDS, as illustrated in Figures 6.13 and 6.14, which give a comparison of EDS and WDS spectra on a heavy element glass (NBS K-251, the composition of which is given in Table 6.3).

(1) The high resolution of the WDS allows the separation of almost all overlaps which plague the EDS spectrum. Thus, in Figure 6.13, bismuth is extremely difficult to detect in the presence of lead in the EDS spectrum (Figure 6.13c) but is readily apparent in the WDS spectrum (Figure 6.14a). Further, the multiplicity of lines which can be observed aids in a positive

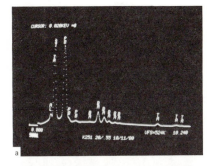
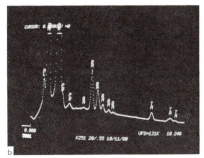
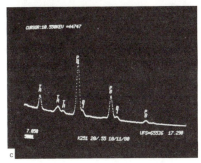

Figure 6.13. EDS spectrum of heavy element glass. (a) Energy = 0–10.24 keV. (b) Energy = 0–10.24 keV, expanded vertical scale. (c) Energy = 7–17 keV.

identification of each element. However, this multiplicity makes the book-keeping that much more difficult and important in order to avoid misassignment of low-intensity peaks.

(2) The improved peak-to-background ratio of the WDS spectrometer provides the capability of detection and identification at a concentration which is about a factor of 10 lower than the EDS (roughly 100 ppm for the WDS compared to 1000 ppm for EDS). (See Chapter 8 for further explanation.)

(3) The WDS spectrometer can detect elements with atomic numbers in the range $4 \leqslant Z \leqslant 11$. While windowless EDS spectrometers can detect elements as low as $Z = 5$, their use is restricted to high-quality vacuum systems, and the detection limits are poor.

The principal drawbacks of WDS qualitative analysis in addition to the multiplicity of lines and higher-order reflections noted above are the following.

(1) Qualitative analysis is much slower than in EDS, with scan times of 10 min to 1 h per crystal typically required. A full qualitative analysis requires the use of four crystals. If multiple spectrometers are available, these crystals can be scanned in parallel. The multiplicity of lines requires much more of the analyst's time for a complete solution of the spectrum. Moreover, this solution must typically be done manually utilizing a full x-ray table (Bearden, 1967).

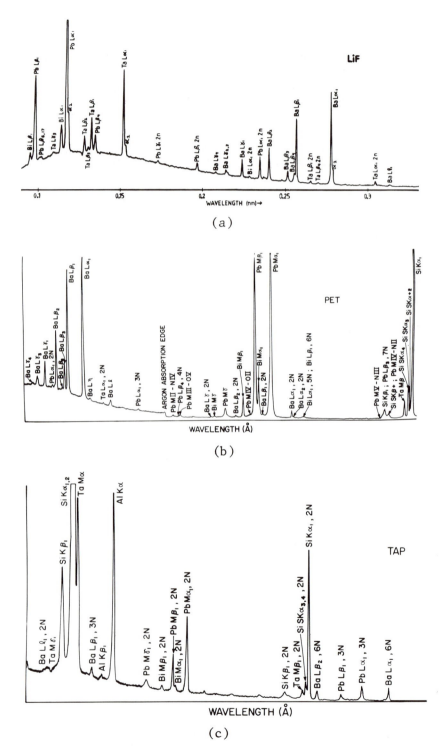

Figure 6.14. WDS spectra of heavy element glass. The EDS spectrum is given in Figure 5.13. (a) LiF spectrum. (b) PET spectrum. (c) TAP spectrum. (d) OdPb spectrum.

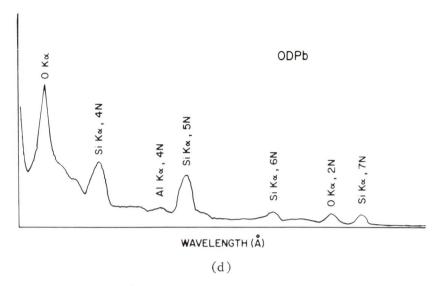

WAVELENGTH (Å)

(d)

Figure 6.14. *cont.*

(2) Much higher beam currents are required for use in the WDS compared to EDS because of the considerably lower geometrical and quantum efficiencies of the WDS. For fragile specimens such as biological targets, these high currents may be unacceptable.

6.3.2. Guidelines for WDS Qualitative Analysis

Given a sample of the complexity of that shown in Figures 6.14a–6.14d, the following procedure should be used.

(1) Because of the possibility of a peak originating from a high-order

Table 6.3. Composition of
NBS K-251

Element	wt%
O	23.0
Al	1.3
Si	14.0
Ba	9.0
Ta	4.1
Pb	44.1
Bi	4.5

reflection, it is best to start at the highest-energy, i.e., shortest-wavelength end of the spectrum for each crystal where the probability is highest for finding parent peaks (first order, $n = 1$).

(2) The highest-intensity peak should be selected in the short-wavelength end of the LiF crystal scan and its wavelength determined. From a complete x-ray reference such as that of Bearden (1967), the possible elements which could produce the peak in question as a $K\alpha_{1,2}$ or $L\alpha_{1,2}$ should be noted. In parallel with the concept of a family of lines introduced in the description of EDS qualitative analysis, when a candidate element is tentatively associated with a peak designated $K\alpha_{1,2}$ ($n = 1$), the analyst should immediately locate the associated $K\beta_1$ peak. Again, the ratio of $K\alpha$ to $K\beta$ should be roughly 10:1. However, because of changes in crystal and detector efficiency, the expected ratios may not always be found. For example, in the Cd spectrum in Figure 6.12, the efficiency of the detector approximately doubles on the low-wavelength side of the argon K absorption edge. Hence the $L\beta_1$ peak, which should be approximately 60% as big as the $L\alpha_1$ peak, is actually larger. The efficiency doubles at the argon K edge because the flow-proportional x-ray detector of this spectrometer utilizes P-10 gas (90% Ar–10% methane). The dimensions of the detector and the pressure of the P-10 gas permit a certain fraction of x-rays of wavelength longer than the edge wavelength to pass through the gas without interaction. For those x-rays with wavelengths shorter than the edge wavelength a greater fraction (approximately twice as many) will interact with the gas and, hence, be detected. Note also that the resolution of the WDS with some crystals such as LiF and quartz is sufficient to show at least some separation of $K\alpha$ into $K\alpha_1$ and $K\alpha_2$, which have a ratio of $K\alpha_1 : K\alpha_2$ of 2:1. Similarly, if an $L\alpha$ peak is suspected, the full L family should be sought. Note that L lines in addition to those listed in Figure 6.1 (i.e., $L\alpha_{1,2}$, $L\beta_1$, $L\beta_2$, $L\beta_3$, $L\beta_4$, $L\gamma_1$, $L\gamma_3$, Ll, $L\eta$) may be found because of the excellent resolution and peak-to-background ratio. In identifying families of lines, it is possible that, because of wavelength limitations of the crystals, only the main peak may be found (e.g., Ge $K\alpha$ on LiF, where Ge $K\beta$ lies outside the crystal range). Applying this approach to the LiF spectrum in Figure 6.14a, the major peak at 0.118 nm could correspond to As $K\alpha_1$ or Pb $L\alpha_1$. Inspection for other family members reveals the presence of the Pb L family but not the As $K\beta$. Thus Pb is identified with certainty and As is excluded.

(3) Once the element has been positively identified and all possible first-order members of the family of x-ray lines have been located, which may require examining spectra from more than one crystal (e.g., Zn $K\alpha$, $K\beta$ on LiF, Zn L family on PET), the analyst should locate all possible higher-order peaks associated with each first-order peak throughout the set of crystals. Continuing the heavy element glass example, the higher orders

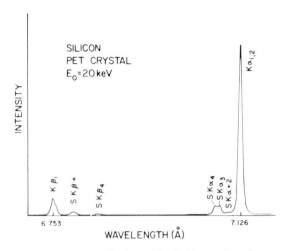

Figure 6.15. WDS spectrum of pure Si taken with a PET crystal at a beam energy of 20 keV. Note presence of satellite lines labeled *S*.

for the Pb *L* family are noted in the remainder of the LiF spectrum as well as on the PET and TAP spectra in Figures 6.14b and 6.14c. No higher-order Pb *L* lines are found in the lead sterate spectrum (Figure 6.14d). Note that the orders can extend to values as high as 7, as shown in Figure 6.14d for silicon.

(4) Only after *all* possible family members of the element and *all* higher-order reflections of each member have been identified should the analyst proceed to the next unidentified high-intensity, low-wavelength peak. The procedure is then repeated.

(5) Wavelength-dispersive spectrometry is sufficiently complicated that a number of special considerations arise: (a) For some elements, the parent *K*α peak ($n = 1$) may lie outside the range of the low-wavelength crystal (usually LiF) whereas higher-order reflections of that *K*α peak may still be found. Example: Zirconium *K*α at 0.078 nm (0.785 Å). (b) Satellite lines will also be observed as low-intensity peaks on the high-energy shoulder of a high-intensity peak. Satellite lines arise from doubly ionized atoms and are illustrated in Figure 6.15. Note that several of the satellite lines of the Si *K*α peak are nearly as high as the Si *K*β peak.

6.4. X-Ray Scanning

The technique of x-ray area scanning provides the investigator with what amounts to a scanning x-ray microscope. The amplified signal from the detector system—energy-dispersive or crystal spectrometer—is made to modulate the brightness of a cathode ray tube (CRT) scanned in synchronism with the electron probe. Thus, on the CRT, a picture is obtained by the variation of x-ray emission from the surface. The same magnification controls, sweep system, and amplifier as in the scanning electron microscope are used here (see Chapter 4). The electron beam can sweep in a line, X or Y direction, to yield an x-ray line scan. An example of a typical line scan for Co and Cr across an oxidized high-temperature alloy is given in Figure 5.14, Chapter 5. The electron beam can of course be swept across an area to yield an x-ray area scan. An x-ray area scan can show tones ranging from black to white depending on experimental conditions. In places of high concentration of the element in the scanned area, the picture will be nearly photographically white; it will be gray where the element's concentration is lower and black where the element is absent. An example showing results for an ore is illustrated in Figure 6.16.

When an energy-dispersive system is used to prepare the area scan photograph, care must be taken to ensure that no other peak interferes with the signal of the desired element. If peak overlaps occur the crystal spectrometer would be needed to separate the two satisfactorily. The desired peak should be carefully isolated by means of a single-channel analyzer; this analyzer is often built into multichannel analyzers. In a programmable system, the desired peak can be similarly isolated. The amplified output of the single-channel analyzer is displayed on a CRT and so provides the elemental distribution desired. Magnification can be varied in the usual way. Separate micrographs of all elements of interest in the scanned area can be built up to give the complete elemental distribution. A detailed discussion of x-ray area scanning has been given by Heinrich (1967).

In the case of an energy-dispersive detector, the time required to build up a satisfactory x-ray area scan may be very long. This hindrance results from the requirements imposed by Poisson statistics to provide at least 20,000 photons per micrograph; frequently 200,000 are needed and occasionally as many as 500,000 photons must be used. For the SEM often run with beam currents on the order of 10^{-10} or 10^{-11} A, the total number of photons produced is fairly low. Since only one peak obtained from the total spectrum is of interest, the time to obtain a satisfactory micrograph is correspondingly long. Times in excess of 15 min are not uncommon to prepare EDS x-ray area scans.

An example of x-ray area scanning information obtainable with the EDS system is shown in Figure 6.17. The specimen is a composite of tungsten wires which were plunged into molten aluminum held at 1100°C under a vacuum of 10^{-4} Torr. The area scans show no significant amounts of Al in W or W in Al. The micrographs were prepared in an SEM operating at 20 kV and a specimen current of 10^{-10} A; exposure time was

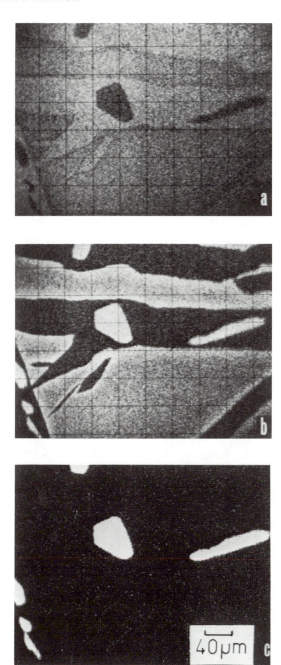

Figure 6.16. X-ray area scans showing distribution of three elements in a galena ore. (a) S $K\alpha$; (b) Si $K\alpha$; (c) Ag $L\alpha$.

Figure 6.17. X-ray area scans of an Al–W composite material photographed on the SEM with an EDS system. Top: W $L\alpha$; middle: Al $K\alpha$; bottom: secondary electron micrograph.

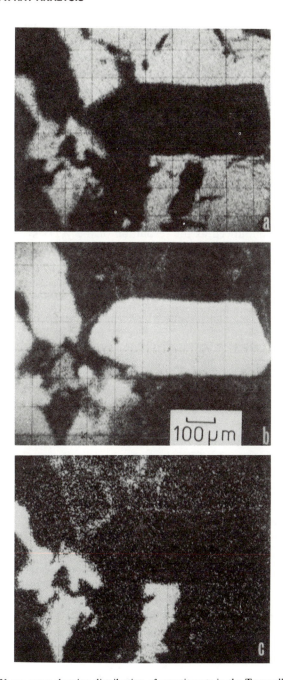

Figure 6.18. X-ray scans showing distribution of constituents in the Tazewell, Tennessee iron meteorite. (a) Fe $K\alpha$; (b) Ni $K\alpha$; (c) P $K\alpha$. Note the nickel enrichment at the phase boundaries; the iron content is about 93% in the kamacite [dark in (b)] and 50%–85% in the teanite [light in (b)]. Differences in iron content of these phases are apparent in the Fe picture.

12 min for each. Figure 6.17 would compare favorably with similar data prepared with a WDS system; however, the time and effort needed to prepare Figure 6.17 would be less in the case of WDS facilities.

If the specimen will not be harmed, higher beam currents are useful—not only for elemental mapping but for all phases of x-ray analysis. The loss of spatial resolution resulting from increasing beam current matters little since the x-ray emission volume is largely determined by electron diffusion and not by probe size (Chapter 3). Since it is difficult to decrease the effective volume of x-ray emission below 1 μm, x-ray scanning pictures are limited to about 3000 × magnification.

Because of the requirement for the x-ray source to be precisely on the Rowland focusing circle of a WDS system, large scan areas may cause intensity drop-off at the edges of the x-ray area scan. This problem becomes more pronounced as the resolution of the WDS system increases. One way to monitor the seriousness of the intensity drop-off is to insert a pure element and to prepare x-ray photographs as a function of raster size. This can be done for each crystal in each spectrometer. Fortunately no such focusing difficulty exists in the case of an EDS system which views a large area of the specimen, even when collimated (see Figure 5.41, Chapter 5).

The eye can observe contrast differences of about 5%. Hence, the composition difference responsible for a 5% observable contrast variation on an x-ray area scan is significant. Figure 6.18 shows kamacite containing about 93 wt% Fe and taenite containing 50%–85% Fe in the Tazewell, Tennessee meteorite. The interface separating these kamacite and taenite phases is evident. This case represents a compositional ratio of about 1.5 Fe (kamacite) to 1.0 Fe (taenite). For compositional ratios of about 1.2 to 1.0, it would probably be very difficult to unambiguously separate phases on the ordinary x-ray area scan. Heinrich (1967) has described methods to improve visibility by electronic enhancement methods.

Black and white photographs showing the distribution of a single x-ray signal are commonly used. It is often difficult, however, to show the correlation of signals from two or more x-ray lines without the use of color. Composite color photographs using x-ray images from the electron microprobe have been produced. One such method uses the black and white scanning images from which color-separation positives are made. With appropriate filters, color prints combining three different signals (Yakowitz and Heinrich, 1969) can be produced. Several techniques have now been developed which use a color TV unit to display three signals at one time (Ficca, 1968).

7

Quantitative X-Ray Microanalysis

7.1. Introduction

With the EPMA and the SEM one can obtain quantitative analyses of $\sim 1\text{-}\mu m^3$ regions of bulk samples using a nondestructive x-ray technique. For samples in the form of thin foils and sections of organic material, the size of the analyzed microvolume is reduced to about one tenth of the value for bulk samples. For metals and alloys the ZAF technique is usually employed. Pure element or alloy standards can be used and the surfaces of the samples and standards must be properly prepared and analyzed under identical operating conditions. For geological samples the a factor or empirical technique is usually employed. For this class of samples secondary x-ray fluorescence is usually not significant and oxide standards of similar atomic number as the sample are used. Biological samples are often adversely affected by the impinging electron beam. It is important to ensure that the standards are in the same form and matrix as the specimen. The purpose of this chapter is to describe in some detail the various methods by which quantitative analyses can be obtained for inorganic, metallic, and biological samples in the form of bulk specimens, small particles, thin films, sections, and fractured surfaces.

Many of the quantitative analytical techniques have been incorporated in microcomputer-based x-ray analysis units. Although the compositional analyses obtained are often quite good, the user should understand what techniques are being employed and their limitations. The accuracy of the analyses is dependent on the standards and the operating conditions (current, operating voltage, x-ray line, etc.) which are chosen.

The ZAF, a-factor, and thin-film techniques have all reached a mature stage of development. Therefore only the most often used procedure for each technique will be presented. In each of these presentations the

305

equations are developed in such a way that the reader can make the calculations directly. It should be pointed out, however, that the ZAF technique is still being improved, especially for low electron energies and light element analysis (Parobek and Brown, 1978; Love and Scott, 1978, 1980). The analysis techniques for biological specimens, as well as for particles and rough surfaces, are less well advanced and are still under development. Therefore several procedures will be discussed for each technique along with some comments to indicate the best method(s) to use.

7.2. ZAF Technique

7.2.1. Introduction

In his thesis, Castaing (1951) outlined a method for obtaining a quantitative x-ray analysis from a micrometer-sized region of solid specimens. Castaing's treatment can be represented by the following considerations. The average number of ionizations, n, from element i generated in the sample per primary beam electron incident with energy E_0 is

$$n = \left(\frac{N_0 \rho C_i}{A_i} \right) \int_{E_0}^{E_c} \frac{Q}{-dE/dX} dE \tag{7.1}$$

where dE/dX is the mean energy change of an electron in traveling a distance dX in the sample, N_0 is Avogadro's number, ρ is the density of the material, A_i is the atomic weight of i, C_i is the concentration of element i, E_c is the critical excitation energy for the characteristic x-ray line (K, L, or M of element i) of interest, and Q is the ionization cross section, defined as the probability per unit path length of an electron of given energy causing ionization of a particular electron shell (K, L, or M) of an atom in the specimen. (See Chapter 3 for a complete discussion of the ionization cross section.) The term $(N_0 \rho C_i / A_i)$ gives the number of atoms of element i per unit volume. The effect of backscattering electrons can be taken into account by introducing a factor, R, equal to the ratio of x-ray intensity actually generated to that which would have been generated if all of the incident electrons had remained within the specimen. The average number of x-rays of element i generated per incident electron, I, is proportional to n. Hence,

$$I = (\text{const}) C_i R \rho \int_{E_c}^{E_0} \frac{Q}{dE/dX} dE \tag{7.2}$$

In practice it is very difficult to calculate the absolute generated intensity, I, directly. In fact the intensity that one must deal with is the measured intensity. The measured intensity is usually even more difficult to calculate. Therefore, as suggested by Castaing (1951), we select a standard

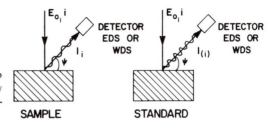

Figure 7.1. Experimental setup for the measurement of intensity ratio $I_i/I_{(i)}$ from sample and standard.

for element i and measure the ratio $I_i/I_{(i)}$, where I_i and $I_{(i)}$ are the measured intensities from the sample and standard, respectively. The experimental measurement procedure is shown in Figure 7.1. The x-ray intensity from element i is measured by the same x-ray detector from both sample and standard in which E_0, the electron probe current and the angle between the specimen surface and the direction of the measured x-rays (take-off angle Ψ) are held constant. By using the ratio of the measured intensities $I_i/I_{(i)}$, the constant term in Equation (7.2) will drop out and the intensity ratio has the following form:

$$\frac{C_i}{C_{(i)}} \propto [\quad] \frac{I_i}{I_{(i)}} \tag{7.3}$$

where the bracket will include terms involving R, ρ, Q, and dE/dX. As discussed by Castaing (1951), to a first approximation, R, ρ, Q, and dE/dX may be assumed equal for both the specimen and standard. If the standard is pure element i, then,

$$C_i = I_i/I_{(i)} \tag{7.4}$$

where $I_{(i)}$ is the intensity measured in pure element i. Equation (7.4) is known as Castaing's first approximation and can be used to obtain an initial estimate of composition.

In most analyses the measured intensities from specimen and standard need to be corrected for differences in R, ρ, Q, and dE/dX and absorption within the solid in order to arrive at the ratio of generated intensities and hence the value of C_i. Recognition of the complexity of the problem of the analysis of solid specimens has led numerous early investigators (Wittry, 1963; Philibert, 1963; Duncumb and Shields, 1966) to expand the theoretical treatment of quantitative analysis proposed by Castaing.

All schemes for carrying out quantitative analysis of solid specimens use a standard of known composition. In many cases, especially for metals, pure elements are suitable. In the case of mineral or petrological samples, homogeneous, compound standards close in average atomic number to the unknown are usually chosen. What the analyst measures is the relative x-ray intensity ratio between the elements of interest in the specimen and the same elements in the standard. Both specimen and standard are examined under identical experimental conditions. The measured relative intensity ratio, commonly called k, must be accurately determined or any quantitative analysis scheme will result in the same inaccuracy. In this chapter we will assume that measurements of $I_i/I_{(i)} = k_i$ can be obtained accurately. In Chapter 8, the measurement techniques to obtain accurate I_i and $I_{(i)}$ values will be considered in some detail.

Once the k_i values have been obtained, they must be corrected for several effects including (1) differences between specimen and standard for electron scattering and retardation, i.e., the so-called atomic number effect expressed by the factor Z_i; (2) absorption of x-rays within the specimen, A_i; (3) fluorescence effects, F_i, and, in a few specific cases, continuum fluorescence. A common form of the correction equation is

$$C_i = (ZAF)_i k_i \tag{7.5}$$

where C_i is the weight fraction of the element of interest. This method is often referred to as the ZAF method. Each of the effects listed above will be discussed in further detail in the following sections.

7.2.2. The Absorption Factor, A

7.2.2.1. Formulation

Since the x-rays produced by the primary beam are created at some nonzero depth in the specimen (see Figure 7.2), they must pass through the specimen on their way to the detector. On this journey, some of the x-rays undergo absorption due to interactions with the atoms of the various elements in the sample. Therefore, the intensity of the x-ray radiation finally reaching the detector is reduced in magnitude. Castaing (1951) has described the intensity, dI, of characteristic radiation—without absorption —generated in a layer of thickness dz having density ρ at some depth z

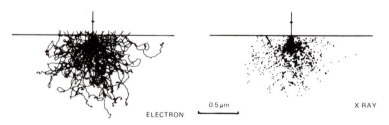

ELECTRON   0.5 μm   X-RAY

Figure 7.2. Schematic representation of 100 electron paths in a copper specimen; primary beam voltage 20 kV. The corresponding x-rays generated by these electrons are shown as well (after Curgenven and Duncumb, 1971).

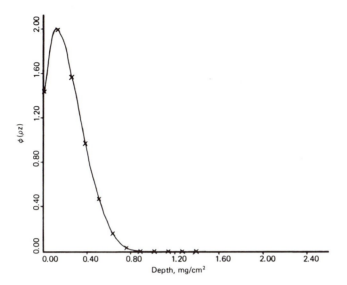

Figure 7.3. Computed $\phi(\rho z)$ curve for copper assuming a primary beam voltage of 20 kV. The value of $\phi(\rho z)$ is the number of $K\alpha$ quanta generated at the indicated depth per incoming electron.

below the specimen surface as

$$dI = \phi(\rho z)\,d(\rho z) \tag{7.6}$$

where $\phi(\rho z)$ is defined as the distribution of characteristic x-ray production with depth. The shape of the $\phi(\rho z)$ curve for Cu $K\alpha$ radiation is shown in Figure 7.3; this shape is typical for all elements. Thus, in the absence of absorption, the total flux generated and detected by the spectrometer system would be

$$I = \int_0^\infty \phi(\rho z)\,d(\rho z) \tag{7.7}$$

But absorption of the generated x-rays does occur and so the actual flux recorded is I' given by

$$I' = \int_0^\infty \phi(\rho z)\exp\left\{-\left[(\mu/\rho)(\rho z)\csc\Psi\right]\right\}d(\rho z) \tag{7.8}$$

where μ/ρ is the x-ray mass attentuation coefficient of the specimen for the characteristic line of interest, $\rho z\csc\Psi$ is the absorption path length, and Ψ is the angle between the x-ray and the specimen surface, the take-off angle. The quantity χ is defined as $\chi = (\mu/\rho)\csc\Psi$. Philibert (1963) referred to I as $F(0)$ when χ was zero, and he referred to I' as $F(\chi)$. The ratio $F(\chi)/F(0)$ was called $f(\chi)$, which is equivalent to I'/I. The term $f(\chi)$ is the standard absorption term referred to by Philibert (1963). Therefore, for the determination of the A term in Equation (7.5) of any element i in any composite specimen, we use the equation

$$A_i = f(\chi)_{std}/f(\chi)_{spec} \qquad (7.9)$$

where std and spec refer to standard and specimen, respectively. Hereafter, all quantities referring to the alloy or specimen will be denoted by an asterisk. All quantities referring to the standard will be left unmarked. Thus we can write

$$A_i = f(\chi)/f(\chi)^* \qquad (7.10)$$

7.2.2.2. Expressions for $f(\chi)$

The absorption correction factor, $f(\chi)$ of any element i, depends upon the respective mass absorption coefficient μ/ρ, the x-ray emergence angle Ψ, the operating voltage E_0, the critical excitation voltage E_c for K, L, or M radiation from element i, and the mean atomic number and mean atomic weight of the specimen, $\overline{Z}$ and $\overline{A}$. Hence we can write

$$f(\chi) = f\left[(\mu/\rho)\csc \Psi, E_0, E_c, \overline{Z}, \overline{A} \right] \qquad (7.11)$$

There have been some direct determinations of the $\phi(\rho z)$ for pure elements. Philibert (1963) fitted a relation for $f(\chi)$ to the experimental $\phi(\rho z)$ curves,

$$\frac{1}{f(\chi)} = \left(1 + \frac{\chi}{\sigma} \right)\left(1 + \frac{h}{1+h} \frac{\chi}{\sigma} \right) \qquad (7.12)$$

where

$$h = 1.2A/Z^2 \qquad (7.13)$$

and A and Z are, respectively, the atomic weight and number of element i. The absorption parameter $\chi = \mu/\rho \csc \Psi$, where μ/ρ is the mass absorption coefficient of element i in itself. The parameter σ is a factor which accounts for the voltage dependence of the absorption or loss of the primary electrons. The σ factor decreases with increasing excitation potential E_0 (Philibert, 1963). At higher excitation potentials electrons penetrate more deeply and the absorption path is lengthened. This is illustrated in Figure 7.4, which shows a representation of the electron and x-ray distribution in copper as a function of primary beam voltage, E_0. As discussed in Chapter 3, x-rays are produced deeper in the specimen as E_0 increases. Figure 7.5 shows schematically the geometry involved in x-ray absorption and the variation of the absorption path length $\overline{P}$ in an aluminum target for varying operating voltages, E_0, and take-off angles, Ψ. Note that the path length increases rapidly with increasing operating voltage and with decreasing emergence angle. The value of $f(\chi)$ will approach 1 [Equation (7.12)] as σ increases, and χ decreases. This occurs at the lowest E_0 values, highest Ψ

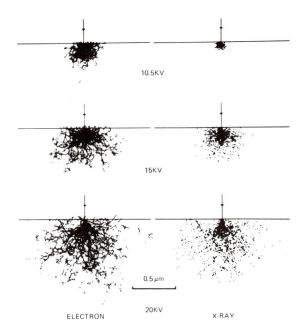

10.5KV

15KV

0.5 μm

20KV

ELECTRON   X-RAY

Figure 7.4. Schematic representation of the variation of 100 electron paths in a copper specimen as a function of primary beam voltage. Note the spatial resolution of the x-rays generated by these electrons (after Curgenven and Duncumb, 1971).

values, and minimum μ/ρ values. As $f(\chi)$ approaches 1, the effect of absorption is minimized. However, we will see later that it is not possible to operate with E_0 too close to E_c.

Duncumb and Shields (1966) proposed that the dependence of E_c should be taken into account in the formulation of σ in Equation (7.12). Later Heinrich (1969), after critical examination of existing experimental

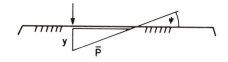

$$\bar{P} = y \csc \theta$$

FOR Al (Z=13)

Figure 7.5. Schematic representation of path lengths for absorption in an aluminum specimen as a function of primary beam voltage, E_0, and x-ray emergence angle, Ψ. The depth Y was obtained using Equation (3.42), Chapter 3.

E_0(kV)	Ψ(deg)	y (μm)	$\bar{P}$ (μm)
10	15	0.3	1.16
10	60	0.3	0.35
30	15	2.0	7.7
30	60	2.0	2.3

$f(\chi)$ data, suggested a formula for σ, namely,

$$\sigma = \frac{4.5 \times 10^5}{E_0^{1.65} - E_c^{1.65}} \qquad (7.14)$$

This development [Equations (7.12)–(7.14)] is known as the Philibert–Duncumb–Heinrich (PDH) equation and is currently the most popular expression for $f(\chi)$. Accordingly, the PDH equation is commonly used in microprobe correction schemes to calculate $f(\chi)$.

In using the PDH equation to calculate $f(\chi)$ or $f(\chi)^*$ of element i in multicomponent samples, the effect of other elements in the sample or standard must be considered. These elements have an effect on the values of h, χ, and σ. After h is evaluated for each element in the specimen, an average h value is obtained using

$$h = \sum_j C_j h_j \qquad (7.15)$$

where j represents the various elements present in the sample including element i and C_j is the mass fraction of each element j. To obtain χ, the mass absorption coefficient for element i in the sample, $(\mu/\rho)_{\text{spec}}^i$, must be calculated. The mass absorption coefficient is given by

$$(\mu/\rho)_{\text{spec}}^i = \sum_j (\mu/\rho)_j^i C_j \qquad (7.16)$$

where $(\mu/\rho)_j^i$ is the mass absorption coefficient for radiation from element i in element j and C_j is the weight fraction of each element in the specimen including i. The σ value is obtained from Equation (7.14) using the E_c value of element i.

7.2.2.3. Practical Considerations

The effects of the errors in input parameters $(\mu/\rho,\ \Psi,\ E_0)$ have been considered in detail by Yakowitz and Heinrich (1968). The major conclusions of their study were as follows:

1. Serious analytical errors can result from input parameter uncertainties.
2. In order to reduce the effects of these errors the value of the absorption function $f(\chi)$ should be 0.7 or greater.
3. To maximize $f(\chi)$, samples should be run at low overvoltage ratios and instruments should have high x-ray emergence angles.
4. To further improve the accuracy of microprobe analysis, particularly of low-atomic-number elements or x-ray lines with energies $\leqslant 1$ keV, more accurate experimental determinations of mass attenuation coefficients and of the function $f(\chi)$ are required.

As an example of the difficulties associated with mass absorption

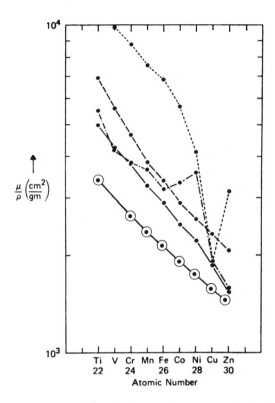

Figure 7.6. Proposed values of $\mu/\rho_i^{iL\alpha}$ where i is an element with $22 \leqslant Z \leqslant 30$ (adapted from Kyser, 1972). Comparison values from: $\odot$ = Henke and Ebisu (1974), ---- = Bracewell and Veigele (1971), ---- = Colby (1968), —·— = Kyser (1972), and —— = Cooke and Stewardson (1964).

coefficient uncertainties, Figure 7.6 shows the range of values of μ/ρ for the $L\alpha$ lines of elements in the range $22 \leqslant Z \leqslant 30$ (Kyser, 1972). Huge discrepancies are obvious.

The case of analysis with an energy-dispersive detector or crystal spectrometer mounted on a typical SEM must be considered. If the specimen is flat and smooth, only a few other obstacles must be considered for quantitative analysis. The most important of these are the correction for background, which will be discussed in Chapter 8, and the exact value of Ψ, the x-ray take-off angle. In the SEM the specimen is usually tilted in order to provide a suitable value of the take-off angle. Uncertainties in the take-off angle Ψ mainly effect the absorption correction calculation; such uncertainties increase as Ψ decreases (Yakowitz and Heinrich, 1968). Thus, values of Ψ in excess of 30° are needed. Furthermore, it is of crucial importance to ensure that the specimen and standard are measured with identical x-ray take-off angles.

7.2.2.4. Calculations of the Absorption Factor, A

The three major variables which influence the absorption factor A_i are the operating voltage E_0, the take-off angle Ψ, and the mass absorption factor for the element of interest i in the specimen, $(\mu/\rho)^i_{\text{spec}}$. Since A_i is determined by the ratio $f(\chi)/f(\chi)^*$, both terms should be similar if the absorption factor A_i is to approach 1. As A_i approaches 1, the measured intensity ratio becomes a better approximation for the concentration ratio of element i from sample to standard.

The significance of the absorption factor can be illustrated by considering two binary systems, Ni–Fe and Al–Mg. In both binaries the atomic numbers of the two elements are so close that no atomic number Z_i correction need be made. We will consider Ni $K\alpha$ absorption in Fe and Al $K\alpha$ absorption in Mg. In both cases secondary fluorescence does not occur so that a F_i correction need not be made. Calculations of $A_i = f(\chi)/f(\chi)^*$ were made for both systems using the PDH correction, Equations (7.9), (7.12)–(7.16).

Tables 7.1 and 7.2 contain the input data for the absorption calculation and also list the various terms χ, σ, h which are evaluated in the calculation. In the case of Ni–Fe, a 10 wt% Ni alloy was considered (Table 7.1). Calculations were performed for two operating voltages, 30 and 15 kV, and two take-off angles, 15.5° and 52.5°. The smallest $f(\chi)$ and $f(\chi)^*$ factors are calculated at $E_0 = 15$ kV and $\Psi = 52.5°$. The amount of absorption is minimized because x-rays are generated close to the surface and the absorption path length is smaller at high take-off angles. In this case the A_{Ni} factor is 1.05, requiring only a 5% correction. On the other hand at 30 kV, and a low take-off angle of 15.5°, the absorption factor is 1.60, requiring almost a 60% correction. The $f(\chi)^*$ factor under these operating conditions is < 0.7, the minimum value suggested by Yakowitz and Heinrich (1968). As pointed out by Yakowitz and Heinrich (1968), A_i will be minimized at low overvoltage and high Ψ angles.

Table 7.2 illustrates the absorption corrections necessary for Al $K\alpha$ in a Mg–10 wt% Al alloy. At $E_0 = 15$ and 30 kV, the $f(\chi)^*$ value is so low that corrections of over 200% are necessary. The very high μ/ρ value of Al $K\alpha$ in Mg (4376.5 cm^2/g) is responsible for the large absorption correction. Since E_c for Al is only 1.56 kV, it is possible to perform an x-ray microanalysis at a lower operating voltage than 15 kV. The excitation region will be closer to the surface and the absorption path length will be decreased. At an operating voltage of 7.5 kV, and a take-off angle of 52.5° (note Table 7.2) the absorption factor is only 1.306. In this case a reasonably small correction can be applied.

It is clear from these calculations that the analyst should be wary of large μ/ρ values and operation at high overvoltages and low Ψ angles. The absorption factor is the most important consideration in most quantitative

Table 7.1. Absorption of Ni $K\alpha$ in a Ni–Fe Binary

(a) Input data Ni–Fe binary		
	Ni	Fe
μ/ρ Ni $K\alpha$ (cm^2/g) absorber[a]	58.9	379.6
Z	28	26
E_c (kV)	8.332	7.111
A	58.71	55.85

(b) Output data Ni–Fe binary								
Ni–Fe Sample	E_0 (kV)	Ψ (deg)	χ	σ	h	$f(\chi)$	$f(\chi)^*$	A_{Ni}
Fe–10% Ni Sample			438	1870	0.0982	—	0.794	
	30	52.5						1.21
Ni Standard			74.2	1870	0.0899	0.959	—	
Fe–10% Ni Sample			1300	1870	0.0982	—	0.555[b]	
	30	15.5						1.60
Ni Standard			220.4	1870	0.0899	0.886	—	
Fe–10% Ni Sample			438	8310	0.0982	—	0.945	
	15	52.5						1.05
Ni Standard			74.2	8310	0.0899	0.990	—	
Fe–10% Ni Sample			1300	8310	0.0982	—	0.853	
	15	15.5						1.14
Ni Standard			220.4	8310	0.0899	0.972	—	
Fe–50% Ni Sample			820.4	8310	0.0945	—	0.902	
	15	15.5						1.08
Ni Standard			220.4	8310	0.0899	0.972	—	

[a] Heinrich (1966).
[b] $f(\chi) < 0.7$.

analyses. Clearly a reasoned choice of SEM operating conditions and x-ray lines with small mass absorption coefficients can help minimize the necessary corrections.

7.2.3. The Atomic Number Factor, Z

The so-called atomic number effect in electron microprobe analysis arises from two phenomena, namely, electron backscattering and electron retardation, both of which depend upon the average atomic number of the target. Therefore, if there is a difference between the average atomic

Table 7.2. Absorption of Al $K\alpha$ in an Al–Mg Binary

(a) Input data Al–Mg binary

	Al	Mg
μ/ρ Al $K\alpha$ (cm^2/g) absorber[a]	385.7	4376.5
A	26.98	24.305
Z	13	12
E_c (kV)	1.56	1.303

(b) Output data Al–Mg binary

	E_0 (kV)	Ψ (deg)	χ	σ	h	$f(\chi)$	$f(\chi)^*$	A_{Al}
Sample Mg–10% Al			5,013	1,657	0.201		0.165[b]	
	30	52.5						4.48
Al Standard			486	1,657	0.192	0.738		
Sample Mg–10% Al			14,884	1,657	0.201		0.04[b]	
	30	15.5						11.7
Al Standard			1,443	1,657	0.192	0.469[b]		
Sample Mg–10% Al			5,013	5,286	0.201		0.443[b]	
	15	52.5						2.04
Al Standard			486	5,286	0.192	0.902		
Sample Mg–10% Al			14,884	5,286	0.201		0.178[b]	
	15	15.5						4.23
Al Standard			1,443	5,286	0.192	0.753		
Sample Mg–50% Al			8,910	5,286	0.197		0.291[b]	
	15	15.5						2.58
Al Standard			1,443	5,286	0.192	0.753		
Sample Mg–10% Al			5,013	17,506	0.201		0.742	
	7.5	52.5						1.306
Al Standard			486	17,506	0.192	0.969		

[a] Heinrich (1966).
[b] $f(\chi) < 0.7$.

number of the specimen given by

$$\bar{Z} = \sum_j C_j Z_j \tag{7.17}$$

and that of the standard, an atomic number correction is required. For example in Al–2 wt% Cu, the value of $\bar{Z}$ is 13.32, so a somewhat larger effect for a Cu ($Z = 29$) analysis would be expected than for an Al analysis ($Z = 13$) if pure element standards are used. In general, unless this effect is corrected for, analyses of heavy elements in a light element matrix generally yield values which are too low, while analyses of light elements in a heavy matrix usually yield values which are too high. Hence, for a Si analysis in Fe–Si the magnitude of the atomic number correction is in the opposite sense to that of the absorption correction.

At present, the most accurate formulation of the atomic number factor Z_i for element i appears to be that given by Duncumb and Reed (1968):

$$Z_i = \frac{R_i}{R_i^*} \frac{\int_{E_c}^{E_0}(Q/S)\,dE}{\int_{E_c}^{E_0}(Q/S^*)\,dE} \tag{7.18}$$

where R_i and R_i^* are the backscattering correction factors of element i for standard and specimen respectively:

$$R_i = \frac{\text{total number of photons actually generated in the sample}}{\text{total number of photons generated if there were no backscatter}}$$

Q is the ionization cross section and defined as the probability per unit path length of an electron with a given energy causing ionization of a particular inner electron shell of an atom in the target (see Chapter 3), and S is the electron stopping power $-(1/\rho)(dE/dX)$ (see Chapter 3) in the region $1 \leqslant E \leqslant 50$ kV. The stopping power S_i given by Bethe (1930) is

$$S_i = (\text{const})\frac{Z}{A}\frac{1}{E}\ln\left(\frac{C_1 E}{J}\right) \tag{7.19}$$

where C_1 is a constant and J is the so-called mean ionization potential. To calculate the term $(C_1 E/J)$, the units commonly used are C_1 equal to 1166 when E is given in kilovolts and J in electron volts.

Values of R lie in the range 0.5–1.0 and approach 1 at low atomic numbers. The backscattering correction factor varies not only with atomic number but also with the overvoltage $U = E_0/E_c$. As the overvoltage decreases towards 1, fewer electrons are backscattered from the specimen with voltages $> E_c$, and consequently less of a loss of ionization results from such backscattered electrons.

There are several tabulations of R values as a function of Z, the pure element atomic number and U (Duncumb and Reed, 1968; Green, 1963; Springer, 1966). Only one experimental study of R, Derian and Castaing (1966), is available. The tabulation given by Duncumb and Reed very nearly agrees with the experimental determinations where comparisons can be made. Other tabulations do not (Heinrich and Yakowitz, 1970). Thus, the R values most often used are those of Duncumb and Reed (1968).

The electron backscattering correction factors R from Duncumb and Reed (1968) were fitted by Yakowitz et al. (1973) with respect to overvoltage U and atomic number Z as follows:

$$R_{ij} = R_1' - R_2'\ln(R_3' Z_j + 25) \tag{7.20}$$

where

$$R_1' = 8.73 \times 10^{-3} U^3 - 0.1669 U^2 + 0.9662 U + 0.4523$$

$$R_2' = 2.703 \times 10^{-3} U^3 - 5.182 \times 10^{-2} U^2 + 0.302 U - 0.1836$$

$$R_3' = (0.887 U^3 - 3.44 U^2 + 9.33 U - 6.43)/ U^3$$

The term i represents the element i which is measured and the term j represents each of the elements present in the standard or specimen including element i. The term R_{ij} then gives the backscattering correction for element i as influenced by element j in the specimen. The fitting factors R_1', R_2', and R_3' have been calculated as a function of U and are listed in the data base, Chapter 14, at the end of the book.

Duncumb and Reed (1968) have postulated that

$$R_i = \sum_j C_j R_{ij} \tag{7.21a}$$

for the standard, and

$$R_i^* = \sum_j C_j R_{ij} \tag{7.21b}$$

for the sample. To evaluate R_i, one must obtain from the Duncumb–Reed tabulations, as given in the data base (Chapter 14) or in Equation (7.20), the value of R_{ij} for each element in the specimen or standard. Thus far no experimental evidence which refutes Equation (7.21) has been found.

There are several relations for Q as discussed in Chapter 3. Despite the fact that Q values differ by several percent depending on the value of the constants, Heinrich and Yakowitz (1970) have shown that discrepancies in Q values have only a negligible effect on the final value of the concentration. A simplifying assumption is often made that Q is constant and therefore cancels in the expression for Z_i [Equation (7.18)].

The value of J to be used in Equation (7.19) is a matter of controversy since J is not measured directly but is a derived value from experiments done in the MeV range. The most complete discussion of the value of J is given by Berger and Seltzer (1964). These authors postulate, after weighing all available evidence, that a "best" J vs. Z curve is given by

$$J = 9.76Z + 58.8Z^{-0.19} \qquad (\text{eV}) \tag{7.22}$$

Other J values have been tabulated by Duncumb and Reed (1968). Figure 7.7 shows the variation of J with Z given by several investigators. Below $Z = 11$, the various curves are very different. The Berger–Seltzer J values as a function of Z are listed in the data base, Chapter 14.

In order to avoid the integration in Equation (7.18), Thomas (1964) proposed that an average energy $\bar{E}$ may be taken as $0.5(E_0 + E_c)$, where E_c is the value for element i and substituted for E in Equation (7.19) for S_i

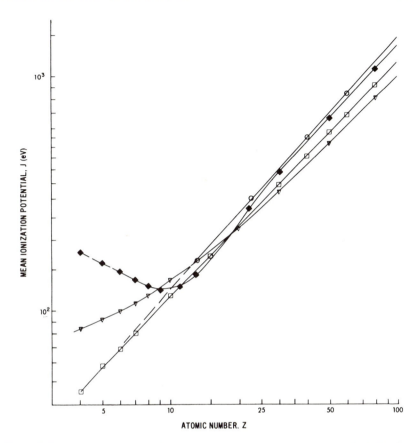

Figure 7.7. Comparison of several expressions for mean ionization potential, J. $\blacklozenge$ = Duncumb–Reed, $\bigcirc$ = Caldwell, $\bigtriangledown$ = Berger–Seltzer. $\square$ = J = 11.5 Z.

with little loss in accuracy. Hence

$$S_{ij} = (\text{const}) \frac{Z_j}{A_j(E_0 + E_c)} \ln\left[\frac{583(E_0 + E_c)}{J_j} \right] \qquad (7.23)$$

where i represents the element i which is measured and j represents each of the elements present in the standard or specimen including element i. The constant in Equation (7.23) need not be evaluated since the constant will cancel out when the stopping power for the sample and standard are compared [Equation (7.25)]. To evaluate S_i and S_i^*, there is experimental evidence that a weighted average of S_{ij} and S_{ij}^*, respectively, can be used, namely,

$$S_i = \sum_j C_j S_{ij} \qquad (7.24a)$$

for the standard and

$$S_i^* = \sum_j C_j S_{ij}^* \qquad (7.24b)$$

for the specimen. The final values of S_i and S_i^* are obtained using Equation (7.24).

If the integration in Equation (7.18) is avoided and Q is constant, we have

$$Z_i = (R_i / R_i^*)(S_i^* / S_i) \qquad (7.25)$$

The use of Equation (7.25) almost always leads to very small errors in Z_i (Heinrich and Yakowitz, 1970).

The major variable which influences the atomic number factor Z_i is the difference between the average atomic number of the specimen and the standard. To illustrate this effect one can consider as an example the analysis of copper in an aluminum-base alloy containing 2 wt% Cu. The average atomic number of the alloy is 13.32 while the atomic number of Cu is 29. A significant atomic number correction can be expected.

A sample calculation of the atomic number correction for this alloy is given in Table 7.3. The atomic number correction Z_{Cu} and Z_{Al} is obtained for two operating voltages $E_0 = 15$ and 30 kV. The values of J_{Cu} and J_{Al} are obtained from Equation (7.22). The values of S_{Cu} and S_{Al} for either Cu $K\alpha$ radiation ($E_c = 8.98$ kV) or Al $K\alpha$ radiation ($E_c = 1.56$ kV) are calculated using Equations (7.23). The constant in Equation (7.23) is set equal to 1 since it eventually drops out in later calculations. The backscattering corrections R_{Cu} and R_{Al} for either Cu $K\alpha$ or Al $K\alpha$ are obtained using Equation (7.20). The S_i, S_i^* and R_i, R_i^* terms are calculated using Equations (7.21) and (7.24). The final corrections Z_{Cu} and Z_{Al} are obtained from Equation (7.25).

The Z_{Cu} correction is decreased by ~50% from 1.16 to 1.08 at 15 kV when a standard of θ phase, $CuAl_2$ ($Z = 21.6$) is used in favor of pure Cu ($Z = 29$). The Z_{Al} correction is essentially negligible, 0.998, in this example since the sample and standard have a very similar atomic number. As discussed earlier unless the atomic number effect is corrected for, analyses of heavy elements in a light element matrix (Cu in Al) generally yield values which are too low ($Z_i > 1$) while analyses of light elements in a heavy matrix (Al in Cu) usually yield values which are too high ($Z_i < 1$).

For both Cu $K\alpha$ and Al $K\alpha$ in the Al–2 wt% Cu alloy the F_i factor is 1.0 since neither element is effectively fluoresced by the other. The absorption correction for Cu $K\alpha$ in the Al–Cu alloy A_{Cu} is 1.0 at 15 kV, $\Psi = 52.5°$ since the mass absorption coefficients are very small, $\mu/\rho_{Cu}^{Cu\ K\alpha} = 53.7$ cm^2/g, $\mu/\rho_{Al}^{Cu\ K\alpha} = 49.6$ cm^2/g. For Cu $K\alpha$ in the Al–Cu alloy, the whole ZAF correction is 1.16 using a pure Cu standard and is due entirely to the atomic number effect. On the other hand Al $K\alpha$ in the Al–Cu alloy is

Table 7.3. Atomic Number Correction for an Al–2%Cu Alloy

(a) Input data Al–2 wt%Cu alloy		
	Al	Cu
Z	13	29
A	26.98	63.55
E_c (kV)	1.56	8.98

(b) Output data Al–2 wt%Cu alloy

Operating conditions: $E_0 = 15$ kV, $\Psi = 52.5°$

For Cu $K\alpha$:
$U_{Cu} = 1.67$
$J_{Cu} = 314.05$ $\qquad$ $J_{Al} = 163.0$
$S_{Cu} = 0.0722$ $\qquad$ $S_{Al} = 0.0894$
$R_1' = 1.641$ $\qquad$ $R_2' = 0.1889$ $\qquad$ $R_3' = 0.7918$
$R_{Cu} = 0.910$ $\qquad$ $R_{Al} = 0.968$
$S_i^* = 0.0891$
$R_i^* = 0.967$

using a Cu Standard: $S_i = 0.0722$

$\left.\begin{array}{l} \\ \\ \end{array}\right\}$ $Z_{Cu} = 1.16$

$R_i = 0.910$

using a θ CuAl$_2$ intermetallic (54 wt%Cu, 46
wt%Al) standard: $S_i = 0.0802$

$\left.\begin{array}{l} \\ \\ \end{array}\right\}$ $Z_{Cu} = 1.08$

$R_i = 0.937$

For Al $K\alpha$:
$U_{Al} = 9.62$
$S_{Al} = 0.119$ $\qquad$ $S_{Cu} = 0.0944$
$R_1' = 2.073$ $\qquad$ $R_2' = 0.332$ $\qquad$ $R_3' = 0.623$
$R_{Al} = 0.910$ $\qquad$ $R_{Cu} = 0.823$
$S_i^* = 0.118$
$R_i^* = 0.909$

using an Al standard: $S_i = 0.119$

$\left.\begin{array}{l} \\ \\ \end{array}\right\}$ $Z_{Al} = 0.998$

$R_i = 0.910$

Operating conditions: $E_0 = 30$ kV, $\Psi = 52.5°$

For Cu $K\alpha$:
using a Cu standard: $S_i^* = 0.0608$ $\qquad$ $S_i = 0.0501$

$\left.\begin{array}{l} \\ \\ \end{array}\right\}$ $Z_{Cu} = 1.11$

$R_i^* = 0.939$ $\qquad$ $R_i = 0.859$

For Cu $K\alpha$:
using a θ CuAl$_2$
intermetallic standard: $S_i = 0.0551$

$\left.\begin{array}{l} \\ \\ \end{array}\right\}$ $Z_{Cu} = 1.05$

$R_i = 0.896$

For Al $K\alpha$: $\qquad\qquad\qquad\qquad\qquad\qquad\qquad\quad$ $Z_{Al} = 0.999$

highly absorbed by Cu, $\mu/\rho_{Al}^{Al\ K\alpha} = 385.7\ cm^2/g$, $\mu/\rho_{Cu}^{Al\ K\alpha} = 5377$. By using the PDH absorption correction described in Section 7.2.2 A_{Al} was calculated for the Al–Cu alloy for the operating condition $E_0 = 15$ kV, $\Psi = 52.5°$. The calculation gave $f(\chi) = 0.902$, $f(\chi)^* = 0.880$, and $A_{Al} = 1.025$. The absorption correction is surprisingly small but is understandable when one realizes that only 2 wt% of the highly absorbing element Cu is present in the sample. For Al $K\alpha$ in the Al–Cu alloy, the whole ZAF correction is 1.023, primarily consisting of the small absorption correction A_{Al} (1.025) and the very small atomic number correction Z_{Al} (0.998).

Heinrich and Yakowitz (1970) have investigated error propagation in the Z_i term. In general, the magnitude of Z_i decreases as the overvoltage $U = E_0/E_c$ increases, but very slowly (5% for a tenfold increase in U). Note Table 7.3, where R_{Cu} decreases from 1.16 at 15 kV to 1.11 at 30 kV. The uncertainty in Z_i remains remarkably constant as a function of U since the R and S uncertainties tend to counterbalance one another. Thus, no increase in the error of Z_i is to be expected at low U values and hence the choice of low operating voltages is still valid for obtaining the highest accuracy.

7.2.4. The Characteristic Fluorescence Correction, F

If the energy of a characteristic x-ray peak E from element j in a specimen is greater than E_c of element i, then parasitic fluorescence must be accounted for in the correction procedure for element i. The correction becomes negligible if $(E - E_c)$ is greater than 5 keV. Such a fluorescence correction is necessitated because the energy of the x-ray peak from element j is sufficient to excite x-rays secondarily from element i. Thus, more x-rays from element i are generated than would have been produced by electron excitation alone.

Electrons are attenuated more strongly than x-rays of comparable energy. Thus, fluorescent radiation can originate at greater distances from the point of impact of the electron beam than primary radiation. (Note Figure 3.49, Chapter 3.) Hence, the mean depth of production of fluorescent radiation is greater than that of primary radiation. Therefore, the intensity of fluorescent emission that can be measured by the x-ray detector relative to that of primary emission increases with increasing x-ray emergence angle.

Since x-ray fluorescence always adds intensity for element i, an equation of the following form can be used:

$$F_i = \left(1 + \sum_j \frac{I_{ij}^f}{I_i}\right) \bigg/ \left(1 + \sum_j \frac{I_{ij}^f}{I_i}\right)^* \tag{7.26}$$

The correction factor I_{ij}^f/I_i relates the intensity of radiation of element i

produced by fluorescence by element j, I_{ij}^f, to the electron-generated intensity of radiation from element i, I_i. The total correction factor is the summation of the fluorescence of element i by all the elements j in the sample. The most popular version of the correction factor I_{ij}^f / I_i was derived by Reed (1965). For element i fluoresced by element j in a specimen containing these or additional elements, we have

$$I_{ij}^f / I_i = C_j Y_0 Y_1 Y_2 Y_3 P_{ij} \tag{7.27}$$

C_j is the concentration of the element causing the parasitic fluorescence, i.e., the fluorescer; Y_0 is given by

$$Y_0 = 0.5 \frac{r_i - 1}{r_i} \omega_j \frac{A_i}{A_j}$$

where r_i is the absorption jump ratio for element i—for a K line $(r_i - 1)/r_i$ is 0.88 and for an L line $(r_i - 1)/r_i$ is 0.75 with very little error; ω_j is the fluorescent yield for element j (see Chapter 3); A_i is the atomic weight of the element of interest and A_j is the atomic weight of the element causing the parasitic fluorescence; $Y_1 = [(U_j - 1)/(U_i - 1)]^{1.67}$ where $U = E_0/E_c$. The term $Y_2 = (\mu/\rho_i^j)/(\mu/\rho)_{spec}^j$, where $(\mu/\rho)_i^j$ is the mass absorption coefficient of element i for radiation from element j, and $(\mu/\rho)_{spec}^j$ is the mass absorption coefficient of the specimen for radiation from element j. The values of ω, A, and μ/ρ are given in the data base (Chapter 14). The other term Y_3 accounts for absorption:

$$Y_3 = \frac{\ln(1 + u)}{u} + \frac{\ln(1 + v)}{v}$$

with

$$u = \left[(\mu/\rho)_{spec}^i / (\mu/\rho)_{spec}^j \right] \csc \Psi$$

where $(\mu/\rho)_{spec}^i$ is the mass absorption coefficient of the specimen for radiation from element i, Equation (7.16) and

$$v = \frac{3.3 \times 10^5}{\left(E_0^{1.65} - E_c^{1.65} \right)(\mu/\rho)_{spec}^j}$$

where E_c is evaluated for element i. Finally P_{ij} is a factor for the type of fluorescence occurring. If KK (a K line fluoresces a K line) or LL fluorescence occurs, $P_{ij} = 1$. If LK or KL fluorescence occurs, $P_{ij} = 4.76$ for LK and 0.24 for KL.

The Reed relation is used in most computer-based schemes for correction. This relation has been found to be accurate under real analysis conditions (Colby, 1965). Heinrich and Yakowitz (1968) tested the Reed model for its response to input parameter uncertainties. They found that ω_j produces the worst uncertainties; the other variables produce negligible errors. The extension of I_{ij}^f / I_i to fluorescence of element i by more than one element j is given by Equation (7.26). The term I^f_{ij}/I_i is calculated by Equation (7.27) for each element j which fluoresces element i. The effects of all these elements are summed as shown in Equation (7.26). If the standard is a pure element or element i is not fluoresced by other elements present in a multicomponent standard, Equation (7.26) for F_i can be written in the more standard form:

$$F_i = 1 \bigg/ \left(1 + \sum_j \frac{I^f_{ij}}{I_i}\right)^*$$ (7.28)

The fluorescence factor F_i is usually the least important factor in the ZAF correction, since secondary fluorescence may not occur or the concentration C_j in Equation (7.27) may be small.

The significance of the fluorescence correction F_i can be illustrated by considering the binary system Fe–Ni. In this system, the Ni $K\alpha$ characteristic energy, 7.478 keV, is greater than the energy for excitation of Fe K radiation, $E_K = 7.11$ keV. Therefore an additional amount of Fe $K\alpha$ radiation is produced. In this system the atomic number correction Z_i is $< 1\%$ and can be ignored. Calculations of F_{Fe} in a 10 wt% Fe–90 wt% Ni alloy were made using the expression given in Equations (7.27) and (7.28).

Table 7.4 contains the input data used for the F_i calculation. Calculations of F_{Fe} were performed for two operating voltages, 30 and 15 kV, and two take-off angles, 15.5° and 52.5°. The amount of Fe fluoresence given

Table 7.4. Fluorescence of Fe $K\alpha$ in a 10wt%Fe–90wt%Ni Alloy

(a) Input data for Fe–Ni binary

	Fe	Ni
μ/ρ Fe $K\alpha$ (cm²/g) absorber	71.4	90
μ/ρ Ni $K\alpha$ (cm²/g) absorber	379.6	58.9
ω	—	0.37
A	55.847	58.71
E_c (keV)	7.11	8.332
C (wt fraction)	0.1	0.9

(b) Output data for 10wt%Fe–90wt%Ni binary

Ψ (deg)	E_0 (kV)	$\dfrac{I^f_{Fe-Ni}}{I_{Fe}}$	F_{Fe}	A_{Fe}	A_{Ni}
52.5	15	0.263	0.792	1.002	1.005
15.5	15	0.168	0.856	1.008	1.015
52.5	30	0.346	0.743	1.011	1.023
15.5	30	0.271	0.787	1.030	1.065

by $I^f_{\text{Fe-Ni}}/I_{\text{Fe}}$, is listed in Table 7.4 and ranges from 16.8% to 34.6%. It increases with increasing take-off angle and operating voltage. To minimize the amount of the fluorescence correction, low kilovoltage operation is suggested. Low-kilovoltage operation will also minimize the absorption correction A_{Fe} and A_{Ni} (Table 7.4). Although the fluorescence correction is minimized at low Ψ angles, the error associated with the term F_i does not increase with the take-off angle (Heinrich and Yakowitz, 1968). Therefore low E_0, high Ψ operation, which is recommended to minimize A_i, is satisfactory even in cases requiring large fluorescence corrections.

If the concentration of C_j, Equation (7.27), decreases the amount of fluorescence, I^f_j/I_i also decreases. For example if the Ni content changes in a binary Fe–Ni alloy from 90% to 50%, a 50% Fe–50% Ni alloy, the amount of fluorescence at $E_0 = 15$ kV, $\Psi = 15.5°$ decreases by over a factor of 2 from 16.8% to 6.5%. Clearly the relative effect of fluorescence increases markedly as C_i decreases and C_j increases.

7.2.5. The Continuum Fluorescence Correction

Whenever electrons are used to excite x-ray spectra, a band of continuous radiation, with an energy range from 0 to E_0, always accompanies the characteristic peaks used as analytical lines. This continuum arises since the incoming electrons decelerate in the electrical field of the atoms to produce quanta of different energies as they proceed through the solid.

The continuum band contains quanta of energy sufficient to excite any characteristic radiation which can be directly excited by the beam electrons since there is always some continuum radiation in the range E_c to E_0. Calculation of the intensity of the continuum-induced radiation is complicated because of the following considerations:

(1) The effect must be integrated over an energy range from E_c to E_0. The cross section for excitation varies with the continuum energy.

(2) Knowledge of the functional dependence of the x-ray continuum on specimen atomic number and beam energy is required.

(3) The difference in the continuum fluorescence effect between sample and standard depends greatly on the relative mass absorption coefficients for the line of interest.

Henoc (1968) has derived a functional expression for the intensity of continuum-induced fluorescence, I'_c.

$$I'_c = f\left(\overline{Z}, \omega, r, (\mu/\rho), \Psi\right) \tag{7.29}$$

where $\overline{Z}$ is the average atomic number, ω is the fluorescence yield, r is the absorption jump ratio, (μ/ρ) is the mass absorption coefficient, and Ψ is the take-off angle.

Accurate computation of I'_c requires extensive program space and computer time. Because the effect can be safely ignored in many cases, most ZAF programs do not include a correction for continuum fluorescence. It is included in such "complete" correction procedures as COR (Henoc *et al.*, 1973) and that due to Springer (1972, 1973).

Myklebust *et al.* (1979) investigated the continuum correction in order to pinpoint cases where the correction can be safely ignored. They came to the following conclusions:

1. The magnitude of the continuum fluorescence correction cannot be neglected in cases where $f(\chi)$ of the material is 0.95 or greater. This corresponds to analysis using x-ray lines from an element in a light matrix, e.g., analysis of the "heavy" component in an oxide.
2. Voltage variation and errors in the take-off angle, fluorescence yield factor, and jump ratio have no significant effect on the continuum fluorescence correction.
3. Continuum fluorescence corrections can be ignored when

$$f(\chi) < 0.95$$
$$C_i > 0.5 \tag{7.30}$$
$$\bar{Z}_{\text{std}} \cong \bar{Z}_{\text{spec}}$$

4. All of the results support the choice of the operating parameters suggested for quantitative analysis: Low overvoltage and high take-off angle (Duncumb and Shields, 1966).

Perhaps the most important conclusion is (1) that uncertainties may result if one employs hard x-ray lines from heavy elements such as Zn, Hg, etc. found as traces in light matrices such as B. For trace amounts of copper in biological tissues, the continuum radiation contribution provides most of the observed intensity. Hence, when analyzing for small amounts of heavy elements in light matrices, the softest possible characteristic x-ray line should be chosen as the analytical line. If it is necessary to use a hard x-ray line, such as Cu $K\alpha$ in a biological matrix, then the standard should consist of a dilute copper constituent in a similar light matrix, such as lithium borate glass.

7.2.6. Summary Discussion of the ZAF Method

We now consider the way in which the individual corrections are utilized to actually calculate the composition of the specimen. In the usual experiment, one determines a set of k values, but the factors Z, A, and F depend upon the true composition of the specimen, which is unknown. This problem is handled by using the k values as the first estimate of composition. These first estimate concentrations are normalized to sum to 100%.

The resulting mass fractions are used to compute the initial ZAF factors for each element. Iteration proceeds until convergence of results occurs.

The iteration procedure most often used was suggested by Criss and Birks (1966) and is based on the idea that the curve relating C and k is a hyperbola (Castaing, 1951; Ziebold and Ogilvie, 1964). This hyperbolic approximation will be discussed in detail in Section 7.3. For each step in the iterative process, the next estimate for the mass fraction, C_m, corresponding to the measured intensity ratio, k_m, can be calculated from

$$C_m = \frac{k_m C(1 - k)}{k_m(C - k) + k(1 - C)} \tag{7.31}$$

since the product $[(1 - k)/k][C/(1 - C)]$ is a constant if the analytical curve is indeed a hyperbola. Thus, to apply Equation (7.31), one assumes a reasonable estimate of the value of C. The k corresponding to that C can be calculated since Z, A, and F can be obtained. Then one measures an actual k_m and solves for C_m. In practice, the k ratio measured is usually taken as the first estimate of C and a k value is calculated from it for use as C and k in Equation (7.31).

This iteration procedure has been extensively tested. It converges rapidly since the hyperbolic approximation is, in fact, almost always a very good representation of the analytical curve of C against k. Rarely are more than three iterations necessary.

Before an x-ray microanalysis is performed, it is wise to calculate the magnitude of the ZAF factors. When the corrections for a given element are significant, the analyst can attempt to minimize them by changing the operating conditions before the analysis is undertaken. The analyst can often control E_0 and sometimes Ψ and can employ standards of similar composition to the sample if necessary. The following two examples show calculations of ZAF corrections and point out how these corrections can be minimized.

The first example is the microanalysis of a seven-component M2 tool steel, composition 0.82 wt% C, 6.11 wt% W, 4.95 wt% Mo, 4.18 wt% Cr, 1.88 wt% V, 0.26 wt% Mn and 81.8 wt% Fe. $K\alpha$ lines were used for all elements except W and Mo, where the $L\alpha$ line was employed. Pure element standards were used for each element. Calculations of ZAF and the expected k ratio were made for two take-off angles, 15.5° and 52.5°, and two operating voltages, 15 and 30 kV. The absorption correction was calculated by the PDH equations, the atomic number correction by the Duncumb and Reed formulation, and the fluorescence correction by the Reed technique as discussed previously. The results are listed in Table 7.5.

The calculations show that there will be a significant atomic number correction for W and Mo since these elements have a much higher atomic number than the matrix. As expected for elements with atomic numbers

Table 7.5. ZAF Calculation for a Tool Steel as a Function of E_0 and Ψ

	Concentration (wt%)	Z_i	A_i	F_i	$[ZAF]_i$	k_i
(a) $\Psi = 52.5°$, $E_0 = 15$ kV	0.82 C	0.872	3.13	1.0	2.73	0.0030
	6.11 W	1.407	1.014	1.0	1.428	0.00428
	4.95 Mo	1.12	1.14	0.996	1.277	0.00388
	4.18 Cr	0.972	1.012	0.836	0.823	0.0051
	1.88 V	0.988	1.019	0.888	0.894	0.021
	0.26 Mn	0.995	1.009	0.998	1.002	0.0026
	81.8 Fe	0.981	1.009	0.997	0.987	0.829
(b) $\Psi = 15.5°$, $E_0 = 15$ kV	0.82 C	0.872	5.2	1.0	4.53	0.0018
	6.11 W	1.407	1.041	1.0	1.465	0.0417
	4.95 Mo	1.12	1.349	0.998	1.508	0.0328
	4.18 Cr	0.972	1.034	0.883	0.889	0.0470
	1.88 V	0.988	1.053	0.924	0.962	0.0195
	0.26 Mn	0.995	1.024	0.998	1.017	0.00256
	81.8 Fe	0.981	1.023	0.998	1.002	0.816
(c) $\Psi = 52.5°$, $E_0 = 30$ kV	0.82 C	0.907	5.32	1.0	4.83	0.0017
	6.11 W	1.258	1.074	1.0	1.351	0.0452
	4.95 Mo	1.085	1.374	0.993	1.482	0.0334
	4.18 Cr	0.983	1.043	0.786	0.807	0.0518
	1.88 V	1.001	1.065	0.843	0.899	0.0209
	0.26 Mn	1.002	1.030	0.996	1.029	0.00253
	81.8 Fe	0.986	1.032	0.994	1.012	0.808
(d) $\Psi = 15.5°$, $E_0 = 30$ kV	0.82 C	0.907	7.62	1.0	6.91	0.0012
	6.11 W	1.258	1.196	1.0	1.505	0.0406
	4.95 Mo	1.085	1.77	0.995	1.912	0.0259
	4.18 Cr	0.983	1.116	0.839	0.921	0.0454
	1.88 V	1.001	1.173	0.885	1.041	0.0181
	0.26 Mn	1.002	1.084	0.996	1.083	0.0024
	81.8 Fe	0.986	1.088	0.995	1.068	0.766

greater than the average atomic number of the sample, Z_i is greater than 1.0 and for elements with atomic numbers less than the average Z of the sample, Z_i is less than 1.0. In the high-kilovoltage, low take-off angle case, the absorption correction is greater than 8% for all elements. However, operation under low-kilovoltage, high take-off angle conditions minimizes this correction. For 5 of the 7 elements, the correction will be less than 2% relative. Both Cr $K\alpha$ and V $K\alpha$ are fluoresced by Fe K lines with the Cr $K\alpha$ more strongly excited. In general low-kilovoltage, high-Ψ conditions are desirable in that they tend to make the ZAF correction more nearly approach 1.0.

The case of carbon, C, is the most dramatic of the elements of interest. The absorption correction A of 300% to 800% is too large for accurate analysis. In this case a standard which more closely approximates that of the sample should be used. If the compound Fe_3C is used as the standard for C, instead of pure carbon, the ZAF correction is decreased. At $E_0 = 15$ kV, $\Psi = 52.5°$, the Z correction decreases from 0.872 to 0.974, the A correction from 3.13 to 1.07, and the total ZAF correction from 2.73 to 1.04. Clearly the compound carbon standard should be employed. A more complete discussion of the practical aspects of C analysis in the SEM is given in Chapter 8.

The second example is the microanalysis of a seven-component meteoritic pyroxene of composition 53.5 wt% SiO_2, 1.11 wt% Al_2O_3, 0.62 wt% Cr_2O_3, 9.5 wt% FeO, 14.1 wt% MgO, and 21.2 wt% CaO (Hewins, 1979). The $K\alpha$ lines were used for all elements and the wt% O was calculated using the cation–anion ratio of the oxides. Pure oxide standards, SiO_2, Al_2O_3, Cr_2O_3, FeO, MgO, and CaO were used for each of the metallic elements. Calculations of ZAF and the k ratio were made for a take-off angle of 52.5° and two operating voltages, 15 and 30 kV, Table 7.6.

The calculations show that there is a significant atomic number effect, $\sim 10\%$, for Cr and Fe, the two elements which have a significantly higher Z than the average atomic number of the sample. The atomic number correction is somewhat less at $E_0 = 30$ kV. The fluoresence effect is small with a maximum correction of 2% to 3% for Cr. Although FeK strongly excites Cr $K\alpha$, the Fe content is relatively small, only ~ 7.5 wt%, minimizing the fluoresence effect. The absorption correction is clearly the most important of the three. Corrections up to 20% at $E_0 = 15$ kV and up to 48% at $E_0 = 30$ kV are necessary. If the take-off angle Ψ were lower than 52.5°, the absorption correction would be much larger. Clearly operation at low E_0 is desirable since the A_i correction is minimized. Therefore, for almost all silicate analyses high take-off angle low-kilovoltage operation is recommended.

Table 7.6. ZAF Calculation for a Meteoritic Pyroxene at $\Psi = 52.5°$ and $E_0 = 15$ and 30 kV

	Concentration (wt%)	Z_i	A_i	F_i	$[ZAF]_i$	k_i
(a) $\Psi = 52.5°$, $E_0 = 15$ kV	53.3 SiO_2	0.980	1.11	0.998	1.091	0.489
	1.11 Al_2O_3	0.995	1.19	0.987	1.17	0.00947
	0.62 Cr_2O_3	1.106	1.019	0.98	1.106	0.0056
	9.5 FeO	1.14	1.01	1.0	1.15	0.0828
	14.1 MgO	0.971	1.20	0.993	1.16	0.122
	21.2 CaO	1.028	1.033	0.996	1.06	0.200
(b) $\Psi = 52.5°$, $E_0 = 30$ kV	53.3 SiO_2	0.985	1.302	0.997	1.278	0.417
	1.11 Al_2O_3	1.0	1.484	0.983	1.460	0.0076
	0.62 Cr_2O_3	1.086	1.07	0.974	1.13	0.0055
	9.5 FeO	1.11	1.033	1.0	1.14	0.0832
	14.1 MgO	0.977	1.47	0.991	1.43	0.0989
	21.2 CaO	1.018	1.11	0.994	1.124	0.189

The calculations in the two examples of ZAF use just discussed were performed using a minicomputer. Although hand calculations are admittedly slow, it may be useful for the new investigator to calculate some of the corrections himself. Such a procedure will help in understanding just what is happening during the ZAF calculation sequence. The necessary equations to calculate ZAF have been given in the previous sections. All the necessary input parameters, ω, J, R_1', R_2', R_3', μ/ρ, Z, A, etc. are available in the data base, Chapter 14.

The ZAF method is often used as a means of obtaining quantitative results from measured relative intensity data. This method is supposed to be applicable to any class of specimen. However, ZAF may not be successful in an analysis using x-ray lines less than 1 keV, mainly because of the lack of knowledge of required input parameters and approximations in the correction models themselves. For these reasons the analysis of low-energy x-ray lines is less accurate than the analysis utilizing higher-energy x-ray lines. When standards of nearly the same composition as the specimen are available and correctly prepared (see Chapter 8), an analysis using low-energy x-ray lines can be performed.

An example of the accuracy that can be expected when good specimens are available is shown in Figure 7.8. This figure is a histogram

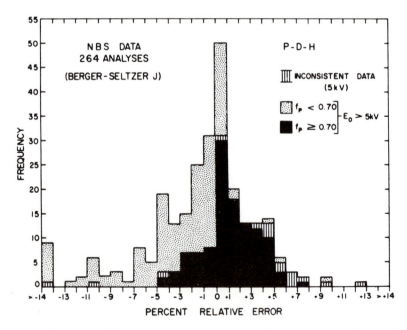

Figure 7.8. Histogram of results for binary alloy quantitative electron probe microanalysis using the Philibert–Duncumb–Heinrich (PDH) absorption expression. (Henoc *et al.*, 1973; Myklebust *et al.*, 1973.)

showing the relative error distribution in a group of 264 homogeneous, well-analyzed binary metal alloys. These alloys were examined under a variety of operating conditions in a commercial electron probe microanalyzer; crystal spectrometers were used. The relative intensities were corrected by means of a theoretical ZAF program called COR2 (Henoc et al., 1973). Some of the experimental measurements were made at higher operating voltages than the optimum or with softer x-ray lines than optimum: i.e., Cu L instead of Cu K or Au M instead of Au L. The results obtained from cases where less than optimum experimental conditions were used provide an indication of what occurs when the nature of the specimen makes it difficult for the investigator to ensure that $f(\chi) > 0.7$. The absorption correction factor is the only term in Equation (7.5) that the operator can control easily; fluorescence effects are inherent in the specimen and the atomic number correction is virtually insensitive to operating conditions (Heinrich and Yakowitz, 1970). The only way to minimize the atomic number effect is to provide a standard whose average atomic number is nearly the same as that of the specimen. Providing such a standard is not always easy.

One way to sum up the implications of the results in Figure 7.8 is in gambling terms; the odds are 2 to 1 against an analysis being within 1% relative (1% of amount present), even money on the analysis being within $2\frac{1}{2}$% relative, and 7 to 1 for the analysis being within 5% relative. These odds are accuracy estimates based on nearly optimum specimens, i.e., properly prepared and placed normal to the electron beam in a commercial electron probe microanalyzer.

7.3. The Empirical Method

In the early 1960s, the ZAF method was not nearly as well developed as it is presently. Hence, histograms, such as those of Figure 7.8, showed a much wider distribution of errors. Furthermore, computer data reduction was much less generally available. In response to this state of affairs, Ziebold and Ogilvie (1964) developed what is known as the hyperbolic or empirical correction method, first described in 1964. In fact, Castaing (1951) laid the basis for this development in his thesis in what he referred to as "the second approximation." This second approximation was a statement that the true weight fraction, C, and the measured relative intensity, k, were related such that a plot of C against k would be a hyperbola.

Ziebold and Ogilvie (1964) expressed this relationship in the form

$$\left(\frac{C}{1-C} \right)\left(\frac{1-k}{k} \right) = a \tag{7.32}$$

Clearly, if a specimen which is homogeneous at the micrometer level and for which C is known, can be procured, k can be measured and a so determined. For example, the National Bureau of Standards certifies SRM-480 to be homogeneous and to consist of 0.215 Mo and 0.785 W. The k values determined using an operating voltage of 20 kV were reported as 0.143 for Mo and 0.772 for W. Solving Equation (7.32) for the respective a values, a is 1.649 for Mo in Mo–W and a is 1.078 for W in Mo–W. These values are valid only for a 20-kV operating voltage. The solution for C in the case of a binary is

$$C = \frac{ka}{1 + k(a - 1)} \tag{7.33}$$

Therefore, any composition of W–Mo can be analyzed by carefully determining k with an operating voltage of 20 kV and solving Equation (7.33) with a desk calculator.

There are two major drawbacks, however. First, it is often difficult to obtain the desired standards, and second, the extension to more components than two is not immediately obvious. The extension to more than a binary and the accuracy of the hyperbolic approximation have been considered in detail by Bence and Albee (1968). These authors were concerned with providing a rapid, accurate means to analyze specimens for geological studies. Mineralogical and petrological specimens may be heterogeneous and frequently contain 6 to 8 elements with concentrations in excess of 1% by weight. For reasons of simplicity and economy, many such specimens are not analyzed by means of the ZAF method. Real-time computer reduction of data is desirable since knowledge of the composition and the calculated formula of the phase is often a necessary prerequisite to the operator's deciding what the next stage in the analytical procedure will be. Bence and Albee noted that a plot of C/k against C must be a straight line of slope $1 - a$ for small C in any binary system. Parenthetically, if a series of alloys in a system is analyzed, a plot of k/C vs. k must be a straight line or the data are internally inconsistent (Bence and Albee, 1968). Figure 7.9 illustrates such curves. A plot of k/C vs. k for Ag $L\alpha$ in four silver–gold alloys and at six operating voltages is shown. Pure silver is the standard for the measurement of the k ratio. The data at each operating voltage except at 5 kV plot as straight lines. Furthermore, Ziebold and Ogilvie (1964) had shown that, for a ternary containing elements 1, 2, 3 with compositions C_1, C_2, C_3 and intensity ratios k_1, k_2, and k_3,

$$\left(\frac{C_1}{1 - C_1} \right)\left(\frac{1 - k_1}{k_1} \right) = a_{123} \tag{7.34}$$

and

$$a_{123} = \frac{a_{12}C_2 + a_{13}C_3}{C_2 + C_3} \tag{7.35}$$

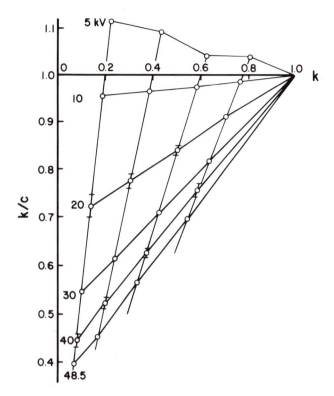

Figure 7.9. Plot of k/C vs. k for silver in alloys of silver–gold. The Ag $L\alpha$ line was used for analysis; primary beam voltages were 5, 10, 20, 30, 40, and 48.5 kV, respectively, as noted on the figure. The data taken at 5 kV are internally inconsistent as indicated by the deviation from linearity.

In other words, the empirical coefficient for the ternary can be determined from the individual binaries (12, 13, and 23). Therefore, standards for each binary are needed. Extending this formulation, Bence and Albee (1968) showed that, in a system of n components, for the nth component

$$C_n = k_n \beta_n$$

where

$$\beta_n = \frac{k_1 a_{n1} + k_2 a_{n2} + k_3 a_{n3} + \cdots + k_n a_{nn}}{k_1 + k_2 + k_3 + \cdots + k_n} \qquad (7.36)$$

Here a_{n1} is the a value for the $n-1$ binary, i.e., for the determination of element n in a binary consisting of element n and of element 1. Thus a_{n2} is the a value for a binary consisting of element n and of element 2, and so on to a_{nn}, which is for n in n and therefore is unity. Similarly, the notation a_{1n} would represent the a value for element 1 in a binary of 1 and n.

The value of k_1 is the relative intensity ratio found for element 1 in the specimen to the standard used for element 1. This notation can be carried along so that k_n is the relative intensity ratio for element n to the standard used for element n.

To solve completely a system of n components, one needs $n(n-1)$ values for a (the a value for each of the n pure elements is unity as noted above). Thus, $n(n-1)$ standards are required. Bence and Albee (1968) procured a number of suitable standards of interest in geology and determined their respective a values at various operating voltages and at one take-off angle. Bence and Albee (1968) used end member oxides rather than pure elements. As an example a_{NiFe} represents the a value for the NiO–FeO binary and is used for the determination of element Ni in a binary consisting of NiO and FeO.

Given the necessary matrix of a values, the analyst chooses his analytical standards and determines $k_1, k_2, \ldots, k_n$. Then $\beta_1, \beta_2, \ldots, \beta_n$ can be calculated. From these, a first approximation to $C_1', C_2', \ldots, C_n'$ is found. Then a new set of β values using the new concentrations $C_1', C_2', \ldots$ are calculated as $\beta_1', \beta_2', \ldots, \beta_n'$:

$$\beta_n' = \frac{C_1' a_{n1} + C_2' a_{n2} + \cdots + C_n' a_{nn}}{C_1' + C_2' + \cdots + C_n'} \tag{7.37}$$

Iteration continues until differences between calculated values of β are made arbitrarily small. In practice, the result really depends on how good the standard is; in geological work, standards of nearly the same average atomic number as the specimen are usually chosen. Hence, atomic number effects are reduced. Fluorescence effects are small in silicate rocks and similar materials. Hence, most of the final correction is for absorption.

Tests of the validity of this method indicate that it yields results comparable to those obtainable with the ZAF method. The chief obstacle to its general use is the need for many homogeneous, well-characterized standard materials. For this reason, the a method is sometimes combined with the ZAF method. In this case, one computes the necessary a matrix by assuming a C value and using the ZAF method to compute the corresponding k. Then, of course, a can be obtained from Equation (7.32). Albee and Ray (1970) have calculated correction factors for 36 elements relative to their simple oxides for $E_0 = 15$ and 20 kV and for several take-off angles using the ZAF technique. A more complex calculation of a coefficients has been developed by Laguitton *et al.* (1975), but its application has been limited. Nevertheless, a calculated a value is subject to all of the uncertainties associated with the ZAF method as outlined previously. The use of such corrections, however, permits x-ray microanalysis using a small suite of simple crystalline oxides as standards and the calculation procedure is much simpler than the ZAF correction.

As an example of the use of the empirical method one can consider the

Table 7.7. Empirical a Factors for a Meteoritic Pyroxene Analysis
($\Psi = 52.5°$, $E_0 = 15$ kV)[a]

Element	Oxide					
	MgO	Al$_2$O$_3$	SiO$_2$	CaO	Cr$_2$O$_3$	FeO
Mg	1.00	1.03	1.09	1.26	1.50	1.76
Al	1.62	1.00	1.01[b]	1.14	1.28	1.37[b]
Si	1.29[b]	1.34[b]	1.00	1.05	1.12	1.19[b]
Ca	1.08	1.06	1.08	1.00	0.91	0.92
Cr	1.13	1.11	1.12	1.14	1.00	0.88
Fe	1.15	1.13	1.15	1.14	1.08	1.00

[a] Albee and Ray (1970).
[b] Modified by T. Bence, personal communication (1978).

microanalysis of the seven-component meteoritic pyroxene discussed previously (Table 7.6) for the ZAF analysis. The $K\alpha$ lines for Si, Al, Cr, Fe, Mg, and Ca are used and the analysis was calculated for $\Psi = 52.5°$, $E_0 = 15$ kV. The appropriate a factors $[n(n-1) = 30]$ for these operating conditions taken from Albee and Ray (1970), are listed in Table 7.7. The β_n factors and intensity ratios k calculated from Equation (7.36) for the meteoritic pyroxene are given in Table 7.8. In addition the calculated ZAF and k values given in Table 7.6a are listed in Table 7.8. The β_n and ZAF factors differ by at most 2% relative. Since the Albee and Ray (1970) values were calculated using the ZAF technique, the close correspondence between β_n and ZAF should not really be surprising.

7.4. Quantitative Analysis with Nonnormal Electron Beam Incidence

In most electron probe microanalyzers, the beam is set normal to a flat specimen. In most SEMs the specimen is tilted in order to provide a

Table 7.8. Calculation Using the Empirical Technique for a Meteoritic Pyroxene at $\Psi = 52.5°$, $E_0 = 15$ kV

Concentration (wt%)	β_n	k_i	ZAF$_i$	k_i
53.3 SiO$_2$	1.074	0.496	1.091	0.489
1.11 Al$_2$O$_3$	1.16	0.00957	1.17	0.00947
0.62 Cr$_2$O$_3$	1.102	0.00563	1.106	0.0056
9.5 FeO	1.133	0.0838	1.15	0.0828
14.1 MgO	1.179	0.120	1.16	0.122
21.2 CaO	1.047	0.203	1.06	0.200

suitable value of the take-off angle. The effect of specimen tilt on the accuracy of quantitative analysis was investigated by Monte Carlo calculations of electron–solid interactions which clearly indicated changes in x-ray distribution and emission as a function of specimen tilt (Curgenven and Duncumb, 1971; Duncumb, 1971). An example of the effect of tilt is shown in Figure 7.10. Experience with tilted specimens indicates that if specimen and standard are handled identically and tilts are not extreme ($\theta < 60°$), large quantitative analysis errors do not occur (Bolon and Lifshin, 1973a; Colby *et al.* 1969).

The chief difficulty in handling tilted specimens is that the ZAF correction model is predicated on normal beam incidence. The effect of a tilted specimen on the parameters of the ZAF method has not been extensively investigated. Reed (1971) has determined that backscattering factors, R, change considerably when the sample is tilted 45° with respect to the beam. If the sample and standard are measured at the same tilt angle, the effect of tilt on R is similar in both. For this configuration Reed (1975) indicates that the change in the backscattering correction with tilt is negligible for $Z < 30$. Even with large differences in atomic number between sample and standard, the maximum effect is only 7%.

The absorption correction, on the other hand, is more likely to be affected for the tilted sample. As shown in Figure 7.10, the x-ray generation

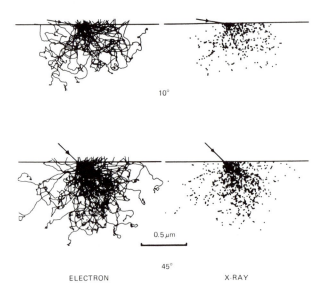

Figure 7.10. Schematic representation of 100 electron paths in copper; primary beam voltage was 20 kV. The effect of tilting the specimen on the paths and the corresponding x-ray generation are apparent. (Compare with Figure 7.2.)

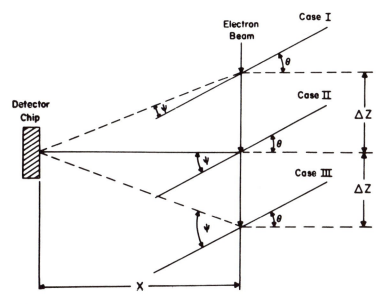

Figure 7.11. Geometry for deducing x-ray emergence angle Ψ using a tilted specimen and fixed energy-dispersive detector chip.

volume is closer to the surface in a highly tilted sample. Since absorption is exponentially dependent upon path length, the tilt effect can lead to a sharp reduction in absorption. There is no consensus in the literature on a method to compensate for tilt effects on the absorption correction. The simplest correction involves modifying χ by a $\sin(90° - \theta)$ term, where θ is the tilt angle (Figure 7.11):

$$\chi(\theta) = (\mu/\rho)_i \csc \Psi \sin(90° - \theta) \tag{7.38}$$

Other workers (Bishop, 1965; Reed, 1975) suggest a more elaborate form:

$$\chi(\theta) = (\mu/\rho)_i \csc \Psi \left[1 - 0.5 \cos^2(90° - \theta) \right] \tag{7.39}$$

More theoretical and experimental work is needed to determine which of these expressions is to be preferred.

The most important factor in analyzing tilted specimens is the proper measurement of the take-off angle. In many SEMs, the lithium-drifted silicon detector chip is placed relative to the specimen as shown in Figure 7.11. For the cases illustrated in Figure 7.11, the specimen is tilted at an angle θ as indicated on the SEM stage control; the center of the detector chip is at a distance, X, from the impact point of the electron beam on the specimen. The distance of the impact point above (case I) or below (case

III) the detector chip center line is ΔZ. Then

$$\text{case I:} \qquad \Psi = \theta - \arctan(\Delta Z / X) \qquad (7.40a)$$

$$\text{case II:} \qquad \Psi = \theta \qquad (7.40b)$$

$$\text{case III:} \qquad \Psi = \theta + \arctan(\Delta Z / X) \qquad (7.40c)$$

It is usually advantageous to try to adjust the height of the specimen such that case II is obtained. However, if case II cannot be obtained, careful measurements of X and ΔZ can be used in conjunction with Equations (7.40a) and (7.40c) to obtain the take-off angle for use in the ZAF correction. The closer the detector chip (smaller X), the greater the uncertainty of Ψ. Furthermore, since the solid angle accepted by the detector increases as the distance X decreases, using the position of the center of the detector as a basis for computing Ψ becomes only an approximation. In the case where the specimen does not face directly toward the spectrometer, a further correction must be applied for the azimuthal angle (Moll *et al.*, 1980).

7.5. Analysis of Particles and Rough Surfaces

7.5.1. Geometric Effects

Quantitative x-ray microanalysis of bulk specimens is restricted to samples which are flat and placed at known angles to the electron beam and the x-ray spectrometer. Under these conditions, the x-ray intensities measured on the unknown specimen differ from the x-ray intensities measured on the standards only because of the difference in composition between the unknown and standard. By employing the methods described previously, e.g., the ZAF or the *a*-coefficient empirical technique, the composition of the unknown can be determined relative to the known composition of the standard.

There exists a broad class of irregularly shaped specimens which do not meet the geometrical requirement of the ideal specimen. This class includes microscopic particles and rough macroscopic samples such as fracture surfaces. The x-ray intensities measured from such irregularly shaped specimens differ from those of the flat standards both because of compositional differences and because of "geometrical effects" described below. Conventional data reduction schemes which have no compensation for geometrical effects lead to inaccurate analyses of such specimens. While in some cases irregularly shaped specimens can be mounted, sectioned, and polished to yield a flat specimen for analysis, in many cases such destructive preparation may eliminate the very information of interest. It thus becomes necessary to develop methods to compensate for "geometrical effects."

The "geometrical effects" can be divided into two major types, the mass effect and the absorption effect, and one minor type, the fluorescence effect (Small *et al.*, 1979). The origin of each of these effects will be considered separately.

7.5.1.1. Mass Effect

When the dimensions of a particle approach those of the interaction volume in a bulk solid, electrons can escape from the sides and bottom of the particle, as illustrated in Figure 7.12, thus reducing the generated x-ray intensity compared to a bulk target. The measured intensity response, normalized to a bulk specimen, as a function of particle diameter for spherical particles is illustrated in Figure 7.13 (curve for Fe $K\alpha$). The mass effect always leads to a lower intensity measured in the particle and becomes significant for particle diameters of 5 μm or less at a beam energy of 20 keV or more.

For rough bulk samples, a mass effect is also observed, which results from the dependence of the size of the interaction volume on the angle of tilt. As shown in Chapter 3, the interaction volume lies closer to the surface in a tilted specimen and the backscatter coefficient increases, resulting in a reduction of generated x-ray intensity compared to a target set normal to

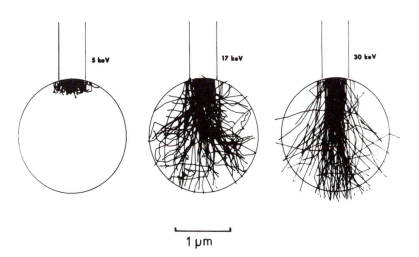

1 µm

Figure 7.12. Monte Carlo electron trajectory calculations of the interaction volume as a function of beam energy in spherical particles of aluminum (from Newbury *et al.*, 1980).

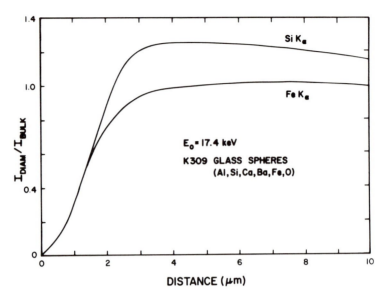

Figure 7.13. Normalized intensity ($I_{\text{diam}}/I_{\text{bulk}}$) as a function of diameter for spherical particles of NBS K-309 glass (15% Al_2O_3; 40% SiO_2; 15% CaO; 15% Fe_2O_3; 15% BaO). Monte Carlo electron trajectory calculations.

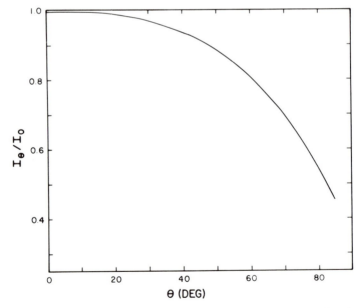

Figure 7.14. Normalized intensity (I_θ/I_0) for a flat, bulk target of iron as a function of tilt (θ). Monte Carlo electron trajectory calculations. $E_0 = 20$ keV.

the beam. This behavior produces the intensity response shown in Figure 7.14.

7.5.1.2. Absorption Effect

Because of the size and shape of a particle, the x-ray absorption path will be different from that of a bulk target. This phenomenon is illustrated schematically in Figure 7.15, where the absorption paths in a spherical particle and a bulk target are compared. Since absorption is exponentially dependent on path length [Equation (3.43), Chapter 3], the particle geometry can have a strong influence on the emitted x-ray intensity. Two cases are especially worth noting:

(1) **Beam Placement.** For particles which are slightly larger than the interaction volume, the location of the beam impact point relative to the x-ray detector can profoundly affect the measured spectrum, as shown in Figure 7.16. When the beam is placed on the side of the particle facing the detector (Figure 7.16, position 1), the absorption path is a minimum and the complete x-ray spectrum, including low-energy x-rays below 3 keV (except for those absorbed in the window), is detected efficiently. If the

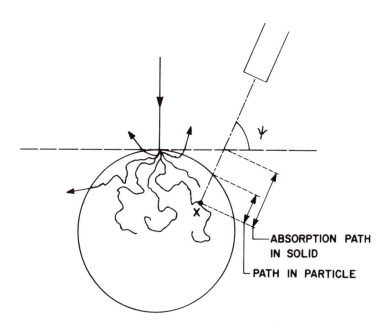

Figure 7.15. Schematic diagram comparing the absorption path in a particle with equivalent path in a bulk target.

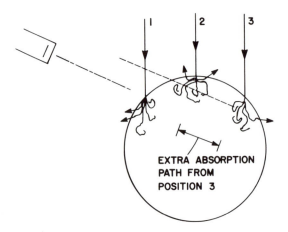

Figure 7.16. Schematic diagram showing the dramatic differences in the absorption path when the beam is located on the front face (toward the x-ray detector), on the top, and on the back face of a particle.

beam is placed symmetrically at the top of the particle and the detector is at a high take-off angle (Figure 7.16, position 2), a situation similar to a bulk target is obtained, and the spectrum is still reasonable. However, if the beam is placed on the side of the particle away from the detector, the extra absorption path through the particle will severely attenuate the low-energy x-rays (Figure 7.16, position 3). Examples of spectra which illustrate these points are shown in Figure 7.17. Note that the ratio Si/Ca changes drastically in Figures 7.17a and 7.17b when the spectra are measured on the front and back surfaces of a homogeneous particle, because of enhanced absorption of the low-energy Si $K\alpha$ line. In analysis of unknowns, the relative heights of characteristic lines are not useful for diagnosing a high-absorption situation which might be encountered accidentally. The high-absorption situation can be easily recognized in energy-dispersive x-ray spectra if the sharp roll-off in measured intensity occurs above the usual 2-keV ($E_0 = 20$ keV) position, as illustrated in the spectra shown in Figures 7.17c, 7.17d, and 7.17e. To minimize problems in beam location during small-particle analysis, the particle is often bracketed by a rapid scanning raster so that all locations are excited.

The effect of the extra path length on attenuating the spectrum is exacerbated by the usual choice of x-ray detector geometry in the SEM. The detector axis is usually placed at a right angle to the beam axis (Figure 7.11). For a flat specimen set normal to the beam, the take-off angle is zero and therefore, the flat specimen must be tilted to obtain an adequate

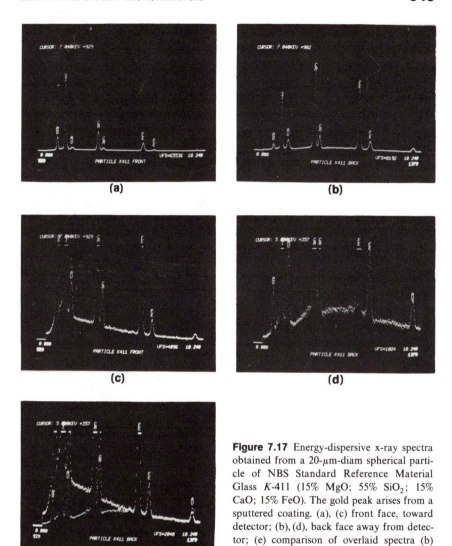

Figure 7.17 Energy-dispersive x-ray spectra obtained from a 20-μm-diam spherical particle of NBS Standard Reference Material Glass K-411 (15% MgO; 55% SiO$_2$; 15% CaO; 15% FeO). The gold peak arises from a sputtered coating. (a), (c) front face, toward detector; (b), (d), back face away from detector; (e) comparison of overlaid spectra (b) and (d), normalized to equal intensities at Fe $K\alpha$.

take-off angle. If, in the analysis of spherical particles, the beam is placed in the center of the particle, the situation illustrated in Figure 7.18a occurs. Although an adequate take-off angle is obtained on the substrate, placing the beam on the center of the particle still results in an effective take-off angle of 0°. For small particles which are completely excited by the beam (Figure 7.18c), an adequate x-ray intensity is measured. For larger particles

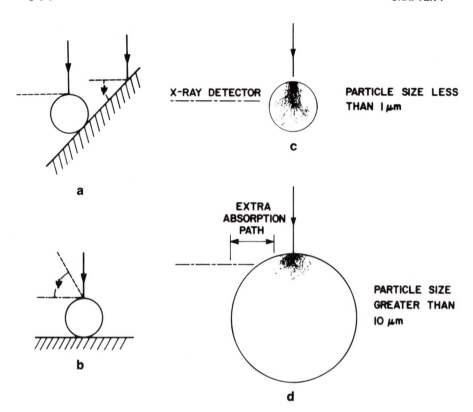

Figure 7.18. Schematic diagram illustrating the effect of particle size and detector position or x-ray path length. (a) Low take-off angle detector, beam set symmetrically on particle; (b) high take-off angle detector, beam located symmetrically on a particle; (c) low take-off angle detector, particle fully excited; (d) low take-off angle detector, interaction volume smaller than particle.

in the size range above 5 μm (Figure 7.18d), the interaction volume is in the center of the particle and the escaping x-rays must pass through the side of the particle to reach the detector. The normalized intensity ($k = I_{\text{diam}}/I_{\text{bulk}}$) curve as a function of particle size which is observed in this case ($\Psi = 0$), shows a peak followed by a rapid decrease with increasing diameter, Figure 7.19. This situation can only be avoided if the beam is placed on the side of the particle facing the detector or if a bracketing raster is employed. For a high take-off angle detector, illustrated in Figure 7.18b, the sharp decrease with increasing diameter is not observed (Figure 7.19, $\Psi = 52.5°$).

(2) **Particle Size.** The second absorption effect is the influence of absorption on the measured k value as a function of particle diameter.

There exists a range of particle sizes for which the generation volume is nearly identical to that in the bulk target. Because of the surface curvature of the particle, however, the absorption path length is substantially reduced compared to the bulk, as illustrated in Figure 7.15. If the absorption of the x-rays of interest is high in the bulk, as is the case for a low-energy x-ray, then the reduction in absorption in the particle may result in more x-ray intensity measured from the particle than from the bulk sample of the same composition, i.e., $k > 1$. This situation is illustrated in Figure 7.13, where the reduced absorption for Si $K\alpha$ x-rays ($E = 1.49$ keV) in the particles leads to $k > 1$ for the size range 2–6 μm. Eventually, as the particle size is decreased further, the mass effect dominates and k decreases below unity.

For rough, bulk samples the absorption effect can be even more dramatic. For a surface oriented randomly relative to the x-ray detector, the x-ray absorption path can vary greatly from the path in a standard set normal to the beam. This situation is illustrated schematically in Figure 7.20. Again, low-energy x-rays will be much more affected by this extra path length than high-energy x-rays.

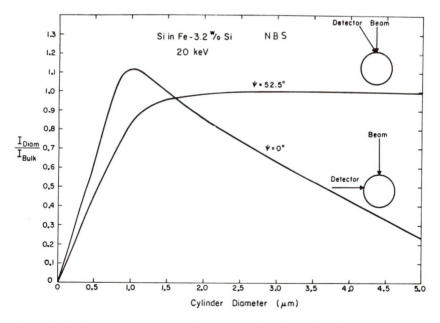

Figure 7.19. Normalized x-ray intensity as a function of particle diameter for particles measured with a high take-off angle (detector-axis–beam-axis angle 47.5°) and a low take-off angle (detector-axis–beam-axis angle 90°). For the low take-off angle detector, note the rapid decrease in intensity for larger particle sizes due to abnormal absorption.

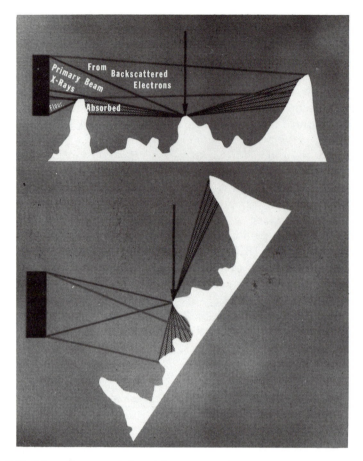

Figure 7.20. Schematic representation of the processes which may occur when rough surfaces are examined.

7.5.1.3. Fluorescence Effect

The phenomenon of fluorescence of characteristic x-rays, induced by the higher-energy characteristic and/or continuum x-rays which are produced directly by the beam electrons, takes place over a large volume compared to the electron interaction volume. This fluorescence volume, which may have a radius of 10–100 μm to contain 99% of the fluorescence effect, results from the low value of the x-ray mass absorption coefficients as compared to the high value of scattering which leads to electron attenuation. A bulk target by definition will be large enough to contain the volume of fluorescence excitation while a particle, depending on its size, may lose a significant fraction of the fluorescence-produced x-rays. The measured k value, compared to a bulk standard, will therefore be lower than expected.

As discussed in Section 7.2.4 of this chapter, characteristic fluorescence is only significant when a specimen contains elements which produce x-ray lines which lie near the absorption edges of other elements. Fluorescence by the continuum is most significant for heavy elements in a light matrix, where the fluorescent x-rays may constitute 20% of the observed intensities. The matrix of an environmental particulate is often composed of light elements, e.g., C, O, Al, Si. The loss of fluorescence by the continuum in a particle may be a significant source of error if minor heavy element constituents are of interest. Rough, bulk targets more closely approximate the flat bulk standard with regard to having dimensions to contain the fluorescence volume. The loss of fluorescence is therefore not as significant in such samples.

7.5.2. Compensating for Geometric Effects

Several techniques are available for dealing with the geometric effects on x-ray intensities encountered in the analysis of particles and rough surfaces. These include (a) ignoring the geometric effects; (b) normalization; (c) use of particle standards; (d) analytic solutions for special particle shapes; and (e) peak-to-background method. These techniques are discussed in the following sections.

7.5.2.1. Ignoring the Effects

Some analysts choose to directly report the results of the analyses of particles or rough samples with conventional quantitative matrix corrections for bulk samples such as the ZAF or a-factor methods. While this may sound facetious in view of what has been indicated in the previous sections on geometrical effects, the uncorrected analyses give a direct indication of the magnitude of the deviation from a flat bulk situation through the discrepancy of the total analysis from 100%. As will be demonstrated in subsequent sections, significant errors can be introduced by casual attempts to correct for geometrical effects. It therefore may be better to report the "raw" analysis since subsequent processing may mask the magnitude of the correction and instill a false sense of confidence.

7.5.2.2. Normalization

Recognizing that geometrical effects usually lead to a total analysis which differs from unity, a simple normalization of the results is often used. Thus, if the total mass fraction as calculated by the ZAF or a-factor method, including oxygen by stoichiometry, is f, the results can be normalized to 1 if all values are multiplied by a factor $1/f$.

Simple normalization is most effective in the analysis of particles in the size range below about 3 μm. For such particles the mass effect dominates and affects all elements similarly. For example, in Figure 7.13 the initial portion of the I_{diam}/I_{bulk} vs. particle diameter curve is identical for both the silicon and iron $K\alpha$ radiation. Normalization of the results can be successful in this situation. An example of an analysis using normalization in which the errors are small is given in Table 7.9 (example—pyrite). However, large errors may be encountered in normalization, particularly when both low-energy and high-energy x-rays are measured.

Normalization produces inadequate, and in fact quite misleading, results when the absorption effect is operating, as in the case of large particles and rough, bulk samples, since low-energy lines are strongly affected while high-energy lines are weakly affected. Examples of the large relative errors which result from simple normalization in such cases are given in Table 7.9 for olivine and Table 7.11 for Au–Cu alloys. Since the absorption effect is almost certainly present in the analysis of rough surfaces, normalization should not be used in this case.

Gavrilovic (1980) uses an interesting variation of the simple normalization procedure in the analysis of micrometer-sized particles. He noted that the mass effect is proportional to the area fraction of a scanning raster which a particle of interest occupies. A normalization factor is applied to the measured intensity I_m which has the form

$$I_0 = I_m S_{max}/S_{part} \qquad (7.41)$$

where I_0 is the normalized intensity, S_{max} is the area of the scanning raster, and S_{part} is the projected area of the particle in the scanning raster.

Table 7.9. Errors Observed in Analysis of Particles by Conventional Methods with Normalization and by an Analytic Method[a]

Constituent	Actual concentration	Conventional analysis, normalized	Percentage relative error	Analytic particle analysis	Percentage relative error
Pyrite (average of 140 particles)					
S	53.45	54.8	+ 2.5%	53.5	+ 0.1%
Fe	46.55	45.2	− 2.9%	46.5	− 0.1%
Olivine (average of 104 particles)					
MgO	43.94	50.7	+ 15%	43.8	− 0.3%
SiO$_2$	38.72	39.3	+ 1.5%	38.9	+ 0.5%
FeO	17.11	9.7	− 43%	17.1	0%

[a] From Armstrong (1978).

7.5.2.3. Particle Standards

A completely different approach to the measurement of a k value against a flat, bulk standard is to employ instead a standard in the form of a particle of known composition and known size. Ideally, the standard particle should have a simple shape, such as a sphere or cylinder (fiber), although randomly shaped particles produced by grinding a bulk standard can be employed. The a-coefficient method can then be employed with the k values determined against particle standards. A number of suitable multielement glass standards in bulk, fiber, and spherical form are available (National Bureau of Standards Catalog, 1979). Alternatively, homogeneous mineral crystals can be analyzed first in the bulk and then ground to produce fine particles.

7.5.2.4. Analytic Solutions

Armstrong and Buseck (1975) have demonstrated that analytic solutions can be derived to predict the k-value vs. diameter curves which can then be used for quantitative analysis of unknowns. The details of the calculations are too complex to discuss here and the reader is referred to their paper. An example of a series of calculated response curves for a specific composition and for different particle shapes is shown in Figure 7.21. In Figure 7.21, the R factor on the vertical axis is the ratio of k_{Ca}/k_{Si}, where the k values are measured for each element relative to a bulk standard. In employing this method for the analysis of unknowns, the analyst must make a good estimate of the size and shape of the unknown in order to use the proper response curve. Table 7.9 illustrates the use of the analytic solutions for minerals of known shape and composition. The relative errors are much smaller than those obtained by the normalization method.

7.5.2.5. Peak-to-Background Method

The peak-to-background method (Small *et al.,* 1978, 1979; Statham and Pawley, 1978; Statham, 1979) is based on the observation that al-

Figure 7.21. Normalized x-ray intensity R as a function of particle size D calculated by the analytic method of Armstrong and Buseck (1975). The two curves mark the extreme limits of the shape dependency. The shapes considered included spheres, hemispheres, rectangular, tetragonal, cylindrical, and right triangular prisms, and the squared pyramid (from Armstrong, 1978).

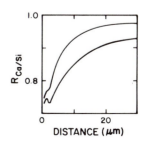

though the characteristic x-rays and bremsstrahlung x-rays of the same energy are produced by completely different processes (inner-electron-shell ionization versus Coulombic interaction), the two types of radiation are produced in nearly the same volume. Moreover, both types of radiation will suffer similar absorption effects upon exiting the specimen. Consequently, at a given energy, the mass effect and the absorption effect will be similar for both the characteristic and the bremsstrahlung. The bremsstrahlung intensity I_B can therefore be used as an internal normalization for the major geometrical effects. Thus, although $k = I_{part}/I_{bulk}$ is a strong function of particle size, $(I_{part}/I_{B\ part})/(I_{bulk}/I_{B\ bulk})$ is nearly independent of particle size, except for very small particles (Newbury et al., 1980).

This experimental observation, which has been confirmed by theoretical calculations, can be employed in several ways (Statham, 1979). One useful technique is to incorporate the following correction scheme into a conventional ZAF method (Small et al., 1978; 1979). Given that

$$\frac{I_{part}}{I_{B\ part}} = \frac{I_{bulk}}{I_{B\ bulk}} \tag{7.42}$$

a modified particle intensity, P^*_{part}, can be calculated which would be equivalent to a measured intensity from the particle if the particle could be scaled to a polished bulk specimen:

$$P^*_{part} = I_{bulk} = I_{part} \times \frac{I_{B\ bulk}}{I_{B\ part}} \tag{7.43}$$

In Equation (7.43), I_{part} and $I_{B\ part}$ are directly measured from the particle spectrum. $I_{B\ bulk}$ will in general not be measured since a standard identical in composition to the unknown is generally not available. $I_{B\ bulk}$ is obtained with the following equation:

$$I_{B\ bulk} = \sum_i C_i I_{i(B\ bulk)} \tag{7.44}$$

where $I_{i(B\ bulk)}$ is the pure element bremsstrahlung at the energy of interest and C_i is the weight fraction of the element. The values of C_i in Equation (7.44) are obtained from the estimate in the interactive loop of the ZAF calculation.

Examples of analyses with the peak-to-background method (P/B, ZAF) are included in Table 7.10. Because this method can deal with the absorption effect, the technique offers considerable promise for the analysis of rough, bulk samples (Table 7.11) which represents the most difficult case. In Table 7.11, simple normalization in a fracture surface analysis produces errors, whereas application of the P/B method reduces the relative errors to a more acceptable level.

Table 7.10. Errors Observed in the Analysis of Particles by Peak-to-Background Method[a]

Constituent	Actual concentration	Conventional ZAF	Percentage error	P/B ZAF	Percentage error
Pyrite					
S	53.4	39.9	− 25%	52.9	− 1.0
Fe	46.6	35.8	− 23%	46.4	− 0.5
Talc					
Mg	19.3	10.3	− 47%	18.5	− 4%
Si	29.8	15.8	− 47%	29.0	− 3%

[a] From Small *et al.* (1979); single analyses on known standards.

Table 7.11. Errors Observed in the Analysis of Rough Objects (Fracture Surfaces of Homogeneous Au–Cu Alloys) by Peak-to-Background Modified ZAF Method[a]

Constituent	Actual concentration	Conventional ZAF	Relative error	Normalized ZAF	Relative error	P/B ZAF	Relative error
Analysis 1							
Au	60.3	28.2	− 53%	49.6	− 18%	58.0	− 4%
Cu	39.6	28.7	− 28	50.4	+ 27	44.0	+ 11
Analysis 2							
Au	60.3	1.08	− 98	29.1	− 52	52.0	− 14
Cu	39.6	2.63	− 93	70.8	+ 79	41.0	+ 3
Analysis 3							
Au	80.1	95.8	+ 20	73.8	− 8	76.9	− 4
Cu	19.8	34.0	+ 72	26.2	+ 32	19.1	− 3

[a] Small *et al.* (1979).

7.5.3. Summary

The analysis of particles and rough surfaces is an area of continuing development. The peak-to-background method holds considerable promise, but additional work needs to be done, particularly a determination of the error histogram from the analysis of rough samples with known composition. Even if a "perfect" analysis method is devised, pitfalls remain in the analysis of rough and particle samples over which the analyst has little control. Returning to Figure 7.20, it is obvious that situations can arise in which electron scattering from the region of interest can excite a nearby region of different composition so that x-rays from two dissimilar compositions contribute to the spectrum. Such a spectrum cannot be deconvoluted, since the relative contributions are unknown. Similarly, in the analysis of inhomogeneous particles, which are frequently observed in practice, the electron interaction volume may intrude from the region of interest into surrounding regions of different composition, again producing a composite spectrum which is intractable to analysis.

When particles are extremely small, less than about 0.2 μm in thickness, a different strategy of analysis is appropriate. Such thin particles are best analyzed by the techniques developed for thin foil analysis, which will be discussed in the next section.

7.6. Analysis of Thin Films and Foils

A sample is considered a "thin" film if its thickness dimension is less than the depth dimension of the electron interaction volume in a bulk solid of the same composition. The lateral dimensions of the film are effectively infinite compared to the lateral spread of the beam. By convention, a thin layer upon a thick electron opaque substrate is defined as a "film" while a free-standing layer is designated a "foil." If a film or foil is analyzed at conventional beam energies (15 to 30 keV) using bulk standards and the composition is calculated using ZAF or a coefficients, the total analysis will be less than 100%. Normalization of the results to 100% can lead to significant relative errors and is therefore not advisable. The use of a lower beam energy, e.g., 10 keV, can be effective for foils or films of intermediate thickness, since the beam penetration may be reduced sufficiently to retain the interaction volume in the region of interest. For very thin samples, conventional bulk sample methods must necessarily fail, and appropriate analytical techniques are needed. The methods of analysis differ greatly depending on whether the sample is a foil or a film.

7.6.1. Thin Foils

Thin foils represent the simplest analytical problem. To a certain degree, thin foils are easier to analyze by x-ray microanalysis than flat, bulk samples. When the sample is very thin elastic scattering and energy loss are reduced to the point that atomic number effects are eliminated, or are secondary influences at best. Since both the elastic and inelastic cross sections decrease as the beam energy increases, thin foils are best analyzed in the analytical electron microscope (AEM), which is typically a combination transmission electron microscope–scanning transmission electron microscope operating at 100 keV or more and equipped with an energy-dispersive x-ray spectrometer. If an AEM is not available, an SEM or EPMA operating in the range 40–60 keV can be used, although atomic number effects may become more important, depending on the composition of the foil and its thickness. Regardless of the beam energy, and depending only on foil thickness, both absorption and fluorescence also become insignificant for thin foils. Thus, all of the matrix effects, atomic number, absorption, and fluorescence, which must be corrected in the analysis of bulk samples, can be neglected in thin foils. As a result, thin foil analysis can be performed by means of a simple relative sensitivity factor technique (Cliff and Lorimer, 1975; Goldstein, 1979).

A relative elemental sensitivity factor, k_{AB}, ("Cliff–Lorimer factor") for element A relative to element B is defined as

$$k_{AB} = \left(\frac{C_A}{C_B} \right) \left(\frac{I_B}{I_A} \right) \qquad (7.45)$$

where C_A is the weight fraction of element A and I_A is the background corrected x-ray intensity. B is a reference element, which is usually taken to be silicon, although any other element can be used. Values of k_{AB} for all elements of interest are measured on thin foil standards of known composition. This suite of k_{AB} values is then employed to convert the intensity ratios, I_A/I_B, measured on unknown thin foil to an elemental ratio, C_A/C_B, by rearrangement of Equation (7.45):

$$\frac{C_A}{C_B} = k_{AB} \left(\frac{I_A}{I_B} \right) \qquad (7.46)$$

Moreover, if C_B is known,

$$C_A = k_{AB} \left(\frac{I_A}{I_B} \right) C_B \qquad (7.47)$$

If this procedure is carried out for all elements in the specimen, then assuming the total concentration $C = 1$,

$$C_B + \sum_i \left[k_{iB} \left(\frac{I_i}{I_B} \right) C_B \right] = 1 \qquad (7.48)$$

Since all k_{iB} values are known and I_i/I_B ratios are measured, Equation (7.48) has only one unknown, C_B, and this value can therefore be determined. From the value of C_B and Equation (7.47), the values of C_i can be determined for all elements in the specimen. If oxygen is present, stoichiometric calculations can be included in the total for Equation (7.48). However, if there are other unmeasured elements present, such as carbon, the forced normalization of Equation (7.48) will obscure the presence of the unmeasured element.

The relative sensitivity factor, k_{AB}, necessarily contains specimen-specific factors, such as ionization cross sections, stopping powers, and sample absorptions for the elements A and B, as well as instrument-specific factors, such as x-ray window absorption and detector efficiency. Although these k_{AB} factors can be calculated, uncertainties in spectrometer window parameters lead to unacceptable errors in k_{AB} values for x-ray energies below 2 keV (Goldstein, 1979), and thus k_{AB} values should be determined on the specific instrument used for analysis.

Sample absorption eventually sets a limit on the maximum thickness of sample to which the analytic procedure in Equations (7.45)–(7.48) can be applied. Goldstein *et al.* (1977) have determined that the thickness limit beyond which absorption corrections must be applied is given by

$$(\chi_B - \chi_A)\rho t / 2 < 0.1 \tag{7.49}$$

where $\chi_i = (\mu/\rho)_i \csc \Psi$, with $(\mu/\rho)_i$ the mass absorption coefficient for element i in the sample and Ψ the take-off angle. Goldstein (1979) has discussed the form of an appropriate absorption correction for thin foils. In cases where fluorescence in the bulk is high, e.g., Fe–Cr alloys, a fluorescence correction is also needed in foils where thickness exceeds the absorption criterion in Equation (7.49).

7.6.2. Thin Films on Substrates

For a thin film on a substrate, the situation which exists for x-ray generation is illustrated in Figure 7.22. Electrons may generate x-rays in the film (1) directly or (2) after backscattering from the substrate. Electrons may backscatter from the film, transmit through the film, and backscatter from the substrate. Electrons may cross the film/substrate interface repeatedly before losing all their energy. The frequency of each of these processes will depend on the thickness and composition of the film as well as the composition of the substrate. Monte Carlo electron trajectory simulations are ideal for the study of x-ray emission from films on substrates since all

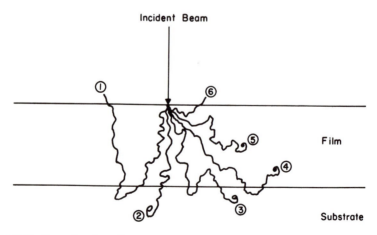

Figure 7.22. Examples of possible electron paths in a thin film on a substrate. Paths 1 and 6 lead to backscattering, while paths 2–5 remain in the film or substrate.

of the cases illustrated in Figure 7.22 can be easily simulated. Several authors have successfully applied the Monte Carlo technique to the thin film case (Kyser and Murata, 1974; Bolon and Lifshin, 1973b).

Other authors (Sweeney et al., 1960; Reuter, 1972; Colby, 1968; and Oda and Nakajima, 1973) have derived mathematical functions for the x-ray intensity ratio of a thin film in a substrate to bulk standard. These expressions have the form

$$k_A = \frac{I_{film}}{I_{bulk}} = \frac{C_A R^* \int_{E_L}^{E_0} (Q/S^*) \, dE + \eta_{ss} \int_{E_{L'}}^{E} (Q/S^*) \, dE}{R_A \int_{E_L}^{E_0} (Q/S_A) \, dE} \quad (7.50)$$

where R^* and S^* are the backscatter factor and stopping power for the foil, η_{ss} is the backscatter from the substrate, E_L is the mean electron energy at the film–substrate interface, and $E_{L'}$ is the mean energy of electrons backscattered from the substrate. In addition, an appropriate absorption correction must be applied to Equation (7.50) (Oda and Nakajima, 1973).

Each of the above methods has been tested by the proponents and found to give useful results in certain analytical situations. These methods are complex and are difficult to adequately describe in a textbook so that the reader can make a straightforward calculation. The interested reader should consult the Monto Carlo and analytical papers in the references listed above for details.

Yakowitz and Newbury (1976) have developed an empirical approach based on fitting of the x-ray depth production curve, $\phi(\rho z)$. This method allows for rapid calculation of thin film thickness and composition on a hand calculator or small computer and is easier to apply than the earlier methods. This method is discussed in the following section and is suggested for analysis of thin films on substrates.

The Yakowitz–Newbury method is based on a simple model of the $\phi(\rho z)$ curve, as shown in Figure 7.23. The $\phi(\rho z)$ curve is modeled as a parabola extending from ϕ_0 (the intercept for $\rho z = 0$) through a peak with coordinates (h, k) to a ρz value of $1.5h$. Thus for the region $0 \leqslant \rho z \leqslant 1.5h$:

$$\phi(\rho z)_{z \leqslant 1.5h} = h^{-2}(\rho z - h)^2(\phi_0 - k) + k \quad (7.51)$$

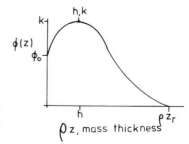

Figure 7.23. Schematic diagram of approximation to $\phi(\rho z)$ curve consisting of a parabola from $(0, \phi_0)$ to $1.5h$ and an exponential decrease from $1.5h$ to ρz_r.

An exponential decrease is used to model the region for $\rho z > 1.5h$:

$$\phi(\rho z)_{z \geqslant 1.5h} = 0.25(\phi_0 + 3k)\exp\left[\frac{3h \sec \theta}{\rho z_r - 1.5h}\right]\exp\left[-\frac{2\rho z \sec \theta}{\rho z_r - 1.5h}\right] \quad (7.52)$$

where θ is the specimen tilt.

The parameters ϕ_0, k, h, and ρz_r are determined as follows (Reuter, 1972):

$$\phi_0 = 1 + 2.8\eta(1 - 0.9W) \quad (7.53)$$

where η is the backscattering coefficient and $W = E_c/E_0$. For tilted samples,

$$\eta(\theta) = 1/(1 + \cos \theta)^P$$
$$P = 9/Z^{1/2} \quad (7.54)$$

The value of the x-ray production range is given by Heinrich's expression (Heinrich, 1980):

$$\rho z_r = 0.007(E_0^{1.65} - E_C^{1.65}) \quad \text{mg/cm}^2 \quad (7.55)$$

The value of h is given by

$$h = \rho z_r(0.49 - 1.6\eta + 2.4\eta^2 - 1.3\eta^3) \quad (7.56)$$

The value of k is given by

$$k = \phi_0(1 + 0.35 \cos \theta \ln U) \quad (7.57)$$

where $U = E_0/E_c$.

The correspondence of the empirical $\phi(\rho z)$ curve obtained from the above equations with an experimental curve is shown in Figure 7.24.

The total generated x-ray intensity for the bulk target is found by integrating Equations (7.51) and (7.52):

$$I_G = \int_0^{\rho z_r} \phi(\rho z)\, d(\rho z) = 0.125(\phi_0 + 3k)(M + 3h) \quad (7.58)$$

where $M = 0.865(\rho z_r - 1.5h)$.

To modify the expressions of Equations (7.51) and (7.52) for the thin film case, a modified backscatter coefficient is needed:

$$\eta_F = \eta_{BF} + \eta_{ss}\eta_{TF}^2 \quad (7.59)$$

η_{BF}, the film backscattering coefficient, is given by

$$\eta_{BF} = 4\eta_z(\rho z/R_{KO}) \qquad \rho z \leqslant 0.25 R_{KO} \quad (7.60)$$

where η_z is the pure element bulk backscattering coefficient and R_{KO} is the Kanaya–Okayama (1972) range:

$$R_{KO} = 2.76 \times 10^{-2}E_0^{1.67}A/Z^{0.889}\rho \quad (\mu m) \quad (7.61)$$

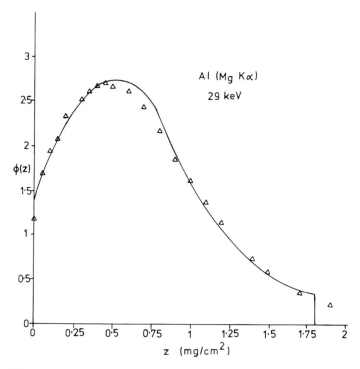

Figure 7.24. Comparison of experimental $\phi(\rho z)$ curve for aluminum (magnesium tracer) (Castaing and Henoc, 1966) and the empirical curve predicted by Equations (7.51) and (7.52) (solid line).

The transmission coefficient through the thin film, η_{TF}, is given by

$$\eta_{TF} = \exp\left\{ -(0.03 + 2 \times 10^{-4}E_0)Z(\rho z / R_{KO})\left[1 + 4(\rho z / R_{KO})\right]\right\} \quad (7.62)$$

Upon substituting these expressions into the equations for the bulk $\phi(\rho z)$, a $\phi(\rho z)$ curve of the form shown in Figure 7.25, as compared to the bulk, is calculated. The intensity generated in a thin film is given by

$$I_G^F = I_{G1}^F + I_{G2}^F$$

$$I_{G1}^F = \int_0^{\rho z \,\leqslant\, 1.5h^F} \phi^F(\rho z)\,d\rho z$$

$$= \left[\rho z\left(\phi_0^F - k^F\right)/h^{F^2}\right]\left[\rho z^2/3 - h^F\rho z + h^{F^2}\right] + k^F\rho z \quad (7.63)$$

$$I_{G2}^F = \int_{1.5h}^{\rho z_r} \phi^F(\rho z)\,d\rho z = 0.125\left(\phi_0^F + 3k^F\right)\left(3h^F + M\right) \quad (7.64)$$

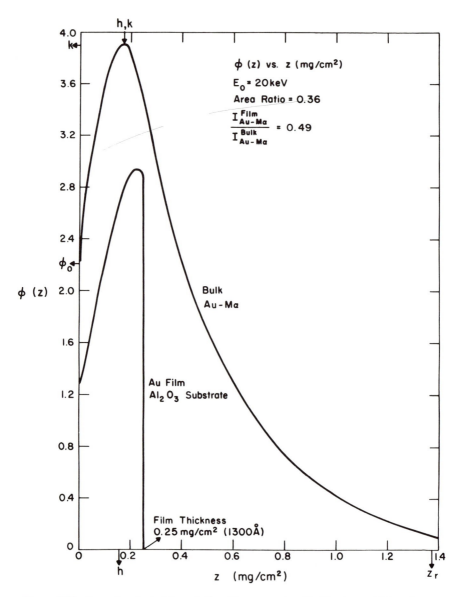

Figure 7.25. Examples of $\phi(\rho z)$ for a bulk gold target and a thin film on an Al_2O_3 substrate predicted by Equations (7.51) and (7.52) with appropriate parameters.

where

$$M = (\rho z_r - 1.5h^F)\{1 - \exp[-(2\rho z - 3h^F)/(\rho z_r - 1.5h^F)]\} \quad (7.65)$$

The emitted k ratio of the pure thin film to a pure bulk target is given by

$$k_F = \frac{I_G^F}{I_G} \frac{f^F(\chi)}{f(\chi)} \quad (7.66)$$

where

$$f(\chi) = 1/(1 + 1.2 \times 10^{-6}\chi\gamma)^2 \quad (7.67)$$

with $\chi = (\mu/\rho)\csc\Psi$, (μ/ρ) is the mass absorption coefficient, Ψ is the take-off angle, and $\gamma = E_0^{1.65} - E_c^{1.65}$ (Heinrich and Yakowitz, 1975). For a thin film $f^F(\chi)$ is approximated as

$$f^F(\chi) = 1 + \{(\rho z/\rho z_r)[f(\chi) - 1]\} \quad (7.68)$$

Equation (7.66) can be used to generate k^F vs. thickness curves for pure elemental films compared to bulk targets. Such calculated k^F vs. ρz curves are shown in Figure 7.26 compared to experimental data.

To analyze a film of unknown multielement composition and thickness, curves such as those in Figure 7.26 are first calculated. This procedure will be illustrated in Figure 7.27 for binary manganese–bismuth films. For thin films, interelement effects will again be presumed to be negligible (atomic number, absorption, and fluorescence). From the experimentally measured k_i^F value for each element i, the mass thickness for that component is determined from the plot of k^F vs. ρz (e.g., Figure 7.27). The total film thickness is then given by

$$\rho z_{\text{tot}} = \sum_i (\rho z_i) \quad (7.69)$$

The concentration of each element i in the unknown film is the fraction

$$C_i = I_i^F / I_i^{F,\rho z} \quad (7.70)$$

where I_i^F is the intensity actually produced for element i and $I_i^{F,\rho z}$ is the intensity which would be produced if the unknown were pure element i. Then

$$C_i = (I_i^F / I_i^{\text{bulk}})/(I_i^{F,\rho z} / I_i^{\text{bulk}})$$
$$= k_i^F / k_i^{F,\rho z} \quad (7.71)$$

Thus, in the graphical sequence shown in Figure 7.27, the measured k^F values for each element are used to obtain the unknown film thickness, ρz_{tot}. The ρz_{tot} value is used to determine $k_i^{F,\rho z}$ values for each element, and the concentrations are determined with Equation (7.71). The concentrations

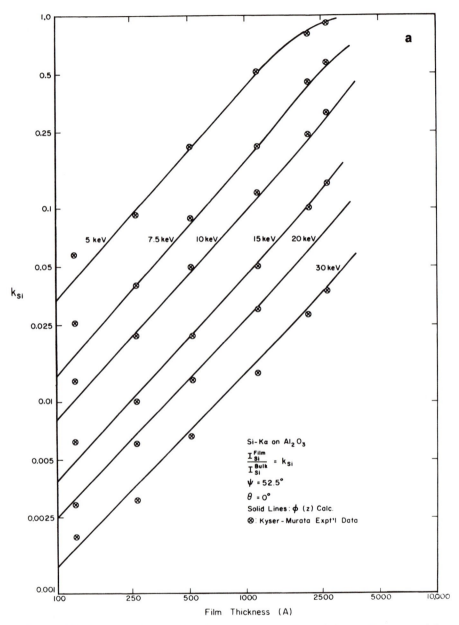

Figure 7.26. Examples of k_F vs. film thickness for (a) Si $K\alpha$ and (b) Au $M\alpha$ on an Al_2O_3 substrate calculated with the Yakowitz–Newbury empirical method compared with data of Kyser and Murata (1974).

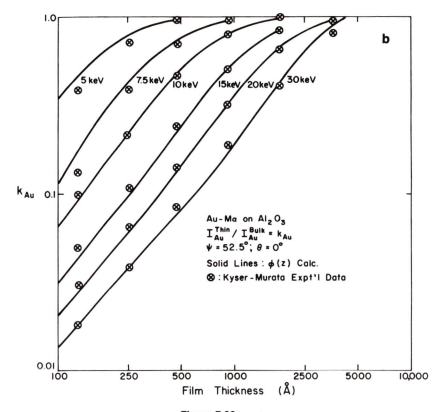

Figure 7.26. *cont.*

determined with this method are listed in Table 7.12 for the Mn–Bi and Co–Pt examples along with independent measurements for comparison.

The Yakowitz–Newbury method is reasonably accurate for films with mass thicknesses up to about 30% of the bulk interaction volume. For thicker films, errors arise because of the development of interelement effects, particularly absorption. To correct for these effects, a second procedure would be applied, in an iterative fashion, to make an absorption correction based on all the mass absorption coefficients through the use of Equations (7.68) and (7.69). Careful testing in this thickness regime has not yet been carried out for multielement films.

The Yakowitz–Newbury method provides a direct empirical approach which can be employed in practical analysis of thin films. If the analyst is confronted with a great deal of thin film analysis on diverse systems with a need for the best accuracy available, a Monte Carlo electron trajectory simulation will provide the best method for analysis, particularly of thicker films.

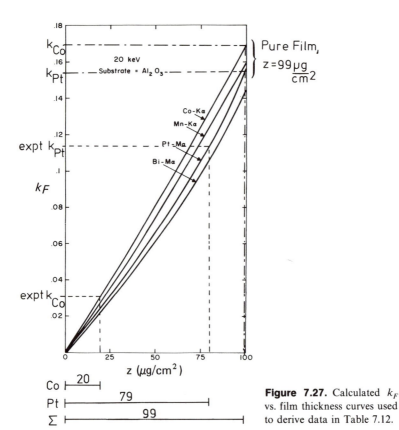

Figure 7.27. Calculated k_F vs. film thickness curves used to derive data in Table 7.12.

Table 7.12. Analysis of Binary Films: Co–Pt on SiO_2, Mn–Bi on SiO_2 Kyser and Murata Experimental Values Compared with NBS Calculations[a]

k_{Co}	k_{Pt}	z_{KM}	z_{NBS}	C_{Co}^{KM}	C_{Co}^{NBS}
0.0177	0.0581	53	51	0.185	0.199[b]
0.0305	0.114	89	99	0.163	0.179
0.0148	0.0675	56	60	0.140	0.154
0.0067	0.0465	40	41	0.101	0.103

k_{Mn}	k_{Bi}	z_{KM}	z_{NBS}	C_{Mn}^{KM}	C_{Mn}^{NBS}
—	0.0766	54	60	0	0
0.0226	0.0515	53	59	0.255	0.263
0.0293	0.0220	38	41	0.510	0.518
0.0329	0.00668	29	30	0.800	0.796
0.061	—	42	41	1.0	1.0

[a] z values in $\mu g/cm^2$, $\theta = 0°$, $\psi = 52.5°$.
[b] Nuclear backscattering gave $z = 53$ $\mu g/cm^2$, $C_{CO} = 0.194$.

7.7. Quantitative Analysis of Biological Material

7.7.1. Introduction

Many of the results of biological x-ray microanalysis may be handled in a semiquantitative fashion without resorting to any numerical schemes, and useful information can be obtained simply by comparing different x-ray spectra. Such spectra may be in the form of $X-Y$ plots or photographs of the TV display of energy dispersive spectra which are a central feature of most modern x-ray analytical systems. Alternatively it is possible to run line scans across the specimen or construct simple x-ray maps. However, these semiquantitative methods do no more than indicate that a particular element might be present in the sample, and at best that certain regions of a sample contain more of a particular element than other regions.

It is now well established that x-ray microanalysis can reliably measure elemental concentrations in microvolumes of cells, tissues, and tissue slices although the term "quantitation" is used less rigorously in biology than in the material sciences. Under ideal conditions it is possible to detect between 10^{-18} and 10^{-19} g of an element in an ultrathin section with a spatial resolution of approximately 25 nm (Shuman and Somlyo, 1976). To arrive at this point of quantitative analysis requires the use of mathematical formulas and an acute awareness of all the things that can go wrong in specimen preparation and analysis and subsequent manipulation of the data.

Although x-ray microanalysis may be accurate and precise, the characteristics of biological material frequently restrict the accuracy of the method to be between $\pm 10\%$ relative of the true value. This uncertainty occurs because biological material is rarely an ideal sample, having variable surface geometry and roughness; is frequently prepared by questionable techniques; and may represent a rich source of contamination in an otherwise clean environment. Another problem peculiar to quantitative biological microanalysis is that the majority of the elements in the specimen, i.e., carbon, oxygen, nitrogen, and hydrogen are difficult to measure accurately. Unlike analyses in the material sciences, most quantitative biological microanalysis is required to measure the concentration of elements $(Z > 10)$ which are thinly distributed in a poorly defined organic matrix. It should also be remembered that the x-ray spectrometers only detect emitted x-rays and this quantity may not necessarily be entirely representative of the x-rays generated within the sample. This problem is exacerbated by the fact that electrons penetrate more deeply into biological material and as a consequence x-ray absorption is increased. The subsequent attempts to correct for absorption are hampered by an incompletely characterized organic matrix and relatively inaccurate values for the mass-absorption coefficients of low-atomic-number elements. Indeed, much of the discussion in this section centers on the corrections one may make to narrow the gap between the numerical value of the x-rays generated by the specimen and the x-rays detected and measured. Hall (1979b) succinctly defines what the microprobe measures in biological systems and considers that the term "concentration" is better expressed as the local mass fraction. The local mass fraction is defined as the mass of an element in the analyzed microvolume divided by the total mass of the specimen in the same microvolume. It may be convenient to express this local mass fraction either in a weight/weight form, i.e., milligrams of element per kilogram local dry mass or as millimoles of element per kilogram of specimen. This latter expression is of more use to physiologists although it must, however, be realized that the results refer to the specimen at the time of analysis. Thus local mass fraction expressed as mM/kg is a measure of the total amount of the element in the analyzed microvolume, making no distinction between the bound and unbound form. Whereas the microprobe can give measurements in terms of mM/kg, it is frequently necessary to combine microprobe measurements with measurements made directly (ion-selective microelectrodes) or indirectly (extraction by micropipette, followed by atomic absorption analysis) on the aqueous phase in order to express the measurements in terms of mM/liter. Although accurate measurement of unbound elements is relatively easy in aqueous extracellular spaces where the dry mass fraction is zero, it is much more difficult to make such measurements in intracellular compartments.

In discussing methods of quantitative analysis, it is convenient to consider separately different types of specimens. Bulk specimens are those which are much thicker than the range of incident electrons; thick sections on bulk substrates are those whose thickness is only marginally less than the range of incident electrons; and thin specimens on very thin supports are those whose thickness is much less than the range of incident electrons. To these three catgories one may add two further types of specimen which require brief consideration: thick specimens on very thin supports and microdroplets. It should be remembered, however, that the majority of biological x-ray microanalysis is carried out on sectioned material on thin supports.

7.7.2. Mass Loss and Artifacts during Analysis

Before entering into a discussion of the different methods of quantitation, it is appropriate to first consider some of the major errors and artifacts which may occur during the process of acquiring the x-ray spectrum.

Problems associated with overlapping spectra, of spurious x-ray signals from the background environment of the specimen support and instrument, and the absorption of the x-rays by the specimen itself have already been considered. We will consider in this section ways in which these extraneous signals can be accounted for and provide simple correction procedures to enable a reasonably accurate estimate to be made of the x-ray signal arising from the specimen.

It is now well established that the high-energy beams of electrons used for electron microscopy and microanalysis can at the same time have a disruptive effect on the specimen. This beam damage is generally more severe in organic and biological specimens (Isaacson, 1979) and it is important to be aware of these beam-induced changes which include gross disruption of the sample, loss of organic material and removal of volatile elements. Although analysis can now be carried out at low beam currents (0.1–5 nA), substantial loss of material may still occur. The amount of material lost from a specimen obviously varies with the sample and the beam current, but it is typically about 30% (Hall, 1979a), and losses as high as 90% have been reported (Bahr *et al.*, 1965). This loss of organic mass is a serious problem, particularly where quantitative measurements are being carried out using the continuum method (see Section 7.7.6) which depends on an accurate measure of local mass during each analysis. Organic mass loss in all forms of electron beam instrumentation can be reduced by cooling the specimen. Shuman *et al.* (1976), and more recently Hall (1979a) and Isaacson (1979), have shown that mass loss is considerably reduced by holding the specimen at low temperatures. This is another of the advantages of using frozen hydrated specimens, although recent studies have suggested that even liquid nitrogen temperatures may not be sufficiently cold to entirely eliminate mass losses.

It is probably impossible to predict in advance whether a given system is likely to suffer mass loss during analysis and it is better to calculate whether mass loss has occurred on the basis of the stability of the continuum radiation count before, during, and after an experimental run. Indeed the stability of the continuum radiation count is a sensitive indicator of mass loss (decreased continuum count) as well as another artifact, contamination (increased continuum count).

Another artifact, and one closely associated with mass loss, is the loss of specific elements from samples during microanalysis. There is very little experimental evidence from biological specimens, but there is good evidence from model systems which shows that losses of chlorine, sodium, potassium, phosphorus, and sulfur do occur; and there is every reason to suspect that the phenomenon also occurs in biological specimens. Morgan *et al.* (1978) have recently reviewed the loss of analyzable elements from specimens, and the data they present suggest that severe problems exist.

The only glimmer of hope is that the elemental losses, like mass loss, are significantly reduced although not entirely removed at low temperatures. In addition a thin coating of a conductive film can reduce the thermal input into the specimen, as well as trapping mobile fragments of organic material which would otherwise vaporize in the microanalyzer (Isaacson, 1979). The coating layers must be applied judiciously because the deposited conductive layer can absorb emitted x-rays, attenuate the primary beam, and in many instances give rise to x-ray lines which interfere with the signal of interest.

7.7.3. Bulk Samples

If bulk samples are to be used for analysis it is necessary to apply a number of corrections to the classical equation of Castaing for quantitative analysis:

$$C_i = I_i / I_{(i)} \tag{7.72}$$

where C_i is the local mass fraction of element i in the analyzed micro-volume of the specimen, $I_{(i)}$ is the measured intensity of the characteristic radiation from the standard consisting solely of element i. The intensities I_i and $I_{(i)}$ are measured under identical instrumental parameters. There are three general methods which may be applied to bulk biological systems; the ZAF correction, the method proposed by Cobet (1972) and Cobet and Millner (1973), and the peak-to-background ratio method of Statham and Pawley (1978) and Small et al. (1978).

7.7.3.1. ZAF Method

This has been described in Section 7.1 of this chapter and will not be repeated here but is given in the equation

$$C_i = (\text{ZAF})_i I_i / I_{(i)} \tag{7.73}$$

where Z is a correction factor related to differences in the mean atomic number between specimen and standard; A takes account of internal absorption of the x-rays generated in the specimen; F is a correction for x-rays generated in the specimen by other x-rays. Continuum-induced fluorescence must also be considered for a biological matrix and is given the term F_c. A recent paper by Barbi (1979) discusses the feasibility of applying the ZAF correction procedures to biological samples.

7.7.3.2. The Cobet Method

An alternative form of analysis for bulk biological material is given by Cobet (1972). He derived the equation

$$C_i = k_{iB} \frac{I_i}{I_B} \tag{7.74}$$

where I_B is the continuum produced from the specimen close to the energy of I_i and k_{iB} is a constant derived from a standard.

The Cobet method minimizes the need to use the correction factors Z, A, and F and continuum-induced fluorescence, because the closeness of the energies for I_i and I_B makes absorption corrections (A) unnecessary, and the use of the ratio of I_i/I_B removes the need for atomic number correction (Z). The correction for characteristic fluorescence (F) is negligible in most biological microanalysis except in the few cases where a high-Z element is present in a low-Z matrix (continuum-induced fluorescence). The constant k_{iB} is determined by means of standards of known concentration.

The term I_B, the continuum or white radiation, is a measure of the electron interaction with the specimen and is proportional to the density and thickness of the specimen (mass thickness). The continuum intensity may also be represented by the following equation derived from Kramers' relationship:

$$I_B \propto \sum (N_i Z_i) \qquad (7.75)$$

where N_i is the number of atoms of element i in the analyzed microvolume and Z_i is the atomic number. In order to use this expression accurately, one must know all of the elements present in the biological material. In the context of the Cobet method, the term I_i/I_B is, to a first approximation, equivalent to the peak minus background ($I_i - I_B$) divided by the background (I_B), $[(I_i - I_B)/I_B]$, at the energy of the selected characteristic line of element i.

The term k_{iB} which can be calculated for a pure standard may be obtained from the equation

$$C_{(i)} = k_{iB} I_i / \text{ZAF} \, F_c \qquad (7.76)$$

where $C_{(i)}$ is the known local mass fraction in the standard (in a pure standard this would be 1.0). Performing the correction procedures Z and A on a pure standard allows k_{iB} to be calculated (F may be ignored for a pure element, because with the exception of a small contribution from the continuum radiation, fluorescence effects are negligible).

7.7.3.3. Peak-to-Background Ratios

The new approach to the quantitative analysis of particles of Statham and Pawley (1978) and Small et al. (1978) discussed in Section 7.5.2.5. can also be applied to bulk materials. The method is based on the use of the ratio between the characteristic and continuum x-ray intensities at the same energies. The peak-to-background ratio (P/B) is defined as $(I_i - I_B)/I_B$. Although the method was devised primarily for the analysis of small (1.0–10.0 μm) particles dispersed on a thin support film, Small et al. (1979) have shown that the method is effective in determining the elemental composition of bulk samples. The Statham and Pawley and the Small *et al.* methods are similar in many respects and are based on the hypothesis that the characteristic and continuum x-rays produced in the same region of the sample are subject to the same absorption and backscatter corrections, and that the peak-to-background ratios from particles are, to a first approximation, equivalent to the peak-to-background ratios from a bulk sample of the same composition. They differ in the fact that the Statham–Pawley method expresses the results as concentration ratios, whereas the method of Small *et al.* expresses results as elemental concentrations. Full details of the procedures are given in these papers, but the essential features of the procedure are as follows.

The most critical step in the procedure is the accurate measurement of the P/B ratio from the x-ray spectrum. With well-separated peaks this presents little problem and the background can be measured directly on the low-energy side of each peak. The difficulty comes when the peaks overlap. The methods described in Chapter 8 must then be applied. Once accurate P/B ratios are obtained, they can be used for quantitation in a number of different ways.

The P/B value for one element can be compared with the P/B value for another element in the same sample and to a first order:

$$\frac{C_i}{C_j} = k_{ij} \frac{(P/B)_i}{(P/B)_j} \tag{7.77}$$

where C_i and C_j are the percent concentrations of elements i and j and k_{ij} is a correction factor which can either be calculated for the two elements concerned or obtained from measurements on standards of known composition. Alternatively, the P/B value for an element in the sample can be compared with the P/B value for the same element in a standard provided there is no significant difference in the matrix composition between sample and standard. This procedure can be applied to the analysis of frozen-hydrated bulk material where the specimen may typically contain 75% water and the standard consists of a frozen salt solution of known concentration and containing 25% carbonaceous material such as a polymeric cryoprotectant. Statham and Pawley (1978) suggest another method, which could be used if the mean atomic number of particles is always the same, a common feature of most biological material. From Kramers' relationship (Chapter 3) the background radiation is proportional to atomic number. If the mean atomic number of the material is always the same, then

$$C_i = k_i(P/B)_i \tag{7.78}$$

where k_i is a constant for each element. If it is possible to analyze all the elements and calculate the concentration of elements such as C, H, O, and N by stoichiometry, then relative concentrations can be readily converted to absolute concentrations.

The advantage of the P/B ratio method for biological material is that the various correction factors used in the ZAF method are less important. Since it is assumed that the percentage of characteristic x-radiation absorbed by the sample is the same as the percentage of background which is absorbed, the absorption factor A cancels out. In biological material, the atomic number effect (Z) is low and in any event cancels out because it is assumed to affect the peak and background in the same way. Because biological material has a low atomic number, the secondary fluorescence effect (F) is low and can be treated as a second-order correction. Both Statham and Pawley, and Small *et al.* have shown that P/B measurements are insensitive to detector efficiency, beam current fluctuations, and live time correction inaccuracies. In addition P/B measurements are less sensitive to the variations in surface geometry frequently associated with bulk samples, particularly in an analytical system with a high take-off angle. With relatively flat specimens, i.e., smooth fracture faces, it has been estimated that a precision of $\pm 10\%$ can be obtained.

7.7.3.4. Summary

In spite of the difficulties involved, several investigators have attempted to carry out quantitative analysis on bulk biological material utilizing one or more of the three correction procedures which have been described. However, the inevitable surface roughness of the sample, the increased depth of beam penetration, the low spatial resolution (5–10 μm) and relative inaccuracy (10%–20%) of the method, coupled with the doubtful validity of the correction technique for light elements in an organic matrix, usually makes analysis of bulk biological samples less useful than other quantitative methods which are described later. The one exception to this may be found in the application of these procedures in the analysis of frozen-hydrated tissue using frozen salt solutions as standards.

7.7.4. Thick Sections on Bulk Substrates

Analysis of sections on thick substrates requires special attention if x-ray intensities from the section plus substrate are to be compared with x-rays from a bulk standard. Warner and Coleman (1973, 1975) derived two correction procedures, BICEP and BASIC, which may be used for the quantitative analysis of sections on bulk substrates. BICEP is based on the thin film model of Colby (1968) and is designed to give ratios of elements present, whereas the BASIC procedure can be used to give local elemental mass fractions. However, these techniques have a limited usefulness in biological analysis because of the difficulty of cutting specimens of even thickness and the fact that there are considerable variations in density of biological material in different parts of the specimen. Other disadvantages are that the organic matrix is assumed to be composed entirely of carbon, which is not the case with frozen hydrated sections, and that all elements in concentrations greater than 1%–2% must be measured. Both techniques are complicated to use, require access to a large computer, and have tended to be superseded by methods using sectioned material on thin supports.

7.7.5. Thin Samples

The use of thin sections greatly simplifies quantitative analysis of biological material. The electrons only lose a small proportion of their energy as they pass through the sample, and electron backscattering from the sample is so small that it may be ignored. Secondary fluorescence of x-rays is negligible, and with the exception of very light elements such as sodium, absorption of x-rays is small.

The only x-ray generation effect which needs consideration is the probability of ionizing an atom of a given element which is described by the ionization cross section:

$$Q \propto \frac{1}{E_0 E_c} \log \frac{E_0}{E_c} \tag{7.79}$$

where E_c is the critical excitation energy of a given x-ray line and E_0 is the incident beam energy. The maximum efficiency of characteristic x-ray production occurs when the electron energy is about 3 times greater than the critical excitation energy associated with a given x-ray line. Because the critical excitation energy for most elemental x-ray lines in thin sections of biological material falls between 1 and 10 keV, the available electron accelerating voltage on an SEM (30–50 keV) and a TEM (75–200 keV) is considerably greater than $3E_c$.

It is convenient to consider the quantitation techniques for thin biological specimens under two headings. In the first we will consider methods which give the ratio of elements in a specimen. The second will deal with techniques which allow measurement of elemental concentrations.

7.7.5.1. The Elemental Ratio Method

For a number of biological investigations it is usually sufficient to measure the relative mass fractions of the elements in the specimen rather than their absolute concentration. In thin specimens, the ratio of x-ray intensities from two elements is proportional to their relative concentration, so that

$$\frac{C_A}{C_B} = k_{AB} \left(\frac{I_A}{I_B} \right) \tag{7.80}$$

where C_A and C_B are the local mass fractions of elements A and B in the same sample and I_A and I_B are the simultaneously measured intensities of characteristic radiation for the two elements. The factor k_{AB} is a constant which accounts for the relative detection and x-ray generation efficiency of the analytical system for the two elements concerned and may be measured from a thin standard composed of a known proportion of the two elements.

If the relative detection efficiency of all the elements in the system being analyzed is known, then the relative mass fractions of a number of elements in the specimen may be compared by

$$C_A : C_B : C_C = \frac{I_A}{p_A} : \frac{I_B}{p_B} : \frac{I_C}{p_C} \qquad (7.81)$$

where p_A, p_B, and p_C are the proportionality factors for each of the elements. Note that k_{AB} in Equation (7.80) is the ratio p_B/p_A. No other standards are needed and it is not necessary to measure the white radiation because, for sufficiently thin specimens, the analyzed volume is the same for each element. Russ (1974) has shown that the constant p may be derived from theory. Further details of the elemental ratio method are given by Cliff and Lorimer (1975) and in Section 7.6.1. of this chapter. While the method can provide the relative concentrations of elements in homogeneous thin samples, the technique has certain limitations when applied to biological material. Biological specimens are rarely homogeneous or uniformly thin, which would make it difficult to compare the relative concentrations of elements from different parts of the specimen. The accuracy of the method when applied to biological samples is typically between 10% and 20%.

7.7.5.2. Measurement of Elemental Concentration

The simplest of the thin sample quantitation techniques is the one described by Ingram and colleagues (Ingram *et al.*, 1972, 1973) and is based on the simple equation of Castaing given in Equation (7.72). Ingram and colleagues have used this simple method to measure the sodium and potassium concentrations in red blood cells. Standards were prepared from solutions of gelatine containing known amounts of sodium and potassium. The red blood cells and gelatine were sectioned at the same thickness and the intensities compared. The success and accuracy of this approach is very dependent on there being no selective loss of elements from either the specimen or the standard.

The Ingram method has the advantage of requiring very little computation and does not require an extensive theoretical background. Indeed, in its purest form there would be no need to make any background correction, as this would be the same for both sample and standard. The disadvantages are that the method is limited in application to what might be described as monophasic systems, i.e., red blood cells, isolated plant vacuoles, and expressed microdroplets, and that the formulation in Equation (7.72) is relatively inaccurate when compared to more recently derived methods. Much of the accuracy is based on the close similarity between sample and standard; most biological samples are likely to be inhomogenous and to contain a wide range of elements. Furthermore it is difficult to find standards whose matrix is identical to that of biological material.

A useful extension of the Ingram method is to encapsulate the specimen with the standard in a suitable organic matrix (gelatin, albumin, dextran, polyvinylpyrrolidone, etc.) and prepare and section the two together (Rick *et al.*, 1978). The mass of elements per unit volume (C_i) is calculated by multiplying the ratio of elemental intensities in the specimen (I_i) and standard ($I_{(i)}$) by the known wet weight concentration of the element in the standard solution ($C_{(i)}$):

$$C_i = \left[I_i / I_{(i)} \right] C_{(i)} \qquad (7.82)$$

Dorge *et al.* (1978) have shown that the intensity of x-ray signals from frozen-dried sections of soft biological material is directly proportional to section thickness up to a thickness of between 1 and 2 μm. As with the earlier studies by Ingram, the accuracy of the method depends on the specimen and standard being of equal thickness. This may be difficult to achieve in frozen sections, which tend to chip, fracture, or even melt during sectioning, and are unlikely to shrink uniformly during freeze-drying.

7.7.6. The Continuum Method

This is undoubtedly one of the most useful methods for biologists, and has been developed primarily by Hall and coworkers (Hall and Werba, 1969; Hall, 1971, 1975, 1979a). The principle of the continuum method is based on an extension of Kramers' relationship, which suggests that there is a proportionality between the mass thickness and the intensity of continuum radiation [see Equation (7.75)]. This relationship, which appears to hold for light elements, is very important because it means that measuring the continuum radiation will provide a built-in monitor of changes in density and thickness as one moves from place to place in the sample. Most of the methods which have been developed primarily for the analysis of thin metal foils cannot be used satisfactorily on biological material because the number of major constituents, i.e., C, H, O, N, S, P, is large, and, with the exception of S and P, the characteristic x-rays are too low in energy for accurate measurement. To carry out the continuum method, thin sections of resin-embedded, frozen-dried, or frozen-hydrated biological material are placed on thin (100–200 nm) support films whose continuum radiation is very low. For elements up to $Z = 20$, the method has an accuracy of between 5% and 10% of the measured concentrations.

It is convenient to consider the continuum method under three separate headings: (1) the measurement of the relative amounts of an element; (2) the measurement of the mass fraction of an element; (3) the application of these methods to frozen-hydrated specimens.

7.7.6.1. Relative Amounts of Elements

By combining the Kramers' relationship with the Castaing equation, which assumes that the intensity of characteristic radiation is proportional to the concentration of the element of interest present in the sample, it is possible to derive an equation which allows one to calculate the relative mass fractions of elements:

$$I_i/I_B \propto C_i/\rho t \tag{7.83}$$

I_i is the intensity of characteristic radiation from element i in the analyzed microvolume of the specimen; I_B is the intensity of continuum radiation from the specimen in a specified region of the spectrum; C_i is the local mass fraction of element i in the analyzed microvolume, and ρt is the mass thickness (ρ is the density and t is the section thickness). The assumption is that the proportionality between I_B and ρt will be the same if sample and standard are similar in average atomic number. For sections of biological material up to about 4 μm thick and composed of light elements, the Kramers' relationship appears linear. For heavier elements above $Z = 20$, i.e., those introduced into the specimen as part of a histochemical or staining exercise, Hall (1971), Shuman et al. (1976), and Sumner (1978) have shown that the proportionality is no longer linear and that the continuum counts vary also with the mean value of Z^2/A where Z is the atomic number and A is the atomic weight.

However, when this analytical method is used to measure ratios of x-ray intensities in the same analyzed region, the effects of mass thickness cancel out. In this case, Equation (7.80) can be used directly.

7.7.6.2. Mass Fractions of Elements

To measure the mass fraction of the element in a microvolume it is necessary to have a measure of the total specimen mass per unit volume. The intensity of the continuum radiation (usually measured with an energy-dispersive spectrometer) can be used to measure the total specimen mass per unit volume, and the mass fraction of a given element can be calculated using the following equation:

$$C_i = \frac{I_i/I_B}{I_{(i)}/I_{(B)}} C_{(i)} \frac{G_i}{G_{(i)}} \tag{7.84}$$

where C_i and $C_{(i)}$ are, respectively, the mass fractions of element i in the analyzed microvolume of a thin specimen and standard; I_i is the measured intensity of characteristic radiation; I_B is the measured intensity of continuum radiation and the ratio of the G factors serves to correct for the fact that the standard and specimen may not necessarily generate the same continuum intensity per unit mass per unit volume. The intensity of continuum radiation (I_B) should be measured over an energy band away from the characteristic lines of the elements of interest, and there is still some debate as to the optimum energy range for the continuum window. For example Somlyo et al. (1977) measure the continuum at between 1.34 and 1.64 keV using an 80-keV transmission microscope. Hall (1979a) favors an energy band between 10 and 16 keV using a 50-keV microprobe, and Saubermann et al. (1981), using a 30-keV scanning electron microscope, measure the continuum between 4.6 and 6.0 keV. The choice of energy band seems to depend more on the amount and energy distribution of extraneous background.

The quantities I_i and I_B must be corrected for extraneous background contributions. It is possible to measure the extraneous contributions coming from the supporting film and specimen holder and from the instrument itself (for details see Appendix A to this chapter).

The factor G (sometimes referred to as Z^2/A) for both specimen and standard may be calculated using the equation

$$G = \sum_i \left(C_i Z_i^2 / A_i \right) \qquad (7.85)$$

where Z and A are, respectively, the atomic number and atomic weight of element i, and C_i is the mass fraction. The sum is taken over all the constituent elements and in the case of the standards where all the constituent elements are known, Equation (7.85) is more conveniently expressed as

$$G = \frac{\sum_i (N_i Z_i)^2}{\sum_i (N_i A_i)} \qquad (7.86)$$

where N_i is the number of atoms of element i in the chemical formula. It is more difficult to calculate G for the specimen, and Hall (1979a) advises making estimates of the composition of the analyzed microvolume, excluding the elements to be assayed. For biological tissue, which is largely made up of C, H, O, and N, this procedure is not too difficult; Hall (1979a) gives the following typical values for G: 3.28 for protein, 3.0 for lipid, 3.4 for carbohydrate, 3.41 for nucleic acid (less the P radiation, which is recorded separately), and 3.67 for water. Because these values are so close to each other it is usually more practical to either take an average value for G, or, if the biological material is known to contain more of one particular constituent, to weight the value accordingly. Where a biological tissue contains a disproportionately high amount of an element, Hall (1979a) suggests that the most accurate procedure is to calculate the value for G from an organic analysis of the sample, but in the absence of the element, and to then correct this G value on the basis of the measured characteristic x-ray radiation from the element concerned. If the standards are made up in a matrix similar to the matrix of the specimen, the factor $G_i/G_{(i)}$ is close to 1 and may be dropped from the equation.

The dimensions of Equation (7.84) are usually expressed as mM/kg provided that the standard concentrations are expressed in the same terms. It is, however, sometimes more useful to express the mass fraction in terms of mM/liter, in which case the mass fractions must be related to the fully hydrated state.

7.7.6.3. Mass Fractions of Elements in Frozen-Hydrated Sections

Because of the increased interest in carrying out analysis on frozen-hydrated sections, it seems appropriate to include a procedure for this type of specimen. The following procedure has been developed by Hall and colleagues, and details may be found in an appendix to a recent paper by Hall and Gupta (1979).

For any measurement on either the specimen or the standard, the continuum count generated within the specimen (I_B) or standard $(I_{(B)})$ is obtained and corrected for any contributions from the supporting film, specimen holder, and instrument. In this procedure, the specimen is surrounded by a peripheral standard and it is assumed that in any one section they are both the same thickness and/or of approximately similar matrix composition.

The terms I_B and $I_{(B)}$ are proportional to the mass in the analyzed volume, while the term I_i, which is the characteristic count for the element of interest, is proportional to the mass of the element in the same analyzed volume. By comparing measurements on areas of specimen and standards, the unknown constants of proportionality can be eliminated and the elemental mass fraction (C_i) can be obtained in mM/kg from

$$C_i = \frac{I_i/I_B}{I_{(i)}/I_{(B)}} C_{(i)} \tag{7.87}$$

If the specimen and its peripheral standard are fully hydrated, and one uses the value of elemental concentration in mM/kg wet weight in the standard, then the concentration in the specimen will also be in mM/kg wet weight. However, this relationship becomes less accurate as the specimen dehydrates since the fractions of retained mass in the specimen and standards are likely to become widely different.

The concentrations measured in the microprobe in terms of mM/kg wet weight may be converted to mM/kg H_2O. This transformation is achieved by measuring changes in the ratio of characteristic to continuum counts for a given element in different parts of the specimen as it is dehydrated. This measurement is compared to the changes in the ratio of characteristic to continuum counts in the peripheral standard which contains a known amount of water as it too is dehydrated (Hall and Gupta, 1979). The values which are obtained for potassium and chlorine using the microprobe (expressed in mM/kg wet weight) are close to the values obtained for the same elements in the same compartments with ion-selective microelectrodes (expressed as mM/liter).

Appleton and Newell (1977) have derived similar procedures whereby wet weight concentrations can be derived from frozen-dried specimens. However, both these procedures are only valid if there is no lateral displacement of elements during sectioning or freeze-drying, if the sections are of uniform thickness before drying, and if drying occurs uniformly throughout the specimen and standard. Only then will the wet weight concentrations be directly proportional to the characteristic counts. While these assumptions may well be correct for tissues and cells with a high organic matrix, there is no reason to believe that any of these assumptions should be correct for cells and tissues which have intracellular compartments containing varying amounts of water and organic matrix and intercellular compartments containing little or no organic matrix.

Further studies by Hall and Gupta (1979) have clarified the relationship between dry weight and aqueous mass fractions. The dry-weight mass fraction $C_i(\text{dry})$ is given by

$$C_i(\text{dry}) = \frac{S}{S + Y\left[(I_B(\text{wet})/I_B(\text{dry})) - S\right]} \tag{7.88}$$

where $I_B(\text{dry})$ and $I_B(\text{wet})$ are the continuum counts in the dry and hydrated state, respectively. The factor S is a shrinkage factor which can be calculated by

$$S = I_i(\text{wet})/I_i(\text{dry}) \tag{7.89}$$

where $I_i(\text{wet})$ and $I_i(\text{dry})$ are the characteristic x-ray intensities for an element before and after drying. The shrinkage factor is included in the equation because there can be a substantial (up to 40%) change in the dimensions of the tissue during drying. This drying will increase the mass per unit area and consequently increase the continuum signal.

The factor Y corrects for any differences in the G value (Z^2/A) between the hydrated and dried state and may be calculated by dividing the G value for the dried organic matrix, i.e., typically 3.2, by the G value for water, i.e., 3.67. The different measurements are made before and after dehydration on adjacent fields rather than on identical fields because it has been found that there can be severe beam damage when reanalyzing a previously irradiated area. The aqueous mass fraction $C_i(\text{wet})$ is obtained by $C_i(\text{wet}) = 1 - C_i(\text{dry})$. A calculation scheme for using the continuum method is given in Appendix A. Several worked examples are given in Appendix B.

7.7.7. Thick Specimens on Very Thin Supports

For thin specimens on a very thin support, with the possible exception of some light elements, absorption of x-rays by the specimen is usually negligible because the mean atomic number and x-ray absorption cross sections of soft tissue are low. Shuman *et al.* (1976) found that the low-energy continuum radiation is absorbed in sections of frozen-dried material thicker than 1.0 μm and make a small correction for this in their quantitative procedure. This finding, coupled with the fact that frozen-dried sections are only about one third the thickness of their frozen-hydrated counterparts would put an upper limit of about 4 μm for negligible absorption in frozen-hydrated sections and about 1.5 μm for frozen-dried and resin-embedded material. Hall and Gupta (1979) find it necessary to multiply all the recorded sodium counts from a fully hydrated 1-μm section by a factor of 1.3 to correct for absorption in the section. They outline a simple procedure for calculating absorption corrections and estimating the mass per unit area using a thin film of known composition and thickness. The standard which is used is a 2-μm film of a polycarbonate plastic Makrofol supplied by Siemens, coated with a measured thickness of aluminum. Separate counts are made on the specimen and the standard with constant probe currents and counting times. The mass per unit area in the specimen M_i can be measured using the following equation:

$$M_i = \frac{I_B}{I_{(B)}}(M)\frac{G_{(i)}}{G_i} \tag{7.90}$$

where I_B and $I_{(B)}$ are the corrected continuum radiation from specimen and standard; (M) is the mass per unit area of the standard and $G_{(i)}$ and G_i are defined in Equation (7.82) as the correction factors which account for the differences in continuum in the standard and the specimen. An alternative procedure for calculating mass thickness is given by Warner and Coleman (1975). A somewhat simpler procedure is given by Eshel and Waisel (1978) and by Halloran *et al.* (1978) and is based on measuring a characteristic peak count rate and the degree of primary beam attenuation after it has passed through the sample. In the beam attenuation method, the transmitted intensity I_T of an electron beam passing through an object of mass thickness ρt is related to the initial intensity I_0 by the following expression:

$$I_T = I_0 \exp(-S_T \rho t) \tag{7.91}$$

where S_T is the total electron scattering cross section, which is constant over a limited range of mass thickness. Provided one can obtain an independent means of measuring S_T, a measure of the incident and transmitted beam intensities will allow the mass thickness to be calculated from

$$\rho t = \frac{\ln(I_0/I_T)}{S_T} \qquad (7.92)$$

These intensities can be readily measured using the exposure meter/photometer fitted to most transmission microscopes and/or the transmission detectors found on scanning electron microscopes. A recent paper by Halloran and Kirk (1979) describes how this quantitative measure of mass thickness may be carried out. This method of measuring section thickness complements the x-ray continuum method, for whereas the latter is useful for sections up to 3 μm thick (depending on beam kilovoltage) and is not as good for very thin sections, the beam attenuation method works well for the ultrathin sections.

Hall (1975) considers that x-ray yields are linearly proportional up to a thickness of about 4 μm for biological material. Above this thickness the linearity of emission no longer holds and it may be necessary either to apply the ZAF corrections mentioned earlier or to use some of the correction procedures described by Warner and Coleman (1975). The use of thick sections (2–10 μm) on very thin supports is no longer a popular way of carrying out microanalysis because of the limited spatial resolution and problems of quantitation.

7.7.8. Microdroplets

X-ray microanalysis has been applied to very small volumes of fluid which have been obtained from tissues by micropuncture and placed either on a polished beryllium surface or a thin support film and subsequently frozen dried. Recent papers by Lechene and Warner (1979) and Garland et al. (1978) give details of the method and the quantitation procedures associated with the technique. There are usually no problems with quantitation as the physical and chemical properties of the frozen-dried fluid specimen and standard are sufficiently similar so that corrections are unnecessary. The calibration curves of the standards are usually plots of the count rate of characteristic x-ray intensity against concentration and are invariably straight lines over the concentration ranges examined. All the analyst has to do is to compare the characteristic x-ray counts from the specimen and standard and read off the concentrations from the working curve. The presence of small amounts of organic material such as protein can have an effect on the quantitation. The protein may affect the accurate delivery of the microdroplets, the pattern of ice crystal formation during specimen preparation, and when dried, can absorb a significant proportion of the sodium signal. When analyzing fluids containing organic material these substances are best removed by ultrafiltration prior to analysis (Lechene and Warner, 1979).

7.7.9. Standards

The success of any quantitation depends on there being suitable standards whose characteristic x-ray counts may be compared with those from the specimen. Although quantitative analysis can be carried out using standards which do not closely resemble the composition of the specimen, it is usually desirable in biological applications to use standards of a nature similar to the specimen. As far as possible the standards should be of similar surface texture and the same thickness and internal homogeneity as the specimen. In addition one should aim at having a standard with approximately the same organic matrix and hence the same dry weight fraction as the specimen. Because of all their attendant problems of corrections for absorption, fluorescence, and atomic number, there has been a move away from using some of the inorganic bulk standards, and biologists are now tending to use thin film standards. These are of four general types: (a) inorganic material within a natural organic matrix, i.e., gelatine, albumin, etc.; (b) electrolytes incorporated into an artificial organic matrix, i.e., macrocyclic polyethers, epoxy resin, etc.; (c) electrolytes incorporated into a freeze-dried or frozen-hydrated artificial organic matrix, i.e., dextran, hydroxyethyl starch, or polyvinylpyrrolidone, which form a peripheral standard around the specimen; and (d) rapidly frozen aqueous solutions of inorganic ions. All four types of standards can be prepared by one means or another as thin sections. An excellent review on the preparation of standards is given in the book by Chandler (1977) and in the papers by Spurr (1975) and Roomans (1979). An overview of standards in biological microanalysis is given in Table 7.13 (Chandler, 1977).

7.7.10. Conclusion

A discussion has been presented of some of the methods which may be used for the quantitative x-ray microanalysis of biological material. The success and relative accuracy of these methods depends very largely on specimen preparation as well as obtaining good peak-to-background ratios for the characteristic x-ray lines. Specimen preparation is the direct responsibility of the experimenter, and, while good peak-to-background ratios are readily obtained using wavelength spectrometers, this requirement is less easily realized, particularly for the light elements, using energy-dispersive spectrometers. Mention is made elsewhere in this book of the various

Table 7.13. Standards for Biological Microanalysis[a]

	Type	Specimen	Materials	Preparation
1.	Inorganic	Thick/thin	Pure metals, formulated glasses minerals, salts	Various
2.	Organic matrix salt mixture	Thin	Albumin or gelatin, agar–agar or mixtures, salts	Shock freeze, freeze-dry, mixtures treated like thin sections
3.	Pellets	Bulk	Organic salts, organometallics	Compress into pellets or into standard holders
4.	Macrocyclic polyether salt complexes in epoxy resins		Polyethers, anhydrous salts, epoxy resin	Dissolve macrocyclic polyether salt complex in epoxy resin and polymerize
5.	Tissue homogenates	Thick	Tissues (kidney etc.), salts	Treat homogenates and salts like tissue
6.	Salt solutions	Bulk	Aqueous solutions of electrolytes	Shock freeze to obtain droplets
7.	Substrates infused with salts	Bulk	Filter paper/membrane, salts	Apply droplets of salt solution onto filter paper, shock freeze and freeze dry
8.	Single crystals	Thick/thin	Organic salts and organometallic salts	Measure minute crystals by optical microscopy, carbon coat
9.	Microdroplets	Bulk	Pico- to nanoliter droplets, salts	Dispense droplets on holders, various methods to ensure uniform spot thickness
10.	Thin films	Thin	Solution of alkali halides in ethanol and colloidon	Apply drops onto C-coated coverslip and evaporate by ethanol
11.	Thin metallic layers	Thin	Metals on flat substrates	Vacuum evaporate metals onto substrate
12.	Cl epoxy resins	Thin	Resin containing Cl	As normally for thin specimens

[a] Adapted from Chandler (1977).

computer techniques which may be used to sort out the problems of spectral overlap and correction of background radiation which can occur when using energy-dispersive spectrometers. It should be remembered that such programs do not provide results which are any more accurate than those which may be obtained by tedious calculation, but that they provide them much faster. This increase in speed of data reduction means that more data can be obtained, which in turn may be subjected to rigorous statistical analysis and thus increase the probability that a particular set of quantitative data is truly representative of the elemental concentrations in the biological material on which the measurements were made.

Appendix A: Continuum Method

A calculation scheme based on the continuum method (refer to Section 7.7.6) is provided for calculating relative mass fractions and absolute mass fractions for elements from thin sections of biological material. The calculations involved can be done using a small hand-held calculator. A Notation listing terms used in the calculation scheme is given after these Appendixes. The procedure is based on the following assumptions regarding specimen preparation, instrumentation, and method of analysis.

(a) Sections which may be frozen-hydrated, frozen-dried, or plastic-embedded have a maximum thickness of 1.0 μm. No heavy metals have been used in either fixation or staining.

(b) The probe current, the raster or spot size, and counting times remain constant throughout a series of experiments.

(c) The instrument has a demonstrably low rate of contamination.

(d) A preselected portion of the continuum or white radiation is measured using an energy-dispersive x-ray detector and the characteristic peaks are measured using either wavelength- or energy-dispersive detectors.

(e) The peak-minus-background figures are correctly measured using the appropriate peak minus background methods, spectral stripping techniques, and background correction routines (Chapter 8).

A.1. Measurement and Correction for Variable Extraneous Background Radiation from the Instrument

If light elements, i.e., Al, Be, and C are used to fabricate the specimen holder and specimen support, these two components will make only a small contribution to the background radiation. However, the regions around the specimen, i.e. specimen stage, specimen chamber, pole pieces, etc. may make a significant contribution to the background, particularly if the x-ray sources are poorly collimated. An important feature of the continuum method is the use of the continuum counts taken from a selected region of the spectrum containing no elemental peaks in order to measure the specimen mass. Extraneous characteristic radiation can contribute to this selected region of the spectrum; therefore, it is necessary to calculate the extent of this contribution and make the necessary corrections.

Instrument Background

To obtain the instrument background, place the beam on a disk of spectroscopically pure carbon, approximately 2 mm thick, and record a complete spectrum using an energy-dispersive spectrometer. The diameter of the disk should be a few millimeters larger than the area likely to be examined and analyzed in the microscope. The accelerating voltage and beam current should be the same as that used for the experiment, i.e., 10–20 kV and 0.1–5.0 nA, and the live counting time should be 100–200 s. Careful examination of the spectrum will reveal the characteristic peaks of the elements which are contributing to the background radiation. The elements most commonly present are copper and iron, the principal constituents of the materials used in the manufacture of the instrument. In most modern instruments which are well collimated these peaks will be absent or very low, but if the x-ray lines are present, rigorous attempts must be made to remove these signals from the recorded spectrum. The details of how this may be done are given in Chapter 5.

If the specimen is to be examined in the transmission mode, it is advisable to make the same measurements and corrections with a carbon disk containing a hole the same diameter as the specimen support. This will give a measure of background radiation from the parts of the stage below the specimen and above the transmitted electron detector.

However, in spite of rigorous attempts to remove extraneous x-ray coming from the instrument, some extraneous background radiation may still exist. It is possible to calculate the extent to which this radiation contributes to the signal from the continuum radiation by using the following formula:

$$I_T = I_i + I_B(S) + \left(I_B^1 + I_B^2 + I_B^3 \cdots \right) \qquad (A.1)$$

which simply states that for a given element being analyzed the total x-ray signal (I_T) entering the detector is equivalent to the sum of the characteristic signal from the element concerned (I_i), the continuum radiation from the support film and the specimen (I_B)(S), and the contribution to the continuum radiation due to electrons interacting with parts of the instrument consisting of elements 1, 2, 3, etc. ($I_B^1 + I_B^2 + I_B^3 \cdots$).

Let us assume that the background radiation from the instrument contains a characteristic peak for copper.

(1) Place the beam on a pure copper standard (a copper electron microscope grid will do) and measure the characteristic radiation for Cu $K\alpha$ (peak minus background) using standardized instrumental parameters and obtain a value $I_{(i)}$. At the same time measure the white radiation in the channels selected for making the continuum measurement (for example 4.6–6.0 kV) and obtain a value I_B.

(2) Calculate $I_B / I_{(i)} = k_1$. Repeat the measurement for $I_{(i)}$ and I_B to obtain an average value for k_1.

(3) Repeat steps (1)–(2) using pure elemental standards and obtain values for $k_2, k_3, \ldots$, etc., that is for any other elements which are shown to make a significant contribution to the extraneous background from the instrument.

(4) The sum of these different k_1 values gives a measure of the

contribution to the signal in the channels selected for measuring white radiation from the different elements making a contribution to the background radiation from the instrument.

(5) The constant k_1 will be used in ensuing calculations.

A.2. Background Radiation from Specimen Holder and Supporting Film

Although it is possible to measure the extraneous background from the instrument, and hopefully to reduce background to acceptably low levels, the specimen holder and support film may also make a contribution to the background signal. If beryllium grids or a thin carbon-coated plastic film have been used to support the specimen, this background will be very low. However, if the specimen support is made of copper, titanium, nickel, etc. this background may be unacceptably high and can be measured as follows.

Place the beam on the specimen holder and record a complete spectrum with an energy-dispersive spectrometer using the beam current, voltage, and counting time which is used to measure the specimen. The recorded spectrum will show which elements in both specimen holder and the instrument are contributing to the background spectrum. If a copper grid is being used, the main peak will be from copper, a titanium grid will give a titanium peak, and so on. This extraneous radiation will also make a contribution to the continuum radiation and must be corrected for in the calculations. Let us consider the two conditions which are likely to arise:

A. The background radiation from the instrument is low and the specimen support is a carbon-coated nylon film on a beryllium holder.

B. The background radiation either from the instrument or the specimen support, i.e., a copper or nickel grid, or from both instrument and specimen support is significant.

In condition A it is unlikely that there will be an extraneous background. In condition B extraneous x-ray signals from the instrument and/or the specimen support may contribute to the continuum radiation which is being used to measure the specimen mass.

Condition A

In this case there is negligible instrument background and negligible extraneous radiation from the specimen support.

(1) Calculate k_1 as described previously.

(2) Place the electron beam on the specimen holder and obtain a value

for the white radiation in the channels selected for making the continuum measurement, i.e., 4.6–6.0 kV. Repeat several times to obtain an average value for $I_B(H)$.

(3) Calculate $I_B(H) \times k_1 = I_B(B)$. [If the chemical composition of the specimen holders remains unaltered and the extraneous radiation from the instrument does not vary, the expression $I_B(B)$ will remain constant and need not be calculated every time.]

(4) Place the beam on the support film and obtain a value for the white radiation in the same channels (i.e., 4.6–6.0 keV). Repeat several times to obtain an average value for $I_B(F)$. Provided the support film thickness does not vary, the value for $I_B(F)$ should remain more or less constant. If the value for $I_B(F)$ varies by more than 5%–10% this is probably due to fluctuations in the beam current, which should be checked.

(5) Place the beam on the specimen and obtain a value $I_B(S)$ for the white radiation from the specimen plus support film in the same channels used to measure the continuum.

(6) Calculate $I_B(S) - I_B(F) = I_B(P)$, the continuum intensity from the specimen minus support film.

(7) Calculate $I_B(P) - I_B(B) = I_B$ where I_B is a measure of the background continuum radiation from the specimen corrected for any contributions of extraneous background radiation from the instrument, specimen holder, and support film. The term I_B will be used in further calculations.

Condition B

In this case there is demonstrable extraneous radiation from the instrument and/or extraneous radiation from the specimen support.

(1) If necessary calculate k_1 as described previously.

(2) Before starting an experimental run place the electron beam on the specimen holder and measure the characteristic radiation for the element of concern in the specimen holder, and obtain a value for $I_i(H)$. At the same time measure the white radiation in the channels selected for making the continuum radiation (e.g., 4.6–6.0 kV) and obtain a value for $I_B(H)$.

(3) Calculate $I_B(H)/I_i(H) = k_H$. Repeat the measure for $I_B(H)$ and $I_i(H)$ to obtain an average value for k_H.

(4) Calculate $k_1 = k_H + k_B$, where k_B is a constant for the contribution to the continuum from specimen holder and instrument. (If the chemical composition of the specimen holder remains unaltered and the extraneous background radiation from the instrument does not vary, the expression k_B will remain constant and need not be calculated every time.)

(5) At the beginning (or end) of an experimental run place the beam on the support film and measure the characteristic radiation for the element of concern in the specimen holder, and obtain an average value for $I_i(F)$. At the same time measure the continuum radiation in the same channels

(e.g., 4.6–6.0 kV) and obtain an average value for $I_B(FH)$, the continuum radiation from the support film plus the specimen holder.

(6) Calculate $I_B(F) = I_B(FH) - [k_B \times I_i(F)]$.

(7) Place the beam on the specimen and measure the white radiation from the specimen and support film in the same channels, i.e., 4.6–6.0 kV and obtain an average value for $I_B(S)$. At the same time measure the characteristic radiation for the elements of concern in the specimen holder, and obtain an average value for $I_i(S)$.

(8) Calculate $I_B = I_B(S) - [k_B \times I_i(S)] - I_B(F)$.

A.3. Calculating Relative Mass Fractions and Absolute Mass Fractions of Elements

(1) Place the beam on the area of interest in the specimen and measure the characteristic radiation (peak minus background) for each of the elements of interest and obtain values for I_i. At the same time measure the white radiation in the channels selected for making the continuum radiation (i.e., 4.6–6.0 kV) and obtain values for $I_B(S)$, and if appropriate $I_i(S)$.

(2) Using $I_B(S)$ and $I_i(S)$ calculate the white radiation from the specimen I_B.

(3) Calculate $I_i/I_B = C_i$ for each of the elements of interest, where C_i is the relative mass fraction of the element concerned.

(4) Place the beam on another region of the specimen and repeat steps (1)–(3) and obtain further values of C_i. Each time a set of characteristic peak minus background readings are made for elements of interest, a measure should be made of $I_B(S)$ and, if appropriate, $I_i(S)$. It is important to periodically measure the continuum radiation from the support film, $I_B(F)$, as this will give an indication of the stability of the beam current, assuming the support film is of constant thickness and composition. $I_B(F)$ is usually measured after every third set of readings of elemental intensities and white radiation.

(5) Repeat step (4) until the analysis is completed in a given section.

(6) Change the specimen. Provided the same type of specimen holder is being used, repeat steps of condition A or steps (5)–(8) of condition B and the steps (1)–(5). At that point further values of C_i can be obtained.

(7) Change the specimen for a sectioned standard and repeat steps (4)–(7) of condition A or steps (5)–(8) of condition B and steps (1)–(5). At that point obtain values for $C_{(i)}$.

(8) Calculate the absolute mass fraction of a given element A in the specimen

$$C_A = \frac{C_i}{C_{(i)}} \times C_{(A)}$$

where $C_{(A)}$ is the concentration in mM/kg dry weight of the standard in the case of dehydrated material or mM/kg fresh weight of the standard in the case of specimen and standard which is in a fully frozen-hydrated state at the time of analysis.

(9) If care is taken in choosing a standard whose organic matrix is close to that of the biological material, then the unknown constants of proportionality $G_i/G_{(i)}$ [see Equation (7.84)] can be eliminated. If $G_i/G_{(i)}$ is not close to unity, then the absolute mass fraction C_A will have to be multiplied by $C_i/C_{(i)}$ to give a correct figure.

(10) Wet weight concentrations may be deduced from measurements of dehydrated sections, provided that the section thickness remains constant, that no shrinkage occurs during specimen drying, and that there is minimal redistribution of elements during dehydration.

(11) Alternatively, wet weight concentrations may be calculated from dehydrated sections by using an internal standard of known water concentrations. Specimens may be encapsulated in known concentrations, i.e., 25% of polymeric cryoprotectants such as hydroxyethyl starch made up in physiological compatible salts solutions also of known concentrations. Frozen-hydrated sections are taken from the specimen plus surrounding cryoprotectant and several measurements are quickly made of the white radiation from the region of interest in the specimen and from the surrounding cryoprotectant, before the specimen begins to lose any water, using the procedures described earlier. The sample may then be carefully freeze-dried in the microscope column, and the elemental analysis carried out using the procedures described previously. Because the white radiation is proportional to the total mass of the analyzed volume, the differences in the continuum between the fully frozen-hydrated state and the frozen-dried state are due largely to water. There may be small inaccuracies due to differences in the water binding capacity of different parts of the cell matrix and the cryoprotectant, but such errors are small in comparison to the gross error of analyzing specimens of uncertain water content.

Polymeric cryoprotectants have the added advantage of being physiologically compatible with most tissues, and when used in concentrations of 20%–25% will vitrify when rapidly frozen with minimal phase separation of the dissolved electrolytes. They form ideal internal standards for x-ray microanalysis.

The calculation relating dry weight and wet weight concentration is as follows:

$$\frac{C_i(\text{dry})}{I_B(D)/I_B(W)} = C_i(\text{wet}) \tag{A.2}$$

where $I_B(D)$ and $I_B(W)$ is the white radiation from a frozen-dried and frozen-hydrated specimen, respectively.

The percentage hydration of a sample is given by

$$1 - \frac{I_B(D)}{I_B(W)} \times 100 = \% \text{ hydration} \qquad (A.3)$$

Appendix B: Worked Examples of Quantitative Analysis of Biological Material

Example A: Calculation of Relative Mass Fractions in a Specimen

The specimen is a 0.5-μm frozen-dried section of plant root tissue supported on a 70-nm-thick nylon film coated with 15 nm of aluminum. The support film is stretched across a beryllium holder and the analysis is carried out in an instrument with negligible extraneous background. Analysis for potassium and chlorine in root cap cells was carried out using an energy-dispersive detector (resolution 150 eV). The peak-to-background ratios were measured for chlorine between 2.52 and 2.68 kV and for potassium between 3.24 and 3.40 kV. The white radiation was measured between 4.60 and 6.00 kV. The operating conditions of the microscope were: accelerating voltage 30 kV, beam current 1.0 nA, probe size on the specimen 2 μm^2, and specimen temperature of approximately 100 K.

(1) The average value for $k_1 = 0.293$.

(2) The average value for $I_B(H) = 79$.

(3) Calculate $I_B(H) \cdot k_1 = I_B(B)$: $0.293 \times 79 = 23$.

(4) The data in Table 7.14 were obtained from the analysis of cells in the root-cap region.

(5) The average value for the white radiation of the film $I_B(F) = 2145$.

(6) For cell 1 calculate

$$I_B(S) - I_B(F) = I_B(P)$$
$$4458 - 2145 = 2313$$

(7) For cell 1 calculate

$$I_B(P) - I_B(B) = I_B$$
$$2313 - 23 = 2290$$

(8) For cell 1 calculate

$$\frac{I_{Cl}}{I_B} = \frac{355}{2290} = 0.155 = C_{Cl}$$

(9) For cell 1 calculate

$$\frac{I_K}{I_B} = \frac{2771}{2290} = 1.21 = C_K$$

Table 7.14. Analysis of Cells in the Root Cap Region

	I_{Cl} $K\alpha$ Cl (P − B)	I_K $K\alpha$ K (P − B)	$I_B(S)$ White radiation, specimen plus film	$I_B(F)$ White radiation, film
Cell 1	355	2771	4458	
Cell 2	365	2968	4842	
Cell 3	446	3669	5503	
Film 1				2167
Cell 4	353	2701	4669	
Cell 5	583	5211	6622	
Cell 6	501	3412	5258	
Film 2				2139
Cell 7	336	2640	4584	
Cell 8	237	1820	3739	
Cell 9	763	5104	6851	
Film 3				2151
Cell 10	420	3211	4954	
Cell 11	447	3799	5406	
Cell 12	168	1383	3330	
Film 4				2120
Cell 13	404	2895	4899	
Cell 14	256	2404	4010	
Cell 15	385	2917	4686	
Film 5				2153
Cell 16	355	2726	4384	
Cell 17	523	4022	5606	
Cell 18	277	2208	4122	
Film 6				2140

(10) Repeat steps (6)–(9) for cells 2–18 and obtain the relative mass fractions of Cl and K as shown in Table 7.15.

(11) The relative mass fractions for the two elements in the root-cap cell vacuoles are chlorine, 0.146, and potassium, 1.14.

Example B: Calculation of Absolute Mass Fractions in a Specimen

Specimen is a 1.0-μm frozen-hydrated section of plant root tissue supported on a 70-nm nylon film coated with a 10-nm layer of aluminum. The support film is placed on top of a 75 mesh copper grid which is placed on top of a beryllium holder and the analysis is carried out in an instrument with negligible extraneous background. Analysis was carried out on phloem parenchyma cells using an energy-dispersive detector (resolution 150 eV). The phosphorus peak-to-background ratios were measured between 1.92 and 2.08 keV; the sulfur peak-to-background ratios were measured between 2.24 and 2.40 keV; and the chlorine peak-to-background ratios were measured between 2.52 and 2.68 keV. The white radiation was measured between 4.60 and 6.00 keV and the Cu peak-to-background ratio

Table 7.15. Relative Mass Fractions of Cl and K for Cells in the Root Cap Region

Cell	C_{Cl}	C_K	Cell	C_{Cl}	C_K
1	0.155	1.21	10	0.150	1.15
2	0.136	1.11	11	0.138	1.17
3	0.134	1.10	12	0.145	1.19
4	0.141	1.08	13	0.148	1.06
5	0.131	1.17	14	0.139	1.31
6	0.162	1.10	15	0.152	1.15
7	0.139	1.09	16	0.160	1.08
8	0.151	1.16	17	0.152	1.17
9	0.163	1.09	18	0.142	1.13
			Average	0.146	1.14

was measured between 7.88 and 8.16 keV. The operating conditions of the microscope were an accelerating voltage of 30 keV, beam current 2.0 nA, probe size on the specimen 2 μm^2, and specimen temperature 100 K.

(1) The average value for $k_1 = 0.06$.
(2) The average value for $I_i(H)$ for Cu = 151764.
(3) The average value for $I_B(H)$ for Cu = 51600.
(4) Calculate

$$\frac{I_B(H)}{I_i(H)} = k_H = \frac{51600}{151764} = 0.340$$

Table 7.16. Analysis of Phloem Parenchyma Cells

	I_P	I_S.	I_K	$I_B(S)$ Continuum radiation, specimen and film	$I_B(FH)$ Continuum radiation, film	$I_i(S)$	$I_i(F)$
	$K\alpha$	$K\alpha$	$K\alpha$			$K\alpha$ Cu	$K\alpha$ Cu
Cell 1	4020	1100	7500	52001		72000	
Cell 2	4564	1147	7559	53190		71680	
Cell 3	3819	1067	7620	50170		72346	
Film					21000		41966
Cell 4	4721	1128	7489	49601		78121	
Cell 5	4319	1201	7496	52116		74286	
Cell 6	4628	1183	7516	53001		75863	
Film					22861		42693
Cell 7	4324	1231	7590	53681		76129	
Cell 8	4432	1168	7601	51816		72691	
Cell 9	4518	1114	7570	52455		71908	
Film					20963		40281
Cell 10	3997	1190	7532	52106		74612	
Cell 11	4318	1107	7488	49970		76121	
Cell 12	4419	1153	7521	53141		73118	
Film					21263		42100

(5) Calculate $k_1 + k_H = k_B = 0.06 + 0.340 = 0.400$.

The data in Table 7.16 were obtained from the analysis of P, S, and K in phloem parenchyma cells in the meristematic region of the root tip.

(6) The average value for $I_i(F) = 41760$.

(7) The average value for $I_B(FH) = 21608$.

(8) Calculate $I_B(F) = I_B(FH) - [k_B I_i(F)] = 21608 - (0.4 \times 41760) = 4904$.

(9) For cell 1 calculate

$$I_B = I_B(S) - [k_B I_i(S)] - I_B(F)$$
$$= 52001 - (0.4 \times 72000) - 4904$$
$$= 18297$$

(10) For cell 1 calculate

$$C_i = \frac{I_P}{I_B} = \frac{4020}{18297} = 0.219$$

$$C_i = \frac{I_S}{I_B} = \frac{1100}{18297} = 0.060$$

$$C_i = \frac{I_K}{I_B} = \frac{7500}{18297} = 0.410$$

(11) Repeat steps (9) and (10) for cells 2–12. Table 7.17 shows the results obtained.

(12) The relative mass fractions for the three elements in the cyto-

Table 7.17. Calculation of C_P, C_S, and C_K for Phloem Parenchyma Cells

Cell	C_P	C_S	C_K
1	0.219	0.060	0.410
2	0.233	0.058	0.385
3	0.234	0.065	0.467
4	0.351	0.084	0.557
5	0.247	0.069	0.428
6	0.261	0.067	0.423
7	0.236	0.067	0.414
8	0.248	0.065	0.426
9	0.241	0.059	0.403
10	0.230	0.069	0.434
11	0.295	0.076	0.512
12	0.233	0.061	0.396
Average	0.252	0.067	0.438

plasm of the phloem parenchyma cells are $C_{(P)} = 0.252$, $C_{(S)} = 0.067$, $C_{(K)} = 0.438$.

(13) To convert the relative mass fractions to absolute mass fractions, repeat steps (6)–(11) using suitable standards 1.0 μm frozen hydrated in this case, standard salt solutions in 25% PVP, and obtain values for $C_{(i)}$.

(14) Calculate values for $C_{(i)}$:

$$C_{(S)} = 0.308 \text{ for } 250 \text{ mM/kg FW}^\dagger$$

$$C_{(K)} = 0.325 \text{ for } 150 \text{ mM/kg FW}$$

$$C_{(P)} = 0.273 \text{ for } 150 \text{ mM/kg FW}$$

(15) Calculate

$$C_A = \frac{C_i}{C_{(i)}} C_{(A)}$$

(a) For sulfur: $C_A = (0.067/0.308)(0.250) = 54 \text{ mM/kg FW}$
(b) For potassium: $C_A = (0.438/0.325)(0.150) = 202 \text{ mM/kg FW}$
(c) For phosphorus: $C_A = (0.252/0.273)(1.50) = 138 \text{ mM/kg FW}$

Notation

C_i = Relative mass fraction of an element in the specimen.
$C_{(i)}$ = Relative mass fraction of an element in a standard.
C_A = Absolute mass fraction (concentration) of an element in the microvolume of the specimen which is analyzed.
$C_{(A)}$ = Absolute mass fraction (concentration) of an element in the microvolume of the standard which is analyzed.
I_B = White radiation from specimen.
$I_{(B)}$ = White radiation from standard.
$I_B(B)$ = White radiation from specimen holder and instrument.
$I_B(D)$ = White radiation from a frozen-dried sample.
$I_B(F)$ = The term for the white radiation from the support film.
$I_B(FH)$ = Continuum radiation from the support film plus the specimen holder.
$I_B(H)$ = The term for the continuum radiation from the specimen holder.
$I_B(P)$ = White radiation from the specimen minus support film.
$I_B(S)$ = The term for the continuum radiation from the support film plus specimen.
$I_B(W)$ = White radiation from a frozen-hydrated sample.
I_i = Characteristic radiation of an element being analyzed.

---

† FW = fresh weight

$I_i(F)$ = Characteristic radiation from an element in the specimen holder which is contributing to the signal from the support film due to electron scatter.

$I_i(H)$ = Characteristic radiation from an element in the specimen holder which might make a contribution to the white radiation or continuum channels.

$I_i(S)$ = Characteristic radiation of an element in the specimen holder which is contributing to the signal from the support film plus specimen due to electron scatter.

$I_{(i)}$ = The characteristic radiation from a pure standard of an element.

k_B = Constant representing the contribution to the white radiation from the specimen holder and instrument.

k_H = Constant respresenting the contribution to the white radiation or continuum channels from the specimen holder.

k_1 = A constant representing the background radiation in the instrument.

I_T = The total x-ray signal entering the detector.

k_{iB} = Constant for the Cobet method.

N_i = Number of atoms of element i in the analyzed microvolume.

A = Atomic weight.

Z = Atomic Number.

P/B = Peak-to-background ratio, $(I_i - I_B)/I_B$.

k_{ij} = Constant for elements i and j in the peak-to-background method for relative concentration.

k_i = Constant for element i in the peak-to-background method for absolute concentration of Statham and Pawley (1978).

Q = Ionization cross section.

E_0 = Operating voltage.

E_c = Critical excitation energy for a given x-ray line.

P_i = Proportionality factors for elemental ratio method.

ρ = Density of specimen.

t = Thickness of specimen.

G_i = Continuum correction factor $\sim Z^2/A$ for specimen.

$G_{(i)}$ = Continuum correction factor $\sim Z^2/A$ for standard.

S = Shrinkage factor.

Y = G correction factor.

M = Mass per unit area in the specimen.

(M) = Mass per unit area in the standard.

Practical Techniques of X-Ray Analysis

8.1. General Considerations of Data Handling

It has already been shown in Chapter 6 that qualitative analysis is based on the ability of a spectrometer system to measure characteristic line energies and relate those energies to the presence of specific elements. This process is relatively straightforward if (1) the spectrometer system is properly calibrated, (2) the operating conditions are adequate to give sufficient x-ray counts so that a given peak can be easily distinguished from the corresponding background level, and (3) no serious peak overlaps are present.

Quantitative analysis, on the other hand, involves (1) accurately measuring the intensity of spectral lines corresponding to preselected elements for both samples and standards under identical operating conditions, (2) calculating intensity ratios, and (3) converting them into chemical concentration by the methods described in Chapter 7. Since quantitative analysis can now be performed with 1%–5% relative accuracy, it is obvious that great care must be taken to ensure that the measured response of the x-ray detector system is linear over a wide range of counting rates, and that the useful signal can be easily extracted from the background. In practice, a knowledge of the absolute spectrometer counting efficiency is generally not required since its effect cancels by establishing intensity relative to a standard. It is necessary to make adjustments for changes in the spectrometer efficiency with count rate, hence the need for dead-time corrections for proportional counters and live-time corrections for Si(Li) detectors. As would be expected, background measurements become increasingly important as peak-to-background ratios get smaller. For example, a 100% error in a background measurement of a peak 100 times larger than the background introduces a 1% error in the measured peak intensity, whereas the same error in the case of a peak twice background introduces a 50% error. Large peak-to-background ratios are not always the case, even with wavelength-dispersive spectrometers, and accurate background measurement becomes important particularly at low concentrations.

As illustrated in Figure 8.1, the commonly used method of background measurement with a wavelength-dispersive spectrometer is to detune to wavelengths slightly above and below the tails of the characteristic peak and then establish the value of the background at the peak setting by linear interpolation. In older instruments it is generally not desirable to detune the spectrometer between sample and standard measurements because the spectrometer position cannot be accurately reproduced. Mechanical back-lash normally prevents accurate repositioning to the previous peak value. It is also difficult to obtain $< 2\%$ reproducibility in the intensity by determining maximum signal on a rate meter while manually adjusting the spectrometer control. If we wish to avoid moving the spectrometer when measuring background, it is sometimes possible to locate a material which has approximately the same average atomic number as the material of interest but which does not contain the element being analyzed. Since the background generated at any energy is approximately proportional to average atomic number, other things being equal, it is then possible to obtain a measure of background under a characteristic x-ray peak. As an example, when measuring the background for Ni $K\alpha$, a copper standard could be used. However, this method must be used with caution. It is sometimes not possible to find a material with an average atomic number sufficiently close to an unknown specimen, for which one often has only a vague idea of the composition in the first place. This method should not be used when the peak-to-background ratio is low, such as for the analysis of concentrations below 1%. Fortunately, in the newer instruments it is possible to accurately determine a peak position, detune to obtain a background

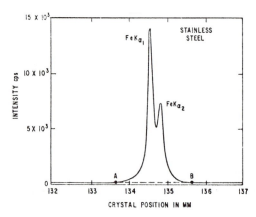

Figure 8.1. Commonly used method of background measurement with a wavelength-dispersive spectrometer using linear interpolation between measured background positions at A and B. Crystal position is equivalent to wavelength in this figure.

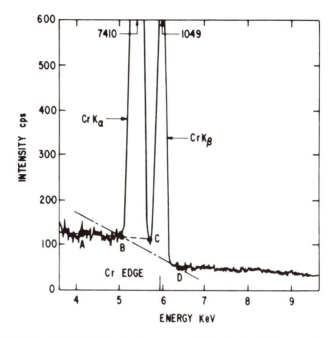

Figure 8.2. EDS background fitting by linear interpolation (points B–C or B–D) or extrapolation (points A–B) in a simple spectrum of chromium (Lifshin 1975).

reading, and return to the peak position by simply dialing in the wavelength setting. In newer instruments a computer-controlled step scan in wavelength can be used for repeaking although some loss of data collection time will result.

Background measurements with energy-dispersive detectors are usually much more critical than with wavelength-dispersive spectrometers because of the lower peak-to-background ratios and the difficulty of finding suitable background areas adjacent to the peak being measured. Figure 8.2 shows a portion of a pure chromium spectrum. The background value at the peak channel determined by interpolation is about 100 counts if points B and D are used or about 130 counts if points B and C are used. The latter pair is the better of the two because the chromium absorption edge at 5.989 keV produces an abrupt drop in the background at higher energies which is obscured by the Cr $K\beta$ peak. If the spectrum had been collected with a detector system with poorer resolution than that used in obtaining Figure 8.2 (165 eV at 6.4 keV) then point C could not be used. In this case, determination of the background by extrapolation of the low-energy points A and B to point C could be used in preference to interpolation of points B and D. The error associated with use of this approach would affect the

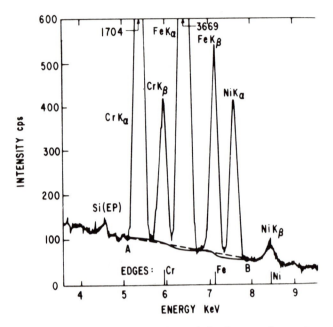

Figure 8.3. EDS background fitting by linear interpolation in a complex spectrum of stainless steel.

measurement of the chromium standard by less than 1%. Figure 8.3 shows a stainless steel spectrum, taken under the same operating conditions as Figure 8.2. There are a sufficient number of $K\alpha$ and $K\beta$ peaks such that the use of extrapolation is questionable for Fe and Ni while interpolation between points A and B must be performed over a 3-keV range. Since the real background is irregular owing to the presence of Cr and Fe absorption edges, an error of about 3% would occur in the Ni intensity measurement in this specimen with background extrapolation by linear interpolation.

Although these errors of a few percent of the amount present in the major constituents are slightly greater than what can be expected with WDS measurements, a much greater difficulty in the example cited would be encountered in the determination of a few percent Mn (Mn $K\alpha = 5.898$ keV) or Co (Co $K\alpha = 6.924$ keV). Not only might the signals of Mn and Co be of the same order of magnitude as the background, but they would be obscured by spectral lines of the major constituents (Cr $K\beta$ and Fe $K\beta$).

8.2. Background Shape

Because compensation for the background, by subtraction or other means, is critical to all EDS analysis, it is worthwhile to spend some time in reviewing what is known about it as well as what reduction procedures are now being used. Basically there are two appoaches to this problem. In one a continuum energy distribution function is either calculated or measured and combined with a mathematical description of the detector response function. The resulting function is then used to calculate a background spectrum which can be subtracted from the observed spectral distribution. This method can be called background modeling. In the other approach, the physics of x-ray production and emission is generally ignored and the background is viewed as an undesirable signal the effect of which can be removed by mathematical filtering or modification of the frequency distribution of the spectrum. Examples of the latter technique include digital filtering and Fourier analysis. This method can be called background filtering. It must be remembered here that a real x-ray spectrum consists of characteristic and continuum intensities both modulated by the effects of counting statistics. When background is removed from a spectrum, by any means, the remaining characteristic intensities are still modulated by both uncertainties. We can subtract away the average effect of the background but the effects of counting statistics cannot be subtracted away! In practice, both of the above methods of background removal have proved successful. The next two sections will discuss these methods.

8.2.1. Background Modeling

Background modeling has been investigated by a number of authors (Ware and Reed, 1973; Fiori *et al.*, 1976; Rao-Sahib and Wittry, 1975; and Smith, Gold, and Tomlinson, 1975). We will describe here only one approach (Fiori *et al.*, 1976) to communicate the essence of the method. Early theoretical work by Kramers (1923) demonstrated that the intensity distribution of the continuum, as a function of the energy of the emitted photons, is

$$I_E \, \Delta E = k_E \overline{Z} (E_0 - E) \, \Delta E \qquad (8.1)$$

In this equation $I_E \, \Delta E$ is the average energy of the continuous radiation, produced by one electron, in the energy range from E to $E + \Delta E$. E_0 is the incident electron energy in keV, E is the x-ray photon energy in keV, $\overline{Z}$ is the specimen average atomic number, and k_E is a constant, often called Kramers' constant, which is supposedly independent of Z, E_0, and $E_0 - E$. The term "intensity" I_E in Equation (8.1) refers to the x-ray energy. The number of photons N_E in the energy interval from E to $E + \Delta E$ per incident electron is given by

$$N_E \, \Delta E = k_E \overline{Z} \left(\frac{E_0 - E}{E} \right) \Delta E \qquad (8.2)$$

This point has been a source of considerable confusion in the literature since Equation (8.1) is often used incorrectly to describe the shape of the spectral distribution of the background generated within a bulk specimen. Equation (8.2) for the intensity of the continuum was discussed in Chapter 3. This ambiguity in the definition of intensity was undoubtedly complicated by the fact that the first ionization detectors integrated total radiant energy collected rather than counting individual x-ray photons. Such an ionization detector would produce the same output for one photon of energy $2E$ as it would for two photons of energy E arriving simultaneously in the detector. The difference in the shape of the two curves described by Equations (8.1) and (8.2) is illustrated in Figure 8.4.

The measured number of photons, $N(E)$, within the energy range E to $E + \Delta E$, produced in time t with a beam current i (electrons per second), observed by a detector of efficiency P_E, and subtending a solid angle Ω, is equal to

$$N(E) = \frac{\Omega}{4\pi} \, itf_E P_E N_E \, \Delta E$$

$$N(E) = \frac{\Omega}{4\pi} \, itf_E P_E k_E \overline{Z} \left(\frac{E_0 - E}{E} \right) \Delta E \qquad (8.3)$$

Here, the term f_E (absorption factor for the continuum) denotes the probability of absorption of a photon of energy E within the target. In Equation (8.3) the term $\Omega/4\pi$, which is the fraction of a sphere which the detector subtends, is equal to the fraction of photons emitted toward the detector only if the generation of the continuum is isotropic. This assumption of isotropy is a reasonable approximation for a thick target although a small degree of anisotropy does exist. If we define a equal to $(\Omega/4\pi)it\,\Delta E$, then we obtain

$$N(E) = af_E P_E k_E \overline{Z} \left(\frac{E_0 - E}{E} \right) \qquad (8.4)$$

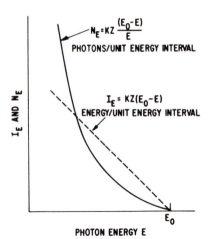

Figure 8.4. The theoretical shape of the x-ray continuum according to Kramers' equation (from Green, 1962).

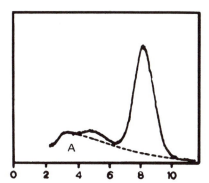

Figure 8.5. Copper pulse distribution curve obtained by Green (1962) using a proportional counter to measure x-ray production efficiencies. $E_0 = 27$ keV.

In order to determine the value of Kramers' constant, k_E, it is necessary to make absolute measurements of the continuum spectral distribution. This is extremely difficult to do with a wavelength-dispersive spectrometer on a microanalyzer because the spectrometer efficiency varies with energy, and, furthermore, the efficiency variation is generally not known. In his dissertation several years before the availability of energy-dispersive x-ray detectors, Green (1962) described a series of measurements of both continuum and characteristic x-ray production efficiencies in which he used a proportional counter of known collection efficiency to directly measure x-ray spectra. Figure 8.5 taken from this work shows a spectrum obtained from copper. Because the broad characteristic and escape peaks mask much of the continuum, Green did not attempt to measure the distribution, but rather assumed Kramers' equation to be correct and calculated k_E by equating the area A under the dashed curve with an integrated form of Equation (8.2), namely,

$$A \cong \frac{1}{2} k_E \overline{Z} E_0^2 \qquad (8.5)$$

As shown in Figure 8.6, the values obtained by various authors for k_E are atomic number dependent. Lifshin et al. (1973) performed the same type of experiment, taking advantage of the better energy resolution and higher collection efficiency of a Si(Li) detector system. Figure 8.7 shows a similar spectrum obtained from copper using a 165 eV (at Fe $K\alpha$) Si(Li) detector, 3 mm thick by 6 mm in diameter. The maximum values of the Cu $K\alpha$ and Cu $K\beta$ peaks have been expanded to emphasize the detailed structure of the background. The decrease in intensity at low energy is due to absorption in the detector window. The sharp discontinuity in the background intensity at about 9 keV is caused by specimen self-absorption effects due to the presence of a discontinuity in the copper x-ray mass absorption coefficient at the energy of the copper K excitation potential. The background at the exact absorption edge energy, 8.979 keV, is obscured, however, by the detector broadening of the Cu $K\beta$ peak. This is a good place to point out

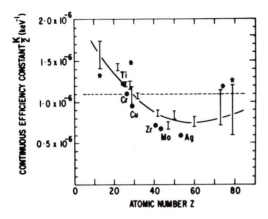

Figure 8.6. Kramers' constant, given as $K/2$, as a function of atomic number: ∗, Dyson (1956); - - -, Compton and Allison (1943); vertical bars, Green and Cosslett (1968); ●, Lifshin et al. (1973).

that any attempt to apply computer-fitting techniques to describe the shape of the observed spectrum for purposes of background subtraction must take into account such details, which in the case of multielement samples can be quite complicated. As mentioned earlier, Kramers' equation does not apply to the observed background distribution, but rather to the continuum as it is generated in the specimen.

The above work of Lifshin (1974) and additional work by Rao-Sahib and Wittry (1972, 1975) and Hehenkamp and Böcher (1974) have shown that the energy and atomic number dependence of $N(E)$, expressed by Equation (8.4), is only approximate. The residual correction has been carried out in at least two ways: (1) Applying more refined theoretical calculations, and modifying the exponents, from unity, on the energy term and atomic number $\bar{Z}$ in Kramers' equation; or (2) using an additive energy-dependent correction term in Equation (8.4) as proposed by Ware

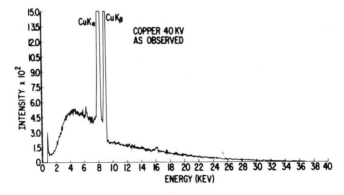

Figure 8.7. A Si(Li) detector spectrum obtained from a pure copper standard.

and Reed (1973):

$$N(E) = a\left[f_E P_E \bar{Z}\left(\frac{E_0 - E}{E} \right) + F(E) \right] \qquad (8.6)$$

where $F(E)$ is taken from an empirically determined table; the authors indicated that for continuum photon energies above 3 keV, the additive energy term was unnecessary. Lifshin (1974) empirically developed and tested a more general relation for $N(E)$:

$$N(E) = f_E P_E \bar{Z}\left[a\left(\frac{E_0 - E}{E} \right) + b\frac{(E_0 - E)^2}{E} \right] \qquad (8.7)$$

in which a and b are fitting factors, and the remaining terms are as defined for Equation (8.4).

Fiori *et al.* (1976) found that it is possible to determine a and b for any target by measuring $N(E)$ at two separate photon energies and solving the resulting two equations in two unknowns. This procedure presumes that both f_E and P_E in Equation (8.7) are known. Since the determination of $N(E)$ is made from measurements on the target of interest only, the inaccuracies of Kramers' law with respect to atomic number do not affect the method. The points chosen for measurement of $N(E)$ must be free of peak interference and detector artifacts (incomplete charge collection, pulse pileup, double-energy peaks, and escape peaks).

8.2.1.1. Absorption Factor, f_E

For characteristic primary x-ray photons, the absorption within the target is taken into account by a factor called $f(\chi)$. The factor $f(\chi)$ can be calculated from Equation (7.12)–(7.14) in Chapter 7 (PDH equation), or a simpler form due to Yakowitz *et al.* (1973) can be used:

$$1/f(\chi) = 1 + a_1 \gamma \chi + a_2 \gamma^2 \chi^2 \qquad (8.8)$$

where γ is $E_0^{1.65} - E_c^{1.65}$. E_c is the excitation potential in keV for the shell of interest, and $\chi = (\mu/\rho)^i \csc \Psi$. $(\mu/\rho)^i$ is the mass absorption coefficient of the element of interest in the target, and Ψ is the x-ray take-off angle. The presently used values for the coefficients a_1 and a_2 are $a_1 = 2.4 \times 10^{-6}$ $\mathrm{g\,cm^{-2}\,keV^{-1.65}}$, $a_2 = 1.44 \times 10^{-12}$ $\mathrm{g^2\,cm^{-4}\,keV^{-2.7}}$.

If it is assumed that the generation of the continuum has the same depth distribution as that of characteristic radiation, and if $E \simeq E_c$, then Equation (8.8) can be modified to

$$1/f_E = 1 + a_1\left(E_0^{1.65} - E^{1.65} \right)\chi_c + a_2\left(E_0^{1.65} - E^{1.65} \right)^2 \chi_c^2 \qquad (8.9)$$

where $\chi_c = (\mu/\rho)^E \csc \Psi$. $(\mu/\rho)^E$ is the mass absorption coefficient for continuum photons of energy E in the target and f_E is the absorption factor

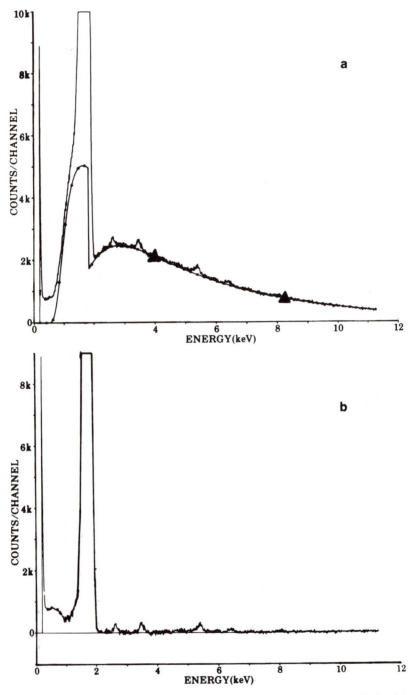

Figure 8.8. Background correction in silicon. (a) Fitted continuum curve calculated by Equation (8.7); background fit points are indicated by triangles. Observed spectrum is superposed. The silicon $K\alpha$ peak contains 53200 counts in the peak channel (b) Difference between observed and fitted continuum spectra. Total spectrum counts equal 9×10^6.

for these photons. Similar assumptions and expressions have been used by other authors, including Ware and Reed (1973), Lifshin (1974), and Rao-Sahib and Wittry (1975).

A criterion for assessing the usefulness of Equation (8.7) is that it should remove the effects of all absorption edges over the energy range of interest. Successful application of this method to a large absorption edge in a pure silicon target measured at 20 keV is illustrated in Figure 8.8.

As pointed out by Ware and Reed (1973), the absorption term f_E is dependent upon the composition of the target. Consequently, when the composition of the target is unknown, it is necessary to include Equation (8.7) in an iteration loop when the ZAF method is being applied.

8.2.1.2. Detector Efficiency, P_E

The continuous radiation emitted from the target toward the detector penetrates a beryllium window, typically on the order of 8 μm thick, a surface-barrier contact ($\sim$20 nm of Au), and an inactive layer of silicon extending 200 nm or less into the detector. The radiation then enters the active (intrinsic) region of the detector which has a thickness typically between 2 and 5 mm. At the energy of the gold M edges, the absorption effects of the gold layer are generally insignificant. Hence, the effects of the gold and beryllium, which become significant at low energies, can be combined into an equivalent thickness, t_{Be}, which is that of a beryllium layer which would cause the same effect as that of gold and beryllium in combination. The absorption losses in the beryllium window, the gold, the silicon dead layer, and the transmission through the active silicon zone can therefore be calculated from

$$P_E = \exp - \left[(\mu/\rho)_{Be}^E t_{Be} + (\mu/\rho)_{Si}^E t_{Si}^D \right] \left\{ 1 - \exp - \left[(\mu/\rho)_{Si}^E t_{Si} \right] \right\} \quad (8.10)$$

In this equation, t_{Si}^D and t_{Si} are the thicknesses (g/cm^2) of the silicon dead layer and of the active detector region, respectively. The mass attentuation coefficients of beryllium and silicon at the energy E, $(\mu/\rho)_{Be}^E$ and $(\mu/\rho)_{Si}^E$, are calculated as described by Myklebust et al. (1979). Since sufficiently accurate values are not usually available from the manufacturer, estimates of the thicknesses of the beryllium window and the dead layer of the detector can be adjusted to optimize the fit between the calculated and the experimental spectra at energies below 2 keV.

An example of applying this technique to fitting the continuum from a complex sample is presented in Figure 8.9. The sample is a mineral, Kakanui Hornblende, with a chemical composition given in Table 8.1. A total of 1.4×10^7 counts were collected, making it possible to detect even the Mn concentration of less than 700 ppm. This method of background modeling and subtraction is used in several ZAF data reduction procedures

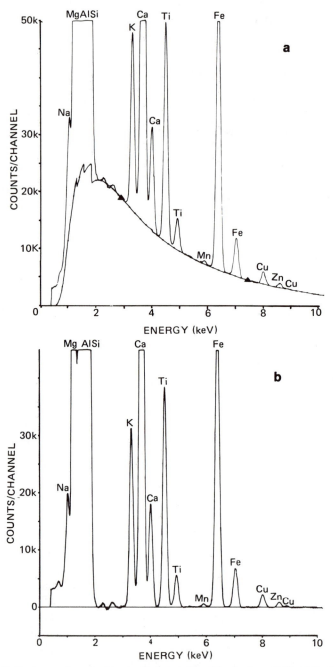

Figure 8.9. Background correction in Kakanui Hornblende (see Table 8.1). (a) Fitted continuum curve calculated by Equation (8.7); background fit points are indicated by triangles. Observed spectrum is superposed. Note the presence of Mn $K\alpha$ peak at 5.9 keV. The concentration of Mn is less than 700 ppm. Total spectral counts equal 1.4×10^7. (b) Background subtracted.

Table 8.1. Chemical Analysis of Kakanui Hornblende

Element	Concentration wt%	Element	Concentration wt%
Na	1.93	Ca	7.36
Mg	7.72	Ti	2.62
Al	7.88	Mn	0.0697
Si	18.87	Fe	8.49
K	1.17	O	Balance

including FRAME B, Fiori *et al.* (1976) and FRAME C, Mykelbust *et al.* (1979).

8.2.2. Background Filtering

In this method of background reduction the continuum component of an x-ray spectrum is viewed as an undesirable signal whose effect can be removed by mathematical filtering or modification of the frequency distribution of the spectrum. Knowledge of the physics of x-ray production, emission, and detection is not required.

Filtering techniques take advantage of the fact that the continuum component of an x-ray spectrum is smooth and slowly varying, as a function of energy, relative to the characteristic x-ray peaks, except at absorption edges. If we mathematically transform an x-ray spectrum from energy space into frequency space the result would be as shown in Figure 8.10. The horizontal axis gives the frequency of an equivalent sine wave component. For example, a sine wave with a full period of 0–10 keV would be plotted at channel 1 while a sine wave with a full period in 10 eV would be plotted at channel 1000. The vertical axis (F) gives the population of each sine wave. In this representation the continuum background which has a long period variation is found at the low-frequency end. The

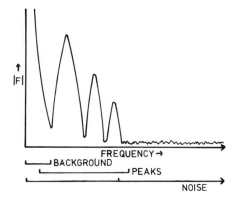

Figure 8.10. Frequency space (Fourier transform) representation of electron excited x-ray spectrum. (See text for explanation.)

characteristic peaks vary more rapidly than the background and are found at higher frequency. The noise is found at all frequencies.

We can now mathematically suppress the frequency components responsible for the slowly varying continuum and additionally suppress a portion of the frequency components responsible for the statistical noise. By then performing a reverse Fourier transform we go back to energy space with a spectrum which should now be comprised only of characteristic peaks devoid of continuum and which is, furthermore, "smoothed" for statistical variation. Unfortunately, close examination of Figure 8.10 reveals that the three main components of the transform overlap. Consequently it is not possible to suppress all the undesirable components without sacrificing part of the peak components. Similarly, it is not possible to keep all of the peak components without including part of the undesirable components. The result in either case will be a spectrum with undesirable distortions.

Although this approach was successfully applied to a number of specific systems, the question of what happens to the instrumentally broadened absorption edges and common frequency components between the background and peaks has not been studied in sufficient detail to ensure accurate analysis for all systems.

Another filtering technique which is now widely used was developed by Schamber (1978). It has been successfully applied to a wide variety of systems and uses what is known as a "top hat" digital filter. The top hat filter is a simple and elegant algorithm—a fact which can easily be obscured by the mathematical formalism required to fully describe it. Simply stated, the top hat filter is a special way of averaging a group of adjacent channels of a spectrum, assigning the "average" to the center "channel" of the filter, and placing this value in a channel in a new spectrum which we will call the filtered spectrum. The filter is then moved one channel and a new "average" is obtained. The process is repeated until the entire spectrum has been stepped through. The filter in no way modifies the original spectrum; data are only taken from the original to create a new spectrum. The "averaging" is done in the following manner. The filter, see Figure 8.11, is divided into three sections: a central section, or positive (+) lobe, and two side sections, or negative (−) lobes. The central lobe is a group of adjacent channels in the original spectrum from which the contents are summed together and the sum divided by the number of channels in the central lobe. The side lobes, similarly, are two groups of adjacent channels from which the contents are summed together and the sum divided by the total number of channels in both lobes. The "average" of the side lobes is then subtracted from the "average" of the upper lobe. This quantity is then placed in a new spectrum into a channel which corresponds to the center channel of the filter.

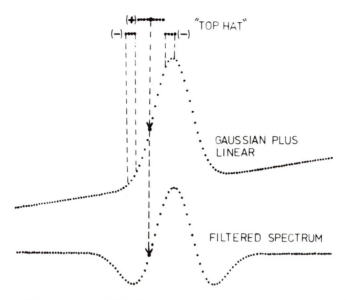

Figure 8.11. Effect of top hat digital filter on spectrum comprised of a Gaussian peak plus a sloped linear background. The filtered spectrum is plotted immediately below the actual spectrum. The channel correspondence for one calculation of the top hat filter is shown.

The effect of this particular averaging procedure is as follows. If the original spectrum is straight, across the width of the filter, then the "average" will be zero. If the original spectrum is curved concave upward, across the width of the filter, the "average" will be negative. Similarly, if the spectrum is curved convex upward the "average" will be positive. The greater the curvature, the greater will be the "average." The above effects can be observed, for a Gaussian superposed on a linear background, in Figure 8.11. In order for the filter to respond with the greatest measure to the curvature found in spectral peaks, and with the least measure to the curvature found in the spectral background, the width of the filter must be carefully chosen. For a detailed treatment of the subject see Schamber (1978) and Statham (1977). In general, the width of the filter for any given spectrometer system is chosen to be twice the full width at half the maximum amplitude (FWHM) of the Mn $K\alpha$ peak, with the number of channels in the upper lobe equal to or slightly more than the combined number of channels in the side lobes.

Because the top hat filter "averages" a number of adjacent channels, the effects of counting statistics in any one channel are strongly suppressed. Consequently, in addition to suppressing the background under spectral peaks, the digital filter also "smooths" a spectrum. Note that in Figure 8.11, the top hat filter converts the sloped background into a flat background.

In conclusion, the effects of passing a top hat digital filter through an x-ray spectrum as recorded by a Si(Li) spectrometer system are to (1) strongly suppress the background and statistical scatter and (2) significantly alter the shape of the spectral peaks. The result strongly resembles the smoothed second derivative; however, this distortion has no adverse statistical or mathematical effects of any consequence. Clear advantages of the method are simplicity and the fact that an explicit model of the continuum is not required. However, since the continuum has been suppressed, the information it carried (i.e., average atomic number, mass-thickness, etc.) is no longer available.

8.3. Peak Overlap

To measure the intensity of an x-ray line in a spectrum we must separate the line from other lines and from continuum. The previous section discussed how to remove the average effect of the background. This section will discuss various ways to isolate an x-ray line from the average effect of other lines when spectral overlap occurs. We will assume in the following discussion that our spectra have been background corrected.

Resolution is the capability of a spectrometer to separate peaks close in energy or wavelength. It is conventionally characterized by the full width of a peak at half the maximum amplitude (FWHM). The resolution of an energy dispersive detector is usually specified at the energy of Mn $K\alpha$ (5.895 keV). Typical detectors have a resolution between 140 and 155 eV. A wavelength-dispersive spectrometer, however, typically has a resolution of 10 eV at the energy of Mn $K\alpha$ (which has a natural width of approximately 2 eV). Therefore, peak interference is rarely a problem with the wavelength-dispersive spectrometer but is often present with the energy-dispersive detector. Consequently, we will consider here only the overlap situation encountered in applications of the energy-dispersive detector. For a detailed treatment of line interference correction in WDS see Gilfrich *et al.* (1978).

When the resolving power of the energy-dispersive spectrometer is insufficient for the case in hand it is necessary to consider ways to isolate a characteristic x-ray peak from the average effects of other peaks by some type of mathematical procedure. It is the purpose of this section to discuss the various techniques in use today. The computer-generated continuous curves in Figure 8.12 illustrate the problem. The upper curve, which is a composite of the two labeled underlying curves, is the sort of spectrum we would see from an energy-dispersive detector if there were no background or statistical modulation and the horizontal (energy) axis was essentially continuous rather than discrete. The underlying curves are what we must

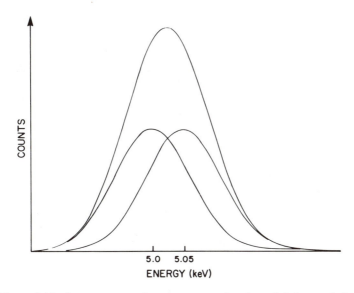

Figure 8.12. Computer generation of two spectral peaks and their convolution.

reconstruct given only the information contained in the upper curve. The area under each of the lower curves represents the number of characteristic x-ray photons recorded by the energy-dispersive detector system and is one of the quantities we require in any method of quantitative data reduction scheme such as the ZAF method.

Procedures which address themselves to the solution of this problem have a variety of names such as curve-fitting, deconvolution, peak stripping, peak unraveling, multiple linear least-squares curve fitting, etc. We will attempt to make the distinctions clear.

All of the procedures just mentioned start with a data set usually obtained from a multichannel analyzer. The data set obtained exists as a one-dimensional array in a computer memory. Consequently, the data are discrete. We need here to reconcile a possible difficulty with terminology. In the nomenclature of the multichannel analyzer each "point" is called a "channel" and a collection of adjacent channels would be called a spectrum. In the nomenclature of the mathematical methods about to be described each channel is referred to as an "element" and a collection of adjacent elements would be called an array, or a vector. However, because of precedent, we will use the terms "channel" and "spectrum" regardless of the context. Consider now a second but corresponding spectrum also existing in a one-dimensional array. We will call this a "calculated" spectrum. This spectrum can be generated in several ways. One way is to utilize a mathematical model to separately describe each peak in the spectrum. At each channel in the calculated spectrum the effects of all the peaks at that channel are summed together. This process is called convolution and it is by this process that we can generate a spectrum. The mathematical model used to describe the shape of each peak usually has a minimum of three parameters, one to specify the amplitude of the peak, one to describe its width, and one to describe its position (energy). The Gaussian (normal) profile is the most often used model for the peaks.

An alternative method of generating a spectrum is to add together simpler spectra (corrected for the average effects of the background) obtained by actually measuring specimens of known composition under known conditions. Each of these spectra is referred to as a "reference" spectrum. For example, if the unknown spectrum was obtained from a copper–zinc brass the two reference spectra would be obtained by measuring pure copper and pure zinc. In this case the only parameter we have to adjust is the amplitude of each of the two reference spectra.

Whether we obtain a calculated spectrum by utilizing a mathematical model for each constituent peak and then adding these or whether we record several simpler spectra which are then scaled and added together, a criterion is needed to determine if the calculated spectrum matches the unknown as closely as possible. We also require a procedure to adjust the various parameters to provide this "best" fit. It is the procedure by which the parameters are adjusted which serves mainly to distinguish the different methods used to account for spectral overlaps. We will discuss in detail several of the more widely used methods and problems common to all of the methods. Before proceeding to particular techniques we will first define linearity and then discuss the criterion most often used to decide if a given fit to a set of spectral peaks is the "best" fit.

8.3.1. Linearity

All of the procedures which address the problem of isolating one peak from the interfering effects of other peaks can be classified into one of two categories: The linear procedures and the nonlinear procedures. We require a definition of linearity as it applies to the above problem. Linearity (or nonlinearity) is a property of the fitting parameters. If all of the parameters which are being adjusted to provide the best fit in a given procedure are used in a simple multiplicative or additive fashion, then the procedure is linear. This definition applies only to the parameters actually being adjusted. There can be other parameters which are used in a nonlinear manner, but as long as they remain fixed during the actual peak fitting then the procedure remains a linear one. An example will help to clarify these points.

The profile of a characteristic x-ray peak obtained from an energy-dispersive spectrometer is closely approximated by a Gaussian (normal) probability distribution function. That is, the contents, Y_i, of any given channel comprising a given Gaussian profile can be calculated from

$$Y_i = A_P \exp\left[-\frac{1}{2} \frac{(E_P - E_i)^2}{\sigma^2} \right] \qquad (8.11)$$

where E_i is the energy (in appropriate energy units, e.g., electron volts) of the ith channel, E_P is the energy (same energy units as E_i) of the profile center, and A_p is the amplitude of the profile at its center. It should be noted that the center of the profile does not have to coincide in energy with the mean energy of any given channel. The parameter σ specifies the width of the profile (again, in the same units as E_i). Of the three parameters, A_P, E_P and σ, only A_P is linear since it is present with a first power exponent. Consequently, any procedure which utilizes Gaussian profiles and adjusts either the width and/or profile center during the fitting is a nonlinear procedure. There is one exception worth mentioning. If only one peak is being fitted it is possible to take the natural logarithm both of the unknown spectrum and of Equation (8.11) and, after some appropriate algebraic manipulation, derive a new set of parameters which are linear. These parameters can be transformed back into the original parameters after the fitting has been accomplished.

8.3.2. Goodness of Fit

We require a criterion with which to evaluate the goodness of fit for a given peak-fitting procedure in a particular application. That is to say, is there a quantity which will tell us how closely our calculated peak(s) matches the peak(s) in an unknown spectrum? A criterion which is widely used is the chi-squared (χ^2) criterion. The set of parameters used in a given fitting procedure which cause χ^2 to reach a minimum value will be the set with the greatest likelihood of being the correct parameters. The χ^2 criterion can be reasonably approximated by the following functional form:

$$\chi^2 \approx \sum_i \frac{(Y_i - X_i)^2}{X_i} \qquad (8.12)$$

where Y_i is the contents of the ith channel of the unknown spectrum and X_i is the contents of the ith channel of the calculated spectrum. For the equation as written to be a proper χ^2 it is required (1) that repeated measurements of Y_i for a given channel i be normally distributed with a true variance of X_i; (2) that the equation we have chosen to describe a peak(s) (or the reference spectrum we have chosen) is a correct representation in all aspects (e.g., a mathematical function would include any peak distortions such as incomplete charge, etc.). Under many conditions of peak fitting the above conditions are approached with sufficient accuracy that a minimum value of Equation (8.12) will mean that the set of fitting parameters which produced the minimum will be the "best choice" parameters. Note that because we square the differences between the two spectra, χ^2 is more sensitive to larger differences than it is to small differences.

A useful variation of Equation (8.12) is the "normalized" χ^2

$$\chi_N^2 = \sum_{i=1}^{M} \frac{1}{X_i} \frac{(Y_i - X_i)^2}{M - f} \tag{8.13}$$

where M is the number of channels used in the fit and f is the number of parameters used in the fit (i.e., A_P, σ, E_P). The utility of Equation (8.13) lies in the range of χ^2 values it will produce. When $\chi^2 \approx 1$ the fit is essentially perfect while for values of $\chi^2 \gg 1$ the fit is poor. For a high-quality, low-noise spectrum, a χ^2 value greater than 100 would indicate a bad fit.

8.3.3. The Linear Methods

As stated previously, linearity requires that only those parameters which are used in a multiplicative or additive manner may be adjusted during the fitting procedure. A simple additive parameter has the effect of moving a peak or spectrum with respect to the vertical (amplitude) axis and consequently is useful mainly for accommodating the average effect of the background if background were to be included in the fitting procedure. While it is possible to include background along with x-ray peaks, it unnecessarily complicates the mathematics, can cause convergence problems, and substantially increases the computational time. Consequently, as mentioned earlier, we will assume that the average effect of the background is first removed from a spectrum before we attempt to unravel the spectral overlaps.

Besides addition, the other possible method by which a parameter can enter an expression, and retain the property of linearity, is multiplication. The practical consequence of multiplication is that only the amplitudes of peaks can be determined in a linear fitting procedure. At first glance this seems to be a severe restriction because it requires us to know beforehand the exact energy calibration of our spectrometer system (that is, the precise position of each peak in the spectrum), and it requires us to know the exact width of each peak. Assuming values for either of these parameters, especially peak position, which differ from corresponding values in our "calculated" spectrum will result in the amplitudes of some, or all, of the peaks to be incorrectly determined by the fitting procedure. Fortunately, the stability, linearity, and count rate performance of modern spectrometer systems are generally adequate when certain precautions are observed (see Chapter 5).

8.3.3.1. Multiple Linear Least Squares

The most popular linear method is the multiple linear least squares technique. Assume that we have a number of channels i which comprise a background-corrected, measured spectrum of N overlapped peaks consisting of Y_i counts in each channel. Further assume a "calculated" spectrum of N corresponding peaks. Each peak is described by a Gaussian profile [Equation (8.11)] for which we must estimate as best we can the width and energy. The calculated spectrum can be mathematically described by

$$Y_i = \sum_N A_N \exp\left[-\frac{1}{2}\left(\frac{E_N - E_i}{\sigma_N} \right)^2 \right] \tag{8.14}$$

where A_N is the amplitude, E_N is the energy, and σ_N is the width parameter for each peak N, and E_i is the energy of each channel i in the calculated spectrum. E_N, E_i, and σ_N each have dimensions of energy (usually eV or keV). We need to find a set of amplitudes, A_N, which will cause the calculated spectrum to match the measured spectrum as closely as possible. The contents, Y_i, of each channel in the measured spectrum will have an associated error, ΔY_i, due to Poisson counting statistics. This error is proportional to $(Y_i)^{1/2}$. Since channels which are near peak centers can contain many more counts than those which are in the tail of a peak it is common practice to weight the effect of each channel such that it has approximately the same effect in the fitting procedure as any other channel. The weighting factor, W_i, is typically chosen to be proportional to $1/(\Delta Y_i)^2$, which is equal to $1/Y_i$. The principle of least squares states that the "best" set of A_N will be those which minimize

$$\chi^2 = \sum_i \left(Y_i - \sum_N A_N G_{Ni} \right)^2 W_i \tag{8.15}$$

where G_{Ni} is the Gaussian profile [the exponential portion of Equation (8.11)]. The desired set of A_N can be found by the following procedure. Successively differentiate Equation (8.15) with respect to each of the A_N. The result will be a set of N equations of the form

$$2\sum_i \left[\left(Y_i - \sum_N A_N G_{Ni} \right)(-G_{Ni})W_i \right] \tag{8.16}$$

Equating each of these to zero serves to couple the equations (which can now be called "simultaneous" equations) and permits us to find the desired minimum. One way to solve the set of equations is by matrix inversion. Equation (8.16) can be written in matrix notation as

$$[B] \times [A] = [C] \tag{8.17}$$

where

$$B_{mn} = \sum_i G_{mi} G_{ni} W_i$$

and

$$C_m = \sum_i G_{mi} Y_i W_i$$

The subscripts m and n refer to the rows and columns of the matrix. Consequently, we can find our "best" set of amplitudes A_N by

$$[A] = [B^{-1}] \times [C] \qquad (8.18)$$

We note that the set A_N has been found analytically. That is to say, we have determined a set of equations which, when solved, give a unique answer which is "most likely" (in the statistical sense) to be correct. The analytical property is a major distinction between the linear and nonlinear methods of peak fitting. In the nonlinear methods we do not have a closed form set of equations which must be solved. We must resort, for example, to iteration techniques, or search procedures to give us the answer we need. We must always keep in mind, however, how we achieved the luxury of linearity. We have made the rather bold assumption that we know very well the width and location of every peak in our observed spectrum. The determination of these parameters, to the required accuracy, is quite feasible but is far from being a trivial exercise.

In this description a "calculated" spectrum was constructed from simple Gaussian profiles. Other functions could have been used, such as a slightly modified Gaussian, to accommodate the small deviation of observed x-ray peaks on the low-energy side from a true Gaussian (described in the next section). Reference (measured on standards) spectra could have been used rather than functions to create the "calculated" spectrum. Finally, digitally filtered spectra both for the measured and the calculated spectra could have been used. The original purpose in applying the digital filter was to remove the average effects of the background. The application of a digital filter to a spectrum modifies the x-ray peaks from nearly Gaussian profiles to approximations of smoothed second derivatives of Gaussian profiles (see Figure 8.11). Neither the application of a digital filter nor the use of functions other than Gaussian affect the "linearity" of the method just discussed. In all cases the parameter being determined is peak amplitude, which is a linear parameter even in the case of digitally filtered spectra.

8.3.3.2. Method of Overlap Coefficients

One of the simplest of the linear techniques is the method of overlap coefficients. In this method we do not utilize an entire spectrum channel by channel as we did, for example, in the method of multiple linear least-squares. Instead, we sum a group of adjacent channels which straddle an x-ray peak. This group of channels we will call a "region of interest." Most multichannel analyzers provide a method (hardware or software) for directly selecting such regions of interest. Calculating the sum of the intensities in a region of interest is equivalent to performing a Simpson numerical integration of a region over a spectral peak. Consequently, each x-ray peak can be completely characterized by three numbers—the lower and upper limits of the region, and the summed contents. These three numbers stand in contrast to the 20 to 40 numbers (channel locations and contents) required to specify a single peak in the multiple linear least-squares method.

The method of overlap coefficients was originally developed by Dolby (1959) to resolve overlapping proportional counter curves and works in the following manner. We first assume, as explained earlier, that our spectra have been corrected for the average effects of the background. If we have i mutually interfering spectral peaks, and N_i^T is the total number of counts due to the effects of all peaks in the ith region of interest, then we can write a set of simultaneous equations of form

$$N_i^T = k_i N_i^P + \sum_{j \neq i} k_j N_j^P C_{ji} \qquad (8.19)$$

In this equation N_i^P is the number of counts in the ith region of interest due to a measurement on pure element i, k_i and k_j are the relative intensities from elements i, j, etc. in our unknown specimen, N_j^P is the number of counts in the jth region of interest due to a measurement on pure element j, and C_{ji} is the "overlap coefficient" or the fraction of the number of counts in the ith region of interest determined from a measurement on pure element j. The k_i, k_j etc. are the k ratios required by a ZAF data correction program. The set of simultaneous equations specified by Equation (8.19) can be solved for this set of k values by algebraic methods. The following example will demonstrate the simplicity and power of the method.

Assume that we have three mutually interfering Gaussian peaks separated from one another by 125 eV. The peak centers are at 5875, 6000, and 6125 eV, respectively, and each peak is the appropriate width for a detector having a resolution at Mn $K\alpha$ of 160 eV. These peaks (dashed lines) and their convolution (solid line) are shown in Figure 8.13. The heavy lines denote the regions of interest. Let the region of interest for each peak be centered over the peak and be 100 eV in width. Assume the measurement of each pure element and the unknown material produce counts in the three regions of interest as shown in Table 8.2. Using the data in Table 8.2 we can write out Equation (8.19) as

(element 1) $6598 = k_1 \cdot 17466 + k_2 \cdot 17344 \cdot 0.257 + k_3 \cdot 17298 \cdot 0.0039$

(element 2) $6704 = k_1 \cdot 17466 \cdot 0.226 + k_2 \cdot 17344 + k_3 \cdot 17298 \cdot 0.289$

(element 3) $9412 = k_1 \cdot 17466 \cdot 0.0035 + k_2 \cdot 17344 \cdot 0.257 + k_3 \cdot 17298$

$$(8.20)$$

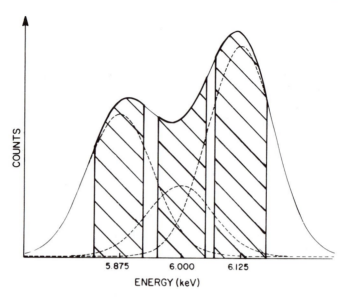

Figure 8.13. Computer generation of three spectral peaks (dashed lines) and their convolution (solid lines). Heavy lines denote regions of interest. (See text for explanation.)

Simultaneous solution of these equations gives $k_1 = 0.333$, $k_2 = 0.167$, and $k_3 = 0.5$.

In practice spectral interference is often more complicated than in the previous example. X-ray lines usually come in families and frequently it is not the peak which we are measuring which interferes with another peak of interest but another family member of the first peak. Furthermore, it is not always possible to obtain pure elements or simple standards from which we can determine overlap coefficients. Fortunately the method can be expanded to accommodate these difficulties (Fiori *et al.*, 1976; Smith, 1975). The method described in the first reference is used in the EDS data reduction procedure FRAME C (Myklebust *et al.*, 1979), which we will summarize here.

Table 8.2. Contents of Regions of Interest Shown in Figure 8.13

	Energy range (eV)		
Counts from	5825–5925	5955–6055	6075–6175
Pure element 1	17466	3941	61
Pure element 2	4454	17344	4454
Pure element 3	68	4998	17298
Unknown	6598	6704	9412

The FRAME C method is based on calculating overlap coefficients rather than measuring them. Calculation has the advantage of flexibility since the width and positions of each region of interest may be changed without the need to remeasure. Most importantly, standards are not required. The coefficients are calculated by formal integration of a Gaussian function which has been modified to accommodate the effects of incomplete charge collection. One x-ray line (which we will call the analytical line) from each element has a region of interest and is used in the ZAF computation to determine the specimen composition. All the other family lines of that element of any consequence are considered to determine if the lines overlap with the regions of interest of the analytical lines of other elements of interest.

Given an x-ray peak near a region of interest of another peak, the fraction of this peak which falls within the region only depends on the relative positions of the peaks and the region of interest and of the standard deviation of the peak. The fraction does not depend on the composition of the specimen. If the interfering peak is also an analytical peak and therefore has its own region of interest, the fraction of this peak within its own region of interest is equally independent of composition. Therefore, the ratio between these two fractions, the overlap coefficient, is computed once early in the procedure and stored for all subsequent measurements. We will develop the argument for two unresolved peaks, A and B (see Figure 8.14). We will only describe the computation of the overlap factor due to peak A

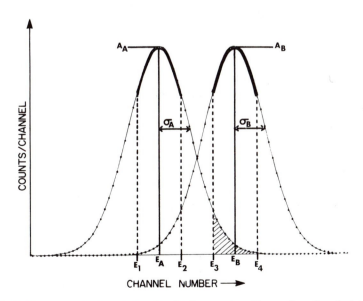

Figure 8.14. Two unresolved spectral peaks showing overlap. The overlap of peak A in the region of interest of peak B is shown as the cross-hatched area.

on the intensity under peak B but the technique could be repeated for peak B.

Let E_A be the center of peak A and E_B the center of peak B. We can calculate the standard deviations σ_A and σ_B as a function of E_A and E_B (Fiori and Newbury, 1978) if the FWHM has been measured for a peak at any energy (e.g., Mn $K\alpha$). The number of counts due to peak A at any channel number E_i is

$$N_i = A_A \exp\left[-\frac{1}{2}\left(\frac{E_A - E_i}{\sigma_A} \right)^2 \right] \tag{8.21}$$

where E_A is the centroid of the peak A, and A_A the counts in a channel of narrow width ($\leqslant 10$ eV) containing E_A.

Consequently, the counts N_A obtained between the limits E_1 and E_2 which define the region of interest of element A are:

$$N_A = A_A \int_{E_1}^{E_2} \exp\left[-\frac{1}{2}\left(\frac{E_A - E_i}{\sigma_A} \right)^2 \right] dE_i \tag{8.22}$$

while the counts of peak A between the limits E_3 and E_4 within the region of interest of element B are

$$N_{AB} = A_A \int_{E_3}^{E_4} \exp\left[-\frac{1}{2}\left(\frac{E_A - E_i}{\sigma_A} \right)^2 \right] dE_i \tag{8.23}$$

The overlap coefficient H of peak A upon B is defined as

$$H_{AB} = \frac{N_{AB}}{N_A} \tag{8.24}$$

The number of counts in peak A between the limits E_1 and E_2 is multiplied by this factor to obtain the number of counts contributed by peak A to the region of interest of peak B.

A different type of interference occurs when a peak from element A which is not used for data reduction such as a $K\beta$ line falls within the region of interest selected for element B (Figure 8.15). In this case, the intensity of the interfering peak from element A must be obtained by using the counts within the region of interest of the analytical peak for element A. If the analytical line of element A and the interfering line of element A are both generated from ionization of the same shell or subshell, their generated intensities are related by their respective relative transition probabilities (or weights of lines). A transition probability φ is the intensity I of the line of interest divided by the sum of the intensities $\sum I$ of all other lines arising from the same absorption edge. The detected intensities are also affected by the respective absorption factors, and the efficiency of the

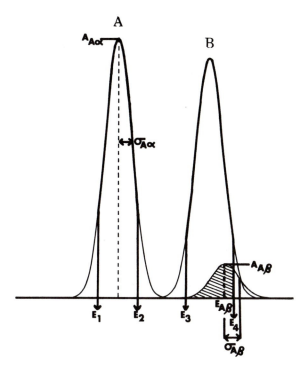

Figure 8.15. Example of interference which can occur when a peak from element A falls within the region of interest (E_3 to E_4) selected for element B.

detector at the respective energies. The ratio of the detected intensities D_I is

$$D_I = \frac{\varphi_{AK}}{\varphi_{AK}} \frac{f(AK\beta)}{f(AK\alpha)} \frac{P_{E(AK\beta)}}{P_{E(AK\alpha)}} \tag{8.25}$$

where $\varphi_{AK} = I(AK\beta)/\sum I(AK)$ is the transition probability of the $AK\beta$ line, $f(AK\beta)$ is the corresponding absorption correction factor, and P_E the detector efficiency at the energy E. Since the absorption factor, f, is a function of specimen composition the overlap correction procedure must be included in the ZAF iteration loop. The number of counts in the region of interest of peak B due to the subsidiary peak from element A, $N_{A\beta,B}$, can be calculated from

$$N_{A\beta,B} = A_{A\alpha} D_I \frac{\sigma_{AB}}{\sigma_{A\alpha}} \int_{E_3}^{E_4} \exp\left[-\frac{1}{2}\left(\frac{E_P - E_{A\beta}}{\sigma_{A\beta}}\right)^2\right] dE \tag{8.26}$$

where the terms have been given previously. The integration limits $E_3 - E_4$ define the region of interest for peak B. The hatched area in Figure 8.15 represents the intensity $N_{A\beta,B}$ which must be subtracted from the intensity in the region of interest of element B. To calculate the overlap coefficient $H_{A\beta,B}$ the value of $N_{A\beta,B}$ from Equation (8.26) is substituted in Equation (8.24).

There are several other cases of overlapping peaks which are encountered in analysis. These cases include: (1) the overlap of lines where the subsidiary line of element A which interferes with element B from a different subshell than the analyzed line of A, e.g., $L\alpha_1 - L\beta_1$; (2) the overlap of lines where a line(s) from another family of element A interferes with element B, e.g., the measured line is Zn $K\alpha$, but Zn $L\alpha$ interferes with Na $K\alpha$; and (3) a silicon escape peak of the analyzed line of element A interferes with element B. The mathematical methods for handling these cases have been described by Myklebust *et al.* (1979).

We noted at the beginning of the discussion on the overlap coefficient method that each x-ray peak can be completely characterized by three numbers—the lower and upper limits of the region of interest, and the contents. This fact and the additional property that the method of overlap coefficients requires a relatively small number of simple arithmetic steps gives the procedure its most appealing characteristics: simplicity and speed. The method of overlap coefficients requires the smallest amount of computer memory and is a minimum of ten times faster than any of the other methods described. As with all of the linear techniques it is required that the spectrometer system be well calibrated. For x-ray peaks which are severely overlapped (i.e., Pb $M\alpha$ and S $K\alpha$) the method is inferior to those procedures which utilize the information in each channel across the overlapped portion of a spectrum.

8.3.4. The Nonlinear Methods

We have seen in the previous section that a major advantage of the linear technique is that it is analytic. We have only to solve a set of equations to get the information we require. The amplitudes returned by a linear peak-fitting procedure are unique and are "most probably" the correct amplitudes. As with everything, very little comes without price and the price paid to gain linearity is that the user must provide the fitting procedure with information about the width and position of all peaks rather than the other way around. The linear procedure then uses this information without change, since it is, by definition, correct. There is a school of thought which suggests that the user cannot always be so sure about his knowledge of peak width and location and these should be included, with amplitude, as quantities to be determined by the fitting procedure. Both linear and nonlinear peak fitting work well in general and each has areas where it is superior. While the nonlinear techniques are capable of determining fitting parameters other than amplitudes the price we must pay for this luxury is dealing with a procedure which is nonanalytic and quite capable of returning an answer which is wrong if certain precautions are not observed. In the following paragraphs we will describe in some detail one of the methods presently in use—the sequential simplex.

A sequential simplex procedure is a technique which can also be used for selection of the set of independent variables in a mathematical expression which cause the expression to be a "best" fit (in a statistical sense) to a set of data points (Nelder and Mead, 1962; Deming and Morgan, 1972).

In this procedure, each of the n independent variables in the function to be fitted is assigned an axis in an n-dimensional coordinate system. A simplex, in this coordinate system, is defined to be a geometric figure consisting of $n + 1$ vectors [in this discussion we will use the purely mathematical definition of a vector, i.e., an ordered n-tuple $(X_i, X_2, \ldots, X_n)$]. In one dimension a simplex is a line segment; in two dimensions a triangle; and in three or more dimensions a polyhedron, the vertices of which are the above-mentioned $n + 1$ vectors. The simplex is moved toward the set of independent variables which optimize the fit according to a set of specific rules. The function used to determine the quality of the fit for any set of independent variables is called the "response function."

We will use a simple example to demonstrate the essence of the method. Consider the problem of finding the "best" straight line through a set of data points. A straight line is defined by the two quantities m, the slope of the line and b, the y-axis intercept of the line. The functional form is

$$y = mx + b \qquad (8.27)$$

The quantities we wish to optimize by a simplex search are m and b. We start the process by estimating, as best we can, three lines each defined by its own slope m_i and y-axis intercept b_i (Figure 8.16). Each of these three lines can be characterized by a vector (m_i, b_i) on a m, b plane. The three vectors form a simplex—in this case a triangle—on the plane. If we now define a third orthogonal axis to represent the "response function" the simplex search can be visualized (Figure 8.17). The search proceeds by calculating the response above each vertex of the simplex, discarding the vertex having the highest (worst) response, and reflecting this point across the line connecting the remaining two vertices. This defines a new simplex and the process is repeated until the global minimum on the response surface is reached.

The problem of fitting a straight line to a set of data points is, of course, a linear problem. However, the method of sequential simplex is an organized procedure to search for a minimum on a response surface and is usable whether the function describing the data set is linear in its coeffi-

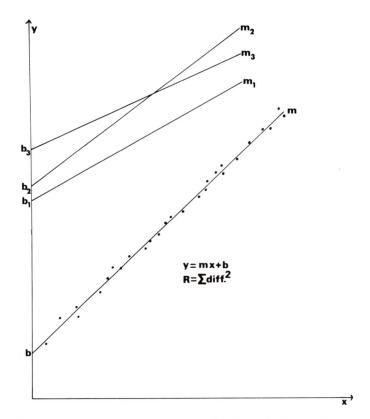

Figure 8.16. Illustration of a simplex procedure. A "true" straight line and three estimates to this line are plotted. Each line is characterized by its own slope, m, and y intercept, b.

cients or not. If the coefficients are linear, or can be made linear by a suitable transformation, the least-squares determination can be accomplished analytically. In this case a search procedure would not be used and is done so here only for the purpose of explanation.

We now require a functional description of an x-ray peak just as required in the linear procedure. Again, we can use an additive combination of N Gaussian profiles [see Equation (8.14)]. The fitting procedure involves the calculation of the function which is then compared with the experimental data points. The purpose of the response function is to provide a measure of the goodness of fit to the data points. A response function which can be used is the normalized chi-square function [see Equation (8.13)].

From Equation (8.11) we see that the independent variables (coefficients) A_p, E_p, and σ must be determined for each peak. Therefore the total number of coefficients, n, is three times the number of peaks N. Here,

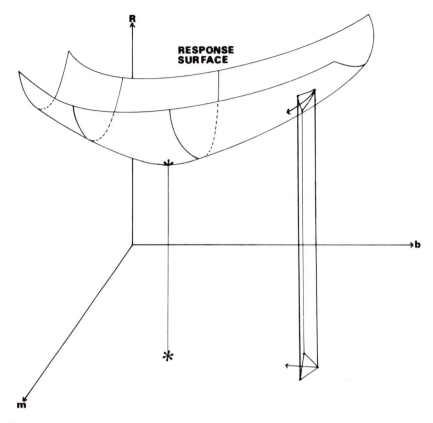

Figure 8.17. Response surface showing simplex and projection onto factor space. The global minimum is shown by the point indicated by an asterisk.

values for $n + 1$ different sets of coefficients are chosen to create the initial simplex and the response, R (Figure 8.17), is determined for each set.

Fiori and Myklebust (1979) introduced several simplifications which reduce the number of coefficients to be fitted. Since the energies of x-ray lines are well known, they are entered as known quantities and it is unnecessary to include all of the peak centroids as coefficients. The energy of only the principal peak is used as a coefficient to correct for small shifts in energy due to miscalibration of the electronics of the energy-dispersive spectrometer. The stability and linearity of present-day amplifiers permits this assumption. In addition, the width σ_P of the principal peak must be included since the widths of the other peaks σ_i are related to σ_P [for a Si(Li) detector, Fiori and Newbury (1978)]:

$$\sigma_i = \left[2500(E_i - E_P) + (2355\sigma_P)^2 \right]^{1/2} / 2.355 \ (\text{keV}) \qquad (8.28)$$

The number of coefficients required is, therefore, reduced from $3N$ to $N + 2$. Consequently, only one more coefficient (amplitude) must be included for each additional peak used in a fit. Since the energies and widths of small unresolved peaks are determined as functions of the principal peak, opportunities for obtaining false minima are considerably reduced. Significant savings in computation time are also realized.

A difficulty common to all minimization methods is the possibility that a local minimum may be found before the global minimum. As the number of peaks to be fitted increases so does the possibility of finding false minima. Consequently, the simplex must be started closer to the unique set of coefficients which determine the global minimum than to a set which determine a local (false) minimum.

8.3.5. Error Estimation

It is the purpose of this section to briefly discuss the various errors which occur when corrections are applied to account for spectral overlap. By "error" we mean the deviation between the true amplitudes and the calculated amplitudes. We have to consider two types of error. First, there are the unavoidable errors due to fundamental statistical limitations and inadequacies in our algorithms. These errors, in general, cannot be avoided but can often be quantified or at least estimated. The second type of error is experimental. This type of error can often be avoided but, in general, cannot be quantified. Examples of the second type include imperfect subtraction of the average effect of the background, distortion of spectral peaks due to electrical ground loops, direct entrance of backscattered electrons into the detector, etc.

The amplitude information we receive is rarely an end product. This information is fed to another procedure which accounts for specimen and standard matrix effects (for example, a ZAF data reduction program). As we have seen in Chapter 7, the matrix correction programs themselves have associated errors which add in quadrature to the peak amplitude errors. Correcting for spectral overlap will always contribute an error in peak amplitude greater than what would be expected from consideration of the counting statistics alone. This additional error due only to the unraveling process can often dwarf all other errors in quantitative x-ray microanalysis. It can categorically be stated that quantitative analysis utilizing spectra without spectral overlap will always be superior to those analyses in which overlap occurs, all other things being equal. For a theoretical treatment of the error due to unraveling see Ryder (1977).

In the discussion on multiple linear least squares, we noted that one way to find the "best" set of amplitudes A_n was to solve the matrix equation

$$[B] \times [A] = [C] \qquad (8.29)$$

by the method of matrix inversion

$$[A] = [B^{-1}] \times [C] \qquad (8.30)$$

The matrix $[B^{-1}]$ is often called the error matrix because its diagonal, designated B_{ii}^{-1}, has the useful property that

$$\sigma_{Ai}^2 \approx B_{ii}^{-1} \qquad (8.31)$$

where σ_{Ai}^2 is the variance associated with the ith amplitude. Consequently, we have a simple means by which to estimate the quality of our fit in terms of each peak. If we have N peaks we will have N estimates of uncertainty, one for each amplitude returned by the fitting procedure. In general, the uncertainity associated with unraveling overlapped spectral peaks increases rapidly as the number of counts under any given peak decreases and/or as the separation between peaks decreases. At a certain point the uncertainty becomes so great that spectral unraveling becomes dangerous.

Consider the problem of determining the true peak heights for two closely spaced, overlapped peaks such as those shown in Figure 8.12. The user who is employing one of the unraveling methods will obtain as results a set of calculated amplitudes and an associated χ^2 value. Since this χ^2 value is supposedly the minimum value in the set of χ^2, it indicates that the peak amplitudes must be the best choice. We can calculate χ^2 for a range of peak amplitudes, and then examine the so-called χ^2 surface to determine how sensitive the χ^2 parameter is to the selection of the true peak amplitudes. Figures 8.18 and 8.19 show such χ^2 surfaces calculated for the case of two peaks separated by 50 eV and centered at 5.0 and 5.05 keV, obtained from a detector with a resolution of 150 eV at Mn $K\alpha$. R is the χ^2 value which is obtained for such selected value of peak amplitudes A and B. In Figure 8.18, the true peak amplitudes were equal at 1000 counts while in Figure 8.19 the true amplitude ratio is 5:1 (1000:200 counts). The χ^2

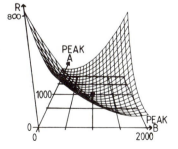

Figure 8.18. Pseudo-three-dimensional plot of χ^2 response surface for deconvolution of two peaks of equal amplitude separated by 50 eV. True solution is indicated by heavy point. Calculation was made by varying peak amplitude while holding peak σ and energy constant.

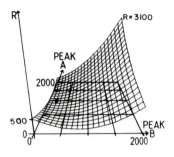

Figure 8.19. Pseudo-three-dimensional plot of χ^2 response surface for deconvolution of two peaks of 5 : 1 amplitude ratio separated by 50 eV. True solution is indicated by heavy point. Calculation was made by varying peak amplitude while holding peak σ and energy constant.

value plotted on the R axis in these figures is obtained by comparing the values of the hypothetical composite curve generated for each selection of the amplitudes of A and B. It can be seen from both plots that a sharply defined minimum in the response surface does not exist. A long, flat-bottomed valley exists near the true solution. Thus, it is not possible to determine the true solution with high confidence using the χ^2 minimization criterion. If the uncertainty due to counting statistics is included, the χ^2 surfaces would be "pot-holed" with local false minima, making the selection of the true solution even more difficult.

This discussion demonstrates that spectral unraveling techniques are not omnipotent. The problem of separating an overlap such as S $K\alpha$ (2.308 keV) and Pb $M\alpha$ (2.346 keV) is extremely difficult and in general unreliable. It is, unfortunately, a common belief that the techniques which we have been discussing have unlimited ability to provide accurate determinations of peak amplitudes no matter how bad the overlap, and/or how few counts are in the peaks. It is for this reason alone that some measure of expected error (such as the error matrix) for individual peaks is so valuable.

The linear methods, in general, all provide a simple means to estimate the fitting error for each peak through the error matrix. The nonlinear methods, on the other hand, do not provide a convenient technique for the error estimates of the individual peaks. It should be noted that both linear and nonlinear methods supply a single measure of the error associated with the entire fit over all of the peaks. This measure is simply the normalized chi-squared, χ_N^2. However, χ_N^2 does not tell us very much about the errors associated with the individual peaks which comprise the overlapped spectrum. We must note, however, that the error matrix of the linear methods is valid only when the conditions which permit a linear assumption are satisfied. That is, we accurately know each peak width and position in our spectra. If we make an error in either or both of these quantities then the amplitudes determined by the fitting procedure will be in error—and the error matrix will not manifest this deviation.

In conclusion, this section has discussed several of the errors associated with unraveling spectral overlap. The several procedures in use today are powerful tools when intelligently used but all have limits to their capabilities. The wise user will be aware of these limits.

8.4. Dead-Time Correction

In order to obtain accurate values of the peak intensities, corrections must be applied for the dead time associated with the measurement of an x-ray. Dead time is the time interval after a photon enters the detector during which the system cannot respond to another pulse. The dead time for an energy-dispersive spectrometer is corrected directly by the pulse-processing electronics during spectral measurement as described in Chapter 5. The analyst should observe the cautions indicated there for testing and proper operation of the dead-time correction system.

In the wavelength-dispersive spectrometer, the dead-time correction is applied after the intensity is measured. For the proportional counter used in the WDS, Heinrich *et al.* (1966) showed that the Ruark–Brammer relation could be employed up to count rates of at least $5 \times 10^4/s$:

$$N = N'/(1 - \tau N') \tag{8.32}$$

where N' is the measured count rate, N is the true count rate which we wish to calculate, and τ is the dead time in seconds. To apply Equation (8.32), the dead time τ must be known. One method for determining τ consists of plotting N' versus the measured beam current, which is directly proportional to N. The deviation of this plot from linearity can be fitted to Equation (8.32) to determine τ. Other methods for determining τ have been discussed by Heinrich *et al.* (1966).

Any intensity measured with the WDS must be corrected for dead time by the use of Equation (8.32) before that intensity is used to form a k ratio. In a typical WDS, τ is approximately 2 μs, so that the correction for dead time is no more than 2% for count rates below 1×10^4 count/s.

8.5. Example of Quantitative Analysis

The analysis of zinc tungstate ($ZnWO_4$) provides an example of a situation in which quantitative analysis by energy-dispersive spectrometry requires correction for peak overlap. The EDS spectrum, shown in Figure 8.20, reveals that the W $L\alpha$ (8.396 keV) and Zn $K\alpha$ (8.638 keV) lines are close enough in energy (242 eV) that the regions of interest for the two analytical lines overlap and the peaks mutually interfere (Figure 8.20a). In addition to the interference of W $L\alpha$ on Zn $K\alpha$, Figure 8.20d shows that the Zn $K\alpha$ peak is interfered with by a second tungsten L line, the W $L\eta$ line at

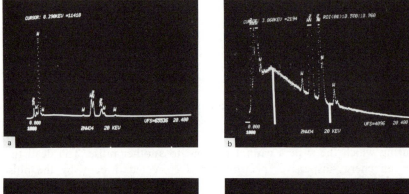

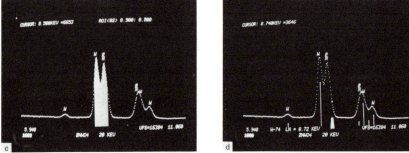

Figure 8.20. Energy-dispersive x-ray spectrum of $ZnWO_4$. (a) Spectrum from 0 to 20.48 keV showing Zn K and L family lines and W L and M family lines; (b) spectrum with expanded fitting by FRAME C procedure; (c) expanded horizontal scale showing W $L\alpha$ and Zn $K\alpha$ regions of interest for quantitative analysis; (d) W L family markers, showing interference of W L (designated "LN" on display) on Zn $K\alpha$.

8.720 keV, within 82 eV of Zn $K\alpha$. Both interferences must be removed to yield an accurate result.

The analysis was carried out with the FRAME C procedure which utilizes the method of overlap coefficients for interference correction and the method of modeling for background removal (Myklebust *et al.*, 1979). The first step in the procedure is to define the regions of interest for the background fit, which are shown in Figure 8.20b. The background modeling method described in Section 8.2.1 is then utilized to provide a first estimate of the background in the analytical regions of interest. The analytical regions are examined for the presence of possible interferences, and appropriate overlap coefficients (W $L\alpha$ on Zn $K\alpha$, Zn $K\alpha$ on W $L\alpha$, and W $L\eta$ on Zn $K\alpha$) are calculated as described in Section 8.3.3.2. The analytical regions are then corrected with these coefficients to provide a first estimate of the characteristic intensities. Similar background and overlap corrections are made in the pure element standard spectra to provide the intensities from the standards. From this set of intensities, k values are calculated. The k values are utilized in a ZAF procedure (see Chapter 7) to provide a first estimate of concentrations. Since the background and the overlaps as well as the ZAF corrections depend on the composition, the procedure is repeated in an iterative fashion to yield a final set of compositions.

The magnitude of the overlap corrections can be seen in Table 8.3. The k values with overlap corrections are about 8% lower than the noncorrected k values. These overlap-corrected EDS k values compare very well with k values determined by wavelength-dispersive spectrometry. Because of the higher resolution of the WDS, the Zn $K\alpha$ and W $L\alpha$ peaks are adequately resolved for analysis, as shown in Figure 8.21 and thus the k values can be measured without overlap correction. In Figure 8.21 the locations of background measurement points are also indicated, since background correction by interpolation is needed in the WDS measurement of intensity.

This example was specifically chosen to illustrate the procedures followed in the case of an analysis in the presence of overlap. For this particular material, $ZnWO_4$, another analytical strategy is possible in which overlap would not occur. As shown in Figure 8.20 alternative analytical lines exist, namely, Zn $L\alpha$ (1.009 keV) and W $M\alpha$ (1.775 keV) which are sufficiently separated in energy. However, the choice of these low-energy lines for analysis would require that a beam energy below 10 keV would have to be used to reduce sample absorption to an acceptable level, and moreover, the sample and standards would have to be prepared identically, including simultaneous carbon coating.

Table 8.3. Comparative Analyses of Zinc Tungstate, $ZnWO_4$[a]

	W($L\alpha$)	Zn($K\alpha$)
Relative intensities (k ratios)		
Wavelength-dispersive spectrometer	0.503	0.230
Energy-dispersive spectrometer		
No overlap correction	0.532	0.255
With overlap correction	0.498	0.236
Calculated mass fractions		
Wavelength dispersive spectrometer	0.587	0.221
Energy dispersive spectrometer		
with overlap correction	0.580	0.227

[a] The operating voltage is 20 keV. The electron beam incidence angle is 90° and the x-ray emergence angle is 52.5°.

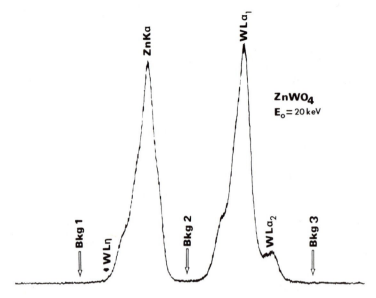

Figure 8.21. Wavelength-dispersive x-ray spectrum of $ZnWO_4$, showing peak positions for intensity measurements of Zn $K\alpha$ and W $L\alpha$. Positions for background readings for interpolation under peaks are also indicated.

8.6. Precision and Sensitivity in X-Ray Analysis

Up to this point we have discussed the corrections for quantitative x-ray microanalysis and the errors associated with the calculation and the data reduction procedure. In addition one must consider the precision or sensitivity of a given analysis. By the precision or sensitivity of an analysis we mean the scatter of the results due to the very nature of the x-ray measurement process. We can take advantage of the statistical equations which describe precision or sensitivity to determine the chemical homogeneity of a sample, the variation of composition from one analysis point to another, and the minimum concentration of a given element which can be detected.

8.6.1. Statistical Basis for Calculating Precision and Sensitivity

X-ray production is statistical in nature; the number of x-rays which are produced from a given sample and interact with radiation detectors is completely random in time but has a fixed mean value. The distribution or histogram of the number of determinations of x-ray counts from one point on a sample vs. the number of x-ray counts for a fixed time interval may be closely approximated by the continuous normal (Gaussian) distribution. Individual x-ray count results from each sampling lie upon a unique Gaussian curve for which the standard deviation is the square root of the mean $(\sigma_c = \overline{N}^{1/2})$. Figure 8.22 shows such a Gaussian curve for x-ray emission spectrography and the standard deviation $\sigma_c = \overline{N}^{1/2}$ obtained under ideal conditions. Here, $\overline{N}$ is considered to be the most probable value of N, the total number of counts obtained in a given time t. Inasmuch as σ_c results from fluctuations that cannot be eliminated as long as quanta are counted, this standard deviation σ_c is the irreducible minimum for x-ray emission spectrography. Not only is it a minimum, but fortunately it is a predictable minimum. The variation in percent of total counts can be given as $(\sigma_c / \overline{N})100$. For example, to obtain a number with a minimum of a 1% deviation in N, at least 10,000 counts must be accumulated.

As Liebhafsky *et al.* (1955) have pointed out, the real standard deviation of the experiment S_c, is given by

$$S_c = \left[\sum_{i=1}^{n} \frac{\left(N_i - \overline{N}_i \right)^2}{n - 1} \right]^{1/2} \tag{8.33}$$

where N_i is the number of x-ray counts for each determination i and

$$\overline{N}_i = \left(\sum_{i=1}^{n} N_i \right) / n \tag{8.34}$$

where n is the number of determinations of i. The standard deviation S_c equals σ_c only when operating conditions have been optimized. In most SEM instruments, drift of electronic components and of specimen position (mechanical stage shifts) create operating conditions which are not necessarily ideal. The high-voltage filament supply, the lens supplies, and other associated electronic equipment may drift with time. After a specimen is repositioned under the electron beam, a change in measured x-ray intensity may occur if the specimen is off the focusing circle of the x-ray spectrometer or if the take-off angle, Ψ, of the specimen varies when using an

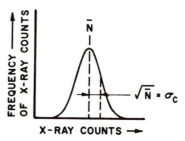

Figure 8.22. Gaussian curve for x-ray emission spectrography and the standard deviation, σ_c, obtained under ideal conditions (from Liebhafsky *et al.*, 1955).

energy-dispersive detector. In practice, for typical counting times of 10–100 s/point, the actual standard deviation, S_c, is often about twice σ_c. If longer counting times are used, S_c/σ_c increases due to instrument drift. Only when counting times are short and the instrument is electronically stable does S_c approach σ_c. Besides the sample signal, sources of variation may also occur if data from reference standards and/or background standards are required (Ziebold, 1967). These, as well as faulty specimen preparation, may also affect the precision of an analysis. Therefore, both instrumental factors and signal variations must be considered when the precision of an analysis is determined.

8.6.2. Sample Homogeneity

An analyst is often asked if a sample and/or a phase is homogeneous. In order to answer this question, the x-ray data must be obtained so that it can be treated statistically. One can either set up criteria for homogeneity and apply them or one can measure the range of composition variation of a sample, at a certain confidence level, and report that number. Either method allows a more quantitative statement to be made than just a simple "yes" or "no" to questions concerning homogeneity. The following material discusses homogeneity criteria that can be used and how to calculate the range and level of homogeneity.

A simplified criterion that has been used to establish the homogeneity of a phase or a sample is that all the data points, n, fall within the $\overline{N} \pm 3\overline{N}^{1/2}$ limits (Yakowitz et al., 1965). If this criterion is satisfied, one assumes then that the sample is homogeneous. The variation,

$$\left(\pm 3\overline{N}^{1/2}/\overline{N} \right)100 \quad (\%) \tag{8.35}$$

for the element of interest in the sample, represents the level of homogeneity in percent that is measured for the sample, remembering that there must be an irreducible minimum level due to the fact that x-ray production is statistical in nature. If 100,000 counts are accumulated at each point in a sample and all these points fall within the limits $\overline{N} \pm 3\overline{N}^{1/2}$, the sample is homogeneous and the level of homogeneity is, according to Equation (8.35) $\pm 0.95\%$. A level of homogeneity of $\leqslant \pm 1\%$ is often desired. If the concentration in the sample C is 10 wt%, the range of homogeneity, that is the minimum variation of concentration that can be validly measured, is ± 0.1 wt%.

A more exacting determination of the range (wt%) and level (%) of homogeneity involves the use of (a) the standard deviation S_c of the measured values and (b) the degree of statistical confidence in the determination of $\overline{N}$. The standard deviation includes effects arising from the variability of the experiment, e.g., instrument drift, x-ray focusing errors, and x-ray production. The degree of confidence used in the measurement states that we wish to avoid a risk, α, of rejecting a good result a large percentage (say 95% or 99%) of the time. The degree of confidence is given as $1 - \alpha$ and is usually chosen as 0.95 or 0.99, that is 95% or 99%. The use of a degree of confidence means that we can define a range of homogeneity, in wt%, for which we expect, on the average, only α (5% or 1%) of the repeated random points to be outside this range.

The range of homogeneity in wt% for a degree of confidence $1 - \alpha$ is

$$W_{1-\alpha} = \pm C \left(\frac{t_{n-1}^{1-\alpha}}{n^{1/2}} \right) \frac{S_c}{\overline{N}} \tag{8.36}$$

where C is the true weight fraction of the element of interest, n is the number of measurements, $\overline{N}$ is the average number of counts accumulated at each measurement and $t_{n-1}^{1-\alpha}$ is the Student t value for a $1 - \alpha$ confidence level and for $n - 1$ degrees of freedom. Student's t values for t_{n-1}^{95} and t_{n-1}^{99} for various degrees of freedom, $n - 1$ are given in Table 8.4 (Bauer, 1971). It is clear from Table 8.4 that at least four measurements, $n = 4$, should be made to establish the range of homogeneity. If less than four measurements are made, the value of $W_{1-\alpha}$ will be too large.

The level of homogeneity, or homogeneity level, for a given confidence level, $1 - \alpha$, in percent is given by

$$\pm \frac{W_{1-\alpha}}{C} = \pm \frac{\left(t_{n-1}^{1-\alpha} \right) S_c (100)}{n^{1/2} \overline{N}} \quad (\%) \tag{8.37}$$

It is more difficult to measure the same level of homogeneity as the concentration, present in the sample, decreases. Although $W_{1-\alpha}$ is directly proportional to C, the value of $S_c / \overline{N}$ will increase as C and the number of x-ray counts per point decreases. To obtain the same number of x-ray counts per point, the time of the analysis must be increased.

Table 8.4. Values of Student t Distribution for 95% and 99% Degrees of Confidence[a]

n	$n - 1$	t_{n-1}^{95}	t_{n-1}^{99}
2	1	12.71	63.66
3	2	4.304	9.92
4	3	3.182	5.841
8	7	2.365	3.499
12	11	2.201	3.106
16	15	2.131	2.947
30	29	2.042	2.750
∞	∞	1.960	2.576

[a] Bauer (1971).

8.6.3. Analytical Sensitivity

Analytical sensitivity is the ability to distinguish, for a given element, between two concentrations C and C' that are nearly equal. X-ray counts $\bar{N}$ and $\bar{N}'$ for both concentrations therefore have a similar statistical variation. If one determines two concentrations C and C' by n repetitions of each measurement, taken for the same fixed time interval, then these two values are significantly different at a certain degree of confidence, $1 - \alpha$, if

$$\bar{N} - \bar{N}' \geqslant 2^{1/2}\left(t_{n-1}^{1-\alpha}\right)S_c/n^{1/2} \tag{8.38}$$

and

$$\Delta C = C - C' \geqslant \frac{2^{1/2}C\left(t_{n-1}^{1-\alpha}\right)S_c}{n^{1/2}\left(\bar{N} - \bar{N}_B\right)} \tag{8.39}$$

in which C is the concentration of one element in the sample, $\bar{N}$ and $\bar{N}_B$ are the average number of x-ray counts of the element of interest for the sample and the continuum background on the sample, respectively, $t_{n-1}^{1-\alpha}$ is the "Student factor" dependent on the confidence level $1 - \alpha$ (Table 8.4), and n is the number of repetitions. Ziebold (1967) has shown that the analytical sensitivity for a 95% degree of confidence can be approximated by

$$\Delta C = C - C' \geqslant \frac{2.33}{n^{1/2}} \frac{C\sigma_c}{\left(\bar{N} - \bar{N}_B\right)} \tag{8.40}$$

The above equation represents an estimate of the maximum sensitivity that can be achieved when signals from both concentrations have their own errors but instrumental errors are disregarded. Since the actual standard deviation S_c is usually about two times larger than σ_c, ΔC is in practice approximately twice that given in Equation (8.40).

If $\bar{N}$ is much larger than $\bar{N}_B$, Equation (8.40) can be rewritten as

$$\Delta C = C - C' \geqslant 2.33 C/(n\bar{N})^{1/2} \tag{8.41}$$

and the analytical sensitivity in percent that can be achieved is given as

$$\Delta C/C(\%) = 2.33(100)/(n\bar{N})^{1/2} \tag{8.42}$$

For an analytical sensitivity of 1%, $\geqslant 54,290$ accumulated counts, $n\bar{N}$ from Equation (8.42), must be obtained from the sample. If the concentration C is 25 wt%, $\Delta C = 0.25$ wt% and if the concentration C is 5 wt%, $\Delta C = 0.05$ wt%. Although the analytical sensitivity improves with decreasing concentration, it should be pointed out that the x-ray intensity decreases directly with the reduced concentration. Therefore longer counting times will become necessary to maintain the 1% sensitivity level.

Equation (8.42) is particularly useful for predicting necessary procedures to obtain the sensitivity desired in a given analysis. If a concentration gradient is to be monitored over a given distance in a sample, it is important to predict how many data points should be taken and how many x-ray counts should be obtained at each point. For example, if a gradient from 5 to 4 wt% occurs over a 25-μm region, and 25 1-μm steps are taken across the gradient, the change in concentration per step is 0.04 wt%. Therefore ΔC, the analytical sensitivity at a 95% degree of confidence must be $\leqslant 0.04$ wt%. Using Equation (8.41) since $\bar{N}$ is much larger than $\bar{N}_B$, $n\bar{N}$ must be at least 85,000 accumulated counts per step. If only ten 2.5-μm steps are used across the gradient, the change in concentration per step is 0.1 wt% and now $n\bar{N}$ need only be $\geqslant 13,600$ accumulated counts per step. By measuring 10 as opposed to 25 steps, the analysis time is cut down much more than the obvious factor of 2.5 since the number of required accumulated counts per step, due to sensitivity requirements, also decreases.

8.6.4. Trace Element Analysis

As the elemental concentration, C, approaches the order of 0.5 wt% in x-ray microanalysis, $\bar{N}$ is no longer much larger than $\bar{N}_B$. This concentration range, below 0.5 wt%, (5000 ppm), is often referred to as the trace element analysis range. For the light elements or x-ray lines with energies $\leqslant 1$ keV, the trace element range begins at about the 1 wt% level (10,000 ppm). The analysis requirement in trace element analysis is to detect significant differences between the sample and the continuum background generated from the sample.

To develop a useful procedure for trace detection we need a criterion that will guarantee that a given element is present in a sample. This criterion can be called the "detectability limit" DL. The so-called detectability limit is governed by the minimum value of the difference, $\bar{N} - \bar{N}_B$, which can be measured with statistical significance.

Leibhafsky et al. (1960) have discussed the calculation of detectability limits. They suggest that an element can be considered present if the value of $\bar{N}$ exceeds the background $\bar{N}_B$ by $3(\bar{N}_B)^{1/2}$. A more sophisticated analysis must consider the measured standard deviation, the number of analyses, and the confidence level desired. By analogy with Equation (8.38) we can also define the detectability limit DL as $(\bar{N} - \bar{N}_B)_{\mathrm{DL}}$ for trace analysis as

$$\left(\bar{N} - \bar{N}_B\right)_{\mathrm{DL}} \geqslant 2^{1/2}\left(t_{n-1}^{1-\alpha}\right)S_c/n^{1/2} \tag{8.43}$$

where S_c is essentially the same for both the sample and background measurement. In this case we can define the detectability limit at a confidence level $1 - \alpha$ (Table 8.4) that the analyst chooses. The 95% or 99%

confidence level is usually chosen in practice. If we assume for trace analysis that the x-ray calibration curve of intensity vs. composition is expressed as a linear function, then C, the unknown composition, can be related to $\bar{N}$ by the equation

$$C = \frac{\bar{N} - \bar{N}_B}{\bar{N}_S - \bar{N}_{SB}} C_S \tag{8.44}$$

where $\bar{N}_S$ and $\bar{N}_{SB}$ are the mean counts for the standard and the continuum background for the standard, respectively, and C_S is the concentration in wt% of the element of interest in the standard. The detectability limit C_{DL}, that is, the minimum concentration which can be measured, can be calculated by combining Equations (8.43) and (8.44) to yield

$$C_{DL} = \frac{C_S}{\bar{N}_S - \bar{N}_{SB}} \frac{2^{1/2} t_{n-1}^{1-\alpha} S_c}{n^{1/2}} \tag{8.45}$$

The relative error or precision in a trace element analysis is equal to C/C_{DL} and approaches $\pm 100\%$ as C approaches C_{DL}.

The background intensity, $\bar{N}_B$, must be obtained accurately so that the unknown concentration C can be measured, as shown by Equation (8.44). It is usually best to measure the continuum background intensity directly on the sample of interest. Other background standards may have different alloying elements or a different composition which will create changes in absorption with respect to the actual sample. Also such background standards may have different amounts of residual contamination on the surface, which is particularly bad when measuring x-ray lines with energies $\leqslant 1$ keV. The background intensity using a WDS is obtained after a careful wavelength scan is made of the major peak to establish precisely the intensity of the continuum on either side of the peak. Spectrometer scans must be made to establish that these background wavelengths are free of interference from other peaks in all samples to be analyzed. It is difficult to measure backgrounds, using the energy-dispersive spectrometer with the accuracy needed to do trace element analysis. Measurements of continuum background with the EDS are discussed earlier in this chapter. If no overlapping peaks or detector artifacts appear in the energy range of interest and the continuum background can be determined, trace element analysis can be accomplished with the EDS detector.

Ziebold (1967) has shown the trace element sensitivity or the detectability limit DL to be

$$C_{DL} \geqslant 3.29 a / (n\tau P \cdot P/B)^{1/2} \tag{8.46}$$

where τ is the time of each measurement taken, n is the number of repetitions of each measurement, P is the pure element counting rate, P/B is the peak-to-background ratio of the pure element, i.e., the ratio of the counting rate of the pure element to the background counting rate of the pure element, and a relates composition and intensity of the element of interest through the Ziebold and Ogilvie (1964) empirical relation (see Chapter 7).

To illustrate the use of this relation, the following values were used for calculating the detectability limit for Ge in iron meteorites (Goldstein, 1967) using a WDS system. The operating conditions were

$$
\begin{aligned}
&\text{operating voltage, 35 kV} \quad &\tau = 100 \text{ s} \\
&\text{specimen current, 0.2 } \mu\text{A} \quad &n = 16 \\
&P = 150{,}000 \text{ counts} \quad &a = 1 \\
&P/B = 200 &
\end{aligned}
$$

With these numbers, Equation (8.46) gives $C_{DL} \geqslant 15$ ppm; the actual detectability limit obtained after calculating S_c and solving Equation (8.45) was 20 ppm (Goldstein, 1967). Counting times of the order of 30 min were found to be necessary to achieve this detectability limit. In measuring the carbon content in steels, detectability limits of the order of 300 ppm are more typical with a WDS system if one uses counting times of the order of 30 min, and the instrument is set up so as to operate at 10 kV with 0.05 μA specimen current.

A criterion which is often used to compare the sensitivity of wave-length- and energy-dispersive detectors is the product $P \cdot P/B$ in the Ziebold relation, Equation (8.46). Comparisons of WDS and EDS detectors have been made on the basis of equal beam current into the specimen (Beaman and Isasi, 1972; Geller, 1977). As an example of this type of comparison Table 8.5 from Geller (1977) lists the C_{DL} (wt%) determined for pure Si and Fe at an operating voltage of 25 kV, 60 s. counting time, and a sample current of 10^{-11} A. The C_{DL} for Si is three times poorer using the WDS than the EDS detector. For Fe, the difference is even larger. The

Table 8.5. Comparison of the Minimum Dectectability Limit of Si and Fe[a] Using an EDS and WDS Detection System on the Basis of Equal Beam Current[b]

		P (cps/10^{-8} A)	P/B	C_{DL}(wt%)
Si $K\alpha$	EDS	5400	97	0.058
	WDS	40.87	1513	0.171
Fe $K\alpha$	EDS	3000	57	0.10
	WDS	12.3	614	0.49

[a] $E_0 = 25$ kV, 60 s counting time.
[b] Geller (1977).

Table 8.6. Comparison of the Minimum Detectability Limit
of Various Elements Using an EDS and WDS Detection System
on the Basis of Optimized Operating Conditions[a,b]

Analysis	Element	P (cps)	B (cps)	P/B	Wet chem. (wt%)	C_{DL} (wt%)
EDS	Na $K\alpha$	32.2	11.5	2.8	3.97	0.195
	Mg $K\alpha$	111.6	17.3	6.4	7.30	0.102
	Al $K\alpha$	103.9	18.2	5.7	4.67	0.069
	Si $K\alpha$	623.5	27.3	22.8	26.69	0.072
	Ca $K\alpha$	169.5	19.9	8.5	12.03	0.085
WDS	Na $K\alpha$	549	6.6	83	3.97	0.021
	Mg $K\alpha$	2183	8.9	135	7.30	0.012
	Al $K\alpha$	2063	16.1	128	4.67	0.008
	Si $K\alpha$	13390	37.0	362	26.69	0.009
	Ca $K\alpha$	2415	8.2	295	12.03	0.009

[a] Geller (1977).
[b] Analysis of D1-JD-35: EDS data collected at 2000 cps for 180 s, dead time, corrected (25%) (1.75 nA probe current at 15 keV). WDS data collected for 30 s for each element, 180 s—total analysis time (30 nA probe current at 15 keV).

P^2/B product is larger for the EDS detector because the geometrical efficiency of the EDS detector is much greater ($\sim 20 \times$) than the WDS. Therefore the peak intensity P is much larger in the EDS detector at constant beam current. The higher P/B of the WDS makes up for some of this difference.

It is clear from earlier discussions, Chapter 5, that the WDS is normally run when the beam current is in the 10^{-8}–10^{-7}-A range. In this current range the WDS detector is optimized for high count rates, P, providing the sample is not degraded by the high beam currents. Therefore it is more realistic to compare the sensitivities of wavelength- and energy-dispersive detectors when each detector is individually optimized for trace element analysis. Geller (1977) has compared WDS and EDS detectors in which each detector was optimized for maximum P^2/B. The specimen used in the analysis was a synthetic mineral D1-JD-35 containing Na, Mg, Si, and Ca. The EDS was operated at a specimen current of 1.75×10^{-9} A at 15 kV. An optimum count rate of 2,000 cps on the specimen was obtained so that summed peaks would not be significant in the analysis. The WDS was operated at a specimen current of 3×10^{-8}A at 15 kV. An optimum count rate was set at $\sim 13,000$ cps on Si so that the maximum count rate caused less than 1% dead time in the detector of the particular spectrometer. Table 8.6 lists the C_{DL} for the elements present in the sample. For the optimized operating conditions the minimum detectability limit using the WDS is almost 10 times better for all elements. Therefore if the solid specimen is not affected by the high beam current, the WDS can offer a factor of 10 increase in sensitivity over the EDS.

8.7. Light Element Analysis

Quantitative x-ray analysis of the long-wavelength $K\alpha$ lines of the light elements (Be, B, C, N, O, and F) as well as long-wavelength $L\alpha$ lines (Ti, Cr, Mn, Fe, Co, Ni, Cu, and Zn) is difficult. The attenuation of the primary radiation is large when these long-wavelength ($\geqslant$ 12 Å), low-energy x-rays ($\leqslant$ 1 keV) are measured, and the correction models developed for quantitative analysis may be difficult to apply in the light element range. A large absorption correction is usually necessary; and, unfortunately, the mass absorption coefficients for long-wavelength x-rays are not well known. These low-energy x-rays are measured using large d spacing crystals with a WDS system. The EDS detector is not able to detect these $\leqslant$ 1-keV x-rays because of the absorption of the x-rays in the Be window of the detector (see Chapter 5).

One can reduce the effect of absorption by choosing to analyze with low operating voltages E_0, and by using high x-ray take-off angles, Ψ. The higher the take-off angle of the instrument, the shorter will be the path length for absorption within the specimen. The penetration of the electron beam is decreased when lower operating voltages are used, and x-rays are produced closer to the surface (see Chapter 3). Figure 8.23 shows the variation of boron $K\alpha$ intensity with voltage, E_0, for several borides (Shiraiwa et $al.$, 1972). A maximum in the boron $K\alpha$ intensity occurs when E_0 is 10–15 keV, depending on the sample. This maximum is caused by two

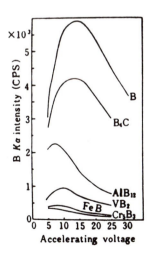

Figure 8.23. Boron $K\alpha$ intensity vs. operating voltage E_0 for several borides (from Shiraiwa et $al.$, 1972).

opposing factors: (1) an increase in x-ray intensity due to increasing voltage, and (2) an increase in absorption due to the fact that x-rays are produced deeper in the sample as incident energy increases. One factor just offsets the other at the maximum intensity. Light element or low-energy x-ray analysis for a given sample is usually carried out using an operating voltage equivalent to this maximum intensity. Even if one uses these procedures, selecting the optimum E_0 and maximizing Ψ, the effect of absorption within the sample is still significant. Other considerations for light element x-ray analysis include overlapping x-ray peaks, chemical bonding shifts, availability of standards, and surface contamination of the specimen. These problems will be considered in some detail in the following material.

Complications in light element analysis may arise because of the presence of the L spectra from heavier metals. Duncumb and Melford (1966) have shown that even if such overlapping occurs, a qualitative analysis can be obtained. For example, titanium carbonitride inclusions, (TiNC), which are probably solid solutions of TiN and TiC, are found in steels. Figure 8.24 shows a comparison of the Ti L spectra obtained from TiN, TiC, and pure Ti. The titanium carbonitride phase gave a more intense peak at the Ti Ll wavelength than that of pure Ti. This peak contains mainly Ti Ll at 31.4 Å (18.3°θ) together with a small amount of nitrogen $K\alpha$ emission indistinguishable from it at 31.6 Å (18.5°θ) (Figure 8.24). The Ti $L\alpha$ line at 27.4 Å (16.0°θ) is heavily absorbed by nitrogen and is about one-third as intense as that from pure Ti. The titanium Ll

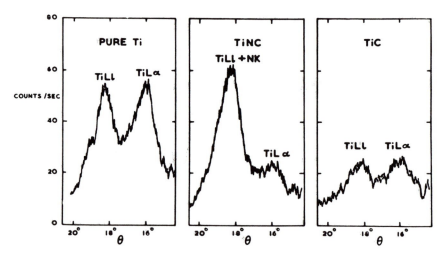

Figure 8.24. Comparison of the Ti L spectra (intensity vs. diffraction angle θ) obtained from pure Ti (left), TiN (center), and TiC (right) at 10 keV operating potential (from Duncumb and Melford, 1966).

emission, however, is only slightly absorbed by nitrogen. The analysis of this type of inclusion may appear to be impossible, but Duncumb and Melford (1966) analyzed the Ti content by using the $K\alpha$ radiation and analyzed the C content by use of the TiC standard (Figure 8.24). The results indicate about 80 wt% Ti and 4 wt% C. Meaningful analysis for nitrogen was not possible for reasons already stated. Nitrogen composition was obtained by difference from 100%. An analysis procedure such as this may have to be used if overlapping lines occur and if they cannot be eliminated by pulse height analysis (Chapter 5).

For the light elements, the x-ray emission spectra consist mainly of a single band produced by the transition of a valence electron to a vacancy in the K shell. As pointed out by Fischer and Baun (1967), the valence electrons are the ones most affected by chemical combination and the emission band can and does reflect the often large effects of changes in chemical bonding between atoms. These changes are signified by wavelength shifts, by increases or decreases in the relative intensities of various lines or bands, and by alteration of shape. Such shifts may cause problems when quantitative light element analysis is desired.

Figure 8.25 shows the C K band from carbon deposited by the electron beam as well as the C K band of electrode grade graphite and various carbides (Holliday, 1967). The wavelength shift for the carbides relative to graphite is significant and can easily be observed with the wavelength-dispersive spectrometer. A similar effect has been observed for B K as shown in Figure 5.13, Chapter 5. This wavelength shift is important since, in order to accomplish a quantitative analysis, the measured $K\alpha$ peak intensity must be made at the position of maximum intensity for both sample and standard alike. Therefore a standard should have a negligible wavelength shift with respect to other standards and unknowns. If this is not possible, the position of the spectrometer must be changed when measuring samples and standards in order to assure that maximum intensity is obtained.

The choice of a primary standard for light element analysis must not only show a negligible wavelength shift but also must give a strong, stable and reproducible peak intensity. For example in carbon analysis, neither spectrographic nor pyrolytic graphite provides reproducible carbon x-ray standards. However, various metallic carbides do provide adequate carbon standards. For steels, cementite, Fe_3C, is an adequate C standard.

Reliable quantitative results can be obtained by comparison with standards whose composition is close to that of the specimen. Figure 8.26 shows the calibration curves for the x-ray analysis of carbon $K\alpha$ at 10 kV in standard alloys of Fe, Fe–10% Ni, and Fe–20% Ni containing specific concentrations of carbon (Fisher and Farningham, 1972). A lead stearate dodecanoate analyzing crystal with a d spacing of 50.15 Å was used. The carbon intensity ratio is the carbon $K\alpha$ line intensity from a given alloy standard less its background divided by the line intensity from a Cr_3C_2 carbon standard less its background. At a given carbon level, the addition of Ni to the steel standard decreases the C intensity ratio. The presence of Ni in iron probably lowers the C $K\alpha$ intensity because it has a greater mass absorption coefficient for C $K\alpha$ than does iron. Fisher and Farningham (1972) used these FeNiC standards to investigate the carbon distribution in carburized nickel gear steels.

Appropriate specimen preparation, as discussed in Chapter 9, must also be considered. During preparation of specimen surfaces, abrasives containing the light elements should either be avoided or, if it is impractical to do so, the sample should be carefully cleaned to ensure total removal of these materials. After the final polish, polishing material can be removed by ultrasonic cleaning. One may not wish to etch a sample since etching can leave a residual contamination layer. Ideally, specimens should be placed in the instrument right after preparation. If this is inconvenient, storage in a vacuum desiccator is usually satisfactory.

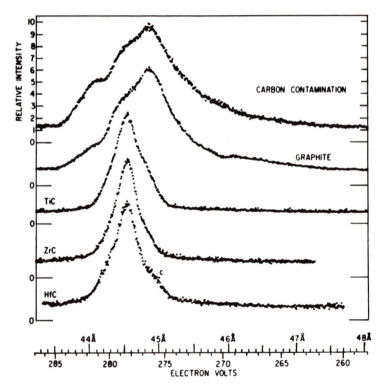

Figure 8.25. C K spectra from various C-containing samples obtained at 4 keV operating potential. The carbon contamination was deposited from the electron beam (from Holliday, 1967).

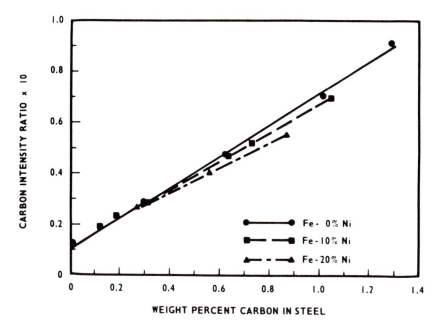

Figure 8.26. Calibration curves for electron microprobe analysis of carbon in nickel steels (from Fisher and Farningham, 1972).

A sample subjected to electron bombardment in a diffusion-pumped vacuum gradually becomes covered with a "contamination" layer due to polymerization, under the action of the beam, of organic matter adsorbed on the surface (Castaing, 1960). The organic molecules come from the oil vapors of the vacuum pumps and the outgassing of any organic material present in the instrument. The effect is not very troublesome unless the deposited layer absorbs the emitted x-rays to a great extent. For low-energy x-rays ($\leqslant 1$ keV) particularly those of Be, B, and C, the absorption of the x-radiation can be severe. The problem, in the case of carbon analysis, is increased because the "contamination" layer contains carbon in large measure. This circumstance leads to the observation of an increasing carbon $K\alpha$ count as a function of beam impingement time.

Two methods have been used to avoid the contamination layer. Castaing and Descamps (1954) showed that directing a low-pressure jet of gas on the specimen at the region bombarded by the electron beam suppresses contamination. When air is introduced into the vicinity of the sample, oxygen oxidizes the hot carbon deposit and the high-energy electron beam acts to produce an ion bombardment-sputtering condition. Air jets have been installed on various SEM instruments and can be built in any laboratory for a nominal cost (Duerr and Ogilvie, 1972). Another method is to provide a surface within the SEM which is cold relative to the surface of the specimen. Organic molecules will then tend to collect on the colder surface rather than on the specimen. The cold surface or cold finger must, however, be placed very close to the specimen. Cold fingers have been installed on various instruments and have effectively reduced the contamination rate to near zero (Borile and Garulli, 1978). In one case (Kohlhaas and Scheiding, 1969) both an air jet and a liquid-N_2 cold finger have been used. Light element quantitative analysis probably ought not be attempted without some kind of decontamination device or operation in a high-vacuum system. The decontamination methods discussed also may be useful in reducing C contamination during scanning in a SEM.

The line through the points in Figure 8.26 for the Fe–Ni–C standards does not pass through the origin at 0 wt% C. The carbon intensity ratio (background already subtracted) does not go to zero as expected even though no contamination layer was observed. Probably, a very thin carbon film was present on the standard before the analysis was made. This specimen-borne contamination is caused by numerous factors such as specimen preparation and exposure to laboratory atmosphere. Even in ion-pumped instruments contamination layers are observed on the specimens. Clearly these layers are not caused by the instrument but are due to the mobilization of specimen-borne contamination. As long as specimen-borne contamination can be characterized and is reproducible, its contribution to the analysis can be subtracted out at an appropriate time.

It is still difficult to obtain accurate light element analyses in multi-component alloys even if overlapping peaks are eliminated, chemical bonding shifts are considered, standards are obtained, and contamination is controlled. Unfortunately the major input parameters for the calculation schemes, e.g., mass absorption coefficients of the light elements in heavy element matrices, are not well known. As an example of the problem, Table 8.7 lists the carbon $K\alpha$ mass absorption coefficients used for a study of carbide compositions in tool steels (Barkalow et al., 1972), those used for a study of binary carbides (Shiraiwa et al., 1972) and the values from the

Table 8.7. Mass Absorption Coefficients for C $K\alpha$

Absorber	μ/ρ for C $K\alpha$ (g/cm^2)		
	Barkalow et al. (1972)	Shiraiwa et al. (1972)	Henke and Ebisu (1974)
Fe	15,000	14,300	13,300
W	18,000	—	18,750
Mo	19,000[a]	—	32,420
Cr	10,000	11,000	10,590
V	15,000	9,300	8,840
C	2,300	2,270	2,373

[a] Estimated.

tables of Henke and Ebisu (1974). The discrepancies are significant, particularly for Mo and V. The mass absorption uncertainties for the $L\alpha$ lines, $\leqslant 1$ keV in energy, have already been discussed and illustrated in Figure 7.6, Chapter 7. In light of this problem the calculation of composition, using low-energy x-rays, is rarely better than $\pm 10\%$ of the amount present in complex alloys. However, the calculation errors can be minimized if standards close in composition are used.

Brown *et al.* (1979) have described a ZAF method for the analysis of carbon based on a generalized function for the $\phi(\rho Z)$ depth distribution of x-ray production. These authors found good agreement in the analysis of known stoichiometric carbides, provided good values of the mass absorption coefficients for C K radiation were employed. Love and Scott (1978) have also described refinements of the ZAF method of Chapter 7 to give improved results in light element analysis.

Materials Specimen Preparation for SEM and X-Ray Microanalysis

9.1. Metals and Ceramics

9.1.1. Scanning Electron Microscopy

One of the great strengths of scanning electron microscopy is the fact that many specimens can be examined with virtually no specimen preparation. Specimen thickness is not a consideration as is the case in transmission electron microscopy. Therefore, bulk specimens can be examined in the SEM with a size limited only by considerations of accommodation in the specimen stage. For the examination of images of topography contrast from metal and ceramic specimens, the only specimen preparation which is necessary is to ensure that the specimen is thoroughly degreased so as to avoid hydrocarbon contamination and, in the case of insulators, to provide a conductive coating. Techniques for cleaning surfaces include solvent cleaning and degreasing in an ultrasonic cleaner, mechanical brushing, replica stripping, and chemical etching. These techniques should be used starting with the least damaging and employing only the minimum cleaning necessary. Usually the first step is to use a solvent wash such as acetone, toluene, or alcohol in an ultrasonic cleaner. Several specific techniques for cleaning metal surfaces are described by Dahlberg (1976).

Since we wish to examine the surface of the material, it is important to remove contaminants which may have an adverse effect on secondary electron emission. The electron beam can cause cracking of hydrocarbons, resulting in the deposition of carbon and other breakdown products on the specimen during examination. Contamination during operation frequently can be detected by making a magnification series from high magnification (small scanned area) to low magnification (large scanned area). The deposit

Figure 9.1. Formation of contamination under electron bombardment. The dark square is a result of hydrocarbon cracking built up during scanning at a higher magnification. The hydrocarbon layer changes the secondary electron emission characteristics. Material: $YFeO_3$; beam: 30 keV.

forms quickly at high magnification because of the increased exposure rate. When the area is observed at low magnification, a "scan square" of contamination is observed, Figure 9.1. It is thus important to avoid introducing volatile compounds into the SEM. Residual hydrocarbons from the diffusion pump oil can also produce contamination under the influence of the beam. This problem can be avoided for the most part by using traps cooled with liquid nitrogen to condense hydrocarbon vapors.

Specimen preparation does, however, become an important consideration under certain circumstances. As explained in Chapter 4, a weak contrast mechanism, such as electron channeling, is frequently impossible to detect in the presence of a strong contrast mechanism, such as topography contrast. It is thus necessary to eliminate specimen topography when we desire to work with electron channeling contrast, types I and II magnetic contrast, and other weak contrast mechanisms. Chemical polishing or electropolishing can produce a mirror surface nearly free from topography in metal specimens. A large amount of literature describing such polishing techniques exists for most metals and alloys (Kehl, 1949). Metallographic mechanical polishing also removes topography and gives a high-quality mirror surface, but such mechanical polishing results in the formation of a shallow layer ($\sim$100 nm; 1000 Å) of intense damage in most metals and ceramics. Such a layer completely eliminates electron channeling contrast, and in magnetic materials the residual stresses in the layer result in the formation of surface magnetic domains characteristic of that particular stress state. If we are interested in domains characteristic of the bulk state of the material, such a residual stress layer must be avoided. Mechanical polishing to produce a flat surface followed by brief electro-polishing or a chemical treatment to remove the damaged layer often gives optimum results. Electropolishing alone can occasionally result in a po-lished but wavy surface. In general, specimen preparation remains an art, with each material presenting a different problem to the investigator.

When the electron beam strikes an insulating material such as a silicate, oxide, or inclusion in a metal specimen, the absorbed electrons accumulate on the surface since no conducting path to ground exists. The accumulation of electrons builds up a space charge region. The problem of charging and its avoidance is discussed in detail in Chapter 10.

9.1.2. X-Ray Microanalysis

9.1.2.1. Surface Roughness and Polishing

Since the x-ray analysis performed is essentially an analysis of the prepared surface, it is requisite that the prepared surface be truly represen-tative of the specimen. Over the years, a number of qualitative criteria for a properly prepared surface have evolved. These are that the specimen should be polished as flat and scratch-free as possible and be analyzed in the unetched condition so as not to alter the topography or surface chemistry. Such criteria were set forth primarily for metallurgical specimens; they can be applied most directly to petrographic specimens. However, for biological specimens and hydrous materials such criteria are virtually meaningless since it is rare that a "polished" specimen is used in such work (see Chapter 12).

Flatness of both specimen and standard is a prime requisite. For pure elements and homogeneous materials, it is feasible to prepare relatively flat surfaces since the hardness will not vary greatly over the specimen. This results in fairly uniform material removal during grinding and polishing. Unfortunately, most specimens submitted fall outside these two categories. In cases where phases of different hardness coexist, sharp steps may occur at the phase boundaries. These steps may cause anomalous absorption and must be taken into account if it is necessary to carry out analysis of regions near the boundary. One way to do this is to rotate the specimen 180° and remeasure the intensities to confirm the presence of a suspected absorption effect. If the absorption effect exists, it can be minimized by rotating the specimen so the x-rays are detected in a direction which is parallel to the step.

When the specimen topography deviates from a flat surface, the differences in local inclinations affect the result of the x-ray microanalysis and are superimposed on the x-ray statistical uncertainty. Yakowitz and Heinrich (1968) showed that one should use high values of the take-off angle Ψ in order to minimize effects of local surface inclinations, $\Delta\Psi$. In addition they showed that effects of differences in surface preparation between specimen and standard will contribute to errors in the measured intensity ratio k.

A somewhat more insidious and more general problem than sharp steps is that of relief due to polishing, etching, and repolishing. This results in variable surface contours over the specimen face, that is, a hilly surface. These contours can occur in the same phase as well as across phase boundaries. Although grain boundaries show the effect most markedly, variation within the grain may often be as large as or larger than at grain boundaries. Relief polishing is often minimized by using lower-nap polishing cloths and higher wheel speeds. If a polish–etch–polish procedure is being used, an etchant giving the lowest relief possible is the most desirable. Quantitative analysis of such specimens may be difficult since the possibility of errors due to topography cannot be precluded.

Another problem is to prepare samples containing inclusions in various matrices. In effect, this is the same as the problem of polishing two phases of variable hardness lying adjacent to one another. The major difference is that the inclusion is small in size and hence may be pulled out during specimen preparation. The standard technique to retain inclusions is to use as little lubricant as possible during the polishing procedure or to electropolish the specimen. The entire junction between matrix and inclusion usually can be examined at high magnification with an optical microscope. If the specimen is properly prepared, both matrix and inclusion should be sharply in optical focus. A detailed treatment of the subject of inclusion polishing and identification is given by Kiessling and Lange (1964).

An associated problem is the introduction of polishing abrasive into the material of interest. One should be suspicious of anomalous inclusions containing elements used in the polishing preparation. This is particularly true when specimens containing cracks or porosity are examined. In such cases a different preparation technique is often warranted in order to determine whether or not such inclusions are artifacts. In extremely soft materials such as lead- or indium-based alloys where local smearing of constituents could lead to erroneous microprobe results, electropolishing may be considered. Again, care and patience should be exercised in order to obtain the flattest surface possible. As for specific techniques for the preparation of metallographic and petrographic sections, there are numerous references in the literature (ASTM, 1960; Anderson, 1961; Cadwell and Weiblen, 1965; Kehl, 1949; Taylor and Radtke, 1965; Tegart, 1959). There are virtually as many techniques for the preparation of a microsection as there are microsections.

The need for truly flat specimens is reemphasized for all line or step scanning operations, or both, in which the specimen is mechanically driven with respect to the electron beam. In such cases, the x-ray flux recorded may vary with both specimen focal level and as a result of surface roughness. With certain spectrometer–specimen geometries, this effect may lead to anomalous x-ray flux variations for the elements of interest resulting in data interpretation difficulties.

9.1.2.2. Preparation of Standards for X-Ray Microanalysis

As discussed in Chapter 7 quantitative x-ray microanalysis is based on determining the x-ray flux emitted by an unknown relative to that of a suitable standard. The most easily obtainable standards are those of the pure elements. Data correction procedures can then be applied to reduce the measured intensity ratio to composition. Models for correction, however, are open to question. Input parameters, such as mass absorption coefficients and fluorescence yield values, are sometimes only poorly known. Furthermore, even if perfect correction procedures were available, some elements such as sulfur, chlorine, potassium, gallium, etc., cannot be obtained in suitable pure form for use in x-ray microanalysis.

For these reasons, the use of intermediate compositional standards has become widespread. There are three basic approaches: (1) to prepare an entire series of standards so that an empirical calibration curve can be established for the system of interest, (2) to obtain a single standard in order to characterize a particular constituent, and (3) to obtain single standards to monitor instrumental performance or as "anchor points" for correction procedures to be applied to similar systems.

The basic requirements for all such standards are that they be homogeneous at micrometer levels of spatial resolution, stable with respect to time and properly prepared for use in the SEM, and carefully analyzed by independent techniques. Usually, the most stringent requirement is that of homogeneity. This factor should be carefully checked by means of line scans, area scans, and point counting. In choosing compounds for use as electron probe standards, it is important to select covalently bonded oxides if available since these materials are least likely to suffer beam damage. In the case of the alkali metals and the halogens, the analyst must usually resort to salts such as NaCl or KI. These materials are susceptible to beam damage and should only be used at low current densities. It is obviously necessary to prepare polished salts in the absence of water and to protect the polished standards from humid atmospheres.

Preparation of entire sets of standards for a single system is usually

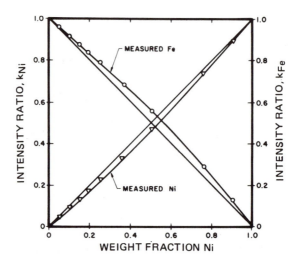

Figure 9.2. Calibration k vs. c curves for the Fe–Ni system. The data points were obtained from nine well-characterized, homogeneous standards (Goldstein *et al.*, 1965).

confined to metallurgical applications. It is warranted when a system is to be studied in great detail, for example, in the case of phase diagram determination. With proper standards, the accuracy of analysis can be made to approach the limits imposed by the x-ray statistics for such systems. Figure 9.2 shows an empirically determined calibration for the analysis of Fe and Ni in binary Fe–Ni alloys at 30 kV operating potential and take-off angle Ψ of 52.5°. The curve was established with the aid of nine well-characterized, homogeneous standards (Goldstein *et al.*, 1965). The Ni $K\alpha$ radiation is heavily absorbed by the nickel and the Fe $K\alpha$ radiation is increased due to x-ray fluorescence by the Ni $K\alpha$ radiation.

The method of obtaining a single standard to characterize a particular constituent may be adopted when no suitable elemental form is available. Before use, the material in hand should be characterized as to homogeneity and composition. Usually, data taken with such a standard must be corrected in the same fashion as those obtained with elemental standards.

The National Bureau of Standards has prepared a few binary and one ternary metal alloy systems suitable for use as standards (NBS Standard Reference Materials Catalog, 1979). At present, the Standard Reference Materials available are a low alloy steel (Michaelis *et al.*, 1964), Au–Ag and Au–Cu alloys (Heinrich *et al.*, 1971), W–20% Mo alloy (Yakowitz *et al.*, 1969), Fe–3.22% Si alloy (Yakowitz *et al.*, 1971), two cartridge brasses (Yakowitz *et al.*, 1966), and an Fe–Cu–Ni alloy (Yakowitz *et al.*, 1972). The principal value of these standards is to provide test specimens for the development, testing, and refinement of analysis techniques. In addition, multielement glasses have now been issued which can serve as standards for the analysis of typical rock-forming minerals (basalt, granite) (Marinenko et al., 1979).

9.2. Particles and Fibers

Particulate samples can usually be examined with little specimen preparation. In general, we will assume that the particles of interest have been collected properly and need only be mounted for SEM examination. A thorough discussion of particle collection techniques is given by DeNee (1978). For mounting a number of dry particles at one time, double-stick adhesive tape which has been previously mounted on SEM specimen stubs is used. It should be noted that dry particles tend to agglomerate when handled. If this agglomeration is acceptable, the process illustrated in Figure 9.3, adapted from Brown and Teetsov (1976), can be used. The advantage of adhesive tape is that the tacky material has a high enough viscosity so that it does not engulf or climb up over the particles when they are mounted, yet is sticky enough to hold the particles. The particles are

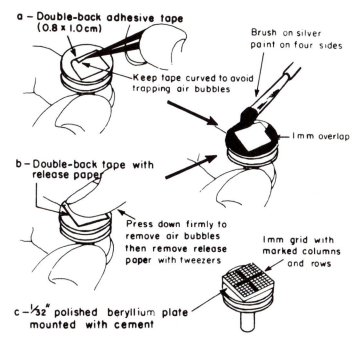

Figure 9.3. Preparing aluminum stubs for small-particle mounting. (a) Double-back adhesive tape method. (b) Tape method with release paper. (Brown and Teetsov, 1976.)

transferred to the tape either by a spatula, to deposit a mound of particles, or by pouring from a bottle. Loose particles are shaken from the tape since they will produce serious charging effects in the SEM even when they are carbon or metal coated. The tape is often masked with carbon paint to ensure good conductivity to the metal SEM stub.

Particles can also be attached to SEM stubs by coating the stubs first with Aquadag (carbon paint) or a thin layer of parlodian or collodion. The particles are transferred just as the dag or polymer coating has begun to dry. If the particles are mounted too early, they will sink in the "liquid" coating and the surfaces will no longer be useful for SEM examination. If the particles are in a liquid suspension, the particles must be separated from the liquid and placed on an appropriate substrate. If the concentration of particles is relatively high, the particle suspension can be directly put on a substrate by (a) placing a drop or several microdrops onto a substrate with a pipette; (b) spray dispersing the mixture onto a substrate; or (c) filtering the liquid (DeNee, 1978). The liquid is removed by heating, evacuation, solvent exchange, or freeze drying.

The mounting of single particles is more sophisticated, particularly because of the need to keep track of each particle as it is analyzed. Particles > 50 μm in size can be mounted directly on normal SEM stubs. They can be selected and placed on the stub by using a stereomicroscope to view the particles during the process. The particles can be handled with stainless steel tweezers or a fine pointed needle. A vacuum pick can also be used for handling the larger particles. Thin coatings of polymer, parlodian, or collodion, can be used to attach the particles to the stub. Also Aquadag or even Duco cement can be employed. Because of the small area of contact, care must be taken not to distort or crush the particles during mounting. Particles greater than 50 μm in size will become firmly attached when any of these media are in a tacky, partially solidified state. Careful documentation by drawings or optical photos of the particle position and relative size is a necessity.

Particles less than 50 μm in size are much more difficult to mount. In addition x-rays are often produced in regions up to 10 μm from the electron beam impact point within the particle. Particularly with particles $\leqslant 10$ μm in size, x-rays may fluoresce the SEM stub, producing unwanted radiation. This spurious x-ray radiation can lead to incorrect identification of the particle chemistry. The use of substrates of heavier metals, which are different from the elements found in the particles, is not recommended because of the higher-order x-ray lines produced (DeNee, 1978). Therefore specimen stubs of polished-smooth carbon or Be which do not contribute a detectable characteristic x-ray signal of their own are recommended. Smooth surfaces are needed so as not to confuse the observer when looking for small particles. The major drawback of using Be is the toxicity of the beryllium oxide which forms when the disk is polished. The polished surfaces are often marked or scratched with a fine tungsten needle to allow for positive identification of the position of the particle. The surface should have a different atomic number than the particles so the samples can easily be observed, for example, in the backscattered electron signal. Obtaining an appropriate mounting medium for coal particles is a particular challenge (Moza *et al.*, 1978).

Fine tungsten needles are used to pick up the small-sized (10–50 μm) particles. These particles are picked up dry with the needle and deposited into a microdrop of solvent such as amyl acetate, ethanol, distilled water, etc., which is placed on the stub with a micropipette, see Figure 9.4. Following drying of the first drop, a second drop of solvent is moved with the needle over a drop of collodion polymer which was placed previously

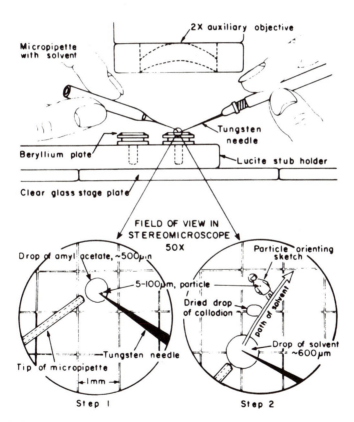

Figure 9.4. Positioning a single particle, 5–50 μm under a stereomicroscope on a SEM stub (Brown and Teetsov, 1976). Step 1: The needle tip with the particle is touched to the center of the drop of solvent. Step 2: A second drop of solvent is moved with the needle over the collodion then over the particle.

on the stub. The mixture of solvent and collodion is carried across the particle. By this process a small amount of collodion, 20–50 nm thick is attached all around the particle and holds the particle firmly to the stub. If done correctly, particle surface detail will not be obscured. Particles 3 to 10 μm in size can also be picked up dry with the needle and deposited within a microdrop of solvent on the SEM stub. Adhesion usually occurs by electrostatic attraction.

For sediment particles less than 50 μm in diameter, the procedure described by Walker (1978) can be used. Some of the wet or dry sediment is placed in deionized water and mixed thoroughly. Approximately 0.2 ml of the solution is placed onto the surface of an SEM stub. The sample is then allowed to dry, preferably in a desiccator.

Particles smaller than 3 μm in size cannot be picked up with a needle. Therefore, a more complicated mounting procedure is followed. First the particle is embedded in a film of collodion and then the film is removed, floated on water, and transferred to an SEM stub. The collodion is then dissolved by a solvent and the particle remains on the stub. This process is described in detail by Brown and Teetsov (1976) and McCrone and Delly (1976). Fine particles can also be mounted dry or in liquid suspension on glass cover-slips attached to specimen stubs. Electrostatic attraction is responsible for the adherence of the dry particles.

An alternate method of mounting small particles (< 10 μm) is to use a 3-mm-diam TEM substrate such as a thin carbon film on a TEM grid. The particles are normally dispersed on the carbon film using a microdrop of solvent such as amyl acetate or ethyl alcohol. In x-ray analysis a significant increase in sensitivity is realized because the x-rays which leave the particle can no longer fluoresce x-rays from the substrate. As discussed by Barbi and Skinner (1976), the amount of continuum generated in the thin-film substrate is far less than that generated in a normal carbon substrate. If the continuum is reduced the sensitivity of the characteristic x-ray lines is improved. Figure 9.5 illustrates the improved peak-to-background ratio achieved by the use of a thin carbon film substrate versus a bulk carbon stub for a 2-μm $CaCl_2$ particle (Barbi and Skinner, 1976). The disadvantage of using a thin-film substrate is the possibility of introducing the characteristic peaks from the grid material into the x-ray spectrum. For work where these spurious x-ray peaks are a problem, carbon films on Be grids can be used. The TEM grid is mounted on a special graphite holder in the SEM to avoid characteristic x-ray generation by beam electrons which may pass through the thin carbon film.

For submicrometer particles, a well-stirred suspension of particles or fibers in water is first obtained. The suspension is then filtered through ~0.1-μm membranes. After filtration, sections of the filter are cut while still wet, fastened to SEM stubs with cellophane tape, and placed in an

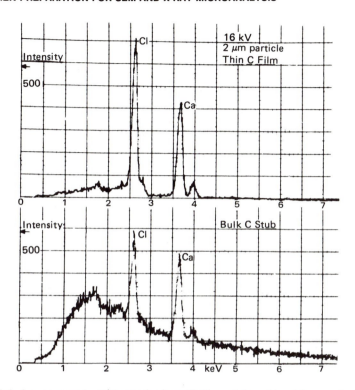

Figure 9.5. Improvement in P/B ratio for 2-μm $CaCl_2$ particles achieved by use of a thin C substrate (upper spectrum) compared to a bulk C stub (lower spectrum). (Barbi and Skinner, 1976.)

oven for 4 h at 40°C (Millette *et al.*, 1978). The specimens are then ready for coating.

If nonmetallic inclusions are < 5 μm in size, and are analyzed directly while still embedded in their matrix material, the x-ray spectrum from the particle will also include information from the matrix. Therefore a specimen preparation technique of major importance in the examination of nonmetallic inclusions in metals is the extraction replica. For metal matrices, the metal is polished and etched so that nonmetallic inclusions stand out from the surface but remain attached to the metal. Carbon is then evaporated onto the specimens. The metal is etched again, leaving the nonmetallic inclusions embedded in the carbon replica in the same positions occupied in the metal. Figure 9.6 illustrates the two-step process (McCrone and Delly, 1973). The next step is to mount the carbon film on a TEM grid and examine the particles in the SEM.

To determine the composition and microstructure of particles > 10 μm

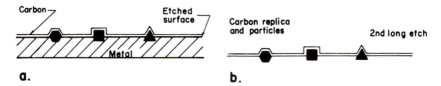

Figure 9.6. Two steps in producing an extraction replica: (a) vacuum carbon coating a heavily etched sample; (b) etching again leaving the particles supported in the carbon film. (Figure 199, McCrone and Delly, 1973.)

in size one can prepare a polished mount for examination in the SEM. To accomplish this, particles are mounted on appropriate SEM stubs as described previously. The SEM stub with the attached particles is embedded with the particle side down in an epoxy mount. Polishing with fine grit sandpaper will expose the embedded particles. Special care must also be taken to locate and document the positions of the particles which are to be analyzed in the SEM.

If a sufficient quantity of particles is available, the particles can be mixed with a small amount of conducting silver epoxy. The resulting mixture is pressed into a blind hole drilled into a carbon or metal block. After hardening, the sample can be ground and polished. This method is useful for particles as small as 1 μm (Small *et al.*, 1978).

9.3. Hydrous Materials

9.3.1. Soils and Clays

The study of soil fabric with the SEM requires that the fluid which is an aqueous solution, be removed from the specimen before placing it in the instrument. If the soil specimen has a high moisture content and/or tendency to shrink upon moisture loss, it is difficult to dry the specimen without disturbing the original structure (Lohnes and Demirel, 1978). Six techniques have been applied for removal of pore fluid (Tovey and Wong, 1973). These techniques are (1) oven drying, (2) air drying, (3) humidity drying, (4) substitution drying, (5) freeze drying, and (6) critical point drying. The first two techniques are straightforward and self-explanatory. Humidity drying is the process of drying the specimen under controlled humidity. Substitution drying involves the replacement of the pore fluid in the soil by a liquid of low surface tension such as methanol, acetone, or isopentane prior to drying (Lohnes and Demirel, 1978). The last two techniques are the same as those used by biologists and are described in Chapter 11. In general air drying is the most widespread technique for low moisture and stiff soils, whereas soils having a fragile fabric can be dried by rapid freeze drying (Lohnes and Demirel, 1978).

After drying the specimen it is necessary to expose an undisturbed surface for study by fracturing and/or peeling. In order to fracture a specimen, it is scored and then bent or pulled to create a tensile failure surface. Peeling involves applying an adhesive material to the specimen surface and then stripping it off to remove disturbed surface particles of the specimen with the adhesive. The stripping process is repeated several times. For a more complete description of soils specimen preparation for SEM the reader is referred to a review article by Lohnes and Demirel (1978). Dengler (1978) has also described an ion milling technique which is used to expose the matrix microstructure.

9.3.2. Polymers

Many fabricated polymers contain water. The study of the surface characteristics of these polymers requires that the SEM preparation does not disturb the surface. Various preparation techniques involve air drying, critical point drying, cryofracture, and freeze drying. The air-drying process is the simplest of the four. The techniques normally used for polymers are very similar to those used for biological materials, and the above techniques are discussed in Chapters 11 and 12.

Coating Techniques for SEM and Microanalysis

10.1. Introduction

Nearly all nonconductive specimens examined in the scanning electron microscope or analyzed in an electron probe microanalyzer need to be coated with a thin film of conducting material. This coating is necessary to eliminate or reduce the electric charge which builds up rapidly in a nonconducting specimen when scanned by a beam of high-energy electrons. Figures 10.1a and 10.1b show examples of pronounced and minor charging as observed in the SEM. In the absence of a coating layer, nonconductive specimens examined at optimal instrumental parameters invariably exhibit charging phenomena which result in image distortion and thermal and radiation damage which can lead to a significant loss of material from the specimen. In extreme situations the specimen may acquire a sufficiently high charge to decelerate the primary beam and the specimen may act as an electron mirror. Numerous alternatives to coating have been proposed and some of these will be discussed in this chapter. Much of what will be discussed is directed towards biological material and organic samples simply because these types of specimens are invariably poor conductors and more readily damaged by the electron beam than most inorganic materials. However, it is safe to assume that the methods which will be described for organic samples will be equally effective for nonconducting inorganic specimens.

This chapter will concentrate on the practical aspects of some of the more commonly used vacuum evaporation and sputter coating techniques which are now standard procedures in most electron microscope and analytical laboratories. It is not proposed to enter into a detailed discussion of the theoretical aspects of thin film technology, but those readers inter-

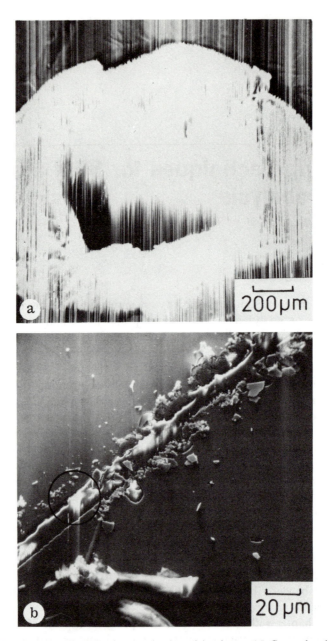

Figure 10.1. Charging effects in the examination of insulators. (a) Gross charging effects during examination of bare Teflon; beam: 30 keV. (b) Minor charging effects. The specimen is carbon-coated glass scratched through the conducting layer. Note local charging effects (circled) due to the exposure of bare glass by the scratch.

ested in this aspect of the subject are referred to the book by Maissel and Glang (1970).

10.1.1. Specimen Characteristics

10.1.1.1. Conductivity

The single most important reason for coating is to increase the electrical conductivity of the sample. Materials of high resistivity, i.e., exceeding 10^{10} Ω-m, will rapidly charge under the incident beam and may develop a potential sufficient to cause a dielectric breakdown in certain regions of the specimen. This leads to variations in the surface potential, giving rise to the complex image artifacts commonly referred to as "charging." These artifacts are manifest as deflection of low-energy secondary electrons, increased emission of secondaries within the crevices of a rough specimen, periodic bursts of secondary electron emission, and deflection of the electron beam, all of which degrade the resolving power and analytical capabilities of the system by introducing astigmatism instabilities, undue brightness, and spurious x-ray signals (Figure 10.2). This undesirable situation is frequently compounded because many of the adhesives used to attach the specimen to the substrate are themselves nonconductors and may prevent any electrical charge leaking away, even from conductive samples. A suitable conducting path may be established with silver or carbon paint. If one is concerned only about electrical conductivity, then a thin layer of gold, silver, or copper will suffice to eliminate the problems associated with charging. Even though metallic samples are usually conductive, there are situations where one may wish to examine nonconducting areas, e.g., inclusions, and in these cases it is necessary to apply a thin coating layer. The conductivity of the thin film should be sufficient to ensure that the specimen current is drained to ground without the development of a significant surface potential.

10.1.1.2. Thermal Damage

Specimen heating is not usually a problem in most samples examined in the SEM, because the probe current is usually in the picoampere range. Although higher currents are frequently used for TV scanning, these are unlikely to seriously degrade the specimen. Thermal effects are potentially more serious for cathodoluminescence and x-ray microanalysis, for which the probe currents are in the nanoampere and even the microampere range.

Figure 10.2. Artifacts during examination in the SEM (a) Beam damage causing cracks on the surface of pollen from *Ipomoea purpurea*. (b) Lines across image due to faulty scan generator. Wood fibers of *Quercus ilex*. (c) Charging causing small flecks on image of *Aesculus hippocastanum* pollen. (d) Charging causing image shift on image of *Lygodesmia grandiflora* pollen.

Excessive heating in the SEM can lead to specimen movement and instability and in extreme situations to breakdown and destruction. The phenomenon described as "beam damage" is most certainly a heating effect and is manifest as blisters, cracks, and even holes in and on the surface of the specimen (Figure 10.2a). In the electron probe microanalyzer the higher beam currents can cause a rapid loss of organic material from plastics, polymers, and biological samples and can even result in substantial elemental losses.

Thermal damage can be reduced by working at lower beam currents, and using thin specimens which are in close contact with a good thermal conductor. Alternatively, the specimens may be coated with a thin film of a good heat conductor such as copper, aluminum, silver, or gold.

10.1.1.3. Secondary and Backscattered Electron Emission

The thin layer of metal which is usually applied to an insulator to make it electrically and thermally conductive is also the source of the bulk of the secondary electrons. A 10-nm layer of metal such as gold would certainly improve the coefficient of secondary-electron emission, δ, for an organic specimen examined at low kilovoltage, but might well diminish δ for a ceramic containing significant amounts of alkaline earth oxides.

Backscattered electrons have also been used in conjunction with standard cytological techniques to localize regions of physiological interest in biological tissue. Thus if one has gone to considerable effort to obtain deposits of lead or silver at specific sites in a piece of tissue it seems inappropriate to mask the atomic-number contrast these locations will give by covering them with a layer of heavy metal. The appropriate technique would be to apply a thin layer of a low-atomic-number conductor such as carbon which would not significantly scatter the incident beam, allowing it to reach the specimen.

For high-resolution, low-loss scanning electron microscopy where the image is dependent on the scattering of high-energy electrons from the specimen surface, it is necessary to coat the samples with a thin layer of a heavy metal which shows no structure at the 1-nm resolution level. Experimental evidence suggests that the refractory metals such as tantalum or tungsten yield such a coating.

10.1.1.4. Mechanical Stability

Particulate matter and fragile organic material are more firmly held in position on the specimen stub after coating with a thin layer of carbon. In many cases it is possible to place such material directly onto the specimen support and stabilize it with a very thin layer of carbon applied from two directions. This simple technique avoids the use of adhesives, most of which are highly nonconductive. Metal coatings, particulary those deposited by a sputtering process, are quite strong and contribute to the increased mechanical strength of otherwise fragile material.

10.1.1.5. Uncoated Specimens

Several methods can be used to examine uncoated specimens in the SEM, including operation at low beam energy, incorporating a second beam of electrons or ions to discharge the specimen, and examining the specimen in the presence of water.

Reduction of charging at low accelerating voltages is due to the characteristics of electron emission from solids, as indicated in the follow-

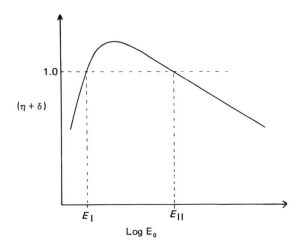

Figure 10.3. Total electron emission (backscattered and secondary electrons) as a function of beam incidence energy E_0. E_I and E_{II} are the first and second crossover points.

ing arguments due to Oatley (1972). If the electron emission coefficient, considering both backscattered primary electrons and secondary electrons, is plotted as a function of beam incident energy, E_0, the relationship shown in Figure 10.3 results. For insulators, a region exists for which the number of emitted electrons exceeds the number of incident electrons, i.e., $\delta + \eta > 1$. This region is defined by two values of the incident energy, E_I and E_{II}, for which $\delta + \eta = 1$; these values are referred to as the first and second crossover points. E_I is of the order of several hundred electron volts and E_{II} ranges from 1 to 5 keV, depending on the material. If the incident beam energy is less than E_I, then $\delta + \eta < 1$ and fewer electrons leave the specimen than enter it, resulting in a buildup of negative charge. This charge lowers the effective energy of the incident beam, producing a further decrease in $\delta + \eta$. This situation continues until the specimen is charged to a sufficient level to totally deflect the beam. If the beam energy is between E_I and E_{II}, then more electrons leave the specimen than enter it, i.e., $\delta + \eta > 1$. This charges the specimen positively, and the positive charge acts to decrease the effective value of δ, since the low-energy secondary electrons are attracted back to the specimen. The effective value of $\delta + \eta$ becomes unity because of these processes, and a dynamic equilibrium is set up with the emitted current equal to the incident current at a small, positive, constant surface charge. Slight variations in the curve of Figure 10.3 at different places in the field of view are not significant, since these will result in slight differences in the equilibrium surface potential. Operation with the $E > E_{II}$ again results in negative surface charging, which decreases the effective value of E, raising $\delta + \eta$ until an equilibrium is established with the effective E equal to E_{II}. Oatley (1972) points out that this equilibrium is unsatisfactory, since E_{II} can vary significantly about the surface, giving rise to large variations in the final surface potential from place to place. Slight leakage through surface conduction can greatly disturb such a situation, producing complicated image behavior as a function of time. It is thus desirable to operate with $E_I < E < E_{II}$, which is often achieved with E set to approximately 1 keV. Under such conditions, insulators can be examined uncoated. However, SEM performance is usually significantly poorer at such low accelerating potentials, since source brightness is greatly decreased under such conditions. With an SEM equipped with a field emission gun, the high brightness can be utilized to obtain good image resolution even at low beam energies (Welter and Coates, 1974).

Low beam energy prohibits x-ray microanalysis, since the overvoltage is inadequate except for x-rays of extremely low energies. To avoid this limitation several authors have reported methods which utilize a second beam of low-energy electrons or ions to discharge the sample during bombardment by the high-energy electron beam. Spivak *et al.* (1972) employed a pulsed beam of low-energy electrons to bombard the specimen during the scan line flybacks. The principle of the sample discharge is the same as that shown in Figure 10.3. Crawford (1979), described a charge neutralization method based on a low-energy beam of positive alkali ions which are attracted to the sites of negative surface charges.

If the specimen contains water, it will have sufficient conductivity to discharge. Considerable success has been obtained with examining and analyzing uncoated biological material at low termperatures (Echlin and Saubermann, 1977). In conditions where water is retained in the frozen state, so too are ions and electrolytes (Echlin, 1978) which provide conductivity. If specimens can be examined in an environmental cell, water can also be retained in the liquid state to discharge the primary beam (Robinson and Robinson, 1978).

10.1.2. Alternatives to Coating

One of the more useful techniques which has been devised is to increase the bulk conductivity of the sample as distinct from the *surface conductivity.* An increase in bulk conductivity may be achieved by metal impregnation from fixative solutions of osmium and manganese, with or without the use of organic metal ligands or mordants such as thiocarbohydrazide, galloglucose, paraphenylenediamine, by exposing specimens to OsO_4 vapor or by bulk staining the specimens after fixation with metallic salts. Kubotsu and Veda (1980) have devised a useful modification to the osmium vapor technique by also exposing the sample to hydrazine hydrate vapors which result in the deposition of metallic osmium. The relative merits of these methods are discussed in the recent review papers by Munger (1977) and Murphey (1978, 1980).

Alternatively, specimen charging may be reduced by spraying or impregnating with organic antistatic agents derived from polyamines, i.e., Duron, Denkil, or sodium alkyl-benzene sulfonate, soaking in conducting colloids of noble metals or graphite, or covering the sample with a thin ($\sim$1.0–20.0-nm) polymer film such as Formvar or styrene-vinylpyridine (Pease and Bauley, 1975).

With the possible exception of the techniques which increase the bulk conductivity of the sample, none of these methods gives anything like the resolution and information content which may be obtained from properly coated samples. Indeed, these alternative methods have diminished in usefulness now that it has been shown (Panayi et al., 1977) that one can adequately coat even the most delicate and thermally sensitive specimens in the sputter coater. Further, it can be argued that if a sample cannot survive the moderate vacuum found in a sputter coater it is unlikely to survive the high vacuum of an electron beam instrument.

The techniques devised to increase bulk sample conductivity are useful especially when they are used in conjunction with coating techniques. Bulk sample conductivity methods are frequently used in connection with the examination of fractured surfaces of three-dimensional specimens whose internal morphology is being investigated. For example, the substructure of fractured surfaces may be revealed by atomic number contrast from osmium incorporation, rather than topographic contrast derived from surface irregularities.

10.1.3. Thin-Film Technology

Thin films can be produced in a variety of ways (Maissel and Glang, 1970), but of these methods only thermal evaporation and sputtering are useful for coating specimens for SEM and x-ray microanalysis. In discussing these methods below, it is important to consider the properties of the ideal film. Such a film should not exhibit any structural features above a scale of 3–4 nm resolution to avoid introducing unnecessary image artifacts. The ideal film should be of uniform thickness regardless of the specimen topography and should not contribute to the apparent chemical composition of the specimen or significantly modify the x-ray intensity emitted from the sample.

10.2. Thermal Evaporation

Many metals and some inorganic insulators when heated by one means or another in a vacuum begin to evaporate rapidly into a monoatomic state when their temperature has been raised sufficiently for the vapor pressure to reach a value in excess of 1.3 Pa (10^{-2} Torr). The high temperatures which are necessary to permit the evaporation of the materials can be achieved by three different methods.

In the resistive heating technique an electric current is used to heat a container made of a refractory material such as one of the metal oxides or a wire support made of a high-melting-point metal such as tungsten, molybdenum, or tantalum. The material to be evaporated is placed in or on the container, which is gradually heated until the substance melts and evaporates. In the electric arc method an arc is struck between two conductors separated by a few millimeters. Rapid evaporation of the conductor surface occurs. This is the usual way by which some of the high-melting-point metals are evaporated. For most high-melting-point materials such as tungsten, tantalum, and molybdenum, the most effective way of heating the substance is to use an electron beam. In this method the metal evaporant is the anode target and is heated by radiation from a cathode maintained at 2–3 keV. This is a very efficient means of heating, as the highest-temperature region is the vapor-emitting surface and not the evaporant source material. An advantage is that the metal evaporant is deposited with very small grain size. Electron gun evaporation can also be used for evaporating some of the lower-melting-point metals such as chromium and platinum which have a very small particle size.

We can conveniently consider evaporation methods under two headings, high- and low-vacuum techniques.

10.2.1. High-Vacuum Evaporation

In this context high-vacuum is considered to be the range between 10 μPa and 100 mPa ($\sim 10^{-7}$–10^{-3} Torr). High-vacuum evaporation techniques are commonly used in electron microscope laboratories.

The formation of a thin film is a complex process and proceeds through a series of well characterized steps: nucleation and coalescence to form a continuous film. The first atoms arriving at the surface of the specimen will only stay there if they can diffuse, collide, and adhere to each other on the surface to form nucleation sites of a critical size. The stronger the binding between the adsorbed atoms and the substrate, the higher the nucleation frequency and the smaller are critical nuclei. Most biological and organic samples are likely to have variable binding energy, which would result in variation of critical nuclei across the surface, producing uneven film deposition. For this reason precoating with carbon at low vacuum to cover the specimen with a homogeneous layer results in smaller, even-sized critical nuclei when subsequently coated with metal. For example the nucleation density of gold can be substantially increased by 5–10 nm carbon precoating. As deposition continues, the nucleation centers

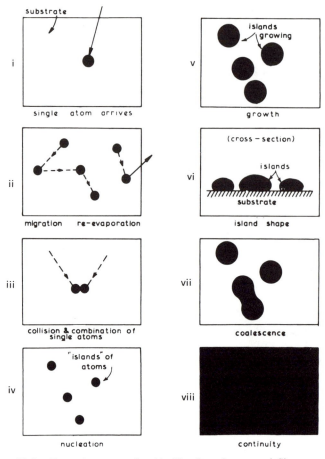

Figure 10.4a. Formation stages of a thin film (from Leaver and Chapman, 1971).

grow in three dimensions to form islands, which gradually coalesce to give a continuous film (Figures 10.4a–10.4e). The rate of continuous film formation and the average thickness at which a given film becomes continuous are influenced by a large number of variables. These include the nature of the evaporant and the substrate, the relative temperatures of the evaporant and substrates, the rate of deposition and final film thickness, and the surface topography of the specimen.

10.2.1.1. The Apparatus

There are four basic requirements for a high-vacuum evaporator. (1) It should have high pumping speeds at low gas pressure to ensure that there is

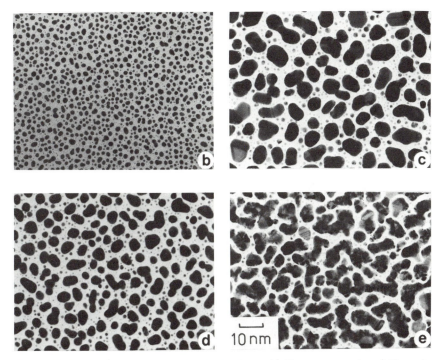

Figure 10.4b–e. Electron micrographs of ultrathin gold films evaporated at 0.05 nm/s. Thicknesses are (b) 1 nm, (c) 4 nm, (d) 6 nm, and (e) 15 nm. (From Kazmerski and Racine, 1975.)

a rapid removal of gases liberated from the evaporant source and specimen during coating. (2) There should be minimum backstreaming into the coating chamber of vapors from the pumps. (3) The system should be readily taken apart for cleaning and maintenance. (4) Adequate provision should be made for electrical connections for multiple evaporation schemes and sample manipulation.

Most units are pumped out using a diffusion pump backed by a rotary pump. The larger the throat size of the diffusion pump, i.e., 150 mm versus 75 mm, the faster the pump-down time of the evaporation chamber, although there will only be a marginal improvement in the ultimate pressure which is obtained. Some units are equipped with turbomolecular pumps backed by rotary pumps. Such units give faster pump-down times and a cleaner vacuum, although the ultimate pressure is no better than that obtained using a diffusion pump. The rotary pump should be a two-stage pump capable of being ballasted. Gas ballasting is one of the most effective means of preventing vapor condensation in a rotary pump. The method consists of leaking a small quantity of air into the rotary pump during the compression cycle so that the exhausted vapor is mixed with a noncondensable gas. This decreases the compression necessary to raise the exhaust valve and prevents vapor condensation. Rotary pump vapor condensation can also be decreased by raising the pump temperature and by placing a desiccant such as P_2O_5 in the exhaust line. To avoid backstreaming, the system should not be pumped below 10 Pa (10^{-1} Torr) on the rotary pump alone. The rotary pump should not be exhausted into the laboratory but vented into a fume cupboard or to the outside. There seems little necessity to resort to ion pumps, sublimation pumps, or exotic cryopumping systems as ultrahigh vacuum (130 nPa to 130 pPa) is not required for the commonly used evaporation methods.

Whatever system is used it is necessary to ensure that the backing and roughing lines from the pumping unit to the chamber are fitted with baffles and/or an activated alumina foreline trap to minimize backstreaming of pump oil. Water-cooled baffles are quite effective, although liquid-nitrogen baffles are better. A liquid-nitrogen trap in the roughing line will minimize diffusion of forepump oil into the evaporation chamber during roughing operations and maintain the forepressure of a diffusion pump at approximately 10 mPa. It is also useful to have a liquid-nitrogen trap above the diffusion pump between it and the coating chamber. Care should be taken to note where the vacuum pressure is read in the system. If the pressure is read near the pumps it is likely to be ten times better than that found in the evaporation chamber. The evaporation chamber should be made of glass and as small as is convenient for the work to be carried out. A safety guard should be fitted over the evaporation chamber to minimize danger from implosion fragments. The chamber should contain at least four sets of electrical connections to allow evaporation of two different materials, specimen rotation and thin-film measurement. The electric power for the evaporation sources should be variable, and it is useful to have a pushbutton control to allow maximum power to be applied in short bursts. The unit should be brought up to atmospheric pressure by means of a controllable needle valve which can be connected to a dry inert gas.

10.2.1.2. Choice of Evaporant

The choice of material to be evaporated and the manner by which it is to be applied is very dependent on the particular application in hand. A tabulation of selected elements and their properties, which are useful for coatings, is listed in Table 14.11 of the data base, Chapter 14. For most SEM work, gold, gold–palladium, or platinum–carbon is used. Silver has a high secondary-electron coefficient and is one of best substances for faithfully following the surface contours. Unfortunately silver suffers from the disadvantages that it easily tarnishes and has a larger grain size than other metals. Gold has a high secondary emission, is easily evaporated from tungsten wire, but has a tendency to graininess and agglomerates during coating, requiring a thicker coating layer to ensure a continuous film. A 60:40 gold–palladium alloy or palladium alone shows less granularity than gold and yields one of the thinnest continuous films. Unfortunately both metals easily alloy with the tungsten holder. Platinum–carbon when evaporated simultaneously produces a fine grain size but of rather low conductivity. Maeki and Benoki (1977) have found that low-angle (~30°) shadowing of specimens with evaporated chromium before carbon and gold–palladium coating improves the image of samples examined in the SEM. The finest granularity is obtained from high-melting-point metals, but they can only be evaporated with electron beam heaters. Figures 10.5 and 10.6

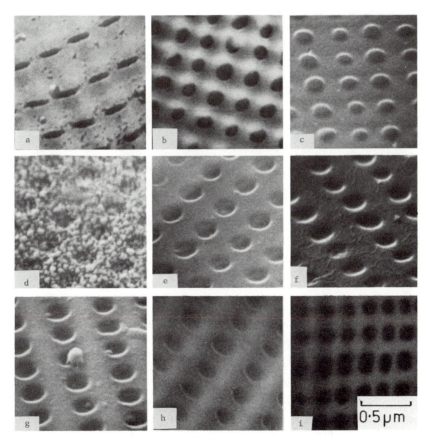

Figure 10.5. Cleaned diatom frustules coated with 10 nm of different metals and examined at 20 keV. (a) gold; (b) aluminum; (c) copper; (d) silver; (e) chromium; (f) gold–palladium; (g) titanium; (h) tin; (i) calcium.

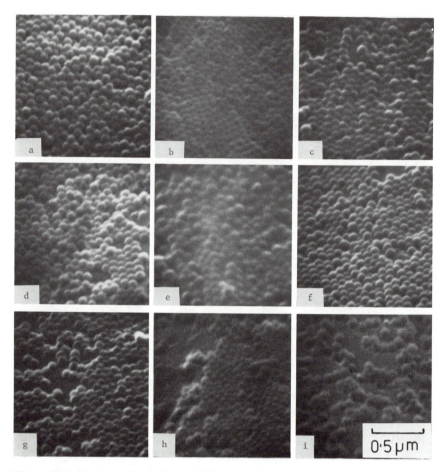

Figure 10.6. Polystyrene latex spheres 0.109 μm in diameter coated with 10 nm of different metals and examined at 20 keV. (a) gold; (b) aluminum; (c) copper; (d) silver; (e) chromium; (f) gold–palladium; (g) titanium; (h) tin; (i) calcium.

show the differences in surface features of some of the metal thin films applied to nonconducting samples. The usual "rule of thumb" for x-ray microanalysis is that the thinnest coating yielding stable specimen current and x-ray flux is desired. This is true since the thinner the coating, the less x-ray absorption within it and the less the energy loss of the primary electron beam entering the specimen. Furthermore, the thinner the coating, the smaller will be the excitation of x-rays from the coating itself. For gold and gold–palladium coatings, which are often used in scanning electron microscopy to provide enough secondary electron flux, the characteristic and/or continuum radiation produced could interfere with the x-ray lines of interest. Particular problems can occur if the element of interest is present in small or trace amounts. The usual thicknesses used range between 5 and 50 nm (50 and 500 Å). For 5–10-nm (50–100-Å) films of carbon, aluminum, gold, and gold–palladium, the energy loss of the primary beam appears to be of small consequence even at low nominal voltages. However, the beam and backscattered electrons obtained from the specimen could excite x-ray radiation from the film. This process may be particularly serious for gold and gold–palladium coatings on specimens with average atomic numbers greater than 10.

Examination of mass attenuation coefficients (μ/ρ) shows that for x-ray lines from 8 to 40 Å, aluminum is lowest of all four, followed in order by carbon, gold, and gold–palladium. In the region below 8 Å, the carbon (μ/ρ) value is lower than those for aluminum, gold, and gold–palladium. However, gold is best when electrical and thermal conductivities are considered, with aluminum about one-third as good, and carbon poor. It would seem that, for general purposes, aluminum is favored by its physical properties for use with x-ray lines of 0.8–4 nm (8–40 Å), while carbon is favored outside this region.

Most of the commonly used evaporants are available in the form of wire. It is recommended that thick wire, i.e., 0.5–1.0 mm is used, as short pieces can be easily looped over the appropriate refractory wire. Substances not available in wire form are available as powders or chips and can be conveniently evaporated from refractory crucibles or from boats made of high-melting-point metals. Care should be taken to ensure that the evaporant does not alloy or form compounds with the refractory. Most metals with a melting point below 2000 K can be evaporated from a support wire or boat made from a refractory metal such as tungsten, molybdenum, or tantalum. Such supports should be good electrical conductors, have a very low vapor pressure, and be mechanically stable.

10.2.1.3. Evaporation Techniques

When a pressure of about 10 mPa has been reached, the refractory support wire and, where appropriate, the carbon rods, can be heated to a dull red color. Gentle heating will cause a sharp rise in the vacuum pressure brought about by outgassing and removal of residual contamination. Once the outgassing is completed, the current is turned down and pumping continued. If a carbon coating is to be applied to the specimen before metal coating, this is best done at a pressure of about 10 mPa.

During coating the specimens are rotated and tilted to give an even coating on all surfaces. Following the deposition of the carbon layer the vacuum pressure is decreased to a pressure between 1.0 mPa and 100 μPa. An electric current may now be passed through the tungsten wire holding the metal to be evaporated. This should be done carefully and it is best to gradually increase the current until the tungsten wire just begins to glow and then back off a small amount. This allows the metal evaporant to heat up slowly and melt as the tungsten wire current is slowly increased.

The metals commonly used in evaporation form a molten sphere in the V of the tungsten wire. This stage should be allowed to remain for a few moments to remove any residual contamination. The current should then be further increased until the sphere of molten metal appears to shimmer and "rotate." When this point is reached evaporation has started and the shutter may be opened to expose the rotating specimen to the evaporation source. In order to achieve a uniform coating on complexly sculptured specimens it is essential that they be rotated rapidly (6–8 rps) during coating. The ideal arrangement is to rotate the specimen in a planetary motion while continuously tilting it through 180°. This slow heating, melting, and evaporation of the metal evaporant is most important, and more particularly so with aluminum which alloys with tungsten. If the source is heated too rapidly the metal evaporant melts at its point of contact with the tungsten wire and falls off. However, the speed of coating is important in obtaining films of good quality, and the faster the speed of evaporation the finer the structure of the layer. This optimal high speed of coating must be balanced against the higher thermal output from the source, and the increased chance of the wire support alloying with the substrate and melting. The thickness of the evaporated film may be measured by a number of different techniques which will be described later, the most convenient of which is a quartz crystal thin-film monitor mounted inside the vacuum chamber. The thickness deposited also depends on the particular specimen being studied and the type of information required from the sample. It has usually been considered necessary to apply sufficient metal to give a continuous film since it has been assumed that only a continuous film would form a surface conductive layer on a nonconductive sample. Although the physical nature of charge transport in particulate films is not clearly understood, recent work suggests that particulate layers may be a useful method for SEM specimens because discontinuous metal films exhibit a significant dc conductivity. Paulson and Pierce (1972) have shown that discontinuous films can conduct a limited current by electron tunneling between evaporated island structures and suggest that such discontinuous films may be useful for the examination of nonconductive specimens viewed at very low current. Specimens have been successfully examined at 20 keV which have only been coated with 2.5 nm of gold (Figure 10.7).

The color of the layer on a white card or glass slide can give a quick practical estimate of thickness. For most specimens a carbon layer visible as a chocolate color and a gold layer which is a reddish-bronze color by reflected light and blue-green to transmitted light will be sufficient. For aluminum coating sufficient metal is deposited when the layer is a deep

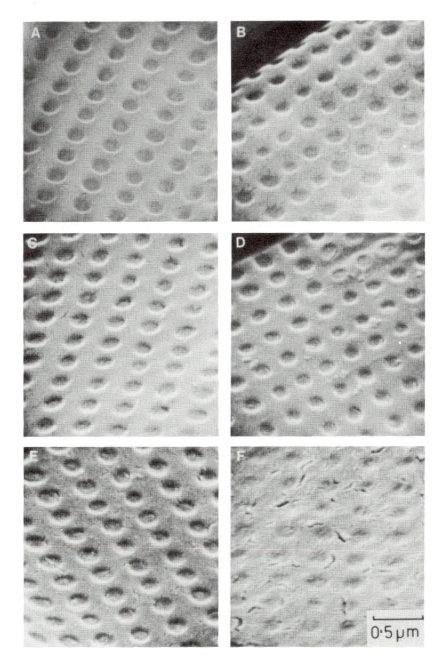

Figure 10.7. Cleaned diatom frustules coated by evaporation, with increasing thickness of gold. (a) 2.5 nm; (b) 5.0 nm; (c) 10 nm; (d) 20 nm; (e) 50 nm; (f) 100 nm and examined in the secondary mode of operation at 30 kV. Note that the very thin layers give optimal information, whereas the thicker layers obscure the specimen surface.

blue to transmitted light. Once these parameters have been worked out for a particular evaporation unit with a fixed geometry of specimen and source, it is only necessary to cut off a standard length of evaporant metal wire for each coating.

10.2.1.4. Coating Artifacts

If the coating procedure has been carried out properly, artifacts are rarely seen. Some of the possible causes of artifacts are discussed below.

(a) **Heat of Radiation.** The intensity of thermal radiation reaching the specimen depends on the source temperature and the source-to-specimen distance. The heat of radiation may be diminished by using a small source and/or moving the specimen further away. The best practical solution is to use a small source at high temperature and have at least 150 mm between the source and the sample. Provided the specimen is adequately shielded from the target, and the shutter only opened at the working temperature, little damage is likely to occur.

Thermal artifacts appear as smooth micromelted areas on inclined fracture faces of biological materials which are bombarded vertically, minute holes, and surface distortions. If specimens are being damaged by heat radiation, the damage can be diminished by placing a cold plate with an aperture over the specimen.

(b) **Contamination.** Contamination is due primarily to dirt and volatile substances in the vacuum system being deposited on the specimen, and it is for this reason that care must be taken to clean the system properly before use. The most effective way to reduce this problem is to surround the specimen with a cold surface. However, this is probably unnecessary except for very-high-resolution studies. Contamination may be recognized as uneven coating and hence charging, as small randomly arranged particulate matter, and in the most extreme situations as irregular dark areas on the specimen.

(c) **Decoration Effects.** Decoration or agglomeration of the evaporated material occurs to some extent with most metal coatings and is a result of uneven deposition of the evaporated metal. Agglomeration occurs if the cohesive forces of the film material are greater than the forces between the film molecules and the substrate. Because of geometrical effects, rough surfaces are particularly difficult to coat evenly, and it is inevitable that those parts which protrude will receive more coating than crevices and holes.

(d) **Film Adhesion.** Poor film adhesion is associated with hydrocarbon and water contamination, and, in the case of plastics, with the presence of a thin liquid layer of exuded plasticizer. Discontinuous and poorly adhesive films are recognized by a "crazed" appearance and have a tendency to flake easily. In the microscope there are variations in the image brightness, and charging occurs on isolated "islands" of material which are not in contact with the rest of the film.

10.2.2. Low-Vacuum Evaporation

In low-vacuum evaporation, carbon is evaporated in an atmosphere of argon at a pressure of about 1 Pa. The carbon atoms undergo multiple collisions and scatter in all directions. This technique is useful for preparing tough carbon films and for coating highly sculptured samples prior to analysis by x-rays, cathodoluminescence, and backscattered electrons. However, its general usefulness for SEM specimens is questionable, particularly as the yield of secondaries from carbon is very low. There is little doubt that the multiple scattering and surface diffusion of the carbon allows it to more effectively cover rough specimens, and for this reason would be a useful method to use if a sputter coater was not available.

10.3. Sputter Coating

Although sputter coating has been known for a long time it is only recently that it has become more widely used for producing thin films. In the process of sputtering an energetic ion or neutral atom strikes the surface of a target and imparts momentum to the atoms over a range of a few nanometers. Some of the atoms receive enough energy in the collision to break the bonds with their neighbors and be dislodged. If the velocity imparted to them is sufficient, they are carried away from the solid (Wehner and Anderson, 1970).

The glow discharge normally associated with sputter coating is a result of electrons being ejected from the negatively charged target. Under the influence of an applied voltage the electron accelerates towards the positive anode and may collide with a gas molecule, leaving behind an ion and an extra free electron. The glow discharge is located some distance from the target. The positive ions are then accelerated towards the negatively biased target where they cause sputtering. At high accelerating voltages many electrons are ejected per impinging ion, and these electrons have sufficient energy to damage delicate targets.

A number of factors affect the deposition rate. The sputtering yield increases slowly with the energy of the bombarding gas ion (Figure 10.8). The current density has a greater effect than the voltage on the number of ions striking the target and thus is the more important parameter determining the deposition rate (Figure 10.9). Variations in the power input can have a dramatic effect on the properties and composition of sputtered

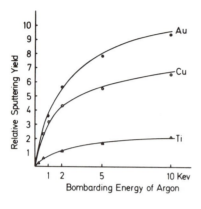

Figure 10.8. Relative sputtering yield of different metals at different bombarding energies of argon.

films; for example, with increasing power levels, aluminum films became smoother with fewer oxide particles. As the pressure of the sputtering system is increased, the ion density also increases. There is a linear increase in current, and hence sputtering rate, between pressures of 3 and 20 Pa. But because there is an increased tendency at higher pressures for the eroded material to return to the target, a compromise pressure of between 3 and 10 Pa is commonly used.

Impurities in the bombarding gas can appreciably reduce the deposition rate. Gases such as CO_2 and H_2O are decomposed in the glow discharge to form O_2, and the presence of this gas can halve the deposition rate. Deposition rate decreases with an increase in specimen temperature, although this phenomenon may not be peculiar to sputter coating. Finally, the deposition rate is higher the closer the target is to the specimen, but this also increases the heat load on the specimen. The ejected particles arrive at the substrate surface with high kinetic energy either as atoms or clusters of atoms, but not as vapor. There is some evidence that the sputtered atoms

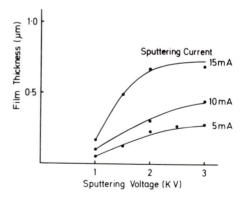

Figure 10.9. Variations in film thickness in relation to the sputtering voltage and current.

have sufficient energy to penetrate one or two atomic layers into the surface on which they land.

There are several different ways to produce the sputtering process, including ion beam sputtering, plasma sputtering, radiofrequency sputtering, triode sputtering, diode (DC) sputtering, and cool diode sputtering. As the technique has evolved for SEM and electron probe specimen coating, only ion beam, diode, and cool diode sputtering are now commonly employed.

10.3.1. Ion Beam Sputtering

The technique of ion beam sputtering is illustrated in Figure 10.10a. An inert gas such as argon is ionized in a cold cathode discharge, and the resulting ions from the ion gun are accelerated to an energy of 1–30 keV.

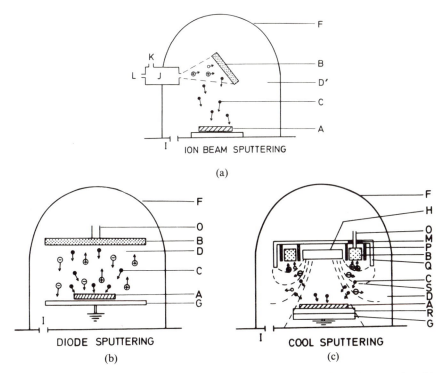

Figure 10.10. Sputter coating techniques: (a) ion beam sputtering; (b) diode sputtering; (c) cool diode sputtering. Key to symbols: A, specimen; B, target or evaporative source; C, target atoms; D, low-pressure argon discharge; D', high vacuum; E, electrode; F, glass vacuum vessel; G, anode; H, permanent magnet; I, line to vacuum pumps; J, ion gun, K, inert gas supply; L, high-voltage power supply; M, insulator; O, direct current power supply (1–3 keV); P, aluminum shield; Q, iron pole piece; R, cooling module; S, lines of magnetic flux.

The ion beam may be formed by collimation or focused with a simple lens system to strike the target. The energetic ions strike the target atoms and impart momentum in elastic collisions; atoms lying near the surface of the target are ejected with energies in the range 0–100 eV. These sputtered atoms then deposit on the sample and all surfaces which have a line of sight to the target. An advantage of this arrangement is that a field-free region exists between the target and substrate so that negative ions and electrons are not accelerated toward the substrate. Multiple coatings can be applied from different targets, providing care is taken to prevent cross-contamination of the targets during sputtering. If a nonconducting target is used, the buildup of positive charge can be suppressed by an electron flood gun. Ion beam sputtering has been used by Adachi *et al.* (1976) and Hojou *et al.* (1977) for high-resolution shadowing. By using sputtered films of tungsten–tantalum, tungsten, and carbon, they were able to see details smaller than 1.0 nm.

10.3.2. Diode or Direct Current Sputtering

This is the simplest, most reliable, and most economical type of sputtering and is the basis of a number of commercially available dedicated instruments as well as sputtering attachments for thermal evaporators. These instruments, which operate between 1 and 3 keV, are sometimes referred to as diode sputtering units as well as DC sputtering units. DC sputter coating units consist of a small bell jar containing the cathode target and water-cooled specimen holder anode, which sits on top of a control module containing the vacuum gauge, high-voltage power supplies, leak valve, and a small timer (Figure 10.10b). The detailed mode of operation and application of this type of instrument has been previously described (Echlin, 1975a). One of the potential problems of this type of coater is that delicate specimens can be thermally damaged.

10.3.3. Cool Diode Sputtering

The heat problem of diode sputtering has been overcome by redesigning the diode coater to incorporate devices to keep the specimen cool throughout the coating procedure (Panayi *et al.*, 1977) as shown in Figure 10.10c. The electron bombardment of the specimens is significantly reduced by replacing the disk-shaped target of the diode coater with an annular target. Thermal damage to the specimen is further reduced by fitting a permanent magnet at the center of the target and an annular pole piece around the target. This arrangement deflects the electrons outside the periphery of the specimen holder. As a further precaution, the specimen holder is cooled by a small Peltier effect cooling module. Using this improved equipment it has been possible to coat crystalline hydrocarbon waxes with melting points as low as 305 K, and thermolabile sensitive plastic films, neither of which had been previously coated successfully by any of the methods available.

10.3.4. Sputtering Techniques

It is important to ensure that the coater is fitted to an adequate supply of clean, dry argon or, where appropriate, another noble gas. A small trace of nitrogen in the argon probably does not matter, but it is important that the gas is free of water, carbon dioxide, and oxygen. Traces of water vapor, which are recognizable by the blue tint in the glow discharge, can easily be removed by passing the gas through a column of desiccant. The final traces of volatile material may be removed from the target by bombarding it with a current of 20 mA at 2.5 kV at a pressure of between 2 and 8 pA. The target can be considered clean when there is no degradation of vacuum when the high-voltage discharge is switched on.

Specimens are placed on the specimen table and the unit pumped down to about 10–15 Pa (10^{-1} Torr) using a two-stage rotary pump fitted with an activated alumina foreline trap to prevent oil backstreaming. It is most important not to let the unit pump for a long time at the ultimate pressure which may be obtained by the rotary pump, because this will cause backstreaming, resulting in contamination. Because of this it is advisable to have the argon leak valve slightly open, to ensure a continual flow of inert gas through the system giving a pressure of about 6–7 Pa. If the unit is fitted with a water-cooled or, better still, a Peltier module cold stage, this should be turned on and the specimens cooled to the working temperature.

One of the commonest sources of poor pumping performance in both sputter and evaporative coaters is the continual outgassing of the specimen and the adhesive used to attach it to the support. It is recommended that biological and organic material, after it has been dried and attached to the specimen support, be placed in a 310 K vacuum oven overnight to ensure that all volatile substances are removed before coating.

Once the specimen has been cooled and an adequate pressure has been reached, the sputter coating may proceed. The system is pumped down to about 2 pA and the high voltage turned to 2.0 kV. The argon leak valve is opened until a plasma discharge current of 12–15 mA is recorded at a pressure of 6–7 Pa. The timer is set and coating continued until the desired thickness has been deposited on the specimen.

It is possible to use nitrogen as an alternative to argon. However, extended discharge times and/or higher plasma currents are necessary to give the same thickness of coating. Sputter coating in air should be avoided.

because the presence of water vapor, carbon dioxide, and oxygen gives rise to highly reactive ions which can quickly degrade the specimen.

10.3.5. Choice of Target

Platinum or the gold–palladium targets have been found to be satisfactory for the routine specimen preparation for the SEM. Targets are available in most of the other noble metals and their alloys as well as elements such as nickel, chromium, and copper. There are differences in the sputtering yield from different elements and these must be borne in mind when calculating the coating thickness. There are difficulties with a carbon target, for although it is possible to very slowly erode the target with argon, the sputtering rate falls off rather rapidly. This decrease is due either to the presence of forms of carbon which have a binding energy higher than the energy of the argon ions, or the poorer conductivity of the carbon gives rise to charging and a decreased erosion rate. The claims that carbon can be sputtered at low kilovoltage in a diode sputter coater are probably erroneous. The deposits of "carbon" which are achieved are more likely to be hydrocarbon contaminants degraded in the plasma than material eroded from the target. There seems to be little likelihood that a simple technique will be devised for sputter coating aluminum. The oxide layer which rapidly forms on the surface of aluminum is resistant to erosion at low keV, and the rather poor vacuum makes it difficult to deposit the metal. For details of other targets, particularly those made by nonconducting materials and bombarding gases, the reader is referred to the book by Maissel and Glang (1970).

10.3.6. Coating Thickness

For a given instrumental setting there is a linear relationship between the deposition rate and the power input. In practical terms this means the thickness of the coating layer is dependent on gas pressure, the plasma current, voltage, and duration of discharge. It is difficult to accurately deposit thin films, i.e., below 10 nm, and it is necessary to rely on evaporative techniques for such films. However, most biological material requires about 15–25 nm of metal coating, and once one has established the ideal film thickness for a given specimen it is easy to accurately reproduce this thickness on other samples.

10.3.7. Advantages of Sputter Coating

One of the main advantages of the technique is that it provides a continuous coating layer even on those parts of the specimen which are not in line of sight of the target. Figure 10.11 compares the major coating

Figure 10.11. Comparison of uncoated, evaporative coated and sputter coated nonconductors. Left, Al_2O_3; center, cotton wool; right, polystyrene latex spheres. Top row, uncoated; center row, evaporative coated with 10 nm of gold; bottom row, sputter coated with 10 nm of gold. Marker = 1 μm.

processes. The continuous layer is achieved since sputtering is carried out at relatively high pressures and the target atoms suffer many collisions and are traveling in all directions as they arrive at the surface of the specimen. Highly sculptured structures or complexly reticulate surfaces are adequately coated. This ability of the target atoms to "go round corners" is particularly important in coating nonconductive biological materials, porous ceramics, and fibers. Complete coating is achieved without rotating or tilting the specimen, and using only a single source of coating material. Provided the accelerating voltage is sufficiently high it should be possible to sputter coat a number of nonconducting substances such as the alkaline halides and alkaline earth oxides, both of which have a high secondary-electron emission. Similarly, it should be possible to sputter substances which dissociate during evaporation. Film thickness control is relatively simple, and sputtering can be accomplished from large-area targets which contain sufficient material for many deposition runs. There are no difficulties with large agglomerations of material landing on the specimen, and the specimens may more conveniently be coated from above. The surface of specimens can be easily cleaned before coating either by ion bombardment or by reversing the polarity of the electrodes. The plasma may be manipulated with magnetic fields, which gives greater film uniformity and reduces specimen heating.

10.3.8. Artifacts Associated with Sputter Coating

Sputter coating of SEM samples has attracted some unfavorable comments, because in the hands of certain users it has resulted in thermal damage to delicate specimens and to surface decoration artifacts. There is little doubt that the earlier diode sputter coaters could cause damage to heat-sensitive specimens, particularly if coating was carried out for a long time at high plasma currents on uncooled specimens. The thermal damage problem has been greatly reduced with the advent of the new series of cool sputter coaters. A careful examination of the examples of decoration artifacts which have been reported in the literature reveals that in most cases scant regard has been paid to cleanliness during coating, and that the artifacts were due to contamination from backstreaming oil vapors and/or the use of impure or inappropriate bombarding gases. Nevertheless it would be improper to suggest that artifacts never occur with sputter coated materials and some of the more common problems are discussed below.

10.3.8.1. Thermal Damage

A significant rise in the temperature of the specimen can be obtained during sputter coating. The sources of heat are radiation from the target and electron boardment of the specimen. There is an initial rapid rise in temperature which then begins to level out and, depending on the nature of the material being coated, may cause thermal damage. Depending on the accelerating voltage and the plasma current, the temperature rise can be as much as 40 K above ambient. However, as mentioned earlier heating effects can be entirely avoided by using the modified diode cool coater where the heat input due to electron bombardment is only 200 mW or somewhat reduced by operating a conventional diode coater intermittently at low power input.

When thermal damage occurs it is manifest as melting, pitting, and, in extreme cases, complete destruction of the specimen. While accepting that thermal damage can be a problem in sputter coating, in nearly all cases where this has been reported it is due to the specimen being subjected to inordinately high power fluxes.

10.3.8.2. Surface Etching

This is a potential hazard in sputter coating and may be caused either by stray bombarding gas ions or, more likely, by metal particles hitting the surface with sufficient force to erode it away. It is possible to find very small holes in the surface of sputter coated specimens examined at high resolution. It is not clear whether these are the result of surface etching or simply thermal damage.

10.3.8.3. Film Adhesion

Film adhesion is much less of a problem with sputter coated films than with evaporated films and is probably due to the fact that the metal particle penetrates the surface of the specimen, forming a strong bond. However, sputter coated samples should not be subject to wide excursions of temperature or humidity, both of which can give rise to expansion and contraction with a consequent rupture of the surface film.

10.3.8.4. Contamination

Because of the low vacuum used in most sputter coaters, the problem of backstreaming from the rotary pumps, and the difficulty of placing effective cold traps in the backing lines, contamination can be a potentially serious problem, particularly if no foreline traps have been fitted. Many of the artifacts which have been described are probably due to contamination, and care should be taken in setting up and using the sputter coater.

10.4. Specialized Coating Methods

Although a description has been given of the general principles of evaporative and sputter coating, there are a number of specialized coating methods which should be discussed because they are applicable to both scanning electron microscopy and x-ray microanalysis.

10.4.1. High-Resolution Coating

As the resolution of scanning electron microscopes increases, greater attention must be given to the resolution limits of the coating layer. For many instruments which can resolve about 10 nm, the granularity of the film is of little consequence, with the possible exception of agglomeration of gold particles. A number of microscopes are routinely operating in the 5–10 nm range and the STEM instruments can give between 2- and 5-nm resolution in the secondary mode. Transmission electron microscopists have long been concerned about high-resolution films, as they are necessary for specimen support, for shadowing, and for making replicas of frozen and ambient temperature fractured material. It is interesting to note that one of the favorite coating materials for scanning electron microscopists—gold—has a coarse granularity when examined at high resolution in the TEM. Chromium, gold–palladium, platinum, and zirconium have much finer grain size, and carbon, platinum–carbon, tungsten, and tantalum all have virtually no grain size when examined in the TEM. The graininess of the deposits generally decreases with an increase in the melting temperature. The studies of Adachi *et al.* (1976) and Hojou *et al.* (1977) show that sputter coated films of tantalum and tungsten have very small grain size and would be most suitable for high-resolution SEM studies.

Bräten (1978) found that thermally evaporated gold–palladium and carbon–gold–palladium gave the best result as far as specimen resolution and surface smoothness was concerned. Thermally evaporated and sputter coated gold gave a far more grainy appearance, and interweaving cracks could be seen on the surface of the specimen. The ultimate particle size is also dependent on the nature of the substrate. Echlin and Kaye (1979) and Echlin *et al.* (1980) found that for high-resolution (2–3 nm) SEM, electron beam evaporation of refractory metals (W, Ta) or carbon–platinum gave the best results. The most convenient means of preparing thin films for medium resolution (5–8 nm) SEM is to sputter coat films of platinum or gold–palladium onto specimens maintained below ambient temperature. Slower sputtering rates also result in smaller particle size. Advantage can also be made of the conductivity of discontinuous metal films which can provide an effective coating layer when only a few nanometers thick. Franks (1980) has used another sputtering technique based on a cold cathode saddle-field source to produce an ion beam. The high-energy beam was used to bombard a target producing coatings at a pressure of 10 mPa. The beam is directed onto the target with the substrate at a fixed or variable angle, and provides very fine grain deposits. Although the coating times are rather long (10–15 min) the films which are produced are ideal for high-resolution SEM studies. Figure 10.12 compares the results obtained with ion beam sputtering with those obtained with cold DC sputtering and evaporative coating. Whereas the sputter coated material and the evaporative coated material show artifacts at high resolution, only random electron noise is evident in the ion beam sputtered material. In a recent study, Clay and Peace (1981) have used ion beam sputtering to coat biological samples. They showed a significant reduction in surface artifacts compared to material which has been diode sputter coated (Figure 10.13). Peters (1980a) has used penning sputtering, which generates high-energy

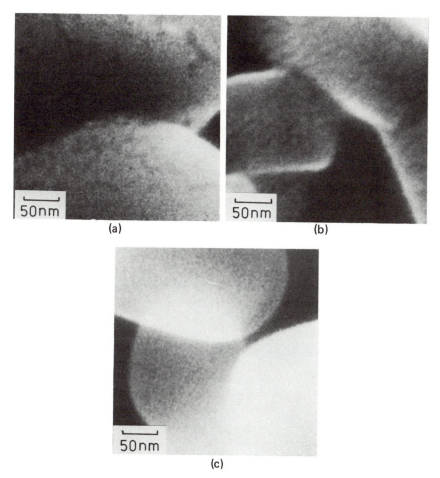

Figure 10.12. A comparison of (a) evaporative coating, (b) diode sputter coating, and (c) ion-beam sputter coating. Each sample has been coated with approximately the same amount of gold. The specimen is copy-paper crystals. Whereas the sputter coated and evaporative coated materials show artifacts at these high magnifications, only random electron noise is visible on the ion-beam sputtered material (Franks, 1980). Original micrographs taken by C. S. Clay, Unilever Research Laboratories, Sharnbrook, U.K.

target atoms resulting in very small crystallites. This technique allows very thin, i.e., 1–2-nm-thick films, to be produced with small particle size. In an earlier study, Peters (1979), has showed that provided great care is taken to minimize contamination, high-resolution films can also be obtained by thermal evaporation of high-melting-point metals.

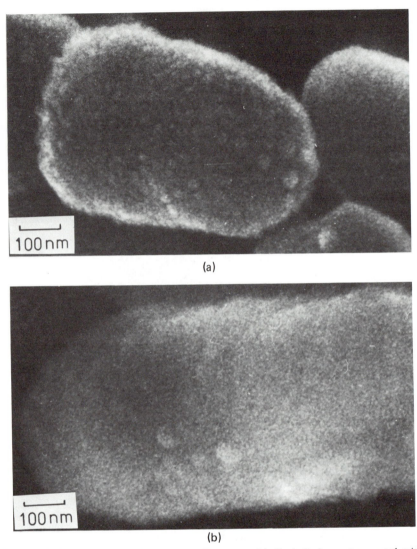

Figure 10.13. The bacteria *Pseudomonas fluorescens*. (a) Cool diode sputter coated with gold–palladium; (b) ion-beam sputter coated with gold. Note the granularity on the diode sputter coated material, which is absent on the ion-sputtered sample (Clay and Peace, 1981).

10.4.2. Low-Temperature Coating

The advantages of examining and analyzing frozen specimens in the scanning electron microscope and the x-ray microprobe have been discussed (Echlin and Saubermann, 1977; Echlin, 1978; Saubermann *et al.*,

1981). The main problems associated with low-temperature coating are contamination and maintaining the specimen below 143 K during the coating procedures. The main source of contamination is residual water vapor, which can readily interact with the metal during the coating process. Saubermann and Echlin (1975) found that unless great care was taken during the coating process, evaporated aluminum was deposited as a grey granular layer onto thin sections of frozen biological material held at below 123 K. A similar phenomenon was occasionally observed during gold coating of fractured, bulk frozen material. These effects are entirely eliminated if the coating is carried in a coating unit interfaced with the microscope via an airlock (Fuchs *et al.*, 1978; Pawley and Norton, 1978; Echlin *et al.*, 1979, 1980).

Low-temperature coating has only been successfully achieved using evaporative techniques, and care has to be taken not to melt the surface of the specimen during the coating procedure. Clark *et al.* (1976), using thin-film thermocouples, recorded an insignificant temperature rise on a thin-film substrate coated from a collimated and shuttered evaporative source. Although attempts have been made to sputter coat frozen samples (Robards and Crosby, 1979), the results bear no comparison with those which can be obtained using evaporative methods.

10.5. Determination of Coating Thickness

There are a number of techniques which can be used to determine coating thickness, and a recent paper by Flood (1980) reviews the methods which are available. However, all the methods give an average value on a flat surface and it is necessary to briefly consider this type of measurement in relation to the coating thickness one might expect on rough surfaces. If the coating thickness is too thin it will be ineffective in eliminating the charge; if it is too thick it will obscure surface detail. A coating layer of a certain minimum thickness is necessary to form a continuous layer, and this thickness varies for different elements and for the relative roughness of the surface. On an absolutely flat surface a layer of carbon 0.5 nm thick will form a continuous coat. A layer ten times as thick is required to form a coat on an irregular surface. As a general rule, thicker coatings are required for irregular surfaces to ensure a continuous film over all edges, cavities, and protrusions.

It is clear that the film thickness we can measure on a flat surface bears little relation to the thickness we obtain on an irregular surface which may have been rotated and tilted during coating. For practical purposes, a given set of rough specimens should be coated with different thicknesses under standardized conditions, each film layer being measured and recorded from a reference flat surface such as a glass slide. The specimens should be examined in the microscope and the one which gives the optimum image in terms of resolution and information content can be considered to have the most suitable coating thickness.

10.5.1. Estimation of Coating Thickness

In high-vacuum thermal evaporation, one can assume that all the vapor molecules leave each part of the evaporant surface without preferred direction and pass to the substrate surface without colliding with residual gas molecules. By allowing for the vapor incident angle at the specimen, and assuming that all the incident vapor molecules have the same condensation coefficient, the coating thickness distribution may be calculated. The formula below may be applied to a flat, nonrotating surface, oriented at an angle θ to the source direction and can be used to calculate the thickness from the quantity of material to be evaporated:

$$T = \frac{M}{4\pi R^2 \rho} \cos\theta \qquad (10.1)$$

where T is the thickness in centimeters; M is the mass of the evaporant in grams; ρ is the density in $g\,cm^{-3}$; R is the distance from source to specimen in centimeters; and θ is the angle between the normal to the specimen and the evaporation direction. With Equation (10.1), the thickness can be calculated to within $\pm 50\%$ if the geometry of the evaporation rig and substrate is known and if the material in question evaporates slowly and does not alloy with the heater. It is assumed that evaporation is from a point source and occurs uniformly in all directions. The uncertainty in T arises in part because of uncertainty in the sticking coefficient of the evaporant or the substrate. The efficiency is material dependent and is rarely better than 75%. The thickness calculated by Equation (10.1) must therefore be reduced by multiplying by the sticking efficiency (approximately 0.75).

The thickness of coating T in nanometers which may be obtained using a diode sputter coater is given by

$$T = \frac{CVtk}{10} \qquad (10.2)$$

where C is the plasma discharge in milliamperes, V the accelerating voltage in kilovolts, t the time in minutes, and k a constant depending on the bombarding gas and the target material. The constant k must be calculated for each sputter coater, target material, and bombarding gas from experimental measurements.

10.5.2. Measurement during Coating

A number of different methods are available which vary in their sensitivity and accuracy. The devices are placed in the coating chamber close to the specimen and measure the thickness during the coating process. It is possible to measure the density of the evaporant vapor stream either by measuring the ionization which occurs when vapor molecules collide with electrons, or by measuring the force which impinging particles exert on a surface. Mass-sensing devices may be used for all evaporant materials; these operate by determining the weight of a deposit on a microbalance or by detecting the change in oscillating frequency of a small quartz crystal on which the evaporant is deposited. The quartz crystal thin-film monitor resonates at a frequency dependent on the mass of material deposited on its surface. The frequency of a "loaded" crystal is compared to the frequency of a clean crystal and the decrease in frequency gives a measure of the film thickness. A typical sensitivity value for a crystal monitor is a 1-Hz frequency change for an equivalent film of 0.1 nm C, 0.13 nm Al, and 0.9 nm Au. The uncertainty of these devices is of the order $\pm 0.1 \ \mu g\,cm^{-2}$. By multiplying this value of uncertainty by the evaporant density, the uncertainty in the film thickness can be determined.

It is also possible to monitor specific film properties such as light absorption, transmittance, reflection, and interference effects by various optical monitors. Film thickness of conductive materials can be measured by *in situ* resistance measurements and the thickness of dielectric materials measured by capacitance monitors. A further sophistication of most of these *in situ* techniques is that they can be used to control the coating process.

10.5.3. Measurement after Coating

The most accurate measurements of film thickness are made on the films themselves. These methods are based on optical techniques, gravimetric measurements, and x-ray absorption and emission. Multiple-beam interferometry is the most precise and depending on the method used can be as accurate to within one or two nanometers. The Fizeau method can be used to check film thickness and involves placing a reflective coating on top of the deposited film step and measuring a series of interference fringes. The film thickness can also be measured by sectioning flat pieces of resin on which a coating has been deposited and measuring the thickness of the metal deposit in the transmission electron microscope. The accuracy of this method is dependent on being able to section the resin and photograph the section at right angles to the metal deposit. A simple method for accurately determining film thickness and grain size has been recently described by

Roli and Flood (1978). They found that linear aggregates of latex spheres only accumulate coating material along their free surface. Transverse to the linear aggregate, the sphere diameter will increase in thickness by twice the film thickness, while parallel to the aggregate the thickness increase will correspond to the film thickness. Using this technique they have measured the film thickness derived from different coating procedures to within ± 2 nm. Film thickness may be estimated by interference colors or in the case of carbon by the density of the deposit on a white tile.

10.5.4. Removing Coating Layers

Having taken all the trouble to deposit a coating layer on a sample it is sometimes necessary to remove it. Provided care is taken, hard specimens can be restored to their pristine uncoated state.

If the sample is a polished section, the coating can be easily removed by returning to one of the final polishing operations (6 μm diamond or 1 μm Al_2O_3). If the sample is rough or a flat sample cannot be repolished, chemical means must be used to remove the coating.

If it is known that the coating layer is to be removed after examination of a rough sample, it is preferable to use aluminum as the coating material because of its ease of removal. Sylvester-Bradley (1969) removed aluminum from geological samples by immersing them in a freshly prepared solution of dilute alkali for a few minutes. Sela and Boyde (1977) describe a technique for removing gold films from mineralized samples by immersing the specimen in a cyanide solution. Gold–palladium may be removed using a 10% $FeCl_3$ in ethanol for 1–8 h (Crissman and McCann 1979).

Carbon layers can be removed by reversing the polarity in the sputter coater and allowing the argon ions to strike the specimen. Care must be taken only to remove the carbon layer and not to damage the underlying specimen. Carbon layers on inorganic samples such as rocks, powders, and particulates can be efficiently removed in an oxygen-rich radiofrequency plasma.

11

Preparation of Biological Samples for Scanning Electron Microscopy

11.1. Introduction

In this and the next chapter we will consider the practical aspects of specimen preparation for both the scanning electron microscope and the x-ray microanalyzer. Although the two types of instrument are very similar and in many respects can be used interchangeably, it is useful, from the biologist's point of view, to consider the preparative techniques separately. The scanning electron microscope gives morphological information, whereas the x-ray microanalyzer gives analytical information about the specimen. It is important for the user to appreciate fully these differences as they have a significant bearing on the rationale behind the specimen preparation techniques. The methods and techniques which are given in these two chapters will provide the optimal specimen preparation conditions for scanning microscopy *or* x-ray microanalysis. It must be realized that anything less than the optimal conditions will result in a diminished information transfer from the specimen. It will also become apparent that it will be frequently necessary to make some sort of compromise between the two approaches, which must result in less information from the specimen.

The practical, do-it-yourself aspects of specimen preparation will be emphasized, and while it is appreciated that a certain understanding of the theoretical aspects might be useful, such considerations will be kept to a minimum and readers will be directed to key review papers on the subject. It is not our intention to provide the biologists with a series of recipes. These already abound in the literature and very few of them can result in optimal specimen preparation for anything other than the specimens for

495

which they were devised. An up-to-date review of specific preparative techniques for a wide range of biological samples may be found in a series of tutorial papers in SEM/1980. The chapters will seek to provide the rationale behind, and suggestions for, the practical execution of specimen preparation but leave the actual choice of recipe to the experimenter. It is advisable to carefully consider what information can be gained from the specimen. For example, the preparative techniques needed to preserve a small part of the cellular ultrastructure for subsequent study at high resolution are quite different from those required to retain the surface topography of the same specimen examined at low magnification and high depth of field. This goal-oriented approach must be related to the resources available and the exigencies of the specimen.

There are enormous technical difficulties in achieving the end result. On the one hand we have an electron beam instrument and on the other hand we have living biological organisms. The interaction of the beam of electrons with the specimen can readily cause both thermal and radiation damage, and the highest signal-to-noise ratios are obtained from flat or thin, thermodynamically stable specimens of high atomic number. Let us now consider the fairly typical materials derived as a product of biological activity. The samples are invariably soft, wet, and three dimensional, composed of low-atomic-number elements with an overall low mass density. They tend to be thermodynamically unstable, needing a continuous influx of energy in order to maintain their form and functional activity. Also, they have low thermal and electrical conductivity and are very sensitive to radiation damage.

Quite obviously the biological material and the electron beam instrument are at odds and some sort of compromise must be made to bring these two extreme situations closer to each other. There are two alternatives. One may either compromise the instrument or one may compromise the biological material. Both alternatives will be considered.

11.2. Compromising the Microscope

We can do two things with the microscope to enable it to be used with unprepared biological material. An environmental stage can be incorporated into the system or the microscope parameters can be set in unusual operating regimes.

11.2.1. Environmental Stages

Recent papers by Robinson (1978) and Valdre (1979) review some of the progress which has been made in building environmental stages for electron microscopes. Such devices seek to provide within the microscope

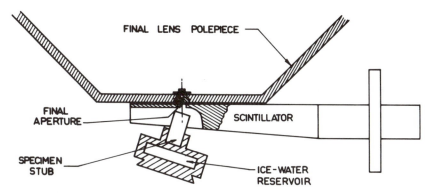

Figure 11.1. Schematic illustration of the detector and specimen configuration employed for wet specimen microscopy in a SEM, Robinson (1978).

column a microenvironment similar to the natural environment of the specimen. An environmental cell can be made by placing the specimen in a small chamber which is provided either with electron transparent windows or closed off from the rest of the column except for small apertures through which there is a constant but small flow of gas. The interior of the chamber is either flooded with an aqueous medium or continuously flushed with an inert carrier gas at a pressure of 30–50 kPa saturated with water vapor. In either situation the electron beam travels through a few millimeters of a less than optimal vacuum before interacting with the specimen surface. Figure 11.1 shows such a apparatus and Figure 11.2 shows an image of wool fibers examined while immersed in water. The mean free path of a high-energy electron in liquid water or water vapor is extremely short and the inevitable scattering of the electrons may result in both specimen damage and image degradation. A critical look at the results which have been obtained using environmental stages shows that very little, if any, new biological information has accrued. The real advantage of this approach would be in the ability to observe dynamic cell processes such as cell division. This has not been achieved, and the evidence presented by Reimer (1975), Glaeser and Taylor (1978), and Glaeser (1980) convincingly demonstrates that the radiation damage which occurs while observing even the fastest-multiplying bacterium would invalidate the biological significance of any observations.

11.2.2. Nonoptimal Microscope Performance

It is possible to run the microscope under less than optimal conditions and still obtain useful information about untreated biological samples. The less than optimal conditions include low-voltage operation, low beam current, short record times, decreased signal-to-noise ratios, and poorer microscope vacuum. Such conditions lessen the energy input into the

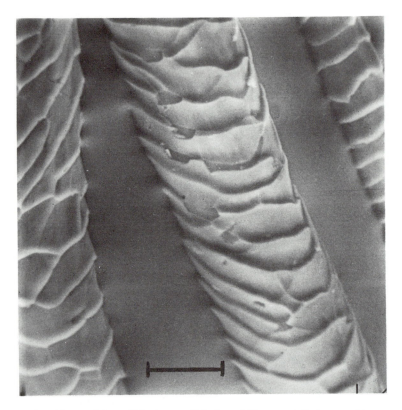

Figure 11.2. Uncoated formaldehyde cross-linked wool fibers partially immersed in water at 273 K. 17 keV accelerating voltage. Marker = 20 μm. (Robinson, 1978.)

specimens and enable the specimens to be examined untreated and in a near natural state, although at much reduced resolution. This approach has been particularly useful in the examination of plant material, the cells of which, although containing a large proportion of water, are invariably bound by a cellulosic wall which is frequently covered by a waterproofing or suberized layer. The net effect is to reduce the rate of water loss from the plant organ, enabling it to be observed for several minutes before obvious dehydration effects become apparent. By the same token, one may observe hard tissues, such as teeth, bone, chitin, seeds, spores, wood, and pollen grains, all of which have a low water content. Examination without preparation is much less successful with most animal tissues, aquatic plants, and microorganisms, where the tissues and cells very quickly shrivel up during drying. However, even this approach should not be dismissed out of hand because useful information can be obtained even from grossly distorted specimens.

It should by now be apparent that attempts to compromise the instrument are not going to achieve very much in terms of obtaining information from biological specimens and we should look at the ways in which we may compromise the specimen to enable it to survive in the microscope column.

11.3. Compromising the Specimen

Much of the effort in developing preparative techniques for biological electron microscopy has centered on devising ways to remove or immobilize water, and considerable success has been achieved. It should be noted that not all preparative problems arise because of water. An attempt has been made to categorize the preparative procedures for SEM into a number of distinct, but interdependent, steps. These procedures are discussed in approximately the chronological order they would be applied during specimen preparation. For a particular problem, it is important to realize that some of the steps may be either unnecessary, mutually exclusive, or out of sequence.

11.3.1. Correlative Microscopy

Any detailed investigation of biological material should attempt to correlate the information obtained from a wide range of instruments. In many ways the SEM is useful to start with because the magnification range encompasses that available on a good hand lens and a high-resolution TEM. The SEM also presents us with a familiar image. Correlative studies are relatively easy to carry out either by processing the specimen for TEM or light microscopy (LM) after examining the specimen in the SEM. An example of a correlative study is given in Figure 11.3 and further details may be found in the papers by Barber (1976, 1972), Wetzel *et al.* (1974), and Lytton *et al.* (1979) and in the book edited by Echlin and Galle (1976). Albrecht and Wetzel (1979) give details of methods which may be used to correlate all three types of images with histochemical techniques, and Junger and Bachmann (1980) give details of correlative studies with light microscopy, autoradiography, and SEM.

The other aspect of the preparative procedure which requires discussion is the realization that the biological objects we see and record in the SEM are artifacts. Preparative procedures have been described as the art of producing creative artifacts, and to a certain extent this is true. Broadly speaking we seek to convert the unstable organic specimen into a stable and largely inorganic form. The net result is a series of amplitude contrast images in which it is very difficult to accurately relate a given signal level to

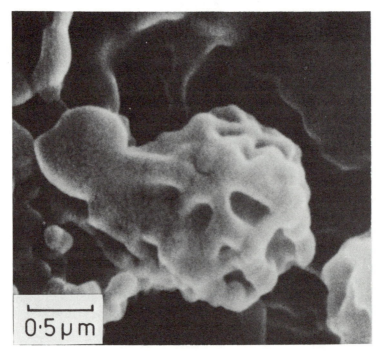

(a)

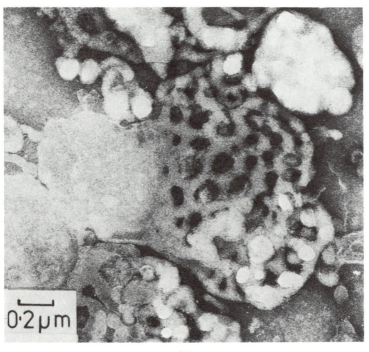

(b)

a feature in the living cell. It is important to realize that one only observes images, which although representative of living biological material are in actuality far removed from it. This aspect of electron microscopy has been forcibly brought home in the studies of unfixed and unstained frozen-hydrated and frozen-dried specimens. In such samples one can no longer see the familiar images obtained by the more conventional techniques, and it may well be that we are going to have to reeducate ourselves to recognize cell structures in frozen-hydrated sections. Acceptance of the fact that everything we observe in the SEM is more or less an artifact in no way lessens the value of these instruments in biological research.

11.3.2. Specimen Selection

Some care must be exercised over the choice of the specimen. The specimen size should be the smallest compatible with the features which are to be examined. Small specimens are more easily handled and are more adequately processed during fixation and dehydration. One of the worst sources of contamination in the SEM is the specimen itself. Small specimens can more adequately be stabilized and dehydrated, which lessens the chance of volatile components being released inside the microscope. Pieces of tissue up to 1–2 mm^3 may be surgically excised from multicellular organisms; somewhat smaller pieces of plant material would be desirable. Tissue culture, single cells, microorganisms, and isolated cell organelles can either be harvested and processed in suspension, encapsulated in agar prior to treatment, or allowed to settle and become attached to a suitable substrate such as glass, freshly cleaved mica, or tissue-compatible metal disks with or without a plastic coat. Alternatively, microorganisms may be studied *in situ* on the surface—naturally or artificially exposed—of the tissue it is inhibiting using techniques recently described by Garland *et al.* (1979). The attached specimens may then be taken through the preparative procedures. Glass and mica substrates have the added advantage that correlative studies can be carried out using the light microscope as shown in Figure 11.4.

Plastic tissue culture dishes should only be used if one is quite certain that they are not soluble in the organic fluids used in the tissue preparation. Many mammalian tissue culture cells are best handled in this way, and it is relatively easy to attach such cells to glass coverslips which have been coated with a monolayer of polylysine (Sanders *et al.*, 1975). The coating method can also be used for other cells, such as microorganisms and unicellular algae, as well as particulate matter. Small particulate matter

---

Figure 11.3. Correlation of SEM and TEM images. (a) SEM image of an isolated Golgi complex from rat liver (Sturgess and Moscarello, 1976). (b) TEM image of a negatively stained Golgi complex from rat liver. (Sturgess and Moscarello, 1976.)

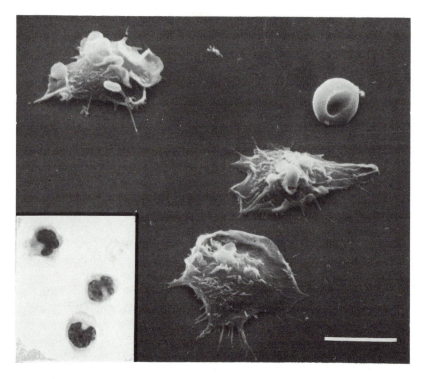

Figure 11.4. Human monocytes on a glass support. Marker = 10 μm. (Wetzel *et al.*, 1974.)

may also be harvested on plastic-covered microscope grids, fine metal gauzes, or, as shown in Figure 11.5, on one of the many different commercially available filter membranes (Millipore, Nuclepore, Flotronic, etc). If filter membranes are to be used, it is important to check their solubility in the various processing fluids and to realize that a membrane which is smooth to the naked eye is frequently very rough and corrugated when observed in the SEM. Such highly irregular surfaces are unsuitable for specimens such as cell organelles and viruses. McKee (1977) and Drier and Thurston (1978) published a useful review of the preparative techniques which can be used with microorganisms such as bacteria and fungi. DeNee (1978) and Chatfield and Dillon (1978) give details of the ways in which particles may be collected, handled, and mounted for SEM.

11.3.3. Specimen Cleaning

It is important to make sure that the surface to be examined is clean and free of extracellular debris. Some extracellular debris such as mucus,

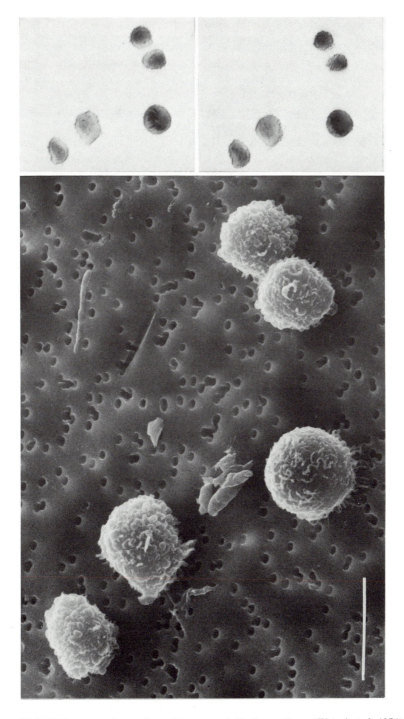

Figure 11.5. CHO tissue culture cells on filter support. Marker = 10 μm. (Wetzel *et al.*, 1974.)

blood, or body fluids may obscure the whole surface to be examined; others such as dust, and cell and tissue fragments may obscure parts of the surface and result in an incomplete image. The cleaning of excised tissue pieces should, within reason, be carried out as rapidly as possible to lessen the chance of post-mortem- and oxygen-starvation- (anoxia) induced changes, which are particularly damaging to mammalian tissues.

The aqueous contaminants usually found associated with multicellular animal tissue can often be removed by gently washing or irrigating the specimen with a buffer that is of the same strength and composition as the cell fluids and is maintained at a temperature compatible with the natural environment of the specimen. The internal spaces may be washed clean by continual internal flushing. Should gentle washing not suffice, then more turbulent washing can be tried with or without a small amount of a surfactant agent. If the biochemical nature of the contaminant is known then enzymatic digestion can be tried. The types of enzymes which have been used are mainly proteolytic, saccharolytic, or mucolytic and include collagenase, hyaluronidsase, neuraminidase, elastase, chymotrypsin, pronase, and some of the amylases. An example of material cleaned with collagenase is shown in Figure 11.6. In the case of some hard tissues only the inorganic matrix is of interest. The organic material may be conveniently removed by gently washing with a 2%–5% solution of sodium hypochlorite at room temperature. Organisms with a characteristic siliceous or calcareous skeleton such as diatoms, foraminifera, and globigerina may be cleaned with oxidizing agents such as potassium permanganate or potassium persulfate. In all cases, great care must be exercised to ensure that only the contaminants are removed and the surfaces remain undamaged. Samples from large multicellular specimens are probably best washed before, during, and after excision to avoid contamination with extracellular fluids.

Aquatic specimens such as unicellular algae, protozoa, and the smaller invertebrates, unless maintained in pure culture, are frequently mixed with a wide range of organisms and other plant and animal debris. Differential centrifugation and repeated washing in an appropriate isotonic buffer will usually precipitate the organisms from the debris (or vice versa) to give a relatively clean suspension.

Figure 11.6. The removal of surface contaminants. (A) Portion of an isolated proximal tubule. Treatment of the kidney with HCl alone allows the microdissection of single nephron. The scanning electron microscope reveals the smooth basal surface of this tubule because the basement membrane is still intact. (B) A portion of an isolated proximal tubule that has been treated with HCl and collagenase. The collagenase selectively removes the tubular basement membrane revealing complex folds or ridges of the tubule (arrows). These folds correspond to the basilar interdigitations seen both in transmission electron micrscopy and in light microscopy. Figures 11.6A and 11.6B from Eisenstadt et al. (1976).

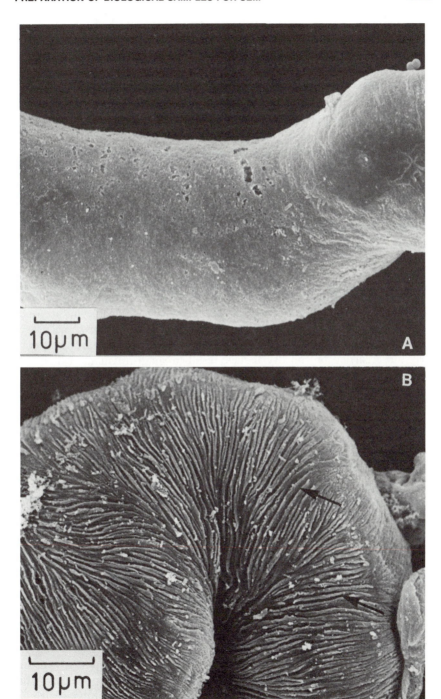

Dry specimens, such as calcified and chitinaceous tissue, seeds, wood, spores, pollen grains, and the aerial parts of plants usually only need dusting, preferably with clean air, although a Freon duster may also be used. Such procedures should not disturb the microflora and phylloplane fungi which are an integral feature of many plant surfaces. Walker (1978) has recently reviewed some of the procedures which may be used in the preparation of geological samples, including fossilized plant and animal specimens.

Robust specimens such as hard plant or animal tissue can be cleaned in a sputter coater or in a cold plasma gas discharge. Some sputter coaters can reverse the polarity of the specimen and the target allowing the plasma to etch and presumably clean the surface. Similarly, a cold plasma discharge of a gas such as oxygen can quickly remove organic material from the surface of an inorganic specimen. Such techniques should be used with extreme caution on suitably stabilized tissue using an erosion rate which will not alter the specimen surface.

Surface cleaning should not be confined to the initial stages of specimen preparation, but should be carried out wherever it appears necessary. For example, fixation and dehydration may also change the nature of a surface contaminant, so that it is more easily removed after these procedures than before. Many of the fracturing techniques (see later), particularly those carried out on dry tissues, result in small fragments which can obscure the specimen. Such fragments should be removed before proceeding to the next stage in the preparation process. The cleaned sample is now ready for the next stage in the process—cell and tissue stabilization.

11.3.4. Specimen Stabilization

Specimen stabilization or fixation is the process whereby the highly mobile solubilized state of the cytoplasm is immobilized by processes of precipitation, denaturation, and cross linking. During fixation, most of the ions and electrolytes, and many of the small-molecular-weight molecules such as sugars and amino acids, are irretrievably lost from the cell, and only the large macromolecular complexes remain.

Many of the fixatives used in SEM have been derived from TEM studies. There are, however, a number of important principles which must be borne in mind when making up a fixative. If one is going to examine the natural surface of a specimen or tissue, or a surface which has been exposed and cleaned prior to fixation, then it is important that the fixative solution should be approximately isotonic with the cell or tissue fluids. In this context the term fixative solution refers to all the components in the fixative other than the fixative itself. This includes the components of the aqueous buffer, balancing ions, and electrolytes and nonelectrolytes such as sucrose. Hayat (1970) has shown that the osmolarity of the fixative solution is as important in obtaining satisfactory fixation as the actual concentration of the fixative. If a specimen or tissue block is to be sectioned or fractured after fixation then the fixative solution should be hypertonic. The fixation time in the former case can usually be quite short, but in the latter case it must be long enough to allow the fixative to penetrate to the center of the specimen. The longer fixation times have the added advantage of hardening the tissue, making it somewhat easier to fracture at a later stage in the preparative procedure.

The other factors which must be considered when making up a fixative are pH, balancing ions, and temperature. The first two should be as close as possible to the natural environment of the cell and tissue fluids. Fixation of mammalian tissues is usually carried out near 277 K. Care should also be taken with the choice of buffer as it is important that its effective capacity is within the range of the pH of the specimen. There are several ways by which fixation may take place: the specimen may be exposed to the vapor phase of the fixative, immersed or floated in or on a solution of the fixative, and in the case of large animals, perfused *in vivo* with a fixative solution. Vapor fixation is useful for preserving delicate aerial structures, while immersion and perfusion fixation ensures that the fixative reaches the appropriate parts of the specimen as rapidly as possible. Flotation fixation has proved effective for leaves, petals, and skin.

Although there is a wide range of chemicals which may be used as fixatives for biological material (Glauert, 1974), it is generally agreed that a regimen suitable for most tissues would be a double fixation using an organic aldehyde followed by osmium tetroxide, both in identical fixative solutions. Glutaraldehyde (2%–5%) is the most commonly used aldehyde, but it can be supplemented with a 1%–2% formaldehyde and/or acrolein which penetrate tissues faster than the dialdehyde. Such a combination is particularly useful for the preservation of plant tissue. Care should be exercised when using acrolein as it is a volatile, inflammable, and toxic liquid. The final concentration of the fixative chemical is dictated by the specimen and the constitution of the fixative solution, but it should rarely exceed 5% of the total fixative. Most tissues are conveniently fixed over-night at room temperature, and it is useful to keep the fixative agitated during this time.

Following the initial fixation, the tissue should be washed in isotonic buffer and postfixed for up to 4 h in 1%–2% osmium tetroxide at the same temperature used for aldehyde fixation. The samples should be washed free of any excess osmium and briefly rinsed in buffer. Most tissues are fixed by immersion, either by placing a small piece of excised tissue into a large volume of fixative, or by initially allowing a large volume of fixative to continuously bathe the specimen *in situ* before it is excised and immersed in the fixative. The alternative approach of perfusion is best suited to mammalian tissues, and details of the procedure are found in the book by Glauert

(1974). For delicate specimens, the exchange method of preparation described by Peters (1980b) avoids specimen damage entirely by preventing sudden changes of concentration and composition.

Cryofixation is assuming a greater importance among the preparative techniques for SEM. The tissue is stabilized by quench freezing and sectioned or fractured in the frozen state. The materials so obtained are examined and analyzed in either the frozen-dried or frozen-hydrated state. Such techniques are particularly useful in the preparation of material for x-ray microanalysis of ions and electrolytes and further details are given in the chapter on specimen preparation for microanalysis (Chapter 12). Cryofixation and the examination of tissues in the frozen-hydrated state also provides a potentially useful tool for the examination of triphasic gas–liquid–solid systems such as pulmonary tissue, air–liquid foams, and developing bakery products or for systems such as oils, fats, and greases which are volatile at ambient temperatures. Preparing such specimens by conventional techniques results in the complete removal of cell fluids making it impossible to distinguish between the structures supporting the gas and liquid phases. An outline of the procedures involved and some of the applications to biological materials is given in the recent book edited by Echlin *et al.* (1978). It must, however, be appreciated that cryofixation is not the best preparative technique for most morphological studies, where it is better to use more conventional methods.

It should be emphasized that each specimen is unique and requires its own peculiar fixation. It is important to experiment with the fixation of a new tissue, varying the specimen size, time, and temperature of fixation, type, concentration and pH of the final fixative and correlating the findings with observation in the light microscope and the TEM. Finally it should be remembered that tissue stabilization is one of a number of different processes in tissue preparation which is best carried out as a continuous sequence of events. It is important to follow fixation either by dehydration or by staining followed by dehydration. The process may be temporarily stopped either when the specimen is in 70% ethanol or when the specimen is completely dehydrated and stored in a desiccator before coating.

Once fixation is complete, a number of different options are available for further treatment of the specimen. One may proceed direct to dehydration, one may attempt to reveal the internal parts of the specimen, or one may attempt to localize areas of specific physiological function.

11.3.5. Exposure of Internal Surfaces

One of the unique features of biological material is that it has a very large functional internal surface area. A number of techniques have been devised which enable these internal surfaces to be examined. The techniques fall into three main classes: sectioning, where there is an orderly procession through a hardened sample by means of tissue slices; fracturing, in which the specimen is broken open, hopefully along a surface of structural continuity but frequently at some random point, to reveal two highly contoured faces; and replication, where a plastic cast is made of the structure concerned and revealed after the biological material is removed by corrosion techniques.

The internal contents may be revealed more or less at any stage during the preparative procedure, even in the microscope by means of microdissection. However, most biological tissues are very soft and are more conveniently sectioned or fractured after stabilization and dehydration.

11.3.5.1. Sectioning

Tissue sections may be prepared using conventional light and TEM techniques and viewed in the SEM in either the transmission or reflective mode. One of the advantages of the SEM is that at a given accelerating voltage there is up to 20-fold increase in the section thickness which may be examined in the transmission mode although at somewhat reduced resolution. Sections examined in the reflective mode show more detail if the embedding medium is removed before examination leaving the biological material behind. For thin sections (20–200 nm) the material is fixed, dehydrated, and embedded in one of the epoxy resins. Sections are cut on an ultramicrotome and the resin removed with organic solvents, sodium methoxide, or controlled erosion in an ion beam or cold plasma gas discharge (Steffens, 1978). Alternatively, the thin sections may be mounted directly onto a solid substrate and examined in the SEM without further treatment. A remarkable amount of detail may be seen in seemingly flat sections, particularly from frozen-dried sections.

Thicker sections (0.2–5.0 μm) can also be cut from resin-embedded material and the resin removed using the technique described by Thurley and Mouel (1974). Sections between 2–10 μm thick may be cut from resin or paraffin wax embedded material on ordinary microtomes and then deparaffinized with xylene or the resin removed by one of the aforementioned methods. The sections are either attached to small glass coverslips using for example Haupt's adhesive for correlative light microscopy/SEM studies or picked up on electron transparent supports for correlative TEM/SEM studies. Humphreys and Henk (1979) have used an oxygen plasma to surface-etch thick (0.1 μm) resin sections of kidney tissue prepared using standard TEM techniques, and revealed ultrastructural details of subcellular orangelles (Figure 11.7).

11.3.5.2. Fracturing

It is sometimes more satisfactory to expose the interior of the specimen using fracturing techniques. Hard tissues can be broken before fixation.

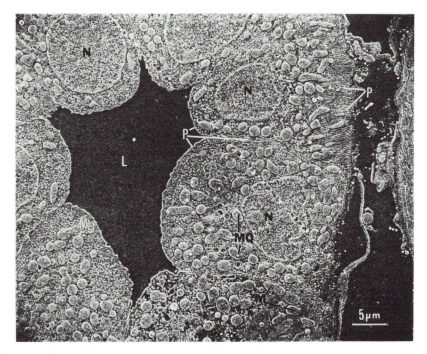

Figure 11.7. Scanning electron micrograph of a 1-μm-thick section of a convoluted tubule of mouse kidney embedded in an Araldite–Epon mixture. The section was surface etched with an oxygen plasma, metal shadowed with platinum–palladium, and sputter coated with gold. Subcellular structures are identified by the etch-resistant residues forming patterns on the surface of the section after etching. Plasma membranes (*P*) are recognizable at cell surfaces lining the lumen (*L*), between adjacent cells, and at the upper right of the micrograph as infoldings from the basal lamina. *N*, nuclei; *M*, mitochondria; *MQ*, mitochondrial matrix granules. (Humphreys and Henk, 1979.)

Clean cut surfaces of wood samples may be readily obtained by using a razor blade. Exley *et al.* (1974, 1977) recommend that low-density woods are best cut fresh or after soaking in cold water, whereas denser woods may need softening in hot water before they cut cleanly. Cytoplasmic debris may be removed with a 20% solution of sodium hypochlorite, and after dehydration and coating, the small blocks of wood are mounted in such a way that the edge between two prepared faces points in the direction of the collector as shown in Figure 11.8. Somewhat softer tissues can be teased apart with dissecting needles before or after fixation, but the most satisfactory fractures are obtained from brittle material.

Flood (1975) and Watson *et al.* (1975) proposed a dry fracturing technique in which a piece of adhesive tape is gently pressed against dehydrated and critical point dried tissues. The piece of tape is removed

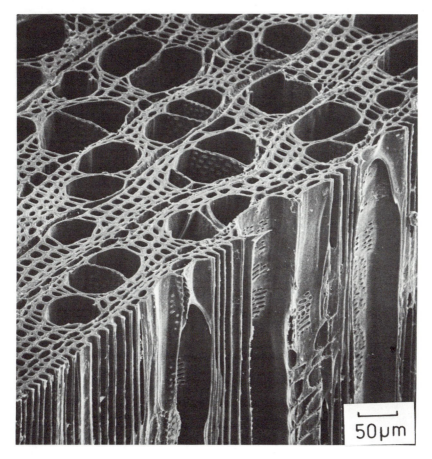

Figure 11.8. Transverse and radial longitudinal surfaces of the wood of *Notofagus fusca* (Exley *et al.*, 1974).

and in the process tears away some of the tissue revealing the internal contents. Complementary fractures may be obtained by placing a small piece of tissue between two pieces of adhesive which is pulled apart, and attached side by side on the specimen stub. Larger pieces of tissue may be dry fractured either by cutting and teasing apart with a small scalpel, or by initiating a cut in the specimen, the two edges of which are then held by fine forceps and pulled apart. Examples of such samples prepared by this technique are shown in Figure 11.9. The sticky tape method can be used equally well on single cells attached to a glass substrate with polylysine, revealing organelles and cellular substructure.

An alternative approach is to use a cryofracturing technique in which unfixed, fixed (Haggis *et al.*, 1976), ethanol infiltrated (Humphreys *et al.*,

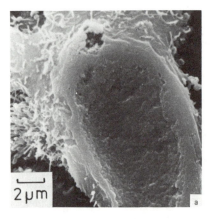
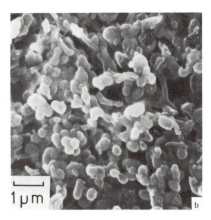

Figure 11.9. (a) Flattened Hela epitheloid cell, dry stripped by adhesive tape method, and showing the undersurface of the raised margins. (b) The surface of a Hela cell stripped by the adhesive tape method revealing rounded organelles. Figures 11.9a and 11.9b from Watson *et al.* (1975).

1977), or resin monomer infiltrated (Tanaka, 1974) material is immersed in liquid nitrogen to convert the liquid to a brittle solid mass. This brittle material is then fractured under liquid nitrogen using a blunt scalpel, the fractured pieces returned to the appropriate place in the sequence of tissue processing, and the preparation completed. Examples of the results of these techniques are shown in Figure 11.10. In a recent paper by Haggis and Bond (1979) the cryofracturing technique has been used with great effect in a study of chick erythrocyte chromatin. The technique involves quench freezing glycerinated specimens which are fractured under liquid nitrogen. After freeze fracture, the cells or tissues are thawed into fixative. Fixation is very rapid, as the fixative quickly diffuses into the fractured cells, but soluble constituents have time to diffuse out, to give a deep view into the cell as shown in Figure 11.11. The cryofracturing methods have a number of advantages over the dry fracturing techniques, not the least of which is a diminution in the amount of debris left on the surface of the specimen which can either obscure the surface or charge up in the electron beam.

Generally speaking cryofracturing is more easily carried out on animal than plant material because of the more homogeneous density of soft animal material and the presence of cellulosic and lignified cell walls in plant material which causes the plant material to splinter more easily and give very irregular fracture faces.

Pawley and Norton (1978) have described a highly sophisticated piece of equipment (Figure 11.12) which is attached to the column of the

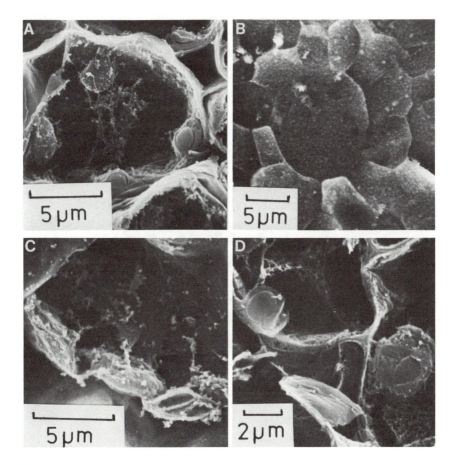

Figure 11.10. Low-temperature scanning electron microscopy. (A) Ethanol cryofracture cells from root of *Lemna minor*. (B) Unfixed, frozen-hydrated fracture surface of cells from root of *Zea mais*. (C) Glycerol cryofracture of cells from root of *Lemna minor*. (D) Resin cryofracture of cells from root of *Lemna minor*.

microscope and permits frozen-hydrated samples to be repeatedly fractured, coated and viewed at low temperatures inside the microscope. With this apparatus Pawley and Norton (1978) have obtained complementary replicas of freeze-fractured yeast membranes at a resolution of 10–15 nm. More recently Echlin, Pawley, and Hayes (1979) have used this apparatus to examine frozen-hydrated fracture faces of developing roots of *Lemna minor*. The resolution is good and one can clearly distinguish subcellular organelles and compartments within the cells of the differentiating phloem parenchyma (Figure 11.13). The fracture faces are usually examined directly in the SEM, although replicas may be made of heat-sensitive material and then coated and examined.

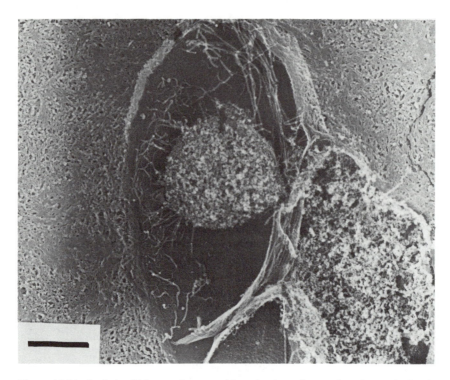

Figure 11.11. A whole chicken erythrocyte with a prominent fractured nucleus, cyloplasmic strands to the left and microtubles to the right. Bar = 1 μm. (G. M. Haggis and E. F. Bond, 1979.)

11.3.5.3. Replication

A recent paper by Pameijer (1979) gives details of a wide range of replica techniques which may be applied to natural external surfaces as shown in Figure 11.14, which compares the original surface and a replica. Replicas can also be made of internal surfaces by infiltrating an organ with a low-viscosity plastic, allowing the plastic to polymerize, after which the tissue is digested away leaving a cleaned solid cast of the surfaces. Such replicas are particularly useful for the interpretation of large internal surface areas such as lung or kidney, and for the investigation of the microvasculature of a wide range of organs (Nowell and Lohse, 1974; Irino et al., 1975; and Frasca et al., 1978). Figure 11.15 shows details of the villous capillary network produced by using the latex method described by Nowell and Lohse (1974). Methyl methacrylate (Batson's casting medium) has also been used; Kardon and Kessel (1979) have used it to make detailed casts of the microvasculature of rat ovary. Nopanitaya et al. (1979)

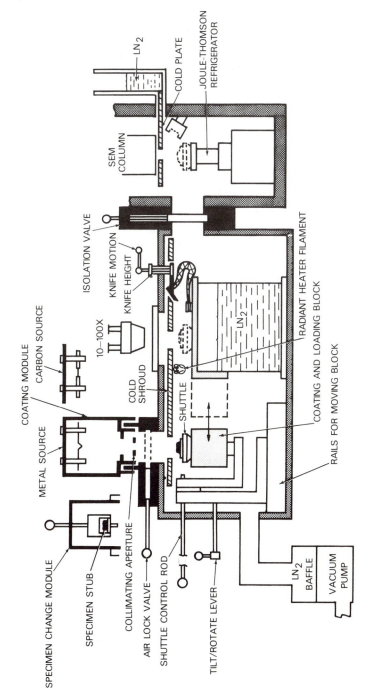

Figure 11.12. Diagram of the AMRAY Biochamber freeze-fracture system. (Pawley and Norton, 1978.)

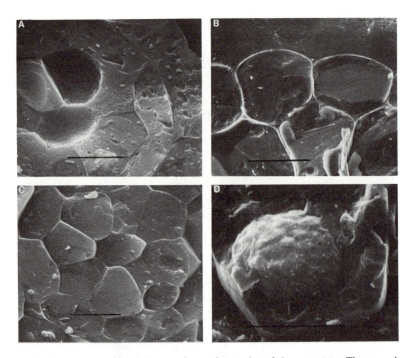

Figure 11.13. Frozen-hydrated fracture faces of root tips of *Lemna minor*. The roots have been encapsulated either in 25% polyvinyl–pyrrolidone (C), or 25% hydroxyethyl starch (A, B, and D), and quench frozen in melting nitrogen. The frozen roots are transferred under liquid nitrogen to the AMRAY Biochamber where they are fractured, etched, and coated with 20 nm of gold at 103 K and a pressure of 50 μPa. The frozen specimens are examined at 93 K and a pressure of 20 μPa in an AMR 1000A SEM. Part A shows a ring of eight phloem parenchyma cells surrounding a central xylem cell. Bar = 10 μm. Part B shows the outer root-cap cells, fully frozen but with no evidence of ice crystal damage. Bar = 20 μm. Part C shows a ring of phloem parenchyma cells, with one cell differentiated into a sieve tube and companion cell. Bar = 10 μm. Part D shows a nucleus from differentiated sieve tube. Bar = 5 μm.

have introduced a modification of the methyl methacrylate medium by the addition of Sevriton, a low-viscosity plastic, to give a mixture about ten times less viscous than the Batson's casting medium. Using this mixture they have obtained stable casts of the gastrointestinal microvasculature of a number of animals. Attempts to use corrosion-casting replica techniques of

Figure 11.14. (a) SEM image of original of a fracture face of a human tooth containing a metal pin. Bar = 2 μm. (b) SEM image of an acetate tape and silicone rubber replica of a fracture face of a human tooth containing a metal pin. Bar = 2 μm. Both from Pameijer (1979).

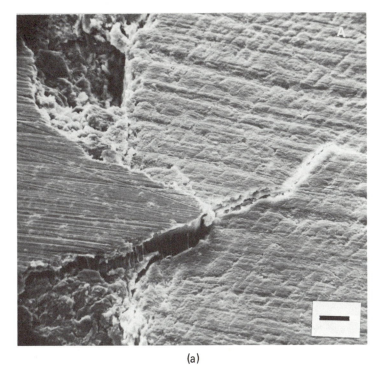

(a)

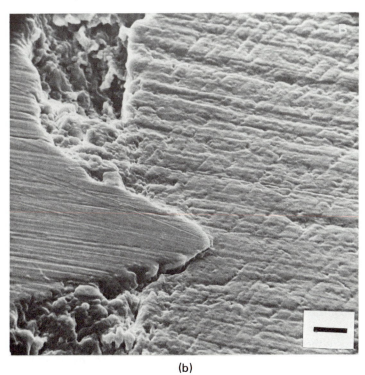

(b)

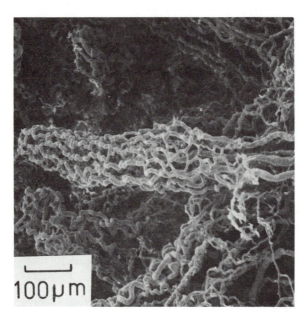

Figure 11.15. Latex injection replica of a villous capillary network from a dog. (Nowell and Lohse, 1974.)

plant tissue have not been successful, presumably because plants do not have a mechanism to pump fluids throughout their structure.

11.3.6. Localizing Areas of Known Physiological Activity

The bulk of biological material is made up of elements of low atomic number whose backscattered electron yield is rather low and so it is necessary to introduce heavier elements, hopefully at specific sites, to increase the signal. This is conveniently achieved by the addition of compounds of elements such as silver, osmium, lead, uranium, and gold, which are moderately specific for certain biological structures when applied either as a block stain or to tissue sections. Abraham and DeNee (1974), and DeNee and Abraham (1976) have reviewed the techniques of biological backscattered imaging, and reference should be made to these papers for specific staining recipes. The staining increases both the secondary-electron and the backscattered coefficients. One of the problems associated with this technique is to distinguish between topographic and atomic number contrast as well as eliminating bright areas due to charging. The complications of topographic contrast can to some extent be limited by examining either tissue sections or flat surfaces of bulk specimens. The atomic number contrast of specimens is enhanced by staining at the expense of topographic contrast, which makes it that much more difficult to unequivocally identify morphological features in the cells and tissues.

The stains may be applied in a number of different ways. Abraham (1977) and DeNee and Abraham (1976) have, as shown in Figure 11.16, stained tissues directly with Wilder's ammoniacal silver and Gomori's methanamine silver, both of which have an affinity for nuclei and nerve reticulum fibers. Geissinger (1972) has used silver nitrate to stain mammalian tissues, and Echlin (unpublished results) has used block staining with saturated aqueous uranyl acetate as a nonspecific stain for plant material examined in the SEM. DeNee *et al.* (1977) describe three techniques for staining tissues with salts of uranium, lead, silver, and tungsten. Alternatively one can, as Becker and DeBryun (1976) have shown (Figure 11.17), apply more specific histochemical techniques where the end product of an enzymatic or biochemical reaction is precipitated as a heavy metal such as salts of lead, uranium, or bismuth. Most of the staining and histochemical techniques which are used in SEM have been modified from the methods used in light and transmission electron microscopy, and the reader is advised to look to these sources for methods which can be adopted for SEM. The autoradiography technique can be used to localize areas of specific physiological activity while the developed silver grains can be

 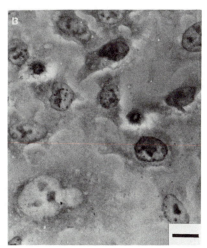

Figure 11.16. (A) Secondary electron image and (B) backscattered image of Hela tissue culture cells stained with Wilder's silver stain. Carbon coated. Marker = 10 μm. (Abraham, 1977.)

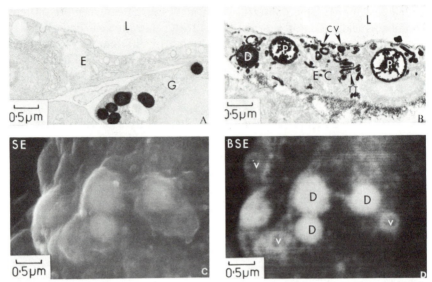

Figure 11.17. (A) TEM image of a control rat bone marrow stained with DAB osmium stain without endogenous peroxidase activity. (B) TEM image of rat bone marrow containing peroxidase activity stained with DAB osmium stain for peroxidase. (C) Secondary electron image of rat bone marrow containing peroxidase activity stained with DAB osmium. (D) Backscattered image of same specimen shown in (C). All from Becker and DeBryun (1976).

imaged in the backscattered mode. Wetzel *et al.* (1977) have successfully combined light microscope, autoradiography, and SEM in a study of tissue culture cell cycle. A review of these techniques is given by Hodges and Muir (1976).

One means of localizing areas of specific physiological activity has been adapted from TEM. These are the cell surface labeling techniques which, when applied to specimens for SEM, result in morphologically distinguishable or analytically identifiable structures on the cell surface. These techniques, coupled with high-resolution scanning electron microscopy, permit one to study the nature, distribution, and dynamic aspects of antigenic and receptor sites on the cell surface. Cell labeling techniques are generally rather complicated and involve immunochemical and biochemical purification procedures. Reference should be made to the papers by Kay (1977), Molday (1977), and Wofsy (1979) for detailed methods but the rationale behind the procedures is as follows. One uses standard immunological procedures to attach antibodies to specific antigenic sites on the cell surface. The trick is to so modify the antibodies that they will also carry a morphologically recognizable label such as latex or silica spheres, a recognizable virus such as TMV, or one of the T-even phages as shown in Figure

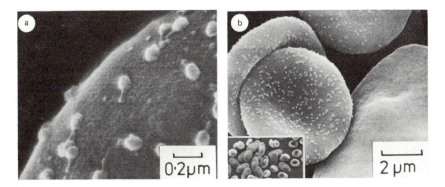

Figure 11.18. (A) Human RBC labeled with T4-lgG. The head and tail piece of the phage is clearly seen. (B) Human RBC labeled with T4-lgG in a mixture of human and chick RBC. Both from Kuman *et al.* (1976).

11.18, or a protein molecule of known dimensions such as ferritin or haemocyanin. Horisberger and Rosset (1977) (Figure 11.19) have used gold granules which have a high secondary-electron coefficient. One part of the antibody has an affinity for the specific antigenic attachment on the cell surface, while another part carries the morphologically recognizable structures. The immunologic techniques have now advanced to the point where they may be used to study both the qualitative and quantitative aspects of the cell surface (Nemanic, 1979; Lotan, 1979).

While these staining, histochemical, and immunological techniques do not have the specificity and resolving capabilities of x-ray microanalysis, they are by and large easier to carry out, and do not required complex and costly equipment.

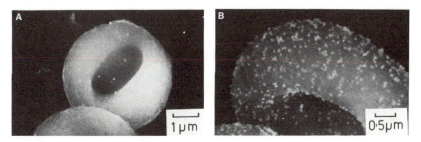

Figure 11.19. (A) Human RBC, control experiment in the presence of an inhibitor to prevent gold granules absorbing on the surface. (Horisberger and Rosset, 1977.) (B) Human RBC marked with gold granules labeled with soya bean agglutinin and wheat-germ agglutinin. (Horisberger and Rosset, 1977.)

11.3.7. Specimen Dehydration

As indicated earlier, one of the central problems in the preparation of biological tissues is the removal or immobilization of water. The procedures for immobilizing the water are based on freezing techniques and are dealt with in the following chapter on microanalysis. The dehydration procedures for SEM can be the same as for TEM and involve either passage of the tissue through increasing concentrations of ethanol, methanol, or acetone followed by critical point drying or by freeze drying at low pressure. Which of the two methods is better is a matter of debate, and it is necessary to balance the tissue extraction and shrinkage brought about by the former against the ice crystal damage of the latter. Whichever of the two methods is used for dehydration, it should be clearly understood that some changes in tissue volume must inevitably take place. Boyde and co-workers have made a series of careful studies on the volume changes which occur in a variety of plant and animal tissues after different dehydration regimes. They found that critical point dried material may shrink by up to 60%, freeze-dried material by 15%, and material air dried from volatile liquids by 80% of the original volume. Although plant material usually shrinks less than animal tissue, each specimen must be considered separately. Provided the measured change in volume is uniform in all directions and the same in all parts of the specimen, corrections may be made to any measurements made on the sample.

Before embarking on a discussion of these two methods it might be useful to consider alternative methods of tissue drying. Air drying is not a satisfactory procedure because it results in such gross distortion of the tissue due to the high surface tension of water, which may set up stresses as high as 20–100 MPa during the final stages of evaporation. These forces become increasingly larger as the structures being dried become smaller (Figure 11.20). Rapid drying of samples from the aqueous phase by placing them in a vacuum has the added disadvantage of ice crystal formation due to the evaporation of the water at reduced pressure. The formation of the ice crystals can severely distort the specimen. Drying tissues from volatile organic liquids with a low surface tension such as diethyl ether or 1 : 2 epoxy propane, or volatile organic solids such as camphor is only marginally better and should only be used in extreme cases.

Chemical dehydration may be achieved by passing the fixed tissues through a graded series of methyl or ethyl alcohol or acetone. This may be done either by complete replacement of one concentration with the next or by slowly but continuously dripping a large volume of 100% alcohol or acetone into a small container holding the specimen. This latter technique has the advantage in that it does not subject the specimen to sudden changes in concentration. Care should be taken not to use denatured alcohols or acetone as this can give rise to small needlelike contaminants on

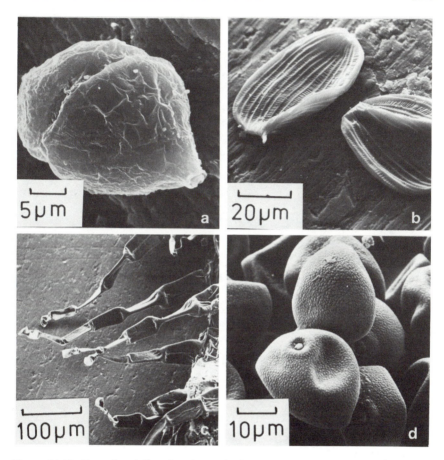

Figure 11.20. Examples of distortion due to air drying specimens. (a) Dinoflagellate, *Prorocentrum micrans*. (b) Euglenoid flagellate, *Phacus sp.* (c) Leaf hairs of *Solanum sarrachoides*. (d) Pollen grains of *Poa pratense*.

the specimen surface. The temperature of dehydration does not appear important, although it is usual to carry out dehydration at the same temperature used for fixation. It is important to remember that 100% alcohol or acetone is very hygroscopic and it may be necessary to carry out the final stages of dehydration either in a dry atmosphere or in the presence of a suitable drying agent. Rapid chemical dehydration can be achieved using a 2–2-dimethoxypropane. Tissue samples are placed into acidulated solutions of 2–2-dimethoxypropane which rapidly converts the tissue water to acetone and methanol. Once the endothermic reaction has ceased the tissues are then transferred through three changes of ethanol, methanol, or acetone. If a rapid preparation is needed, then this is an effective dehydra-

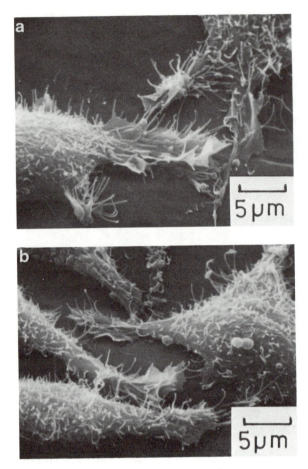

Figure 11.21. Cultured mouse fibroblast cells. (a) Dehydrated with ethanol. (b) Dehydrated with 2–2 dimethoxypropane. (Kahn *et al.*, 1977.)

tion procedure. The slower procedures are preferred because there may be a danger of tissue distortion by the more rapid procedure, particularly with larger samples. Kahn *et al.* (1977) have made a careful comparison of ethanol and dimethoxypropane (DMP) dehydration methods on the preservation of tissue cultures, and as Figure 11.21 shows there is little to choose between the two techniques. Tissue distortion during dehydration may be diminished if in the early stages the dehydrant (ethanol, acetone, etc.) is diluted with the buffer used as the fixative vehicle.

Liepins and de Harven (1978) have shown that it is possible to dry single tissue culture cells in a desiccator from Freon 113 at room temperature. Cells, attached to coverslips, are fixed and dehydrated to 100% ethanol and then passed through gradual changes to 100% Freon 113. The specimens immersed in Freon 113 are transferred to a desiccator containing Drierite, and evacuated to a pressure of 3 Pa (3×10^{-2} Torr). The liquid phase of the Freon 113 evaporates in a few minutes, leaving cells whose ultrastructure is comparable to cells dried by critical point drying (Figure 11.22). Further studies are necessary to show whether this technique can be effectively applied to pieces of tissues. Lamb and Ingram (1979) have devised a method which allows specimens to be dried directly from ethanol under a stream of dry argon without the use of the critical point drier. The technique gives results as good as those obtained by other methods, although with sensitive samples such as ciliated epithelium the critical point drying method gives better results.

It is difficult to give a set time for effective dehydration because, like fixation, dehydration is very dependent on specimen size, porosity, and whether the internal and/or the external features of the cell are to be examined. As a general guide most tissue blocks 1–2 mm³ are effectively dehydrated after 15 min in 15%, 30%, 50%, 70%, 95%, and 100% ethanol or acetone, followed by three 10-min changes in anhydrous solvent, the whole procedure being completed in about 2 h. Rostgaard and Jensen (1980) have shown that there is better structural preservation if the dehydration is continuous rather than being carried out in a series of steps.

The timing of the preparative procedures is rather important, to avoid any unplanned delays during the process. Thus it is usually convenient to select and clean tissue samples on the afternoon of day 1; to fix in aldehyde overnight, wash, postfix in osmium, dehydrate, and mount on specimen holders during day 2; and to coat the samples first thing on day 3 and spend the remainder of the day examining the specimen in the microscope. The final stage of dehydration, unless one is going to examine sectioned or fractured material, is the critical point drying procedure.

11.3.7.1. Critical Point Drying

Scanning microscopists rediscovered the critical point drying method which was first developed nearly thirty years ago by Anderson (1951) for studies on bacterial flagella. The rationale behind the procedure is that the two-phase stage (vapor and liquid) of the majority of volatile liquids disappears at a certain temperature and pressure—the so-called critical point. At the critical point the two phases are in equilibrium, the phase boundary disappears, and there is, therefore, no surface tension. Much has already been written about the theory and practice of critical point drying, and reference may be made to the papers by Bartlett and Burstyn (1975) and Cohen (1979) for the details of the technique.

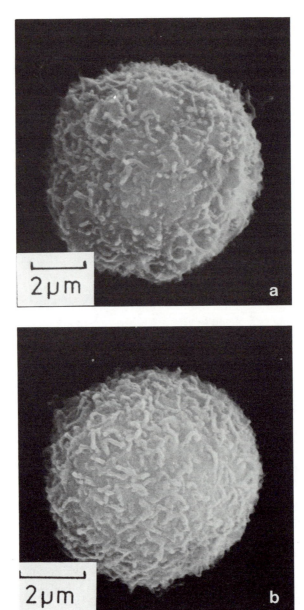

Figure 11.22. A comparison of critical point drying and direct drying using Freon 113. (a), (b) Mastocytoma cells showing surface villi: (a) prepared by critical point drying; (b) prepared by direct drying. (c), (d) Surface of a mastocytoma cell examined in field emission SEM: (c) prepared by critical point drying—note the porosity of the cell surface membrane; (d) prepared by direct drying method—note that the cell surface membrane does not show perforations. From Liepins and de Harven (1978).

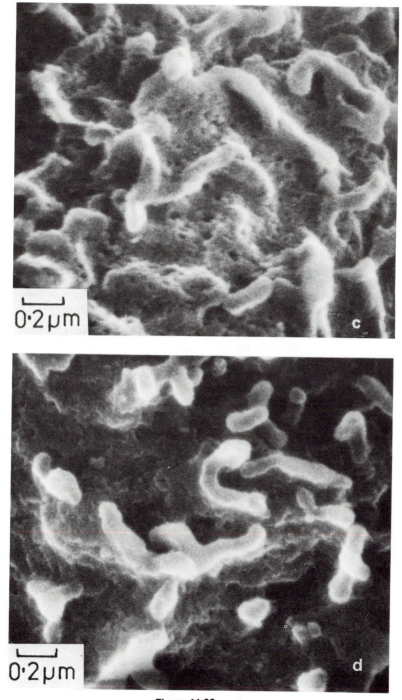

Figure 11.22. *cont.*

The practical aspects of the method involve passing the tissue from the dehydrating fluids (methanol, ethanol, acetone) through the intermediate fluids (amyl acetate, Freon TF) into the transition fluid (liquid CO_2 or Freon 13) in the critical point drying apparatus (or bomb). Ideally, one would like to critical point dry from water, but the critical temperature of water (648 K) is so high that one would effectively cook the specimen.

We have already discussed the significance of the dehydrating fluids. The transition fluids are the chemicals from which critical point drying is carried out. The critical temperature of CO_2 is 304 K and its critical pressure is 7.39 MPa (1073 psi), and for Freon 13 (monochlorotrifluoromethane) 311 K and 3.87 MPa (561 psi), respectively. The intermediate fluids are used if the dehydrating fluids are not miscible with the transition fluid. This presents no problem with ethanol or acetone and liquid CO_2, but it is necessary to interpose an intermediate fluid between ethanol–acetone and the fluorocarbon Freon 13.

Amyl acetate may be used between ethanol–acetone and CO_2 as it is somewhat less volatile and diminishes the chance of the specimen drying out during transfer to the critical point dryer. Another advantage of amyl acetate is the distinctive (but toxic) odor, and its absence from the venting transition fluids is an accurate indicator that the transition has been successfully completed. Pawley and Dole (1976) have shown the importance of complete removal of the intermediate fluid before the drying is carried out.

Although the critical point drying procedure is roughly the same for all pieces of equipment, important differences do exist between the various commercially available critical point drying bombs and it is important to follow the procedures which the particular manufacturer recommends. As long as the operator remembers that the process goes on inside a bomb, with its attendant high pressures, and uses the technique with care, then critical point drying can be carried out safely and effectively. It is important to regularly check the valves, seals, and where appropriate, the viewing windows. Further safety precautions are given in the papers by Cohen (1979) and Humphreys (1977). Most critical point drying is carried out using liquid CO_2. It is cheaper and more readily available than Freon 13. The CO_2 should be anhydrous and free of particulate matter and for convenience should be supplied in a siphon cylinder. If possible the cylinder should be shot-blasted on the inside to remove any traces of rust, which could contaminate the surface of the specimen. The specimen should be placed in a sealed open meshwork container towards the end of the dehydration. There is considerable fluid turbulence in the bomb, and since it is usual to dry several different specimens at once, the containers prevent the specimens from becoming mixed. The only proviso with all the containers and supports is that they do not impede the flow of the transition fluid and remain immersed in the intermediate of dehydration fluid until the bomb is filled under pressure. Once the critical point drying has been completed, the specimens should be immediately transferred to a desiccator, where they may be stored prior to mounting and coating.

11.3.7.2. Freeze Drying

Freeze drying is an effective alternative to critical point drying, although there is always the danger of ice crystal damage. Infiltrating the specimens with a penetrating cryoprotectant, such as glycerol or dimethyl sulfoxide, does to some extent alleviate volume shrinkage but in turn creates other problems. The cryoprotectants are retained in the specimen after drying and will slowly outgas and obscure the specimen and contaminate the microscope column. This outgassing can be reduced by examining specimens at low temperature or by using high-molecular-weight nonpenetrating polymeric cryoprotectants such as polyvinylpyrrolidone or hydroxyethyl starch (see Chapter 12). Practical hints on how best to carry out the procedure are given in Chapter 12, on specimen preparation for microanalysis, and in the recent review by Boyde (1978), but an outline of the procedure is as follows: Small pieces of the specimen are quench frozen in a suitable cryogen such as melting Freon 13 and transferred under liquid nitrogen to the precooled cold stage of the freeze dryer. The chamber of the freeze dryer is evacuated to its working vacuum of 20–100 MPa, at which point the ice sublimes from the sample. Ice sublimation is a function of temperature and vacuum—the lower the temperature, the higher the vacuum required to enable the water to leave the sample. Most freeze drying is carried out at about 200 K and at a pressure of 20–100 mPa. Lower temperatures diminish ice recrystallization but require much higher vacuum and/or longer drying times. It is important to provide a condensing surface for the water molecules which have been removed from the sample. A liquid-nitrogen-cooled trap a few millimeters from the specimen is the most effective way to achieve this, but chemical desiccants such as phosphorus pentoxide or one of the zeolites will also suffice. The actual time taken for freeze drying depends very much on the properties of the sample such as size, shape, the amount of free and bound water, and the relative proportions of fresh to dry weight. The drying procedures should be worked out for each specimen, and it is important to gradually allow the specimen to reach ambient temperature whilst under vacuum before it is removed from the freeze dryer. There is still considerable debate regarding the merits of freeze drying vs. critical point drying. On balance, it would appear that the freeze-drying method is better for preserving cellular detail. The critical

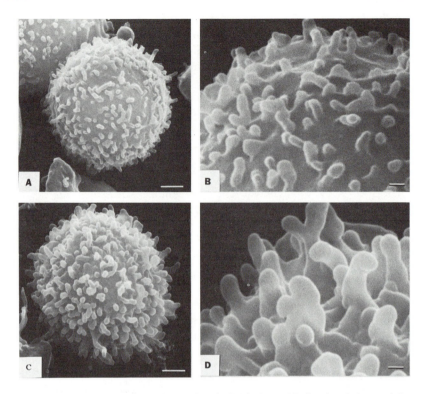

Figure 11.23. A comparison of the results obtained after critical point drying and freeze drying. A and B are of critical point dried lymphocytes showing narrow microvilli and a relatively smooth underlying surface. Bar on A = 1.0 μm, B = 0.2 μm. C and D are of the same specimens which have been freeze dried. The microvilli are much wider and the cell surface is more complex. Bar on C = 1.0 μm, D = 0.2 μm. All from Schneider *et al.* (1978).

point drying method provides more information about the whole specimen. A good demonstration of the results obtained by the two methods is shown in Figure 11.23.

It is also possible to dry tissues which have been dehydrated in ethanol or acetone and then quenched to about 163 K. The cold solvent dehydrated samples are then dried at 243–253 K at a pressure of between 1.0 and 0.1 Pa. Katoh (1978) gives details of a similar scheme in which specimens are fixed, dehydrated in ethanol, and then gradually infiltrated with Freon TF. The specimen is then placed on a large block of aluminum and the two are quench frozen in liquid nitrogen. The cold specimen and metal block are then placed in a vacuum desiccator and the specimen is dried under a vacuum of about 1 Pa while the metal block slowly warms up. Because it is possible to freeze-dry unfixed tissue this procedure is doubly useful in that it provides samples for morphological study and analysis. As with critical point dried material the samples should be stored in a desiccator following drying as they can readily absorb water.

11.3.8. Specimen Supports

One of the prime considerations in choosing a support is that it should provide some form of conducting pathway, for even the most suitably metal-coated sample will quickly charge up if electrically insulated from the microscope stage. As discussed earlier, the specimen may have already been fixed to a substrate such as glass, plastic, mica, or one of the membrane filters. In these instances it is only necessary to attach the support to the specimen stub using some form of conductive paint such as silver dag or colloidal carbon. It is important to paint a small area on the specimen support and to continue this paint through over the edge and onto the specimen stub. The specimen should then be placed in a 313 K oven or low-vacuum desiccator for several hours to ensure that the solvent in the conductive paint has been completely removed before the specimen is coated. Care should be exercised when mounting membrane filters because the conductive paint can infiltrate the filter by capillary action and obscure the specimen and/or the paint solvents may dissolve plastic specimen supports. Because specimens emerge dry from the critical point dryer or freeze dryer they may be attached directly to the metal stub by a wide variety of methods. One of the simplest ways is to use double-sided tape. The specimens are dusted or carefully placed on the adhesive and, in the case of large specimens, a small dab of conductive paint is applied from the base of the specimen, across the adhesive, and onto the metal specimen stub. Double-sided adhesive tape is a poor conductor and it is important to establish a conductive pathway between the specimen and the metallic support.

The specimens may be attached to the stubs by a whole range of adhesives, glues, and conductive paints. Rampley (1976) and Witcomb (1981) have made a careful study of the different glues which are available and have recommended a number of useful ones. Irrespective of which adhesive is used, it is important that it does not contaminate the specimens, does not remain plastic and allow the specimen to sink in, and contains volatile solvents which can be easily removed. It should also be resistant to moderate heat and electron beam damage.

The metal specimen support should be clean and have good surface finish to allow the specimens to be attached more easily. With a little care it is possible to load several different specimens onto the same stub. Identifying marks can be scratched on the surface, and the different specimens should be firmly attached to prevent cross contamination and misinterpretation.

Once the dry specimens have been mounted and the adhesive has dried, it is useful to gently dust the surface of the specimen with a gentle jet of clean air to remove any loose pieces and contaminating dust. The specimen should then be either coated and examined in the SEM or stored in a dry dust-free container.

11.3.9. Specimen Conductivity

Biological material has a high electrical resistivity of the order of 10^8 Ω/m and will rapidly charge up when irradiated with a high-energy electron beam. The conductivity of biological material may be dramatically increased by the addition of heavy metals or nonmetallic conductors either in the form of a thin layer on the specimen surface, or by impregnation of the whole material with heavy metal salts. Coating techniques are discussed in some detail in Chapter 10, but the alternatives to coating will be briefly discussed here.

11.3.10. Heavy Metal Impregnation

The very act of postfixing specimens for a couple of hours with osmium tetroxide increases the heavy metal content of the specimen. As indicated earlier, it is also useful to block stain specimens after fixation with salts of heavy metals such as lead, bismuth, uranium, etc. The image-forming properties of tissue blocks or large specimens are improved by placing them in a freshly made up saturated solution of uranyl acetate for 2 h at room temperature. The blocks are rinsed thoroughly in distilled water before dehydration and may be used for either transmission or scanning microscopy.

The amount of heavy metal in the specimen can be substantially increased by the addition of a ligand-binding agent which increases the loading of heavy metals in the specimen. The method was first introduced by Kelley *et al.* (1973) and later modified by Malick and Wilson (1975). Following fixation in aldehyde and osmium, the specimen is treated with a ligand-binding agent such as thiocarbohydrazide. The thiocarbohydrazide binds to the osmium present in the tissue by one of its terminal amino groups, and another of its terminal groups is then available to bind further osmium, which is added after the excess thiocarbohydrazide is removed. This is usually referred to as the OTO (osmium–thiocarbohydrazide–osmium) method, and an example is shown in Figure 11.24. As long as the

Figure 11.24. (A) SEM image of OTO treated testis tissue followed by uranyl acetate block staining, coated with 5 nm Au/Pd. Photographed at 18 keV. (B) SEM image of OTO-treated liver tissue, coated with 5 nm Au/Pd. Photographed at 20 keV. (Munger *et al.*, 1976.)

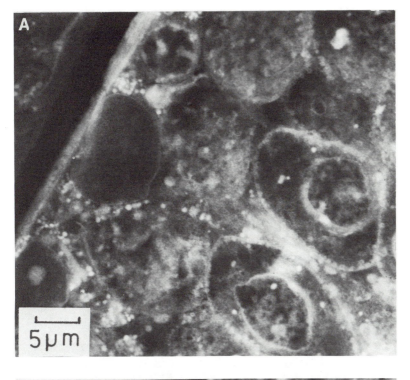

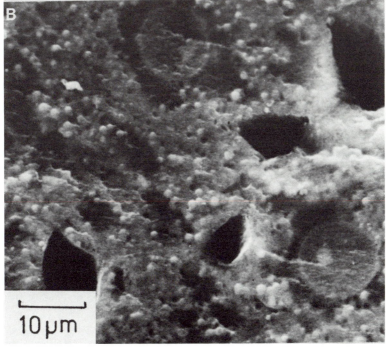

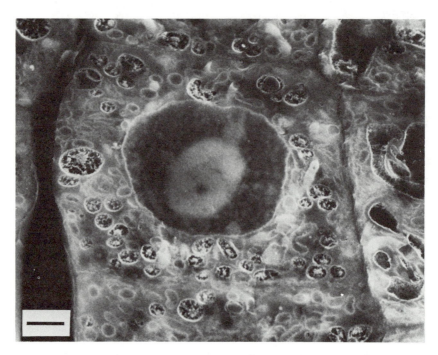

Figure 11.25. SEM photomicrograph of meristematic cell from root tip of *Zea mais*. Specimen processed by osmium ligation technique, dehydrated and embedded in resin monomer. Fractured in frozen resin, washed in acetone, and critical point dried. Uncoated. Bar = 2 μm. (Woods and Ledbetter, 1976.)

osmium remains unoxidized the sequence may be repeated several times to ensure a high loading of osmium. Recent papers by Woods and Ledbetter (1976) and Postek and Tucker (1977) give a useful modification of this method for plant material. An example of a cell prepared by the method of Woods and Ledbetter (1976) is shown in Figure 11.25. Tannic acid has also been used as a ligand-binding agent. Other less specific chemical treatments of specimens to increase their conductivity include the use of osmium vapor, potassium iodide and lead acetate, and silver nitrate. These and other metallic impregnation techniques have been recently summarized by Munger (1977) and Murphey (1978, 1980). The net effect of all these chemical treatments is to increase the bulk conductivity and electron emissivity of the specimens, allowing them to be viewed uncoated in the SEM, although not at the maximum operating capabilities of the instrument.

11.3.11. Interpretation and Artifacts

In spite of the most carefully devised preparation techniques mistakes do occur and it is important to be able to recognize the artifacts in the final image. Damage to the specimen can occur during preparation or during observation and it is frequently very difficult to separate one from the other. It is for this reason that correlative studies are so important, as each of the different methods allows cross monitoring of the whole preparative procedure. A number of the more commonly encountered artifacts are given in Tables 11.1 and 11.2 and a cause suggested for each form of damage. Boyde (1976) has presented similar data in the form of pictures which all SEM users should examine. An example of some of the types of damage which occur during preparation is shown in Figure 11.26. Different types of artifacts appear for different specimens. For details of the physical

Table 11.1. Artifacts Arising from Faults in Specimen Preparation

Bubbles and blebs on surface—incorrect initial washing, hypertonic fixative
Burst bubbles—hypertonic fixative
Cell surface depressions—incipient air drying
Crystals on surface—buffer electrolytes, medium electrolytes
Detachment of cells—poor freezing
Directional surface ridging—cutting artifact of frozen material
Disruption of cells—poor fixation, poor freezing, incipient air drying
Disruption of cell surface—poor washing, poor fixation
Distortion of cell shape—air drying
Fine particulate matter on surfaces—too much, or improper coating
General shrinkage—hypotonic fixatives and washes
Gross damage—poor handling of prepared material, microbial decay
Increased cell dimensions—too much coating
Internal contents clumped and coagulated—poor fixation, dehydration
Large particulate matter—disintegration of specimen, dust, and dirt
Large regular holes—freezing damage
Large surface cracks—movement of coated layer
Microcrystalline deposits—incorrect buffer, fixative, or stain
Obscured surface detail—too much coating
Plastic deformations—incorrect cutting, fracturing
Regular splits and cuts—thermal contraction between cells
Ruptured cell contents—poor critical point drying
Separation of cells—incorrect fixation
Shriveled appendages—hypotonic fixation
Shriveling of cells—overheating during fixation
Slight shrinkage—poor critical point drying
Small holes on surface—incorrect sputter coating
Smooth surface overlaying a featured surface—incorrect drying
Speckled surface—contamination in sputter coater
Stretched cell surface—hypotonic fixation
Strings and sheets of material on surface—incorrect washing
Surface crazing and etching—incorrect sputter coating
Surface folds and/or ridges—fixative pH is wrong
Surface melting—overheating during coating
Surface sloughing—air drying from organic fluid, slow application of fixative
Swelling of sample—hypertonic buffers and fixatives, poor freeze drying

Table 11.2. Artifacts Arising during Specimen
Observation as Seen on the Image

Background distortion—charging
Bright areas—charging
Bright areas in cavities—charging
Bright banding—charging
Bubbles on surface—beam damage
Cracks on surface—beam damage
Dark areas on specimen—charging
Dark raster square—contamination
Image distortion—charging
Image shift—charging
Light and dark raster squares—beam damage
Loss of parts of specimen—beam damage
Raster square on surface—beam damage
Specimen movement—poor attachment, charging
Specimen surface shriveling—beam damage

nature of beam damage reference should be made to the paper by Isaacson (1979). The most obvious indications of specimen damage arising from faulty preparation are shrinkage due to extraction and drying, distortion due to drying and ice crystal damage associated with cryobiological techniques, and changes in surface features brought about by faulty coating techniques (Figure 11.27).

Figure 11.28 shows some of the damage which may occur inside the microscope. The damage can be conveniently divided into a number of different types, although it is sometimes difficult to distinguish one from another. Specimens, particularly those whch have not been coated, may occasionally show vacuum damage. This manifests itself in the form of structural collapse. Beam damage can occur when irradiation with the electron beam causes undue local heating. When this occurs the specimen may bubble, blister, or crack; it may show movement or even disintegrate. In its simplest form it may appear as a light or dark raster square on the specimen surface. This appearance should not be confused with contamination—which can introduce its own set of artifacts and problems of interpretation. Beam damage can be reduced by putting less energy into the specimen, although this may well compromise the optimal performance of the instrument. Specimen charging is another common artifact; although the phenomenon is usually seen as anomalous bright areas on the specimen, the effect runs the whole gamut from small streaks on the image to grossly distorted and unrecognizable images. Pawley (1972) has provided a useful discussion of charging artifacts explaining how they occur and suggesting ways that they may be avoided.

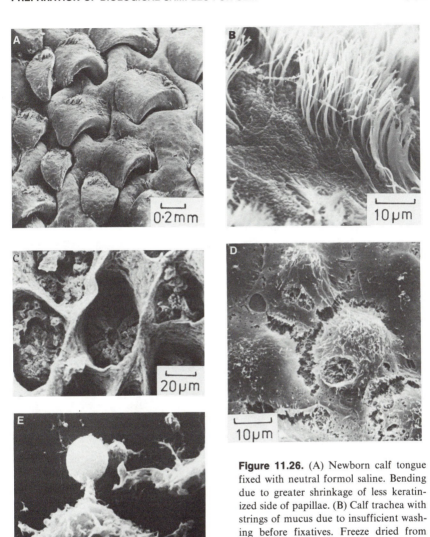

Figure 11.26. (A) Newborn calf tongue fixed with neutral formol saline. Bending due to greater shrinkage of less keratinized side of papillae. (B) Calf trachea with strings of mucus due to insufficient washing before fixatives. Freeze dried from water. (C) Rhesus monkey cardiac stomach. Five minutes treatment with *N*-acetyl, *L*-cystline before fixation causing loss of surface epithelium. Air dried from acetone. (D) K4 cells fixed with hypotonic cacodylate buffer plus 1% formaldehyde + 1.9% glutaraldehyde. Note burst blebs. (E) Kr cell rapidly frozen and thawed before fixation in glutealdehyde followed by CPD. Severe blebbing produced. (From Boyde, 1976.)

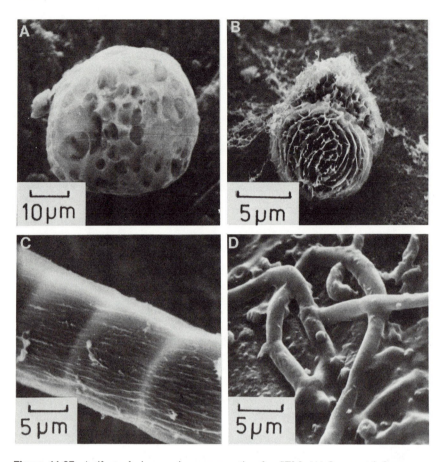

Figure 11.27. Artifacts during specimen preparation for SEM. (A) Ice crystal damage on badly freeze-dried pollen of *Clarkia spp*. (B) Incomplete removal of capsule material from surface of freeze-dried alga *Chroococcus turgidus*. (C) Partial collapse of critical point dried alga *Lyngbia majuscula*. (D) Slime layer obliterating surface of bacteria, *Bacillus megatherium*.

The actual interpretation of the images obtained by SEM is paradoxically both the easiest and the most difficult part of the whole process. It is easy simply because the images are familiar to us and, at low magnifications, we can readily recognize images we see using a stereobinocular microscope. It is difficult, particularly if a new feature appears at high resolution, because one must clearly separate the artifacts which occur in any specimen prepared for the electron microscope from the structure of the biological feature one is attempting to elucidate.

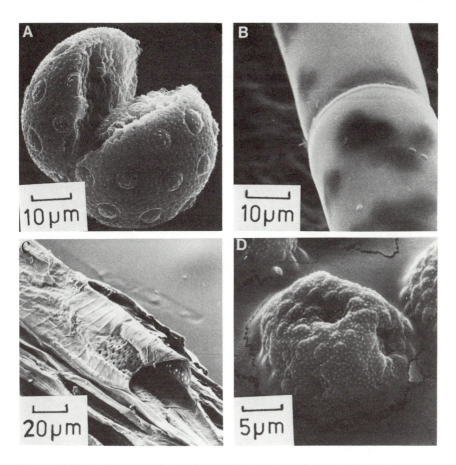

Figure 11.28. Artifacts seen during the examination of specimens in the SEM. (A) Dry fractured pollen gain of *Silene zawadski* showing bright edge due to incomplete coating. (B) Fixed and coated filament of the green alga *Oedogonium* showing black contamination spots. (C) Wood vessels of *Tulipa sp.* showing collapse during drying. (D) Pollen grain of *Plantago media* poorly mounted and showing cracking of the coating layer.

12

Preparation of Biological Samples for X-Ray Microanalysis

12.1. Introduction

Any preparative technique for x-ray microanalysis must result in reasonably recognizable structural detail, while at the same time preserving *in situ* the elemental constituents one wishes to analyze. This is frequently difficult, and some elaborate and ingenious preparative techniques have been devised to achieve this goal. In many respects the principles of the preparative techniques for x-ray microanalysis are not dissimilar to those used in scanning electron microscopy. The important difference is that the methods of preparation, while seeking to preserve ultrastructure, should not achieve this goal at the expense of the soluble cell constituents. In the discussions that follow it is not intended to provide innumerable recipes which could be applied to specific biological systems, but rather to consider the principles underlying the methodology and thus allow experimenters to devise their own methods for the particular sample in question. The subject is not new, and while an attempt will be made to give details of some of the more recent advances, reference should be made to the earlier studies and collected papers by Hall, Echlin, and Kaufmann (1974), Echlin and Galle (1976), Chandler (1977), Echlin and Kaufmann (1978), Morgan *et al.* (1978), and Lechene and Warner (1979), which provide recipes for specific biological samples.

12.1.1. The Nature and Enormity of the Problem

Microanalysis in biological samples can be defined as the detection and measurement of small amounts of elements, molecules, and even macromolecules with high spatial resolution in a biological matrix usually of low mass density. The term "microanalysis" is usually associated with

541

x-ray microanalysis, a technique which locates and measures the elemental composition of specimens. With currently available detectors the minimum amount of an element in a biological tissue that is measurable by x-ray microanalysis is about 10^{-19} g, at a spatial resolution of 20–30 nm and a mass-fraction limit of about 1 mM/kg (Hall 1979b). Biologists prefer to use the term "local mass fraction" as an expression of the concentration of an element. The mass fraction is defined as the mass of an element in the analyzed microvolume divided by the total mass of the specimen in the same microvolume. An explanation of what the microprobe measures including a more precise definition of the terms local mass fraction and measurement by means of elemental ratios are given in the papers by Coleman (1978) and Hall (1979b).

The ideal specimen for x-ray microanalysis would be a thin smooth section of stable low-atomic-number material with good thermal and electrical properties containing inclusions of high local elemental concentration. A thin, $\sim$ 200-nm, aluminum foil containing gold inclusions would probably be the nearest one could hope to approach such a specimen. Unfortunately, biological material is far removed from this ideal.

Biological specimens are three-dimensional, unstable, wet insulators. They are composed primarily of organic molecules and macromolecules bathed in low concentrations of ions and electrolytes. The degree of binding of the elemental components ranges from the relatively strong covalent bonds found in sulfur-containing proteins to the freely diffusible potassium ions in the cytosol. A characteristic feature of living material is that it moves. The movement ranges from the obvious locomotion through cytoplasmic streaming, to the movement of ions, electrolytes, molecules, and macromolecules relative to each other within the cell. Biological material is constantly changing as patterns of metabolism form, destroy, and rebuild the functional architecture of the cell. This very instability mitigates against any x-ray microanalysis being carried out unless ways can be found of instantaneously arresting cell activity and holding it in this state until the investigations are complete. If these were not problems enough, the environment in which the x-ray microanalysis is to be performed is totally alien to life processes. A typical unicellular organism less than 2 μm across synthesizes many hundreds of compounds by a precisely regulated process, is able to replicate itself, and is capable of the genetic evolution and modification of these processes. If one wanted a quick way of destroying that uniquely delicate mechanism there is probably no better tool than a flux of fast electrons which in one second could boil away water many times the weight of the specimen.

As indicated earlier the biologist must accept a compromise situation between the properties of a specimen and the conditions under which the analysis must be performed. There seems little to be gained by compromising the instrument, which means we must carefully consider ways of preparing the biological material. Most of the procedures which have been devised are based on methods used in transmission electron microscopy. This is an unfortunate geneology because the transmission electron microscope relies on adequate preservation of macromolecules, whereas the x-ray microanalyzer detects elements and is thus best suited for inorganic analysis. A careful study by Morgan *et al.* (1978) shows that there is a gross loss and redistribution of nearly all elements during all phases of conventional histological procedures. The loss is also far from uniform; for example, large amounts of potassium are removed, but variable amounts of phosphorus are removed, depending on the tissue compartment. Of equal concern are the elements which can be introduced into the tissue during preparation. The preparative procedures are considered under two main headings, ambient temperature and low temperature, and will be presented in chronological order from the living specimen to the sample inside the x-ray microanalyzer. We shall briefly discuss the high-temperature preparation procedures of microincineration. To gain a proper perspective on the preparation techniques, we shall first consider the types of analytical investigations and applications to biological systems, the types of specimens to be examined, and the strategy and criteria for preparation.

12.1.2. Applications of X-Ray Microanalysis

It is useful to consider briefly the principal types of investigations which are undertaken using x-ray microanalysis. Although these investigations are categorized below, it must be accepted that the categories clearly and frequently overlap.

a. Localization of ions and electrolytes of known physiological function.

b. The study of systems that have been deliberately or naturally perturbed, i.e., by drugs, disease, histochemical stains, and precipitating agents.

c. Localization of enzyme activity.

d. Microchemical analysis of very small volumes of expressed or extracted physiological fluids.

e. The relationship of elemental composition to a particular morphological feature.

f. The general distribution of ions and elements in normal plant and animal cells.

In all six categories, the element(s) being analyzed can exist in a number of different forms; as amorphous or crystalline deposits, covalently bound to molecules or macromolecules, ionically bound to molecules or macromolecules, or as the unbound ion or soluble element freely diffusible in the cell and tissue fluid.

12.1.3. Types of X-Ray Analytical Investigations

The investigations can be carried out at two levels of sophistication. In *qualitative analysis* one attempts to find whether a particular element is present in a cell or tissue. Figure 12.1 shows an example of qualitative analysis on a piece of wood impregnated with a copper-containing preservative. In *quantitative analysis* one is trying to measure whether one part of a cell or tissue contains more or less of a particular element than another part. The ulitmate aim of quantitative analysis is to find out as accurately as possible how much of a particular element is present in a given volume of tissue. Most published work on x-ray microanalysis falls into the first category, and provided adequate care is taken during specimen preparation, the technique can give valid biological information. The second category of accurate quantitative analysis is somewhat more difficult to achieve and involves far more than just the optimal preparation of the specimen. It is necessary, for example, to have accurate standards, high instrumental stability, and computational accessories capable of analyzing a multiplicity of repetitive data.

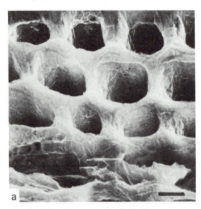
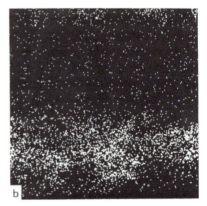

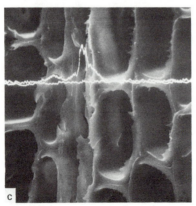

Figure 12.1. A piece of pine wood impregnated with a copper-containing preservative. (a) Secondary-electron image. Bar = 40 μm. (b) X-ray map for Cu $K\alpha$, of same area as (a). (c) Line scan across the piece of wood, with Cu $K\alpha$ peak in the vascular ray region in both (a) and (b).

12.1.4. Types of Biological Specimens

It is convenient to categorize specimens for x-ray microanalysis into six different types. The different types of specimen show varying degrees of x-ray spatial resolution depending on the volume of the sample contributing to the analysis. This volume is a function of both the beam diameter and depth of electron penetration and of electron and x-ray scattering within the specimen.

12.1.4.1. Bulk Samples

These are defined as specimens which are too thick to allow transmission of either photons or electrons and for which the morphological information can only be obtained with a reflected signal. Although bulk specimens are becoming less useful for x-ray microanalysis because of the rather poor x-ray spatial resolution (i.e., 5–8 μm), there are a number of useful applications, some of which are discussed by Marshall (1975), Nagy et al. (1977), and Fuchs et al. (1978). Hard tissues such as bone, shells, wood, and fossils may be fractured, and such specimens are useful for qualitative analysis, but their surfaces are so irregular as to make accurate quantitative analysis practically impossible. This problem can be partly overcome by polishing the surface, but it is important to use pure abrasives to avoid contamination. The fracture faces of soft material can also be analyzed, but only after the tissue either has gone through stabilization, dehydration and hardening, or has been frozen and fractured at low temperature and examined and analyzed in the frozen-dried or frozen-hydrated state. Recent studies (Echlin et al. 1979, 1980) have shown that it is possible to obtain large areas of smooth fracture surfaces on frozen-hydrated bulk specimens, and providing the specimen–detector geometry is optimized, the technique can give useful analytical information on specimens which are difficult to section at low temperatures, i.e., most young plant tissue.

12.1.4.2. Microorganisms, Isolated Cells, and Organelles

There are now many reliable biochemical and manipulative procedures which allow the isolation of single cells and organelles. It is not proposed to discuss these methods, but to consider how such cells and tissue fragments may be studied in the x-ray microanalyzer. As an example of a successful study of a single cell, Figure 12.2, from the work of Chandler (1977) shows data from human sperm cells examined and analyzed in an electron microscope microanalyzer. In some respects isolated cells and organelles which are only a few micrometers thick can be treated as particulate matter, but their very softness and high water content place severe restric-

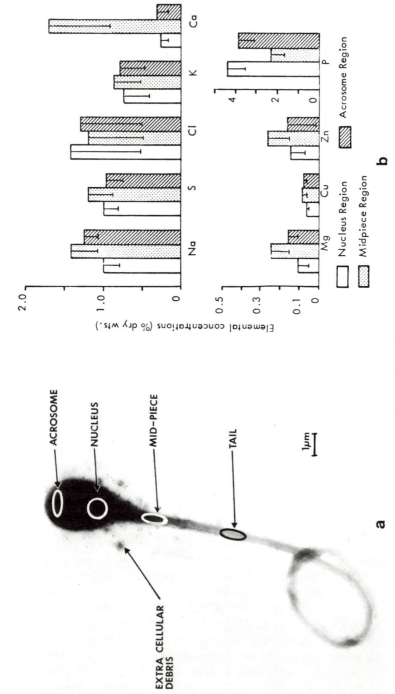

Figure 12.2. (a) Air-dried human sperm cell analyzed in EMMA. The subcellar regions of analysis are indicated by the circles (Chandler, 1977). (b) Elemental composition of air-dried human sperm cells from 20 donors. The bars represent 1 standard deviation of the mean (Chandler, 1977).

tions on the preparative procedures which may be used. Depending on their size, isolated cells and organelles—and, for that matter, inorganic particulate matter—can for analytical purposes be treated either as bulk samples or sections. Some examples of specimen preparation for isolated cells are given in the recent papers by Masters *et al.* (1979) and Cameron and Smith (1980).

12.1.4.3. Inorganic Particles and Fibers

It is sometimes useful to carry out analysis on particulate matter which has either been eluted from tissues with physiological saline, collected as a dry powder by some form of filtration method, or individually isolated from the tissue matrix by micromanipulation. Henderson and Griffiths (1972) have devised a simple extraction replication technique by which a plastic replica is made of the natural or fractured surface. Particles which were in the specimen surface are removed with the replica and may be analyzed. Dry particles can be analyzed *in situ* on a filter as long as they are sufficiently well dispersed, or embedded in a resin which is then either fractured and polished or cut into sections. Alternatively the particles can be suspended in a nonaqueous solvent and spread as a thin layer on a suitable substrate. Particles also occur in a matrix of organic material, and this may be removed by washing in sodium hypochlorite solution, boiling in strong solutions of KOH, or cold plasma ashing. An example of this type of analysis of foreign bodies in tissue is given in the recent paper by Champness *et al.* (1976), who studied the presence of asbestos in lung tissue.

12.1.4.4. Liquid Samples

Much of the interest in biological x-ray microanalysis is centered on the study of physiologically active fluids which may be derived from single cells or from spaces surrounded by a few cells. Methods have been developed to extract such liquid samples by micropuncture and to determine the elemental composition and concentration of liquids with a volume of only a few picoliters. The techniques which have been developed by Ingram and Hogben (1967), Morel and Roinel (1969), and Lechene (1978) involve placing the small drops on a highly polished beryllium surface immersed in a bath of paraffin oil. The aqueous drops do not mix with the hydrophobic paraffin oil and remain as separate drops on the beryllium surface. The paraffin oil is removed by xylene or chloroform and the droplets are frozen dried leaving the nonvolatile material in a microcrystalline form. Hundreds of samples can be processed at a time as the surface of the beryllium block is marked with a grid to facilitate sample identification.

Lechene and his colleagues have automated many of the analytical procedures and the method is now both accurate and reproducible. A slightly different procedure has recently been described by Quinton (1978) and Van Eekelen *et al.* (1980) and involves placing the microdroplets on a thin supported film of parlodion and desiccating the samples by flash evaporation rather than freeze drying. Beeuwkes (1979) has extended the microdroplet technique to the analysis of organic samples by precipitating very small amounts of urea with thioxanthen-9-ol and then carrying out analysis for the sulfur in the precipitate.

12.1.4.5. Thick Sections

Thick sections (i.e., between 0.2 and 2.0 μm) are a useful compromise as reasonably good morphological information can be obtained using scanning transmission images and the specimens are sufficiently thick to contain enough material to be analyzed with good x-ray spatial resolution. As most plant and animal tissues are very soft, they must first be stabilized and strengthened before any sections can be cut. As will be discussed later, most of these preparative procedures can cause serious losses of soluble substances from tissues and should be used with great care. Alternatively one may analyze freeze-dried or frozen-hydrated sections.

12.1.4.6. Thin Sections

Thin sections, i.e., less than 200 nm, usually show the greatest amount of morphological detail and potentially the highest x-ray spatial resolution, but their very thickness may reduce the amount of material to very low levels because of the small microvolume of the section being analyzed. However, one is faced with the same problems of preparation which are found with thick sections, and the only sensible recourse is to use thin sections of freeze-dried or freeze-substituted material.

The advantages and disadvantages of the different types of specimen are summarized in a recent paper on specimen visualization (Echlin 1975b). Specimen visualization is an important feature of x-ray microanalysis as it is necessary to accurately localize the probe on the cells and/or tissues.

12.1.5. Strategy

Before any attempt is made at preparing the specimen, five parameters must be clearly understood: (a) type of investigation, (b) degree of analytical sophistication, (c) type of instrumentation, (d) type of specimen visualization, and (e) nature of the specimen. There is, for example, a quite different approach to specimen preparation for the qualitative analysis for a covalently bound element in a liver biopsy sample using a scanning microscope fitted with an energy-dispersive spectrometer operating in a reflective mode and the quantitative analysis for diffusible ions in a thin section of plant root using an electron probe microanalyzer fitted with wavelength diffracting spectrometers and operating in a transmission mode. A list should be made of these five parameters, and having ascertained the constraints under which the analysis is to be performed, the analyst should decide which are the optimal specimen preparation techniques.

12.1.6. Criteria for Satisfactory Specimen Preparation

There are a number of criteria which should be fulfilled when applying preparative procedures for biological microanalysis. Coleman and Terepka (1974) have suggested five such criteria:

(1) *The normal structural relationships of the specimen should be adequately preserved.* It is difficult to set limits for morphological preservation but a good guideline would be to achieve a morphological spatial resolution one order of magnitude better than the hoped-for x-ray spatial resolution.

(2) *The amount of material lost from, or gained by, the sample must be known.* This can be achieved by carrying out a bulk chemical analysis on one half of an identical sample before preparation and on the other half after preparation. Although this procedure only gives the total elemental concentration of the sample, it does show whether there have been major changes as a result of a particular preparation protocol.

(3) *The chemical identity of material lost from, or gained by, the sample must be known.* Not only is it important to know if there are changes in the elemental concentration in the sample, but whether these changes are selective. Elements in biological specimens are usually partitioned into different compartments, each of varying composition. Within any one compartment, elements may be loosely or strongly bound, and a preparative procedure may selectively affect the concentration of a particular element.

(4) *The amount of elemental redistribution and translocation within the sample must be known.* Although a given preparative procedure may not change the total concentration of elements in the system it may well have caused gross redistribution of the elements within the tissue. Artificial movement of material in the sample is the most difficult phenomenon to assess. The observation of unusually high concentrations of an element, either as crystals or precipitates, gives some indication that redistribution has occurred.

(5) *The chemicals used to prepare the samples should not mask the elements being analyzed.* If one examines the characteristic x-ray spectra of the elements it is possible to see that their K, L, and M radiations overlap

particularly in spectra drived from an energy-dispersive x-ray spectrometer with its poor energy resolution. Unless care is taken, the L or M radiation of one element may overlap and mask the K radiation of the element being analyzed. Table 6.1 in Chapter 6 gives an indication of the type of the overlaps which may occur between some of the elements used in specimen preparation and some of the lighter elements of biological interest.

In many instances it is not necessary to go through elaborate pretesting or preparative procedures, as parallel physiological and biochemical studies will give an indication whether the preparative procedures are causing changes. Because no other reliable analytical procedure exists to verify the accuracy of the microanalysis on this scale, it is difficult to obtain an independent assessment of the accuracy of analysis.

12.2. Ambient Temperature Preparative Procedures

12.2.1. Before Fixation

In general, there should be minimal delay in obtaining the piece of tissue or cells from the experimental organism. Enough is known about the morphological effects of anoxia, stress, and post-mortem changes to suggest that local elemental concentrations are also likely to change. Care must also be taken to remove any contaminating body fluids such as blood, mucus, or extracellular fluids which might infiltrate the specimen and give rise to spurious x-ray signals. Details of the necessary procedures are given in Chapter 11 on specimen preparation for scanning electron microscopy.

12.2.2. Fixation

It is of paramount importance to choose a fixative which does not cause loss, redistribution, or masking of the elements to be analyzed. This is one of the first things which must be checked by carrying out, for example, flame spectrophotometry on the fixation fluids before and after fixation. Very few systematic studies have been undertaken to examine the effect of sample preparation on the elemental concentration of tissues. Yarom *et al.* (1975a) have examined the effects of a number of different fixatives on the ionic concentration of frog muscle, and (Yarom *et al.* 1975b) on rat tissue. Holbrook *et al.* (1976) have carried out a similar study on human skin, details of which are shown in Figure 12.3. These studies provide a convincing demonstration that all fixation techniques have some effect on the natural distributions of elements in tissues. Whereas the cryobiological techniques have a minimal effect on the tissue electrolyte concentration, the wet-chemical methods have a profoundly deleterious effect. *All* fixatives

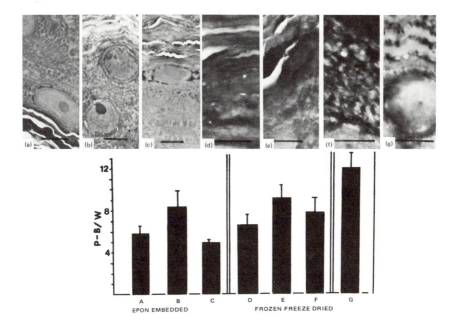

Figure 12.3. Electron micrographs and histograms of peak-to-background ratios comparing the amount of sulfur retained in human stratum corneum cells prepared using different experimental protocols. (a)–(e): Bar = 5 μm, (f)–(g): Bar = 2 μm. [The histograms labeled A–G correspond to the micrographs labeled (a)–(g).] (a) Osmium fixation, Epon embedding; (b) glutaraldehyde fixation, Epon embedding; (c) glutaraldehyde and osmium fixation, Epon embedding; (d) osmium fixation, freeze dried; (e) glutaraldehyde, freeze dried; (f) glutaraldehyde and osmium fixation, freeze dried; (g) unfixed, freeze dried. Note that the better elemental retention is associated with poorer morphological preservative. (Holbrook *et al.*, 1976.)

have an adverse effect on diffusible unbound elements and may have varying effects on bound elements. A recent review article by Morgan *et al.* (1978) contains a summary of the elemental losses recorded during various fixative procedures from a wide range of tissue types. It paints a somber picture and should be sufficient to discourage the use of fixatives, particularly when diffusible elements are the subject of analysis. Commonly used fixatives such as the organic aldehydes have been shown to cause a dramatic loss of soluble material from cells. Figure 12.4, which is taken from the work of Sjostrom and Thornell (1975), shows how fixatives and staining can alter the analytical information from specimens. Brief fixation results in gross ionic changes, and sectioning with a trough liquid led to extraction of elements. If aldehyde fixatives are used it is important to check whether they have been stored over barium carbonate. It is probably

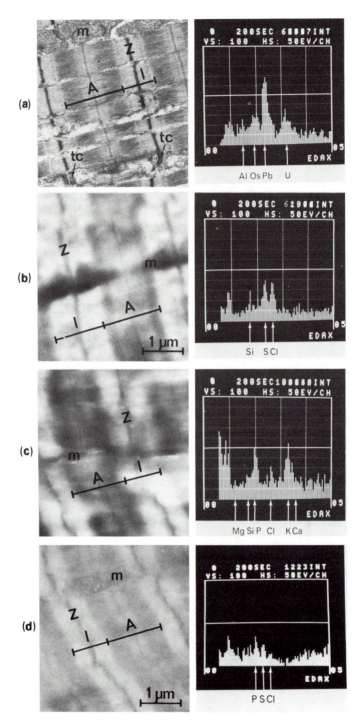

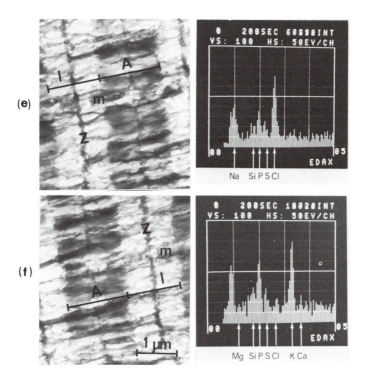

Figure 12.4. Variations in the elemental composition of tissues due to differences in the preparative techniques. Longitudinal sections of skeletal muscle fibers. Mitochondria (*m*), thick and thin myofilaments comprising the *A* (*A*), *I* (*I*), and *Z* bands (*Z*), parts of the sarcoplasmic reticulum (arrows and tc = terminal cisternae). All cryosections were unstained. Elemental spectra obtained by analysis of sections. The analyses were performed over the middle of an *A* band. The beam spot diameter was 0.5 μm. Below the spectra, arrows indicate the expected sites of counts arising from irradiation from a specific element. (a) Conventional wet-chemical preparation. The reagents used for fixation and staining prevented the identification of naturally occurring ions because the energy peaks overlapped. Here an unexplained aluminum peak appeared. (b) Section, cut using DMSO 50% in distilled water as trough liquid, of a glutaraldehyde-stabilized (5 min) and frozen fiber. After cutting, the section was rinsed in distilled water and air dried. Only a few elements were identified. (c) Unfixed, frozen and wet-cut (DMSO 50% in distilled water) muscle fiber. The section was then rinsed in distilled water and air dried. The potassium peak is present as well as a relatively high chlorine peak. Often a sulfur peak was found, although none is present here. (d) Glycerol-treated muscle fiber (30% glycerol in distilled water), frozen unfixed and dry cut. The section was subjected to freeze-drying procedure. No signs of damage due to ice-crystal formation during freezing were seen. Only a few elements were detected. (e) Dry-cut and freeze-dried section of glutaraldehyde-stabilized and frozen fiber. Note the high peaks of sodium and chloride. These elements probably originated from the solutions used. (f) Unfixed, frozen and dry-cut muscle fiber. The section was freeze dried and stored dry until examination. Numerous peaks are seen. The potassium peak is prominent. The occurrence of a calcium peak might be due to a high concentration of this element in components of the sarcoplasmic reticulum present in the column analyzed. (Sjostrom *et al.*, 1975.)

better to use freshly distilled material. The loss of soluble elements is exacerbated by postfixation in osmium and permanganate, which have the added disadvantages of masking the elements being analyzed. If it is quite clear that some sort of tissue stabilization is necessary, and if fixatives have to be used, the fixation time should be as brief as possible, and it is preferable to use fixatives in the vapor phase rather than the more usual liquid state in order to lessen the redistribution of soluble elements. Osmium tetroxide vapor and formaldehyde vapor have been used, but the depth of penetration of the fixative is limited and only small specimens can be successfully treated in this way. Ingram and Hoben (1967) have applied osmium vapor treatment to frozen-dried blocks of tissue, and Somlyo *et al.* (1977) have treated frozen-dried sections. The principal advantage of the osmium treatment is that it improves tissue contrast, but it should only be used if the presence of osmium is not going to interfere with the analysis.

Organic buffers such as bicarbonate, piperazine-N-N'-bis-2-ethanol sulfonic acid, veronal acetate and collidine are preferred to the more commonly used inorganic buffers because there is always a danger that the latter may contribute unwanted elements to the cells (for example, cacodylate buffer contains arsenic). However, an interesting and potentially useful variation on this theme has recently been reported by Fisher *et al.* (1976), who were able to identify high calcium affinity sites in *Paramecium* by fixing the cells in the presence of high calcium concentrations. It is also important to check that the pH of the preparative fluids does not cause changes in elemental solubility. For those investigators who are interested in "wet-chemical" preparative techniques for microanalysis, it is recommended they read the recent review article by Morgan (1979).

An alternative to chemical fixation is heat fixation. Cells or thin tissue splices are placed on a suitable substrate blotted free of excess fluid and then passed through a Bunsen flame. This rather draconian measure is obviously destructive and causes morphological distortion and chemical redistribution, yet it has been used to successfully prepare specimens for x-ray microanalysis.

12.2.3. Histochemical Techniques

It has been long known to light and electron microscopists that it is possible to localize areas of specific physiological and/or metabolic activity by the use of highly specific dyes and histochemical agents. The same idea can, in principle, be applied to x-ray microanalysis. Figure 12.5 from the work of Ryder and Bowen (1974) shows electron dense deposits resulting from a modified Gomori technique for acid phosphatase; the x-ray spectra confirm the presence of lead. Unlike electron microscopy, the end product need not necessarily be a heavy metal as long as it is sufficiently distinct

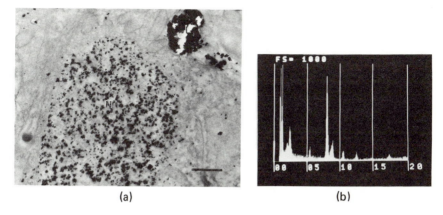

(a)   (b)

Figure 12.5. (a) TEM of section of planarian acidophil cells exposed to Lead-Gomori test for acid phosphatase. Note deposit of lead. $\times 15,000$ Bar = 1 μm. (b) Energy-dispersive x-ray spectrum from stained nucleus seen in (a) showing the $K\alpha$ peak for phosphorus. Note $M\alpha$, $L\alpha$, and $L\beta$ peaks for lead and the K peaks for copper. (Ryder and Bowen, 1974.)

from the bulk of the sample matrix. For example the Alcian blue stain for mucin contains copper, sulfur and chlorine, all of which are readily detected by x-ray microanalysis. Ryder and Bowen (1974) have localized phosphatase activity using an azo-dye which has three covalently linked chlorine "marker" atoms. Martoja *et al.* (1975), Vallyathan and Brody (1977), and Okagaki and Clark (1977) have reviewed the application of histochemical techniques to microanalysis. Van Steveninck and Van Steveninck (1978) have recently published a comprehensive review of the histochemical–analytical techniques which may be applied to plant material. A similar review article dealing principally with animal tissue, has been written by Bown and Ryder (1978).

12.2.4. Precipitation Techniques

A number of workers have attempted to precipitate diffusible elements *in situ* before fixation takes place. The technique is based on the reaction of an ion with a heavy metal compound to produce an electron dense precipitate which is either observable in the transmission electron microscope or can be detected in the x-ray microanalyzer. Table 12.1 shows some of the precipitating reagents which have been used on biological tissue. Alternatively, if the ion itself is a heavy element, organic reagents may be used as the precipitating agents. The same principle can be applied to the intracellular localization of certain enzymes whose reaction products are inorganic ions.

Läuchli (1975) and Van Steveninck and Van Steveninck (1978) have

Table 12.1. Precipitation Reactions for *in Situ* Demonstration of Inorganic Ions and ATPases in Plant and Animal Tissues

Ion or ATPases	Precipitating reagent	Ion or ATPases	Precipitating reagent
(a) Animal specimens		(b) Plant specimens	
Cl^-	Ag-acetate or Ag-lactate	Cl^-	Ag-acetate or Ag-lactate
Na^+	K-pyroantimonate	Na^+	K-pyroantimonate
Na^+, Mg^{2+}, Ca^{2+}	K-pyroantimonate	Na^+, Mg^{2+}, Ca^{2+}	K-pyroantimonate
Na^+, K^+, Mg^{2+}	K-pyroantimonate	Na^+, Ca^{2+}	Benzamide
Ca^{2+}, Mn^{2+}		Ca^{2+}	NH_4-oxalate
Ca^{2+}	K-pryoantimonate K-pyroantimonate or oxalate K-oxalate	PO_4^- ATPases	Pb-acetate $Pb(NO_3)_2$
PO_4^- ATPases	Pb-acetate $Pb(NO_3)_2$ or Ca^{2+}		

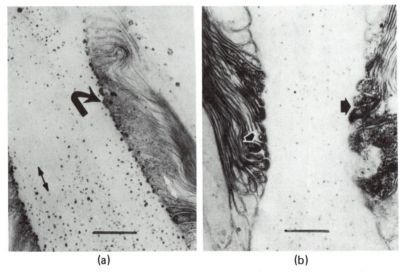

(a)   (b)

Figure 12.6. (a), (b) Localization of sodium and calcium in nerve tissue using the osmium-pryoantimonate precipitating technique. (a) TEM image of an unstained node of Ranvier showing the distribution of pyroantimonate precipitates. Large grains occur in axoplasm (white arrow) and smaller grains are restricted to the cytoplasm of the paranodal loops (black arrows). Bar = 0.5 μm. (b) TEM image of a portion of the paranodal region of the node of Ranvier showing pyroantimonate precipitates. Small-grain precipitates are seen in the cytoplasm of the paranodal loops (white arrows) and large grain precipitates are within the axoplasm (black arrows). Bar = 0.5 μm. X-ray microanalyses by means of wavelength spectrometers show that the granular deposits contain both Na^+ and Ca^{++}. (Ellisman *et al.*, 1980.)

published a critical review of the precipitation techniques for diffusible substances (Table 12.1) and Figure 12.6 from the work of Ellisman *et al.* (1980) shows ways in which the method can be applied in biological tissues. It is important that the specificity of the reaction is adequately tested. Thus the organic silver salt precipitation technique for chloride will also give a precipitate with bromide, and the widely used pyroantimonate technique for sodium also gives a precipitate with potassium, magnesium, calcium, and manganese. Simpson *et al.* (1979) have recently published a review of the pyroantimonate precipitation methods and provide a number of criteria which may be usefully applied when applying this method to the analysis of cations. It is important to know the extent of loss from the specimen during sample preparation. Läuchli *et al.* (1978) were able to show that after preloading the tissue with ^{36}Cl only about 4% of the chloride was lost during the silver precipitation technique. Localization artifacts, occurring either before, during, or after the precipitation technique is carried out, must be considered. Yaron *et al.* (1975b) demonstrated that calcium precipitated with pyroantimonate was lost from the tissue during alcohol dehydration. Van Steveninck *et al.* (1976) were able to demonstrate by x-ray microanalysis that silver-precipitated chloride deposits were relatively unaffected by postfixation in osmium.

12.2.5. Dehydration

Since the specimen is ultimately examined in the vacuum of the microscope, water must either be removed or its vapor pressure reduced by lowering the sample temperature. There is no doubt that chemical dehydration contributes to the loss of diffusible substances from cells and tissues initiated by chemical fixation. Although no critical comparison studies have been made there appears to be little difference in the effect of ethanol, methanol, or acetone as dehydrating agents. However, Thorpe and Harvey (1979) have found that plant material dehydrated in dimethoxypropane showed considerably improved ion retention (Na^+, K^+, Cl^-) compared with dehydration in acetone. The classical dehydration procedures may be circumvented by using the inert dehydration procedures suggested by Pease (1974), water-soluble resins, the glutaraldehyde–urea embedding technique of Pease and Peterson (1972), or by passing glutaraldehyde-fixed material through increasing concentrations of glutaraldehyde up to 50%, after which the tissue is passed directly to Epon 812 (Yarom *et al.*, 1975a). An alternative procedure (Munoz-Guerra and Subirana, 1980) involves infiltrating fixed specimens with increasing concentrations of polyvinyl alcohol solution (MW 14,000) up to a final concentration of 20%. Water is then eliminated by dialysis and the resultant hard gel cross-linked with glutaraldehyde. However, these procedures do not appear to significantly reduce the loss of soluble materials from the specimens tested. Simply air drying the specimen also causes elemental redistribution. Similarly the critical point drying procedure which usually comes at the end of fixation and dehydration is likely to make little difference to the level of soluble substances which have long since been removed during dehydration. Critical point drying may have a severe effect on bound elements, particularly when fluorocarbons are used as the transition fluid. There are, unfortunately, few definitive data on the solubilization effects of the critical point drying method. The solvent evaporation drying procedure described by Boyde and Maconnachie (1979) might also be useful in preparing specimens if it can be demonstrated that minimal elemental extraction occurs during processing. Fixed and chemically dehydrated specimens are infiltrated with Freon 113 and then rapidly exposed to a vacuum pressure of 1–10 Pa. The initial high rate of Freon 113 evaporation cools the remaining Freon which solidifies and sublimes at a slower rate. The absence of any water avoids the problem of ice crystal damage and the morphological results are as good as that produced by critical point drying.

12.2.6. Embedding

Most fixed and dehydrated biological tissues need infiltrating with either resin or paraffin wax to enable sections to be cut, but the resin effectively dilutes cellular constituents by increasing specimen mass. This step in the preparative procedure may be omitted if very thin specimens or fine suspensions are to be examined, or if the specimens are to be examined only by reflected photons or electrons.

The resins can extract and redistribute elements and there is no doubt that the embedding procedure can contribute elements to the materials which are to be analyzed. Some of the epoxy resins can contain relatively high amounts of sulfur, while epoxy resins based on epichlorohydrin contain small amounts of chlorine. The low level of chlorine (0.73%) in Spurr's low-viscosity resin is still too high where critical studies for chlorine are to be carried out. It is advisable to carry out an elemental analysis of all embedding chemicals before they are used for specimen preparation. Methacrylate resins are theoretically free of mineral elements, but cause serious shrinkage during polymerization and are unstable in the electron beam.

12.2.7. Sectioning and Fracturing

It may be difficult to see how sectioning and fracturing could influence the analytical results, but a number of effects should be considered. Knife edges do not stay sharp indefinitely and are progressively worn away by interaction with the specimen. One should consider where the material from the knife edge disappears to during sectioning. Glass typically contains boron, silicon, sodium, and potassium, together with traces of other elements; diamond knives are pure carbon and unlikely to be of any significance in specimen contamination. Steel knives are often used to cut thick sections and could quite possibly contaminate specimens. It is important to thoroughly clean steel knives after they have been sharpened to remove all traces of the honing compound. This cleaning is best achieved using a soft cloth and an organic fluid such as acetone. Attention should be paid to cleanliness of the trough liquid if this is being used and to section treatment after cutting. For example, if resin sections need flattening, it is probably better to avoid chloroform vapors, which can be readily absorbed by resin sections giving rise to high levels of chlorine contamination.

12.2.8. Specimen Supports

Ideally the specimen support should be a good conductor and made of material which will make no contribution to the x-ray signal from the specimen. For bulk specimens or sections to be examined by secondary electrons it is usual to place the specimens on highly polished ultra-high-purity carbon, aluminum, or beryllium disks. Single crystals of silicon doped with boron are also useful. These materials are reasonably good conductors and make only a small contribution to the x-ray background. Materials to be examined using light optics should be placed on quartz or clear plastic slides which can be thinly coated, $\sim$ 5–7 nm, with aluminum to provide a conductive layer. For sectioned material a whole range of specimen supports are available, mostly based on the standard 3.08-mm transmission electron microscope grid. It is possible to buy grids made of copper, titanium, nickel, aluminum, beryllium, gold, carbon, and nylon. These may be used with and without a plastic support film. There is a tendency to use grids made of low-atomic-number elements, such as aluminum, carbon, or beryllium because these materials make significantly less contribution to the x-ray background. Aluminized or carbon-coated nylon films have been used as specimen supports (Saubermann and Echlin, 1975; Gupta *et al.*, 1977), and have the advantage of being strong, and transparent to electrons and photons although they make a contribution to the background. Bulk specimens should be fixed to the specimen support using a high-purity conductive paint such as colloidal carbon, and it is important to check that the adhesive does not have any interfering x-ray lines.

12.2.9. Specimen Staining

The cautionary remarks which are directed towards the use of fixatives are equally applicable to staining. In view of the danger of extracting soluble constituents and the likelihood of introducing heavy elements whose x-ray peaks can mask or intefere with elements of interest, it is probably best to avoid any staining. If the contrast of the specimen image is unacceptably low, then some compromise will have to be made or the specimen will have to be examined by scanning transmission electron microscopy, which gives higher-contrast images, or by using secondary-electron imaging, which gives surprisingly good information from thin sections.

12.2.10. Specimen Coating

It is usual to coat specimens with a thin conductive layer prior to examination to minimize undue heating and prevent local charge buildup. However, provided the specimen is in good contact with the specimen support, the coating procedure is generally dispensed with in the case of samples to be analyzed by one of the energy-dispersive x-ray spectrometers. The higher specimen currents and longer counting times normally needed for analysis with wavelength spectrometers may require that the specimens be coated before being placed in the microanalyzer. The procedures for applying thin coating layers to specimens are considered in Chapter 10 and will not be mentioned here.

Procedures other than coating discussed in Chapter 11, on specimen preparation for SEM, are not generally useful for tissues to be analyzed. Operation at low voltage to reduce charging is not practical for x-ray microanalysis because of the requirement for accelerating voltage about three times greater than the critical excitation potential of the element being analyzed. Most instruments in which sections are being analyzed are run at 20–50 keV. Infiltrating the specimen with chemicals such as thiocarbo-hydrazide or tannic acid, which have a high affinity for metallic charge carriers such as osmium, prolonged fixation in osmium tetroxide or uranyl acetate, and poststaining with salts of silver, gold, or lead are unlikely to be of any use in microanalytical preparations for the reasons mentioned earlier.

12.3. Low-Temperature Preparative Procedures

Many, although by no means all, of the problems associated with "conventional" specimen preparation are considerably diminished by carrying out the procedures using low-temperature techniques. A useful review of low-temperature techniques can be found in the collected papers from the First International Low Temperature Biological Microscopy Meeting, which were recently published as a book (Echlin *et al.*, 1978).

Cryobiological techniques are assuming increasing importance in x-ray analytical studies of biological specimens. They are probably the only way one may hope to analyze soluble ions and elements. Cryofixation is an entirely physical process and confers considerable mechanical strength on otherwise soft biological tissue. The conversion of water from the liquid to the solid state arrests physiological processes and greatly diminishes the movement of dissolved substances. It is probably the only procedure which will result in the preservation of biological tissue in its nearly natural state. Added advantages which accrue from working with specimens at low temperature included a marked reduction in the contamination rate and reduced thermal damage to the specimen under the electron beam. Echlin (1978) has reviewed some of the low-temperature techniques used in scanning microscopy, many of which are applicable to x-ray microanalysis. Reference may also be made to the recent reviews by Saubermann (1978) and Hall and Gupta (1979) and the books edited by Hall, Echlin, and Kaufmann (1974), Echlin and Galle (1976), Echlin and Kaufmann (1978), and Lechene and Warner (1979), all of which contain a number of papers describing low-temperature techniques. The cryobiological procedures will now be considered in order of application during preparation.

12.3.1. Specimen Pretreatment

(a) Fixation. It is probably inadvisable to carry out any form of fixation, as it is now quite clear that even the seemingly most gentle fixation can cause gross membrane permeability changes. If fixation must be used to retain morphology, it is advisable to carry out parallel studies in which one half of the specimen is prepared for morphological detail, while the other half is prepared for analysis.

(b) Encapsulation. If the specimen is to be sectioned or fractured, it may be necessary to encapsulate it in an inert substance dissolved in a physiologically compatible fluid. The encapsulation confers additional mechanical strength on the specimen once it is frozen. This procedure is particularly important for small, soft, biological samples and for plant material with their thick cellulosic walls. The types of materials which can be used include agar, bovine serum albumin, dextran, polyvinylpyrrolidone, and hydroxyethyl starch at a concentration of 10%–30%. It is most important to check the physiological action these substances may have on the functional activity of the tissue being analyzed.

(c) Cryoprotection. It is virtually impossible to freeze any but the smallest pieces of biological tissue without some form of ice crystal damage. This damage can be considerably reduced if the specimen is infiltrated with a cryoprotectant before freezing. Studies have been shown that the commonly used penetrating cyroprotectants, glycerol and dimethyl sulfoxide, while giving good structural preservation, cause gross physiological damage

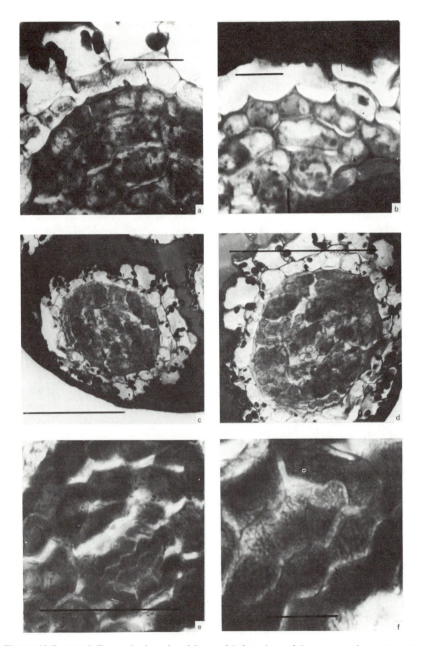

Figure 12.7. (a)–(f) Frozen-hydrated and freeze-dried sections of *Lemna* root tissue, encapsulated in polymeric cryoprotectants, sectioned at 200 K and examined at 113 K on the cold stage of an AMRAY scanning microscope. (a), (b) Freeze-dried section encapsulated in PVP. Bar = 20 μm. (c), (d) Freeze-dried sections encapsulated in HES. Bar = 100 μm. (e), (f) Frozen-hydrated sections encapsulated in HES. Bar (e) = 20 μm; bar (f) = 10 μm.

Table 12.2. Relative Mass Fractions of Elements from Frozen-Hydrated Sections Cut at Different Regions along the Root of *Lemna minor*[a]

Morphological region of root	Disorganized root meristem	Root meristem	Root meristem	Differentiated root
Distance from tip (μm)	40–50	50–60	60–80	120–150
Phosphorus	0.93 ± [b]0.13 (12)	1.46 ± 0.06 (9)	3.66 ± 0.15 (17)	1.16 ± 0.12 (12)
Sulfur	1.63 ± 0.28 (12)	0.25 ± 0.02 (9)	0.59 ± 0.03 (17)	0.44 ± 0.11 (12)
Chloride	1.60 ± 0.23 (13)	0.21 ± 0.02 (9)	0.45 ± 0.03 (17)	1.31 ± 0.10 (12)
Potassium	3.95 ± 0.66 (13)	2.06 ± 0.06 (9)	5.71 ± 0.25 (17)	8.49 ± 0.41 (12)
Calcium	0.34 ± 0.05 (13)	0.06 ± 0.01 (9)	0.23 ± 0.04 (17)	0.37 ± 0.09 (12)

[a]Analysis carried out on the ring of phloem cells, similar to those shown in Figures 12.7e and 12.7f. The changes in elemental concentrations may be correlated with changes in the differentiation of the phloem parenchyma cells.

[b] ± = standard error of the mean. Analysis made on two or three sections in each region.

to cells, and they should not be used in connection with microanalytical studies. High-molecular-weight nonpenetrating cryoprotectants such as polyvinylpyrrolidone (PVP) and hydroxyethyl startch (HES) can give good morphological preservation while at the same time causing minimal perturbation to the functional physiology of the tissue. A series of papers outlining the procedures is given by Franks *et al.* (1977), Echlin *et al.* (1977), Skaer *et al.* (1977, 1978). Studies by Saubermann *et al.* (1977a) and Echlin *et al.* (1978, 1980) show that the high-molecular-weight cryoprotectants can probably be safely used in conjunction with x-ray microanalysis of tissues. This conclusion is corroborated by Schiller *et al.* (1979) in work which demonstrated that solutions of PVP give effective cryoprotection without impairing the normal electrophysiological functions of the frog motor neuronal tissue. A recent study by Barnard (1980) of rat tissue shows that these high-molecular-weight cryoprotectants offer a practical way forward for the preparation of material for x-ray microanalysis of diffusible elements. Figure 12.7 shows the morphological image and Table 12.2 contains the corresponding analytical information which has been obtained from *Lemna* roots encapsulated in PVP and HES. Both polymers considerably reduce the ice crystal damage in the cells, as well as improving the cutting and fracturing of the encapsulated tissues. There is evidence that infiltration of the specimen with artificial nucleators may help to reduce the ice crystal damage. These substances, e.g., chloroform, are quite toxic to cells and should on no account be used in investigations involving x-ray microanalysis.

12.3.2. Freezing Procedures

The different freezing procedures for biological materials have been reviewed by Costello and Corless (1978). The paper by Franks (1980) provides a useful physicochemical background to the process of freezing in biological systems. The important point of all the methods is that the specimen should be cooled as rapidly as possible, although there is still considerable debate how this rapid cooling may best be achieved.

There appear to be four main ways by which specimens can be rapidly cooled:

(1) The specimen may be rapidly propelled against high-thermal-conductivity material kept at low temperature. The copper block at 4 K method described by Heuser and colleagues (1979) is a good example of this type of fast cooling.

(2) Jets of liquid propane at 83 K may be rapidly sprayed onto a stationary sample (Moor *et al.*, 1976).

(3) Microdroplets of suspended cells may be sprayed on a cold surface (Buchheim and Welsch, 1977).

(4) Specimens may be plunged, mechanically or manually, into a melting cryogen such as one of the fluorocarbons, nitrogen, iso-pentane, etc.

The first three methods have been used on small samples, usually single cells or very small pieces of tissue, and from the appearance of the freeze-etched images it is evident that very high freezing rates have been achieved. Most low-temperature biological microanalysis is carried out on larger (1–3 mm^3) pieces of tissue which are frequently sectioned or fractured following quench cooling. Under the circumstances it is doubtful whether the first three methods of cooling have any advantage over the well-established procedures of plunging the specimen into a melting cryogen. Figure 12.8 shows an example of a device which has been used to rapidly propel small tissue samples into melting cryogen. A similar device has been described by Chang *et al.* (1980).

The specimen should be as small as possible compatible with the retention of physiological activity, and as much of the surface fluids should be removed as possible without impairing the internal electrolyte balance of the cells. Although the literature abounds with freezing rates for various cryogens, it should be remembered that these rates have been measured with thermocouples many times smaller than the sample we are trying to cool. While the measured cooling rates for cryogens are a useful indicator, the best measure of the efficiency of a given cooling regime is the size of ice

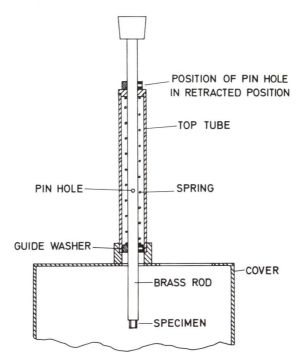

POSITION OF PIN HOLE
IN RETRACTED POSITION

TOP TUBE

PIN HOLE

SPRING

GUIDE WASHER

COVER

BRASS ROD

SPECIMEN

Figure 12.8. A mechanical injector device for fast freezing. The specimen arm is retracted into the top tube and held in position by a small pin. The injector device is placed over a Dewar containing the melting cryogen and the retaining pin released. Injection speeds 3–4 times faster than hand dipping may be obtained, and the specimen is easily released from the injector arm while below the cryogen.

crystals, and hence the amount of damage and redistribution of elements in that part of the specimen being analyzed. Even with the steepest cooling gradient, any specimen with a diameter greater than 250 μm could be vitrified on the outside while the center could possess substantial ice crystals.

For most biological x-ray microanalysis, it is probably adequate to cool samples in melting nitrogen (63 K), melting monofluorotrichloromethane (84 K), or 8% methyl cyclohexane in iso-pentane (100 K). If chlorine is to be included in the microanalysis of frozen-hydrated samples it is probably better to avoid fluorocarbons as traces of halogen may persist in the frozen sample.

Once the tissue is frozen it is important that it should either be stored in liquid nitrogen or remain at as low a temperature as possible to avoid ice recrystallization. In pure water, ice recrystallization can occur at 143 K, although in biological systems there is very little evidence for ice crystal growth until the temperature rises above 173 K. A recent paper by Rall *et*

Figure 12.9. Frozen section of mouse liver tissue showing numerous small electron transparent ice crystals seen in the scanning transmission mode. The main compartments (nucleus and cytoplasm) of the hepatocytes can be distinguished. Bar = 1 μm.

al. (1980) indicates that the absence of intracellular freezing during cooling is no guarantee that freezing will not occur during warming. In tissues slowly warmed from 77 K, intracellular ice first appeared at 188 K, and it is considered that this phenomenon is associated with devitrification, as extensive migratory recrystallization (and cell injury) did not occur until 208 K.

Very little is known about whether the freezing process may affect x-ray microanalytical data. As the tissue cools a freezing front proceeds through the specimen and ice crystals form from pure water at the expense of the cell fluids. An examination of frozen hydrated tissue (see, for example, Figure 12.9) shows ice crystals of varying size. It is certain that the soluble elements are pushed to the edges of the ice crystals or are attached to the macromolecules which abound in the cytoplasm. It would be wrong to suggest that the soluble elements do not move, because the very process of freezing must cause some translocation. During ice formation in the cell, there is a certain period of time when both the liquid and solid phases of water are present together. During this time the concentration of electrolytes in the liquid will increase considerably and cause changes in the electrochemical gradients across membranes which in turn may give rise to some ion redistribution. This problem is exacerbated in cells and tissues with large watery lumens, for ice crystal formation may be more rapid in the extracellular space than in the intracellular space, leading to increased electrolyte concentration in the extracellular compartment. The faster the freezing, the smaller the ice crystals, and the shorter the distance the electrolytes can move from their normal position in the cell (soluble ions are extremely mobile in the cell).

12.3.3. Movement of Elements within a Given Cellular Compartment

This may not, at present, be of any great concern to x-ray microanalysis, as most studies are carried out localizing elements within and without the natural compartments of the cell as in the nucleus in Figure 12.9, where ice crystal growth seems to be limited by the cell membranes. Ice crystal growth is a problem where studies are made on the distribution and concentration of elements within a cellular compartment. It is doubtful whether much reliance can be placed on analytical data obtained from different regions within a given cellular compartment from frozen tissues in which ice crystals can clearly be seen.

12.3.4. Postfreezing Procedures

Once the tissue is frozen and the cell fluids immobilized, a number of options are open regarding both the further processing of the tissue and the possibility of examining and analyzing it directly.

12.3.5. Frozen-Hydrated and Partially Frozen-Hydrated Material

Providing the microscope or x-ray microanalyzer is fitted with a cold stage it is possible to examine and analyze bulk and sectioned specimens in which some or all of the water is retained in a solid state. These types of investigations are fraught with technical difficulties, not the least of which is to ensure that the specimen remains in a frozen-hydrated and uncontaminated state throughout the whole of the preparation and examination procedures. Space does not permit a detailed discussion of these techniques, and reference should be made to the recent reviews by Echlin (1978) and Hall and Gupta (1979), which give both the rationale behind the use of frozen-hydrated samples for x-ray microanalysis and details of some of the procedures which have been developed.

12.3.6. Freeze Drying

Freeze drying consists of the sublimation of ice from cells and tissues under vacuum and as such is an important preparation technique for biological microanalysis. It is by no means the ideal technique and it is necessary to balance the problems associated with the inevitable formation and growth of ice crystals with the advantage of being able to avoid the tissue coming into contact with any chemicals during the preparative process. Moreover, it is not the best method for all specimens. Optimal preservation has usually only been obtained from samples in which there is a tissue matrix remaining after the drying process is complete. Freeze drying in connection with microanalytical studies is probably best applied to sectioned material, cell monolayers, isolated cells, and thin liquid samples. Freeze-dried bulk material can be infiltrated with wax or resin and the polymerized material sectioned. For the analysis of bulk material it is probably better not to use freeze-dried material because of the increased size of the x-ray interaction volume (Echlin *et al.*, 1980). Freeze drying is probably not the best preparative method for the *in situ* analysis of extracellular fluids—such analyses are more properly carried out in the frozen-hydrated state. Freeze drying of biological specimens for microscopy and analysis is by and large an empirical process and it is impossible to set out a protocol which will work for all specimens; each sample will require its own procedure. Such procedures are probably best designated after considering some of the physicochemical aspects of freezing and freeze drying. It is therefore proposed to first consider some of the theoretical aspects of freeze drying and follow with a discussion of some practical aspects which will apply to all specimens. Although much has been written about freeze drying, the recent papers of Meryman (1966), Burstone (1969), MacKenzie (1976), and Franks (1980) will provide a firm theoretical basis of the method.

A deep frozen biological sample consists of a mixture of pure water (ice) and the biological matrix, the so-called frozen eutectic mixture. The ice phase consists primarily of the unbound water which is an important part of all biological systems whereas the biological matrix consists of the heteropolymers which make up living matter together with a variable amount of bound water. Conventional freeze drying can readily remove the unbound water but it is much more difficult to remove the bound water which inevitably forms an integral part of the biological polymers— proteins, carbohydrates, and nucleic acids.

The sublimation of water vapor from the pure ice crystals in the eutectic mixture occurs when the partial pressure of water vapor of the frozen surface is greater than that of the atmosphere adjacent to it. The rate of sublimation from an ice crystal is only a function of temperature. Table 12.3 shows how this rate varies over the range 173–273 K. It should be remembered that the sample may well *cool* during freeze drying with a concomitant decrease in the ice sublimation rate, because of withdrawal of latent heat of evaporation. However, there is usually sufficient radiation

Table 12.3. Vapor Pressure, Rate of Evaporation, and Evaporation Times of Crystalline Ice at Different Temperatures

Temp (K)	Maximum evaporation rate (g/cm²/s)	Evaporation time per g/cm²	Vapor pressure of ice
273	8.3×10^{-2}	1.2 s	595 Pa
233	3.4×10^{-3}	4.8 m	13 Pa
213	2.3×10^{-4}	72 m	2.6 Pa
193	8.3×10^{-6}	33.5 h	0.052 Pa
173	5.0×10^{-7}	20 days	1.3×10^{-3} Pa

and conductive heat from the equipment and the environment to balance the cooling effects of water sublimation.

There are a number of different freeze dryers available, but the basic features are as follows:

(1) There is a drying chamber containing a cold stage which should be at low enough temperature to avoid recrystallization of the ice if one has achieved vitrification during cooling and to prevent ice crystal growth if the cooling only produced small ice crystals. It should be possible to back-fill the drying chamber with a dry inert gas, i.e., nitrogen. In some circumstances it is useful to have a stopcock to allow liquid resin to be passed into the chamber to infiltrate the dried specimen.

(2) A vacuum system is required which is capable of maintaining a pressure of between 10 and 100 mPa in the drying chamber. At high vacuum the mean free path of the subliming water molecule increases and the heat conduction between the tissue and condensing surface is minimized. The vacuum is usually achieved with a diffusion pump and a rotary pump, although when freeze drying is carried out at warmer temperatures the diffusion pump may not be necessary. The vacuum line should contain either a chemical or low-temperature water vapor trap.

(3) A system is necessary to maintain a high water vapor gradient at the solid phase boundary. This is best achieved by molecular distillation of the water vapor onto a liquid-nitrogen-cooled trap within the drying chamber as a 1-cm² surface at 77 K will pump water vapor at a rate of 15 liters/s. Alternatively it is possible to use a chemical desiccant such as phosphorus pentoxide or one of the zeolites, but these are not as effective as a liquid-nitrogen-cooled trap. It is important to maintain the cold trap at least 20° colder than the specimen. The greater the temperature difference between the drying specimen and the condenser, the more effective the retention of water vapor from the specimen. The distance between the drying specimen and the water vapor trap should be less than the mean free path of the residual water vapor molecules. Table 12.4 shows the mean free path of water molecules in air at different pressures.

Table 12.4. Mean Free Path of Water
Molecules in Air at Different
Vacuum Pressures

Pressure (Pa)	Mean free path
13	500 μm
1.3	5 mm
0.13	50 mm
0.013	500 mm
1.3×10^{-3}	5 m
1.3×10^{-4}	50 m

The drying rate depends on a number of factors including specimen temperature, size, and shape, the relative amounts of bound and free water in the sample, and to some extent the vacuum in the drying chamber. The greatest impediment to fast drying is the object itself, and in particular the resistance of the drying shell. Tissues dry from the outside inwards and water molecules subliming inside the specimen must pass through the dried areas once occupied by ice crystals in order to be removed by the vacuum system. This dry organic shell increases in thickness as the freeze-drying front proceeds through the specimen and offers progressively more resistance to the water molecules, which undergo many collisions before reaching the surface. Table 12.3 gives some figures for the drying rates of pure crystalline ice at different temperatures. These rates are faster with vitreous ice but are much slower for ice embedded in a biological matrix.

The temperature at which drying should take place is a question of much debate. If the specimen is maintained at low temperatures, then recrystallization is diminished but the drying times become impracticably long. Nevertheless, small samples such as single cells have been frozen-dried initially at 173 K followed by a three-week period during which the temperature was gradually allowed to warm up. Drying at warmer temperatures increases the risk of ice recrystallization but does allow a faster removal of water, and this is the more practical approach. Freeze drying at warmer temperatures also increases the risk of the "collapse" of the solute matrix, with a concomitant loss of the structural integrity of the sample. The collapse phenomenon is a feature of many aqueous solutions and may be best avoided by only freeze drying small samples at low temperatures (MacKenzie, 1976). Paradoxically there is a greater likelihood of collapse with a finer freezing pattern—the very type of pattern we aim for but rarely achieve when quench cooling biological tissue. Most freeze drying is carried out between 213 and 203 K and under these circumstances a monolayer of cells is dried in a few hours whereas a piece of tissue several millimeters thick may take several days to dry. Although no single freeze drying protocol can be satisfactorily applied to all specimens, there are a number of practical operational considerations which can be applied to all specimens.

The specimens should be as small as practicable and as much as possible of the surface water should be removed. Both these suggestions put severe restraints on the specimen, and it is frequently difficult to remove the surface layer of water from aquatic specimens without causing irreversible damage to the sample. If the specimens are only required for morphological studies then it is useful to prefix in one of the organic aldehydes. This treatment helps to cross-link and strengthen the biological matrix and diminishes the amount of ice crystal damage.

Penetrating cryoprotectants should not be used, for although these substances diminish the size of the ice crystallites they have a much lower volatility than water and are very difficult to remove during the drying process. Nonpenetrating cryoprotectants are, however, most useful, as they readily give up their water during drying and form a stable matrix around the specimen. Neither fixation nor penetrating cryoprotectants should be used with specimens for analysis. The nonpenetrating cryoprotectants can be used in connection with microanalysis as they do not appear to cause any significant redistribution of electrolytes.

When quench freezing the samples it is important to ensure that the specimen does not float on top of the melting cryogen. This is usually not a problem with larger samples, but small specimens should be placed in perforated containers, or held in fine forceps and plunged below the surface. Very fine particulate matter, i.e., single cells, can either be sprayed onto a cold surface or deposited on thin metal foils before quench cooling. Once frozen the samples should be transferred as rapidly as possible to the platform of the freeze drier. This is best achieved by placing the samples in a shallow metal dish about 5 mm deep, which can be filled with liquid nitrogen. The dish containing the samples under liquid nitrogen is quickly transferred to the precooled cold stage in the drying chamber, which is being back-filled with dry nitrogen. Once the nitrogen has bubbled away the chamber can be pumped out to its working vacuum. This procedure minimizes the formation of ice crystals on the surface of the specimen, which under some circumstances can aggregate and lead to elemental redistribution within the specimen. Although no hard and fast rule can be established for the time and temperature of drying it should be carried out at as low a temperature as possible commensurate with the size and shape of the specimen and the exigencies of the experimental system. It is usual to gradually raise the temperature of the specimen during the drying process. This must be carried out slowly or the specimen water may melt and cause redistribution of electrolytes. Towards the end of the drying process the specimen temperature should be allowed to rise a few degrees above ambient temperature, while still maintaining a high vacuum in the drying chamber. This will ensure that the last remnants of free water are sublimed from the specimen. Once the specimen has reached ambient temperature, it is convenient to allow the condenser to warm up, and/or swing the sample away from the condenser. It is most important to make sure the vacuum line water vapor traps are operating efficiently to ensure that the water released from the condenser is trapped there and not allowed to become redeposited on the sample. Once drying is complete, the equipment should be back-filled with a dry inert gas, and the sample placed in a desiccator. Most frozen-dried specimens are very hygroscopic and will rapidly reabsorb water from the atmosphere.

Alternatively, the dried sample may be infiltrated with resin or wax and thin sections cut for analytical studies, and the review by Ingram and Ingram (1980) should be consulted for recent advances in this procedure.

A general procedure which appears to work for most samples is to start the drying at 190 K and a pressure of 1—2 mPa, and then allow the temperature to gradually rise to ambient over a 24–48-h period. The condenser is kept at 77 K. Some freeze driers rely on a Peltier cooling module as a cold platform. Under these circumstances the drying initially starts at about 210 K and a pressure of 1 Pa, and with small specimens the drying is complete within 24 h. With Peltier freeze driers it is more convenient to use phosphorus pentoxide as the trapping agent for the sublimed water.

Boyde (1980) recommends a much faster freeze-drying technique. It should be emphasized that these procedures are only useful for morphological studies of the surface features of samples and it would be necessary to make sure that no melting of the internal ice has occurred. An alternative method of freeze drying is to expose thin sections of frozen material to a slow stream of dry cold nitrogen gas at atmospheric pressure and at a temperature of between 200 and 170 K. Drying is usually completed within an hour.

Freeze drying can give rise to a number of artifacts and it is important that these are recognized. The claim by Lechene et al. (1979) that the material may "explode" during drying is a most unusual observation and suggests that the drying was being carried out at too warm a temperature. This caused the internal ice to melt and quickly evaporate with explosive force. Boyde (1980) has shown that a certain amount of shrinkage accompanies freeze drying but that this shrinkage is much smaller than that seen in critical point dried material.

Whether freeze drying is the best method for preventing redistribution of soluble material is still a matter for discussion. It is probable that during drying, salt crystals may move around within the cellular matrix owing to electrostatic attraction and gravitational forces. The work of Shuman et al. (1976), Somlyo et al. (1977), (1979), Dorge et al. (1975), Rick et al. (1978),

Appleton (1974), Appleton and Newell (1977), and Barnard and Seveus (1978) provides convincing evidence that physiologically significant data can be obtained using quantitative x-ray microanalysis on frozen-dried specimens. In the work of Shuman *et al.* (1976) the analytical data obtained using the microprobe agreed with the data obtained by purely physiological means. These and other studies have been carried out on tissue samples which contain a substantial amount of organic matrix. As the supporting water matrix is sublimed, the once soluble components must be redistributed within the tissues in the same way as salt crystals are deposited on the sides and bottom of a dish of sea water left out in the hot sun. In cells with a large amount of biological matrix, i.e., many animal cells and meristematic plant cells, the redistribution may only be over short distances because the solutes become enmeshed with the drying cellular matrix. With highly hydrated tissues, the migration may be over long distances, which would diminish the accuracy of any analytical studies, and in all cases it is necessary to accept that some redistribution of soluble elements must occur (Dorge *et al.*, 1975).

Finally, an alternative freeze-drying procedure involving cryosorption might have a useful application for preparing biological material for microanalysis. Edelman (1979) describes a simple piece of equipment in which a cooled specimen support, loaded with frozen specimens, is sealed in a closed vessel containing 50 g of the molecular sieve Zeolite 13 X, or Linde 3 A. The sealed container is plunged into liquid nitrogen, and the subsequent freeze drying carried out for 3 days at 203 K and 6 days at 213 K. The dried tissues may be subsequently embedded in resin at 258 K or the container allowed to reach ambient temperature and then back-filled with dry nitrogen. No vacuum is needed, and the system works because of the high trapping capacity of the molecular sieve, which at 77 K can produce a vacuum equivalent to 6.5 nPa.

12.3.7. Freeze Substitution

Freeze substitution is a useful technique for preparing cryofixed specimens for thin sectioning. Although the main force of the investigations has been directed towards preserving the morphological integrity of specimens, there is now sufficient evidence to suggest that freeze substitution might also be a useful preparative technique for biological microanalysis. The reader is directed to the recent reviews by Harvey (1980) and Marshall (1980), which discuss a wide range of methods which have been applied to plant and animal tissue.

The procedure is a relatively slow process and consists quite simply of replacing the ice in a quench-frozen sample with an anhydrous solvent at low temperatures followed by low-temperature infiltration of the tissue with an appropriate embedding medium. The whole substitution, resin infiltration, polymerization, and sectioning must be carried out under strictly anhydrous conditions and at low temperatures. These requirements can create technical problems as atmospheric water vapor is very rapidly absorbed by desiccants at low temperatures. For this reason, the changes and transfers are either carried out in a glove box maintained at a slight positive pressure with dry nitrogen or in sealed containers fitted with entry and exit ports.

To be effective, freeze substitution has to be carried out below the ice recrystallization temperature, which for most biological tissues falls between 163 and 193 K. A number of organic liquids have been used, and the temperatures and times of substitution are given in Table 12.5. As with freeze drying, no strict times can be given and it is necessary to vary the procedures for different samples. The times given in Table 12.5 are averages for specimens 1–2 mm^3, and the temperatures are the range over which the substitution is carried out. To aid structural preservation of the specimen, a fixative such as OsO_4 or uranyl acetate can be added to the substituting fluid, but the presence of a heavy metal can seriously compromise the analytical quantitation.

The lower the substitution temperature the longer the substitution time, which makes the use of butanol, propanol, diethyl and dimethyl ether somewhat impracticable. As a number of workers have shown, the elemental losses are considerably reduced at lower temperatures. It is more usual to use either ethanol, methanol, or acetone. A typical substitution procedure is given by Harvey *et al.* (1976) and is described below.

Small pieces of tissue in an open-mesh wire basket are quench frozen in a melted pool of a mixture of 8% methylcyclohexane in iso-pentane cooled to 100 K in liquid nitrogen. The frozen samples are transferred to

Table 12.5. Organic Liquids Used in Freeze Substitution

Substitution fluid	Temperature (K)	Time	Reference
Acetone	200–233	3–4 days	Harvey *et al.* (1976)
Acetone + acrolein	190–233	1–2 days	Steinbiss and Schmilz (1973)
Butanol	193–233	More than 14 days	—
Diethyl-ether	200–233	20 days	Harvey *et al.* (1976)
Diethyl-ether and acrolein	163–233	6–20	Van Zyl *et al.* (1976)
Dimethyl-ether	158–233	28 days	—
Ethanol	200–233	3–4 days	Harvey *et al.* (1976)
Methanol	178–243	1 day	Muller *et al.* (1980)
Propanol	193–233	More than 14 days	—
Propylene oxide	200–273	10–15 days	Fisher (1972)

anydrous acetone at 193 K, and the cold acetone changed every day for 3 to 4 days. It is convenient to use 50×20 mm glass bottles with clip-on plastic tops, as these can be easily removed, even at low temperatures using a beer bottle opener. It is also useful to keep the bottles in holes in a large metal block as this provides a larger thermal mass when the bottles are away from the refrigerator during transfer. The most convenient drying agent for the substitution fluid is pellets of Linde Molecular Sieve 3A or 4A. This may be dried by heating to ~900 K under vacuum and allowing it to cool in an inert dry atmosphere. The glass bottles are half-filled with the dried molecular sieve and then filled with the anhydrous substitution fluid. During transfer, which is carried out in a glove box pressurized with dry nitrogen, the open-mesh containers may be rapidly and conveniently taken from one bottle to another without any ice forming on the surface.

Following substitution at 193 K, the samples are slowly allowed to come up to 233 K, over a period of 4 h and then transferred to fresh anhydrous acetone at 233 K. The now dried specimens are gradually infiltrated with increasing concentrations of resin in acetone such as the low viscosity Spurrs medium, in which the chlorine-containing component DER 736 is replaced by DYO64 (Pallaghy, 1973). The anhydrous conditions must be retained during the 3-day infiltration, and the temperature of the specimens is slowly allowed to rise to 293 K during this period. The resin is polymerized at 333 K.

Sections may be cut from the polymerized resin block, and any aqueous medium should not be used as a trough liquid, as there can be substantial losses of elements from the thin section. Various liquids have been used in the cutting trough including acetone, dimethyl sulfoxide, hexylene glycol, glycerine, and silicone fluids (Harvey *et al.*, 1976) but all with limited success. It is possible to cut dry sections about 1 μm thick in an atmosphere of dry nitrogen and pick up each section using a fine tungsten needle. Such sections may be sandwiched between evaporated carbon films and are thin enough for examination and analysis in an SEM working in the transmission mode at 20 to 30 kV.

Because these procedures are designed for microanalysis, the specimen contrast is usually very low. Nevertheless, it is possible to distinguish landmarks in the cells and tissues. With a little practice and with the assistance of correlated studies using conventionally prepared material, analysis can be made at recognizable sites in the cell at a spatial resolution of 0.1–0.2 μm. The process of isothermal fixation (MacKenzie *et al.*, 1975; Asquith and Reid, 1980), a form of high-temperature freeze substitution, has been used to fix biological material at relatively high subzero temperatures (253 K) without disturbing the configuration of ice or the hydration state of the crystalline matrix. Although the structural preservation appears quite good, the use of high concentrations of NaCl and DMSO in the fixative fluid to obtain the necessary meeting point depression, would preclude using this technique in the preparation of samples for x-ray microanalysis.

The promising analytical results obtained by a number of workers do not mean, however, that freeze substitution is an ideal method of preservation. There is, for example, considerable variation in the solubility (and hence extraction) of electrolytes in the substituting fluids. There is minimal extraction when the ethers are used (acetone is the next best), while there is considerable tissue extraction with the alcohols. Lechene and Warner (1977) and Chandler and Battersby (1979) have shown that elemental displacement can occur during freezing and substitution. Morgan *et al.* (1978) give a long list of the elemental changes which can occur during the application of different freeze-substitution regimes to a number of different tissues.

In spite of these shortcomings freeze substitution is a useful preparative technique and is probabily best applied to cells and tissues with a dense matrix. This would limit the amount of ice crystal growth during freezing and substitution and minimize the distances over which electrolyte movement may occur.

12.3.8. Sectioning

It is possible to section frozen biological material and examine thick (0.2–2.0-μm) or thin (50–100-nm) sections in the frozen-hydrated or freeze-dried state. The thin sectioning is adequately described in a number of papers (Sjöstrom and Thornell, 1975; Appleton, 1974; Hodson and Williams, 1976; Spriggs and Edwards, 1976), all of which painstakingly outline the procedures to be followed for this technique. Appleton (1974, 1978) and Sjöstrom and Thornell (1975) have carried out x-ray microanalysis on thin sections which have been freeze-dried and have been able to localize a number of soluble elements. Hodson and Williams (1976), Spriggs and Edwards (1976), and Roomans and Seveus (1976) have recently extended the temperature range of cryoultramicrotomy by cutting thin (50–100 nm) sections at between 148 and 123 K. These workers all consider that it is necessary to cut at lower temperatures to be sure that no transient thawing occurs during the cutting procedures. Thicker sections (0.5–1.0 μm) have also been used for microanalysis, but unlike the thin sections, which are usually analyzed in the freeze-dried state, the thick sections are usually prepared and analyzed in the frozen-hydrated state. In the original procedures devised by Moreton *et al.* (1974), Saubermann and Echlin (1975), and Gupta *et al.* (1977) there was considerable uncertainty regarding the degree of hydration of the sections during preparation and analysis. Although thick frozen-hydrated sections could be produced, after the first analysis the sections were substantially dehydrated and the only way one could hope to relate the analytical findings to the wet weight of the tissue was to carefully and quickly calculate the amount of water in the tissue by measuring selected parts of the continuum radiation before and after drying while in the microanalyzer. The only hope of maintaining the specimens fully hydrated for long periods of time is to have the sections in closer contact with a support film of high conductivity.

Recent studies by Saubermann *et al.* (1981a), have refined the cryosectioning technique, specimen support and transfer method, and mode of analysis, so that it can routinely produce sections from a variety of tissues which remain frozen hydrated after 3–4 h in the microscope. In this procedure, small pieces of tissue are placed on the end of copper chunks and quench cooled in melting Freon 12. Cryosectioning is carried out over a range of temperatures using a new cryosectioning device mounted on a microtome in which the large cutting chamber is cooled by a continuous flow of dry nitrogen gas entering the chamber. The temperature of the gas, and hence the temperature of the knife and specimen, can be controlled by passing the nitrogen through an insulated resistor. The continuous upward flow of dry nitrogen gas prevents any ice forming in the sectioning chamber.

Sections are cut singly, and are transferred to the section holder cooled to 123 K with liquid nitrogen. The section holder (Figure 12.10) and sections maintained at ∼ 123 K are transferred from the microtome chamber to the precooled cold stage of the scanning microscope with a transfer rod consisting of a vacuum-tight specimen chamber and a precooled copper heat sink (Figure 12.11). Examination and analysis is carried out at 100 K.

The major factor for successful cryosectioning of thick sections is the temperature of the tissue block. Saubermann and Echlin (1975) consider that thick sections cut at low (∼ 193 K) temperatures are produced by

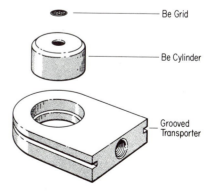

Figure 12.10. Exploded view of a complete specimen holder assembly suitable for frozen-hydrated sections. The tapped hole on the beryllium grooved transporter permits connection to the transfer system heat sink. The grooves on the transporter enable it to make close thermal contact with the microscope cold stage. (Sauberman *et al.*, 1981a.)

Be Grid

Be Cylinder

Grooved Transporter

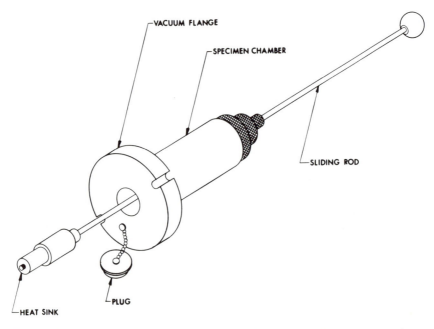

Figure 12.11. Schematic drawing of a sealed transfer system with the heat sink in an extended position. The precooled (77 K) heat sink is screwed onto the grooved transporter, and both are retracted into the specimen chamber, which is then plugged. This enables specimens to be transported to the air lock of the scanning microscope chamber. (Sauberman *et al.*, 1981a.)

multiple fracturing, as the section surfaces are rough. A smooth section surface would imply that transient melting and thawing may have taken place during the cutting process. It is important from the microanalytical point of view to know whether any incipient thawing has occurred during the cutting process. The presence of a thaw zone in either the section and/or the tissue block could lead to redistribution of soluble elements, both during the time the ice is melted and during the recrystallization phase which follows quickly after. The dislocation would be minimal where the cell matrix is dense and homogeneously distributed, but might well be serious in matrix-free compartments which contain only water and small molecules. Saubermann (1980) and Saubermann *et al.* (1977b, 1981a) have considered the physical nature of the cutting process in frozen tissue in more detail and have measured the thermal energy generated during cryosectioning. They consider that the heat generated during sectioning is mainly dissipated by the metal knife and that transient thawing is unlikely to be a problem with thick frozen-hydrated sections.

Saubermann *et al.* (1981a) have examined frozen-hydrated sections in the SEM using both the transmission and secondary mode of imaging. Sections cut at 193 K are rough and show evidence of multiple fracturing. The morphology is distorted and it is difficult to recognize cellular components. Sections cut at 223 K showed the major cell compartments whereas sections cut at 243 K were smooth and flat and the major cell components could be readily recognized as shown in Figure 12.12. In addition, Saubermann *et al.* (1981a) and Bulger *et al.* (1981) have shown quite conclusively that known ionic gradients produced within *in vitro* systems, and electrolyte concentrations from *in vivo* systems, are maintained in frozen-hydrated sections cut at between 233 and 244 K (Figures 12.13 and 12.14 and Table 12.6).

The final cutting temperature for thick sections must remain a compromise. At low temperatures it is impossible to recognize cellular features and any analytical data are probably questionable as the cutting process at these temperatures results in a series of overlapping ice chips. But it is unlikely that any transient melting has occurred. At warmer temperatures the sections are flat and the cellular features can be easily identified, and although there still remains a risk of transient melting, recent evidence would suggest that this may not be such a serious problem.

The problem of transient thawing during the sectioning process for thin sections remains unsolved. The problem is made more difficult by the fact that glass knives are used to cut thin sections, thus removing a potential thermal sink for any heat produced during the cutting process. Hodson and Marshall (1972) consider that the thaw zone is confined to a layer about 60 nm thick, although this is disputed by other workers. A melting zone of these dimensions would not be too serious a problem in thick (1–2-μm) sections, but would certainly create grave problems with thin (50–100-nm) sections. Roomans and Seveus (1976) consider this problem can be obviated by cutting at below 173 K using a dry knife and very slow cutting speeds. Another potential problem associated with cryosectioning is the migratory recrystallization of ice, which in turn could lead to redistribution of electrolytes. Frederik and Busing (1981) have shown that at 193 K ice crystal growth does not significantly contribute to ice crystal damage in frozen thin sections. No data are available for sections cut at warmer temperatures.

Cryosectioning is not an easy procedure, being even more difficult for plant material, and the operational parameters must be optimized for every sample. The image contrast in unfixed material is poor, and it is frequently difficult to recognize familiar cells and tissues. These discouraging notes should not, however, deter people from trying the method, for without doubt analysis of sections, and in particular, thick frozen-hydrated sections,

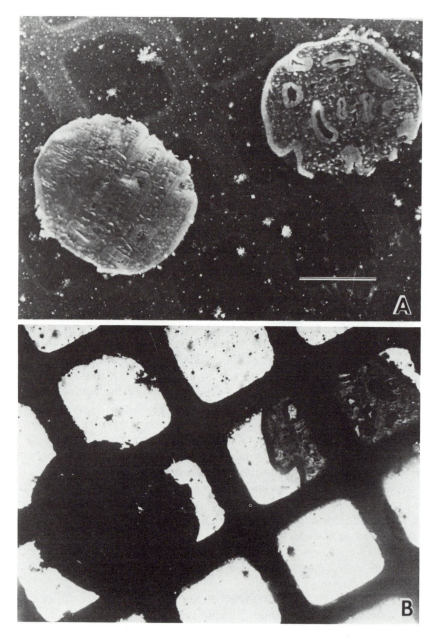

Figure 12.12. Frozen-hydrated (left) and freeze-dried (right) tissue sections of rat renal papillae shown in the secondary (A) and transmitted electron (B) imaging modes. The absence of distinctive morphology and the nearly uniform density of the frozen-hydrated section makes absolute tissue identification difficult. The identity of tissue compartments is confirmed after analysis in the frozen-hydrated state by allowing the specimen to partially dry inside the microscope. Bar = 300 μm. (Sauberman *et al.*, 1981a.)

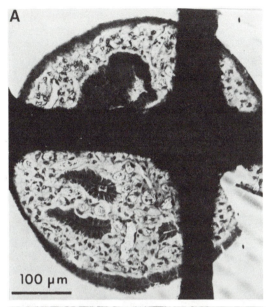

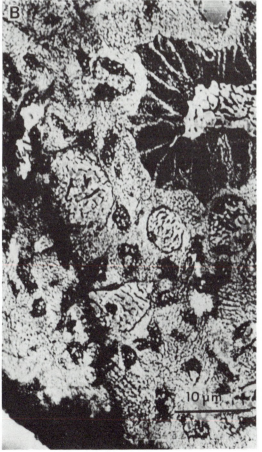

Figure 12.13. STEM images of freeze-dried sections of rat renal papillae. (a) Lower-power view of a section near the papilla tip. (b) At higher magnification, shows a collecting duct at the upper right of the picture and papillar epithelium at the lower left of the picture. The reticulate appearance of the tissue is due to ice crystals formed during the initial freezing step. (Bulger *et al.*, 1981.)

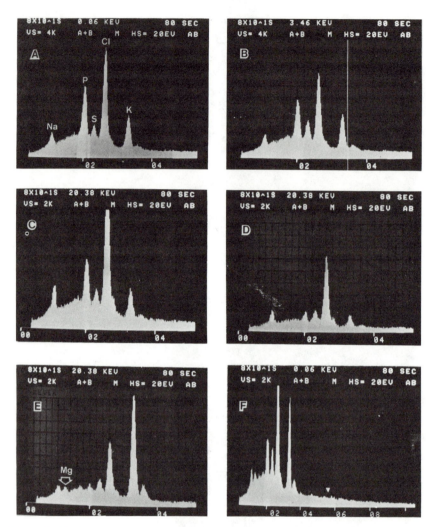

Figure 12.14. X-ray spectra obtained during microanalysis of defined compartments of rat renal papillae. (a) cytoplasm of collecting duct cell; (b) cytoplasm of papillary epithelial cell; (c) cytoplasm of interstitial cell; (d) interstitial space; (e) lumen of collecting duct; (f) plasma in lumen of capillary vessel. The arrow in (f) indicates the position of an iron peak. The absence of a peak in this position indicates little or no hemolysis and smearing had occurred. (Bulger *et al.*, 1981a.)

provides quantitative analytical data most closely resembling conditions existing in the living tissue. It is important to recognize the artifacts associated with transient or complete thawing, the accidental rehydration of freeze-dried sections, the redistribution of electrolytes by migratory ice recrystallization, and contamination of frozen sections.

Table 12.6. Compartmental Water Fractions and Elemental Concentrations in Frozen Sections of Rat Renal Papillae[a]

Compartment	% Water	Na	P	S	Cl	K
Collecting duct cells	56.7 ± 5.6					
mM/kg dry wt.		795 ± 293	581 ± 46	67 ± 15	765 ± 249	360 ± 125
mM/kg wet wt.		344 ± 127	252 ± 20	29 ± 6	331 ± 108	156 ± 54
g/kg wet wt.		7.9 ± 2.9	7.8 ± 0.6	0.9 ± 0.2	11.8 ± 3.8	6.1 ± 2.6
Papillary epithelial cell	59.5 ± 10.8					
mM/kg dry wt.		708 ± 261	597 ± 133	93 ± 39	785 ± 196	348 ± 90
mM/kg wet wt.		287 ± 106	242 ± 54	39 ± 16	318 ± 79	141 ± 36
g/kg wet wt.		6.6 ± 2.4	7.5 ± 1.7	1.2 ± 0.5	11.3 ± 2.8	5.5 ± 1.4
Interstitial cell	59.9 ± 6.5					
mM/kg dry wt.		2236 ± 483	361 ± 65	64 ± 15	1805 ± 348	197 ± 40
mM/kg wet wt.		898 ± 194	145 ± 26	26 ± 6	725 ± 140	79 ± 16
g/kg wet wt.		20.7 ± 4.5	4.5 ± 0.2	0.8 ± 0.2	25.7 ± 5.0	3.1 ± 0.6
Interstitium	77.4 ± 2.3					
mM/kg dry wt.		2609 ± 526	172 ± 52	120 ± 40	1966 ± 508	186 ± 35
mM/kg wet wt.		590 ± 119	40 ± 15	27 ± 9	445 ± 115	42 ± 8
g/kg wet wt.		13.6 ± 2.7	1.20 ± 0.4	0.9 ± 0.3	15.8 ± 4.1	1.6 ± 0.3

[a]Percent water is expressed as the mean ±SD of measurements made in five animals. Elemental concentrations (mean ± SD) are based on dry weight fractions, water content, and calibration factors based on the analysis of standard solutions. Bulger et al. (1981).

12.3.9. Fracturing

An alternative way of exposing the internal surface of the hydrated or freeze-dried biological material for microanalysis is to fracture the sample at low temperature and examine and analyze either the freeze-dried or frozen-hydrated surface. In spite of the limitations imposed by the x-ray microanalysis of bulk samples, a number of workers have used this technique for analysis. Marshall and Wright (1973) have analyzed diffusible ions in bulk frozen insect larval tissue; Fuchs *et al.* (1978) have analyzed the gut contents of *Chironomus* larvae; and Yeo *et al.* (1977) have measured electrolyte concentrations across the roots of halophytic plants. Echlin *et al.* (1980) have carried out analysis on frozen-hydrated differentiating root tissue of *Lemna minor*. Figure 12.15 shows some of the preliminary results, and while it is clear that the x-ray resolution is no better than 5.0 μm, the method provides an easy and certain way of analyzing frozen-hydrated tissues. The specimens are prepared by mounting samples, exposed briefly to a nonpenetrating polymeric cryoprotectant, in a small ($\sim$ 1.0-mm) holder and quench cooling in melting nitrogen. The frozen specimens are transferred under liquid nitrogen to the precooled cold stage of the special SEM biological sample chamber. This is a specially-built device attached to a scanning electron microscope which permits samples to be fractured, etched, and coated under strictly controlled conditions of low temperature and high vacuum (Pawley and Norton, 1978). Specimens can be repeatedly fractured by means of a fixed-angle cold knife, the height of which may be controlled by a micrometer. The fracturing process can be observed by means of a binocular microscope. The fractured surfaces may be etched under controlled conditions using an overhead radiant heater, and carbon (or metal) coated before being transferred to the microscope cold stage where they are examined and analyzed at low temperatures. The whole preparation is carried out at low temperatures, $\sim$ 100 K, in a clean high vacuum (50 μPa) with the assurance that the specimen remains fully frozen-hydrated. The ability to control the level of the fracture plane means that it is possible to obtain the smooth fracture faces necessary for bulk analysis. The additional facility of being able to surface etch the specimens means that a thin, i.e., 0.1–1.0-μm, surface layer of water can be sublimed from the fracture face to enhance the specimen morphology, while the bulk of the specimen remains in a fully frozen-hydrated state.

12.3.10. Specimen Handling

There are special problems associated with handling frozen biological material, particularly in the transfer of samples from one instrument to another. Appleton (1974) has shown that it is most important to freeze-dry sections in a highly controlled manner. He collects sections on EM grids

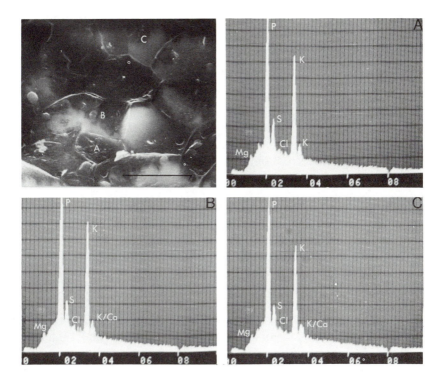

Figure 12.15. Energy-dispersive x-ray microanalysis of a frozen-fracture surface of *Lemna minor* root. Young roots are cryoprotected with 25% hydroxyethyl starch mounted in beryllium specimen holders, quench frozen in liquid nitrogen and fractured and lightly etched in the AMRAY Biochamber at 100 K. The fractured surface was coated with 20 nm of carbon and analyzed at 120 K in an AMR 1000 scanning electron microscope fitted with a Kevex detector. The analysis was carried out using a 1-μm^2 reduced raster at 20 keV, a beam current 0.5 nA, and a counting time of 100 s. (a) Secondary electron image of the frozen-hydrated fracture face showing a ring of phloem parenchyma cells surrounding a central xylem cell. One of the phloem-parenchyma cells has differentiated into a sieve element (A) and a companion cell (B) and analytical data have been taken from these two cells together with data from the phloem-parenchyma cell (C), which is presumed to be undergoing differentiation to form the second pair of sieve element and companion cells normally found in this plant. (a)–(c) Analytical data from these cells. All cells show the high amounts of phosphorus and sulfur normally found in meristematic cells. The levels of potassium, chloride, and magnesium are approximately the same in all three cells, but the sieve element (A) lacks any calcium, which is present in both the companion cell (B) and the phloem-parenchyma (C). It should be noted that the data presented here are only a visual representation of spectra which have extensively deconvoluted and which show that the relative elemental concentrations are significantly different. Marker on micrograph, 10μm.

and slowly dries them at low temperature over a period of several hours. Dorge *et al.* (1975) sandwich the frozen sections between two carbon-coated grids and slowly dry them in a stream of dry nitrogen. If the freeze-drying process is too fast, ice crystals will grow with the consequent redistribution of the soluble elements. Once sections have been dried it is important that they should not rehydrate, as this can cause considerable changes in the cell morphology as well as causing redistribution of electro-lytes.

If frozen-hydrated specimens are to be examined, it is necessary to bring the sections to the microanalyzer in a state where one may be sure that the sections have neither lost water (by melting or sublimation) nor gained water (by condensation from a contaminated atmosphere). A wide range of transfer devices have been developed and they all serve the same purpose—to keep the specimen cold ($\sim$ 123 K) and free from contamina-tion. Hutchinson and Borek (1979) have devised a useful and simple technique whereby they enclose their samples in a Formvar pouch prior to preparation and transfer.

Finally, it is important to consider the difficult question as to whether frozen-hydrated specimens remain in this state throughout preparation, examination, and analysis. Moreton *et al.* (1974), Saubermann and Echlin (1975), Echlin (1978), and Hall and Gupta (1979) have suggested a number of criteria which might be useful in establishing whether a specimen is frozen hydrated at the time of analysis. Clark *et al.* (1976) and Talmon and Thomas (1977a, b) have shown that thin sections do not melt in the electron beam provided they are kept sufficiently cold. It is unlikely that all these criteria can be applied under all conditions.

The quantitative analysis of frozen-hydrated and partially frozen-hydrated sections probably represents the acme of biological microanalysis, as it seems at present to be the only way we may work out the low concentrations of soluble elements within different compartments of the cell and in the extracellular spaces.

12.4. Microincineration

It has been known for some time that it is possible to remove the organic material from biological material by placing it in a muffle-furnace at 773 K or by low-temperature ashing in a reactive oxygen plasma (Thomas, 1969). Although far from an ideal method, microincineration does, however, have a limited number of uses in microanalysis. Albright and Lewis (1965) have detected Fe and Cu in bacteria and Davies and Morgan (1976) have used the method as part of a preparative procedure for the analysis of tissue samples and found it could be controlled sufficiently to retain some volatile elements such as sulfur and chlorine. Meyer and Stewart (1977) have also applied the technique to plant material. Holman (1974) describes ways in which ultramicroincineration can be used for thin-sectioned tissue. Although this is a potentially useful technique for the removal of background material where the mineral content is low, Barnard and Thomas (1978) in a study of microincinerated resin sections, showed that whereas x-ray sensitivity of sulfur, calcium, and phosphorus was increased, there was a loss of chlorine from the sample. As some elements are likely to be displaced during microincineration, the application of this technique should be more towards bound elements and elements of low volatility.

Applications of the SEM and EPMA to Solid Samples and Biological Materials

Several examples of the use of the SEM and x-ray microanalysis are given in this chapter. These case studies were chosen to illustrate various types of SEM and x-ray microanalysis techniques that can be applied to problems of interest to the user. No attempt was made, however, to cover the wide spectrum of applications available in the literature.

13.1. Study of Aluminum–Iron Electrical Junctions

Problem. To study the metallurgical phenomena which occur during electrical arcing at aluminum-wire–iron-screw junctions (Newbury and Greenwald, 1980).

Approach. Electrical arcs were produced between aluminum wire and an iron screw held in an assembly which provided for an adjustable gap between the metals. The free surfaces of the arc-damaged components were then examined in an SEM equipped with an energy-dispersive x-ray spectrometer. In order to study the subsurface region below the arc-damaged regions, metallographic cross sections were taken at right angles through the craters. These regions were analyzed by quantitative x-ray microanalysis procedures with pure element standards and a ZAF matrix correction method.

Results. SEM images of the arc-damaged regions on both the wire and screw revealed similar craters which contained structures indicative of melting, as shown in Figure 13.1a. On the iron screw, these melt structures occasionally appeared in regions which EDS analysis revealed to be nearly

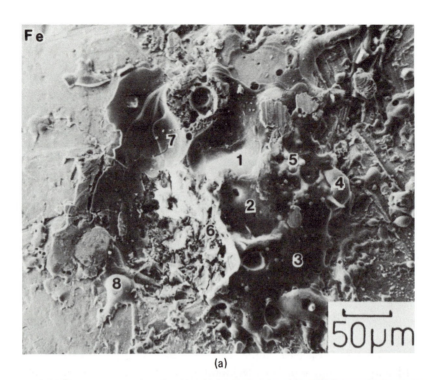

(a)

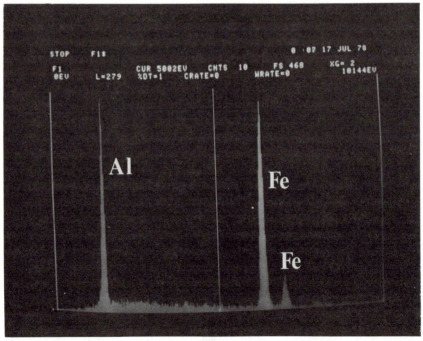

(b)

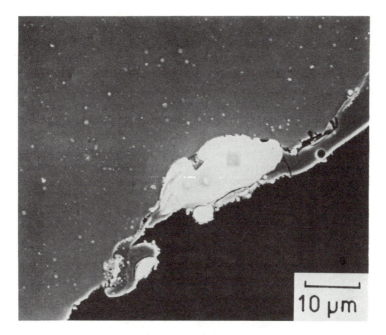

Figure 13.2. SEM image of metallographic cross section of an aluminum wire from an aluminum–iron junction subjected to arcing. The bright region is a layer of Fe_3Al formed by the reaction during the arcing process.

pure iron. This suggests local temperatures in excess of 1800 K. EDS x-ray microanalyses in the center of these craters revealed the presence of aluminum as a major constituent with iron on the screw, and iron as a major constituent with aluminum on the wire (Figure 13.1b). These observations indicated that vapor phase transport occurred between the two metals during the arcing process. The metallographic cross sections revealed the presence of a distinct phase below the surface of each crater, as shown for the aluminum wire in Figure 13.2. Quantitative x-ray microanalysis of these regions indicated that the composition was approximately stoichiometric Fe_3Al, an intermetallic compound.

These studies were then extended to household electrical connectors wired with aluminum. After an induced arcing failure of a connector, SEM

←───

Figure 13.1. (a) SEM image of an arc-damaged region on the surface of an iron screw. X-ray analyses were performed at the numbered locations, revealing major amounts of aluminum transferred by the arcing process from the aluminum wire in the connection being tested. Note the solidification structures at locations 4 and 8. (b) X-ray spectrum obtained at location 2 in (a). Note large aluminum peak indicative of a major aluminum constituent in this region of the crater on the iron screw.

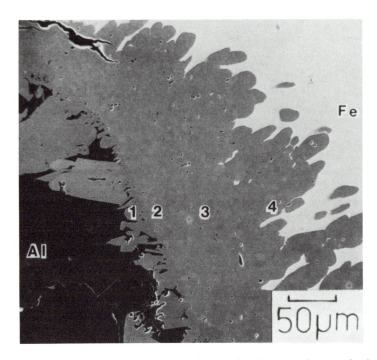

Figure 13.3. SEM image of metallographic cross section of an iron screw from an aluminum–iron junction subjected to arcing. X-ray microanalyses performed at the numbered locations are given in Table 13.1.

images of a metallographic cross section at the iron–aluminum interface revealed the presence of an extensive pad of intermetallic compounds (Figure 13.3). The results of x-ray microanalyses at several points across this structure, given in Table 13.1, indicated the presence of two iron–aluminum intermetallic compounds, identified as Fe_2Al_5 and $FeAl_3$. The formation of these extensive deposits of intermetallic compounds at the current-carrying interface is of special interest. The resistivity of these compounds is much higher than that of the pure metals and thus may contribute substantially to the I^2R heating observed in such failures.

Table 13.1. Analysis of Regions 1–4 at an
Iron–Aluminum Interface (Figure 13.3)

Location	Al (wt%)	Fe (wt%)
1	62.5	38.1
2	57.5	43.3
3	57.3	43.3
4	55.5	44.4

This study illustrates how a combination of SEM/EDS analysis of the free surface and quantitative x-ray microanalysis of metallographic cross sections can provide a complete view of the structure of a sample.

13.2. Study of Deformation *in Situ* in the Scanning Electron Microscope

Problem. Examine the surface of a superplastic metallic alloy during tensile deformation.

Approach. In order to conveniently observe the same region of a sample after various amounts of deformation, the sample was subjected to tensile strain *in situ* in the SEM with a specially constructed tensile stage. The large volume of the SEM sample chamber is ideal for the introduction of special-purpose stages, such as the tensometer illustrated in Figure 13.4. The large depth of focus of the SEM is a particular advantage in such studies where an initially flat sample can develop significant surface roughness during deformation, rendering optical microscopy useless.

Figure 13.4. Tensometer stage for scanning electron microscope chamber which provides the capability for *in situ* deformation experiments.

(a)   (b)

Figure 13.5. (a) Superplastic lead–tin alloy in undeformed condition. Note atomic number contrast between the lead-rich phase (bright) and the tin-rich phase (dark). (b) Same region as (a) after 90% tensile strain. Note grain boundary sliding and void formation. The development of strong topographic contrast suppresses the atomic number contrast visible in (a).

Results. The sample studied consisted of eutectic lead–tin alloy, swaged, rolled, and annealed to produce a microstructure of nearly equiaxed grains of two distinctly different chemical phases. These phases, a lead-rich and a tin-rich solid solution, can be detected via atomic number contrast in the electropolished, undeformed sample, shown in Figure 13.5a. After a tensile deformation of 90%, Figure 13.5b, the grain structure is observed to undergo grain boundary sliding, in which the tensile strain is accommodated mainly by relative motion at the grain boundaries with little deformation of the bulk of the grains. The formation of voids at certain boundaries is observed to accompany grain boundary sliding.

Grain boundary sliding can occur in a direction perpendicular to the plane of the specimen, leading to roughening and the development of topography. Note that as the strain increases, Figure 13.5b, this topography produces such strong topographic contrast that the atomic number contrast is lost.

In situ deformation studies such as these, combined with advanced materials characterization techniques such as electron channeling (Newbury, 1974), can be effectively used to gain a better understanding of the microscopic mechanisms of plastic flow (Dingley, 1969; Joy *et al.*, 1972).

13.3. Analysis of Phases in Raney Nickel Alloy

Problem. Determine the composition of the phases in the microstructure of Raney nickel, a methane production catalyst, in the as-cast form.

Approach. The as-received specimen consisted of a cast hollow vane used for assembly on a tie rod to form the catalyst structure for a methanation system. To prepare a sample suitable for x-ray microanalysis, a single blade was removed from the vane, sectioned with a diamond saw, and polished by standard metallographic procedures. The polished sample was degreased with ethanol, and a coating of approximately 20 nm of carbon was applied by thermal evaporation.

Results. Images formed with the Everhart–Thornley detector revealed at least three distinct phases, as shown in Figure 13.6. The core of the dendrites, position 1, is brightest, and therefore has the highest average atomic number. The interdendritic material, position 3, is the darkest. The interdendritic material may contain more than one phase, but the composition of the fine, dark particles observed (Figure 13.6) could not be obtained by x-ray microanalysis. Qualitative energy-dispersive x-ray microanalyses made at the locations indicated in Figure 13.6 revealed that the only constituents present (with $Z \geqslant 11$) were aluminum and nickel.

Quantitative energy-dispersive x-ray microanalyses were performed at the three locations indicated in Figure 13.6 at a beam energy of 20 keV and a system dead time of 25%. The resolution of the detector was 146 eV at

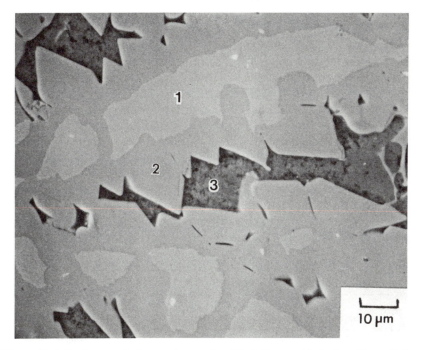

Figure 13.6. Backscattered electron image of polished Raney nickel alloy. Three distinct phases (1, 2, 3) can be observed.

Mn $K\alpha$. Spectral intensities were reduced to quantitative compositional values using the ZAF technique. Pure element standards were employed for the analysis. The beam energy of 20 keV gives an adequate overvoltage for the measurement of nickel. However, the overvoltage for aluminum is quite high, causing Al $K\alpha$ production deep in the sample. The resultant strong absorption for aluminum in the high nickel alloy is significant [$f(\chi)$ for Al equals 0.33 in a 58 wt% Ni–42 wt% Al alloy]. Since the two analytical lines are widely separated in energy, some compromise must be made in choosing the analytical conditions.

The compositional values obtained at locations 1, 2, and 3 (Figure 13.6) are listed in Table 13.2. The high total of the analysis at location 1 may indicate an error, most probably in the aluminum concentration, due to the low value of the absorption parameter for this situation. The compositional values listed in Table 13.2 show that the dendrites are nickel-rich while the interdendritic material is mostly aluminum. The average atomic number for each phase is listed in Table 13.2 and is consistent with the observed backscattered electron image contrast.

13.4. Quantitative Analysis of a New Mineral, Sinoite

Problem. To quantitatively analyze a mineral containing the light elements Si, N, and O.

Approach. In the course of the electron microprobe study of the Jajh deh Kot Lalu, a stony meteorite of the enstatite ($MgSiO_3$) chondrite class, a new mineral containing silicon, nitrogen, and oxygen was discovered (Andersen *et al.*, 1964). The occurrence of the new mineral, sinoite (Si_2N_2O), seems to be of particular significance for the specific and extraordinary environment in which this rock was formed. Apparently this type of meteorite is unusual in that it was formed under extremely reducing conditions, and the occurrence of sinoite indicates that there was not sufficient oxygen available to bind all the excess silicon as SiO_2. The mineral was identified in thin sections of the meteorite and occurs in lathlike crystals up to about 200 μm in diameter. Each thin section was polished and carbon coated for electron probe microanalysis. Quantitative analyses of the mineral were carried out by moving the mineral grains in

Table 13.2. Analysis of Phases in Raney Nickel

Location	C_{Al} (wt%)	C_{Ni} (wt%)	$\bar{Z}$	Image brightness
1	44.2	58.4	22.1	High
2	60.2	40.9	19.3	Intermediate
3	97.9	1.3	13.1	Low

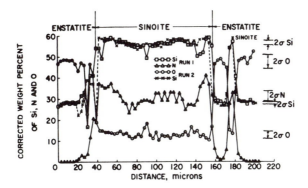

Figure 13.7. Concentration distance plot for Si, Ni, and O across a sinoite (Si_2N_2O) grain in the Jajh deh kot Lalu enstatite chondrite (Keil and Andersen, 1965).

3-μm steps under a stationary electron beam. The analyses of sinoite were performed using WDS. A flow detector with an ultrathin nitrocellulose window was used for both N $K\alpha$ and O $K\alpha$. Enstatite ($MgSiO_3$), quartz (SiO_2), and boron nitride (BN) were used as standards for Si, Ni, and O.

Results. The results of the quantitative microprobe analyses are shown in Figure 13.7. The accuracy of the Si values is about 2% and of the O and N values probably not better than 15%. These analyses were some of the first performed for the light elements nitrogen and oxygen.

13.5. Determination of the Equilibrium Phase Diagram for the Fe–Ni–C System

Problem. To obtain the equilibrium composition of coexisting phases as a function of temperature and composition at the Fe-rich end of the Fe–Ni–C system.

Approach. Electron probe analysis has some important advantages over other conventional methods (x-rays, quantitative metallography) used for phase diagram analysis. If the alloy phases are at equilibrium at the temperature of interest the electron probe can measure the composition of these phases directly. Tie lines can be obtained by measuring the composition of the two coexisting phases at equilibrium. Also in the three phase regions the composition of the coexisting phases can be measured directly and only one alloy is necessary to determine the phase field. Even if the various phases are not totally in equilibrium, phase equilibria data can still be obtained by measuring the interface compositions of coexisting phases (Goldstein and Ogilvie, 1966). This procedure is suitable as long as equilibrium is maintained at the phase interfaces.

In the Fe–Ni–C system various ternary alloys were prepared, homoge-

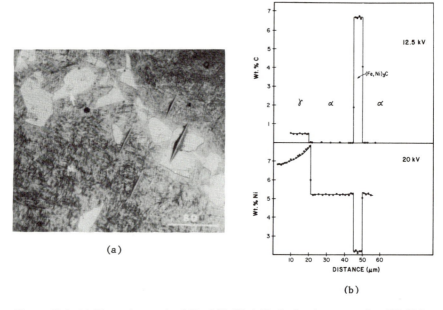

(a)

(b)

Figure 13.8. (a) Photomicrograph of Fe–5.4% Ni–1.1% C alloy heat treated at 923 K for 60 h. Scale bar, 80 μm. *M*, martensite. (Arrows indicate location of cohenite.) (b) Concentration profiles across a three-phase assemblage in an Fe–5.4% Ni–1.1% C alloy heat treated at 923 K. The alloy microstructure is shown in (a).

nized, and heat treated at 773, 873, 923, and 1003 K. The heat-treated samples were mounted, polished, and etched for metallographic examination. In samples where the phase structure could not be seen in the as-polished condition, fiduciary marks were used to define the areas of interest for probe analysis (Figure 13.8a). The sample was then repolished and coated with Al in a vacuum evaporator.

The electron probe analyses were obtained using WDS with an ARL microprobe. For quantitative iron and nickel analyses the electron probe was operated at 20 kV and 0.05 μA specimen current. For quantitative carbon analysis, the electron probe was operated at 12.5 kV and 0.05 μA specimen current. An air jet was also used to minimize carbon contamination as discussed in Chapter 8. A pure Ni sample was used for the Ni standard and a pure Fe sample was used for the Fe standard. Meteoritic cohenite $(FeNi)_3C$, which contains 6.67 wt% C was used for the carbon standard. The background intensity for Ni and Fe was measured on pure Fe and on pure Ni, respectively. The background intensity for carbon was measured on the Fe–Ni ferrite of the meteorite standard. A ZAF matrix correction program using the carbon absorption coefficients of Henke and Ebisu (1974) was used to convert the microprobe data to composition.

Results. The x-ray results were used to obtain the various two-phase $\alpha + \gamma$, $\alpha + (\text{FeNi})_3\text{C}$, $\gamma + (\text{FeNi})_3\text{C}$ tie line compositions and the three-phase, $\alpha + \gamma + (\text{FeNi})_3\text{C}$, compositions. Figure 13.8a shows the typical morphology of the three-phase alloys. The matrices are consistently austenite which has transformed to lath martensite on final quenching. The ferrite has a blocky structure and apparently nucleated at prior austenite grain boundaries. The carbides were found totally within α, totally within γ, and at the α/γ interfaces. A concentration profile from γ to α to $(\text{FeNi})_3\text{C}$ to α phase in a Fe–5.4% Ni–1.1% C alloy heat treated at 923 K is shown in Figure 13.8b. The location of the microprobe analysis is indicated by the arrow on the microstructure shown in Figure 13.8a. The carbon content of ferrite is essentially zero, and cohenite always contains 6.67 wt% C. The interface composition values for each of the three phases did not vary with the morphology of the two-phase interfaces. The composition of the three coexisting Fe-rich phases is α, 5.30 ± 0.1 wt% Ni; γ, 7.95 ± 0.2 wt% Ni, 0.48 ± 0.10 wt% C; and $(\text{FeNi})_3\text{C}$, 2.20 ± 0.10 wt% Ni, 6.67 wt% C. The compositions of other coexisting two-and three-phase regions in the Fe–Ni–C system are given by Romig and Goldstein (1978).

In summary this study shows how the electron probe microanalyzer can be used to obtain quantitative chemical analyses of phases in equilibrium and in nonequilibrium assemblages. Quantitative carbon analyses can be obtained but specific techniques such as lowering the beam energy and using an air jet must be employed.

13.6. Study of Lunar Metal Particle 63344,1

Problem. To determine the microstructure and phase compositions of the largest metal particle found on the lunar surface.

Approach. Metal particle 63344,1 was found in the Apollo 16 coarse soil fines (4–10 mm diameter) and is a unique specimen consisting of metal globules. The metal particle was examined initially using the scanning electron microscope (SEM), electron microprobe, and optical microscope. The sample was then mounted in epoxy, polished, and examined metallographically with the optical microscope and the SEM. Electron microprobe analysis for Fe, Ni, Co, P, and S in the metal was carried out using WDS. Silicate analyses were obtained using an energy-dispersive detector and the data reduction technique of Statham (1974).

Results. About 400 metal globules in a restricted size range from 150 to 600 μm in diameter are welded together into a cohesive mass. Figure 13.9 shows an SEM photograph of the sample. The material which fills in between the globules is a fine dust and consists of 10–100-μm fragments of plagioclase crystals and very fine-grained anorthositic-rich rocks. The dust

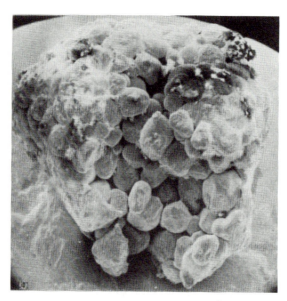

Figure 13.9. Low-magnification SEM photograph of metal particle 63344,1. Field of view is 6.2 × 5.8 mm.

was derived from the lunar soil, which is composed of fragments of highland-type rocks. Figure 13.10 shows the microstructure of the specimen after it was ground down ~1 mm from the top surface. The globules remain loosely packed together and no silicate was observed within or between any of the metal globules. Very few of the globules are really spherical and most of them appear to be sintered together and/or plastically deformed.

Measurements of the bulk compositions of over 12 globules were made. The globules were analyzed by sweeping the electron beam over several $50 \times 40 \ \mu m^2$ areas in each globule. The average composition of all the globules is 6.25 ± 0.2 wt% Ni, 0.37 ± 0.04 wt% Co, 0.80 ± 0.1 wt% P, and 92.5 ± 2 wt% Fe. Since the globules are chemically similar, they were probably produced at the same time and from the same source.

Microstructural examination of individual globules showed a Widmanstätten growth pattern of α bcc phase. Figure 13.11 is an SEM picture of one of the Widmanstätten plates, the boundary of which is decorated with micron-sized phosphides. Microprobe measurements of local compositions in phosphide-free areas give values of the bulk composition of the globule, 6.1 wt% Ni–0.8 wt% P. Measurements of Ni and P interface compositions from phosphides greater than 5 μm in size and from the adjoining α yield 13.5 ± 1 wt% Ni, 15.5 ± 0.5 wt% P for the phosphide, and 5.8 ± 0.5 wt% Ni, 0.30 ± 0.3 wt% P for α. These interface compositions are consistent with equilibration of phosphide and α at 600°C.

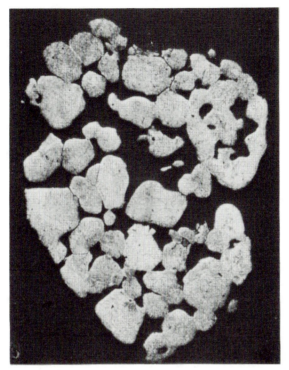

Figure 13.10. Optical photograph of globules ∼1 mm from the top surface of the metal particle. The black bacground is Lucite mounting material. Field of view is 5.4 × 4.0 mm.

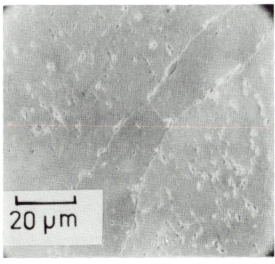

Figure 13.11. SEM photomicrograph of one Widmanstätten plate in a globule of metal particle 63344,1. The α boundary is decorated with micron-sized phosphides. Smaller phosphides are observed in the matrix.

In summary SEM and x-ray microanalysis showed that the globules of 63344,1 were of the same composition and probably cooled slowly from the solidification temperature to 600°C. Various possible mechanisms have been proposed for the origin of the particle but all of these argue for the particle to have formed from an impacting iron meteorite (Goldstein *et al.*, 1975).

13.7. Observation of Soft Plant Tissue with a High Water Content

Problem. To make a study of the cell orientation in the differentiating phloem parenchyma of the growing root tip of *Lemna minor*. This morphological study is a preliminary investigation before carrying out a more detailed microanalytical study on frozen-hydrated material. It was necessary to find out whether it was possible to obtain a recognizable cell morphology in fracture faces of frozen-hydrated material which had not been chemically fixed. The ultrastructure of the root has been elucidated using the transmission electron microscope and the SEM investigations form part of a correlative study.

Approach. The tips of the growing roots were excised, damp dried, and then thoroughly wetted in a 25% solution of hydroxy-ethyl starch made up from *Lemna* growth media. It is important that the roots stay in the hydroxy-ethyl starch (a polymeric cryoprotectant) for only 2–3 min. The cryoprotected roots were mounted in thin silver tubes and quench frozen in liquid nitrogen slush at 63 K. The tubes and their frozen contents were placed in a specimen holder, held under liquid nitrogen, and quickly transferred via an air lock onto the precooled cold stage (103 K) of an AMRAY Biochamber. With the specimens maintained at 103 K in a vacuum of 50 μPa (5×10^{-7} Torr), the frozen root tips were fractured with a knife held at 100 K. The fracture faces were etched for 90 s with an overhead radiant heater, and then coated with 20 nm of evaporated gold while the specimen was rotated and maintained at 103 K. The specimen was transferred via a second air lock to the precooled (93 K) cold stage of the AMRAY 1000 SEM. The specimen was examined and photographed at 93 K and a pressure of 25 μPa (2×10^{-7} Torr), with a 20-keV beam and a beam current of 20 pA.

Results. Figure 13.12 shows a transverse section of the stelar region of the root of *Lemna minor*. This picture confirms what we already know about the relative disposition and ultrastructure of the cells in this region of the root. The small central xylem cell ($\times$) is surrounded by a ring of eight undifferentiated phloem parenchyma cells, outside of which there is a ring

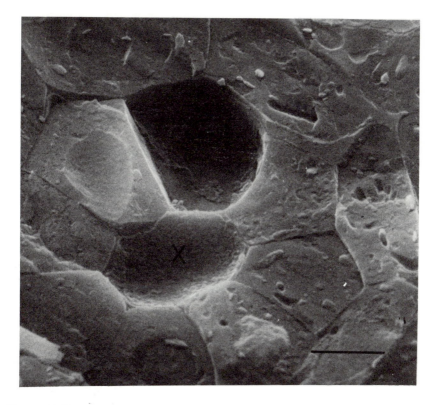

Figure 13.12. Transverse section of the stelar region of the root of *Lemna minor L.* (duckweed). Bar = 5.0 μm.

of larger endodermal cells. The fractures have passed through the cells at different levels and revealed many of the cell contents. Similar images have been obtained from carbon-coated material and a large number of x-ray analyses have been carried out on such fractures, with the assurance that one knows where in the sample the fracture is located. The technique can be used to obtain morphological information about cells and tissues and avoids the use of deleterious wet chemical fixation methods. The combination of rapid freezing and polymeric cryoprotectants has produced a frozen-hydrated specimen which is free of any visible ice crystal damage. The clean fracture face is devoid of any ice crystal artifacts, and the coating layer is sufficiently thick to prevent any charging. The etching has successfully removed a surface layer of ice and has emphasized the morphological features of the tissue. The fracture face is relatively smooth and flat and is suitable for carrying out x-ray microanalysis.

13.8. Study of Multicellular Soft Plant Tissue with High Water Content

Problem. To study the differentiation and development of the pollen grains and tapetal tissue of *Helleborus* by means of transmission and scanning electron microscopy.

Approach. Young, whole anthers were fixed for several days at 277 K in a fixative composed of 2.5% glutaraldehyde, 1.5% para-formaldehyde, and 1% acrolein in a 0.05 M phosphate buffer pH 7.2. Following the aldehyde fixation the specimens were thoroughly washed in 0.05 M phosphate buffer and then postfixed for 3 h at room temperature in 2% osmium tetroxide in 0.05 M phosphate buffer. The anthers were then washed in distilled water and dehydrated through a graded ethanol series ending up in anhydrous ethanol over Linde Molecular Sieve 3A. The samples for transmission microscopy were passed through propylene oxide and into increasing concentrations of Araldite Epoxy resin. Thin sections were cut from cured resin blocks, picked up on uncoated 400 mesh grids, and stained with lead citrate and uranyl acetate. The sections were examined and photographed at 80 keV in an AE1 EM6 transmission electron microscope. The samples for scanning microscopy were passed into amyl acetate and then critical point dried from liquid CO_2. The whole dried anthers were carefully attached to a specimen stub using plastic glue which was allowed to set and harden overnight at 308 K. A small piece of adhesive tape was lightly pressed on the surface of the dried anther, and a dry fracture was made by gently pulling away the adhesive. The exposed interior surface was coated with a thin layer of evaporated carbon and the edges of the sample were painted with conductive paint. The paint was allowed to dry overnight at 308 K and the specimens were coated with 25 nm of evaporated gold. The samples were rotated and tilted during coating. The anthers were examined and photographed at 20 keV in a Cambridge Instrument Stereoscan scanning microscope.

Results. Figures 13.13a and 13.13b are a good example of a correlative study combining transmission and scanning electron microscopy. A direct correspondence can be made between the size, location, and relative position of structures seen at the surface of the specimen using the SEM and in a cross section of the sample as viewed in the transmission electron microscope. The TEM gives information on the way the Übisch bodies are extruded from the tapetal cell into the anther cavity; the SEM provides information on the relative density of these structures at the tapetal surface. The SEM gives information on the surface features of the wall; the TEM

Figure 13.13. Pollen grain and tapetal tissue of *Helleborus foetidus L.* (stinking hellebore). Bar = 5.0 μm on (a) TEM and (b) SEM micrographs.

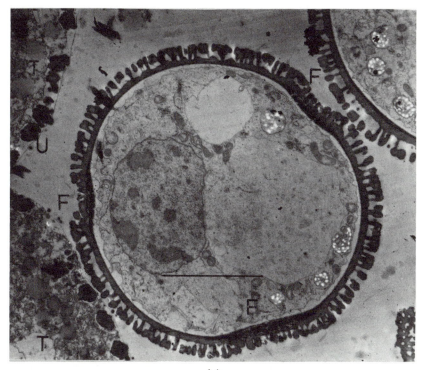

(a)

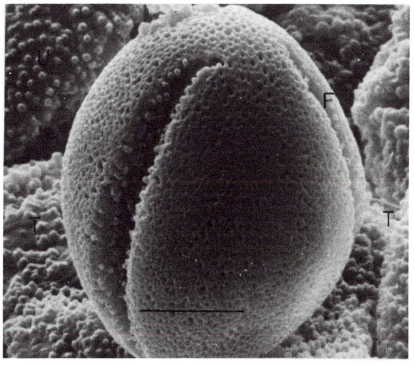

(b)

gives information on the underlying architecture of this structure. The TEM gives little information about the width and extent of the furrow, but this is readily available from the SEM. The SEM gives no information about the substructure of the pollen grain cytoplasm, the TEM readily reveals the localization and interrelationships of the cellular organelles. The prolonged fixation with a mixture of reactive aldehydes is frequently necessary for the adequate preservation of large pieces of tissue from the aerial parts of plants because of the time it takes for the fixative to penetrate to the center of the material. The fixation is doubly difficult with this particular sample because the tapetal tissue (T), Figure 13.13, is in a senescent phase and easily disrupted by sudden osmotic changes. The fixation has retained the structural relationships within the nearly mature anther, even to the point of preserving the Übisch bodies (U) which are forming on the surface of the senescing tapetum. The characteristic shape and the detailed patterning of the pollen grain wall can be clearly seen in both surface and transverse view. Two of the large furrows (F) can be seen on the scanning image; all three are apparent on the transmitted EM image.

13.9. Examination of Single-Celled, Soft Animal Tissue with High Water Content

Problem. To study the surface morphology of red blood cells, as part of a more general investigation on the appearance of blood components in the SEM.

Approach. A sample of blood was slowly dripped into a fixative composed of 0.9% glutaraldehyde in $0.1\,M$ sodium cacodylate buffer pH 7.2 and a total osmolality of 286 m/os. After 10 min fixation the sample was centrifuged and the cells were resuspended in $0.1\,M$ sodium cacodylate buffer adjusted to 268 m/os with sucrose. The blood cells were dehydrated through a graded acetone series and placed in anhydrous acetone. For the remaining steps aliquots of anhydrous acetone and red blood cells were transferred to a small envelope (10×10 mm^2) made of a hard filter paper held together with a paper clip. The envelope remained immersed in anhydrous acetone throughout the loading procedure. The cells in their envelopes were critical point dried with liquid CO_2 and the dried envelopes carefully opened. Portions of the envelope containing cells were fixed to the specimen stub with a very thin layer of colloidal silver and stored overnight in a vacuum desiccator. The specimens were sputter coated with 30 nm of gold at 2.4 kV, a plasma current of 20 mA, and a pressure of 8 Pa. The specimens were photographed in a Cambridge Instrument S-600 microscope operated at 15 kV.

Results. As shown in Figure 13.14 the red blood cells show no sign of collapse and have retained their distinctive bi-concave shape. The sample is

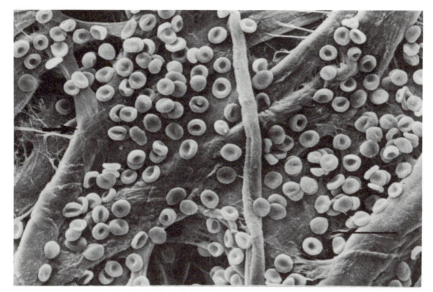

Figure 13.14. SEM photomicrograph of human red blood cells. Bar = 20 μm. Photograph courtesy of Ken Thurley, Department of Anatomy, University of Cambridge.

nicely dispersed and there is no sign of charging. Although the specimen has been examined and photographed at relatively low kilovoltage (15 kV) there is still some beam penetration (→). Because this picture has been taken at low magnification (~1400 ×) it would have been better to have reduced the accelerating voltage to 10 kV or less. The red blood cells have retained their characteristic shape. At this relatively low magnification it is impossible to see any detailed surface irregularities. Although the red blood cells have been adequately fixed and dehydrated, the background filter paper matrix tends to distract attention away from the objects of biological interest. The coarse filter paper has a sufficiently rough surface to allow many of the specimens to be hidden within the crevices. This is not an important problem in this present application where one is examining many identical cells. It would be a serious problem in a situation where there are only a few cells to be examined, and under those circumstances it would be better to use a smoother filter substrate such as one of the finer grades of Millipore filter.

13.10. Observation of Hard Plant Tissue with a Low Water Content

Problem. To study the wood anatomy of the mature stem of the common grape vine *Vitis vinifera*. The possibility exists that the wood may have been infected with microorganisms.

Approach. Small pieces of vine stem (10×5 mm^2) were excised with a clean razor blade and fixed for two hours in 4% osmium tetroxide in 0.05 M sodium cacodylate pH 7.2 at room temperature. The fixed tissue was thoroughly washed in distilled water, chemically dehydrated in acidulated dimethoxy propane, rinsed several times in anhydrous acetone, and critical point dried from CO_2. The dried tissue blocks were mounted in different configurations on slabs using colloidal silver and allowed to dry overnight in an oven at 313 K. The samples were sputter coated at 293 K with 30 nm of gold–palladium alloy 80 : 20 at 2.4 kV and a plasma current of 20 mA and an argon pressure of 8 Pa. The specimen was examined and photographed in a Cambridge Instrument S4-10 microscope at an operating voltage of 30 kV and a beam current of 35 pA. This type of specimen is a good example of a relatively hard biological tissue of low thermal and electrical conductivity. The 4% osmium tetroxide ensures some degree of bulk staining and the rapid dehydration procedure is unlikely to cause any distortion. The coating layer is deliberately thicker than usual to make sure that there is an adequate metal coating in the deeper parts of the cells.

Results. Figure 13.15 demonstrates the large depth of focus (~ 100 μm) available with the SEM, enabling the operator to see into the large vessels which are a predominant feature of the specimen. The picture has not been subject to any image signal processing. The relatively dark vessel at the top of the picture ($\rightarrow$) might have revealed further information if the black level had been suppressed and/or nonlinear amplification had been used. Biologically, all that is observed is the lignified cellulosic cell walls; the cell contents have long disappeared. There is no evidence of bacterial or fungal infection. Because the specimen has been cut with a clean razor blade, and care has been taken to wash away any small pieces of material, the surface is clean and devoid of contaminants. The preparative technique is adequate, and the sputter coating has reached to the bottom of the large vessels since there is no evidence of charging.

13.11. Study of Single-Celled Plant Tissue with a Hard Outer Covering and Relatively Low Internal Water Content

Problem. The pollen grain is one of the principal causative agents of hayfever. This study was made to see if any of the surface features of the pollen grain wall could be correlated with the allergenic properties of the cell.

Approach. The pollen grains were taken from a dried plant on a herbarium sheet and dusted onto nearly dry but tacky plastic glue on the surface of the specimen stub. The samples were stored in a vacuum desiccator overnight at room temperature and then coated with 30 nm of gold in a vacuum evaporator at a pressure of 5×10^{-4} Pa. The specimens

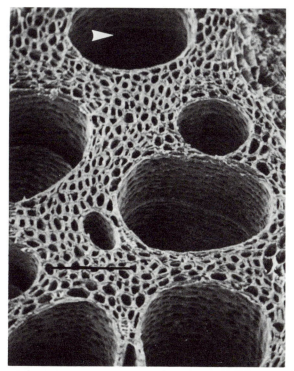

Figure 13.15. SEM photomicrograph of a cross section of a common grape vine, *Vitis vinifera*, stem. Bar = 100 μm.

were rotated and tilted during the coating. Specimens were examined and photographed at an operating voltage of 30 kV and a beam current of 30 pA in a Cambridge Instrument scanning microscope.

Results. The pollen grains have retained their more or less spherical shape as shown in Figure 13.16. It is unlikely that the spines or the slightly furrowed surface bears any relation to the antigenic properties of these pollen grains. The large cracks on the surface (→) of the pollen grain are the pores through which the pollen tube would emerge prior to fertilization. The degree of opening of these pores is related to the humidity of the environment, being much more widely open when the pollen grain is on the sticky stigmatic surface (or the moist nasal membranes of a human being). The allergenic substances are within the pollen grain and in this plant species reach the outside via the pore.

These specimens together with many similar specimens such as seeds and shells are relatively easy to prepare because they have such a low water content. If one had taken fresh pollen grains it would have been necessary to take them through a fixation, dehydration, and critical point drying

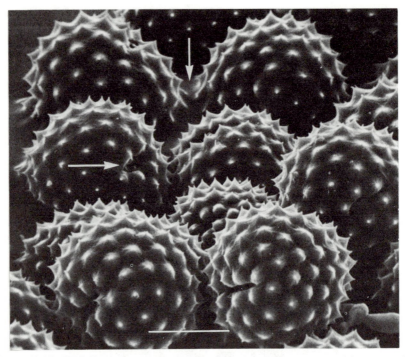

Figure 13.16. SEM photomicrograph of pollen grains of *Ambrosia artemesifolia* (ragweed). Bar = 10 μm.

procedure, to avoid the grains collapsing. One of the disadvantages of sprinkling dried samples on plastic glue is that the sample may sink into the glue and obscure the features one wishes to examine. An arrow (↓) points to a place where this has occurred. The coating procedure is adequate but there is a small amount of charging at the top of each of the spines. The charging could have been avoided if the specimens had been photographed at a lower accelerating voltage.

13.12. Examination of Medium Soft Animal Tissue with a High Water Content

Problem. To examine the early stages of development and differentiation of the tongue tissue.

Approach. A whole sheep foetus was slowly prefused with a fixative containing 1% glutaraldehyde and 1.5% formaldehyde in 0.1 M sodium cacodylate buffer with added sucrose to give a final osmolality of 880 m/os. Small pieces of tongue tissue were excised and immersed in the same fixative for 4 h at room temperature. Specimens were washed overnight at 275 K in 0.1 M sodium cacodylate buffer adjusted to 500 m/os with sucrose. The tissue blocks were trimmed to size with a clean razor blade and dehydrated through a graded acetone series and transferred to anhydrous acetone. Specimens were critical point dried from liquid CO_2 and the tissue blocks attached to the stub using colloidal silver. After overnight storage in a vacuum desiccator the samples were blown clean with dry air to remove any loose pieces. The specimens were sputter coated at 293 K with 40 nm of gold 1.4 keV, a plasma current of 40 mA, and an argon pressure of 7.0 Pa. The samples were examined and photographed at an operating voltage of 15 kV in a Cambridge Stereoscan 600 microscope.

Results. Figure 13.17 shows the surface of a sheep tongue. Although this specimen is from a relatively large piece of tissue, sufficient attention has been paid to the fixative procedure to retain the surface features of the individual surface papillae. With the exception of two pieces of debris (→), the surface is clean. In one or two places (*) there is evidence of surface erosion, which poses the question to the investigator as to whether this is a genuine pathological feature of the tissue or an artifact generated by the preparative procedure. It is clear that the surface of the papilla is composed

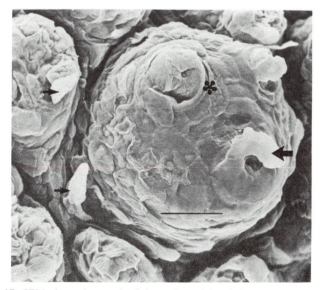

Figure 13.17. SEM photomicrograph of the surface of foetal sheep tongue showing filiform papillae. Bar = 40 μm. Micrograph courtesy of Ken Thurley, Department of Anatomy, University of Cambridge.

of a series of loosely arranged overlapping cells, some of which are becoming detached (←). Because this is a low-magnification image, the coating layer is deliberately made somewhat thicker than usual. There is no evidence of any beam penetration although the two pieces of debris show some charging. At this low magnification the granularity seen in sputtered gold films is not a problem. If the sample had been postfixed in osmium, with or without an additional thiocarbozide treatment, there would have been sufficient increase in the bulk conductivity of the sample to have obtained the same information from uncoated material. A detailed interpretation of this micrograph is only possible from an examination of stereo pairs. The papillae project a considerable way from the surface of the tongue and it is impossible from a single micrograph to obtain any real indication of their structural interrelationships. This case study is a good example of the necessity of adequately preserving the internal parts of a piece of tissue, even if the investigation is to be confined only to a natural surface.

13.13. Study of Single-Celled Animal Tissue of High Water Content

Problem. To study the number and arrangement of the microvilli and the method of surface attachment of individual tissue culture cells.

Approach. Clean glass cover-slips were coated with polylysine by brief immersion into a 0.01% polylysine hydrobromide solution in distilled water. The coated cover-slips were thoroughly rinsed in distilled water and air dried. The polylysine provides an extremely thin but firm attachment site for cells and microorganisms. HeLa cells in suspension were added in a drop of medium to the coated cover-slips. Most cells settled and adhered to the glass within 5 min. The cover-glass and attached cells were incubated at 310 K for 3 h in growth medium. Cells attached to the cover-slips were fixed for 30 min in 2.5% glutaraldehyde in Hank's basal salt solution buffered with 20 mM HEPES to pH 7.2 at 310 K and allowed to equilibrate to room temperature. After fixation, cells were washed in Hank's buffer and kept for 20–25 min at 277 K. Cells were postfixed for 30 min at room temperature in 1% osmium tetroxide in Hank's salt solution, dehydrated in graded ethanol, transferred to amyl acetate, and critical point dried from liquid CO_2. The cover-glass and dried cells were attached to the specimen stub with colloidal silver and sputter coated at 293 K with 50 nm of gold–palladium at 1.4 kV, a plasma current of 30 mA, and an argon pressure of 8 Pa. Specimens were examined in a Cambridge scanning microscope.

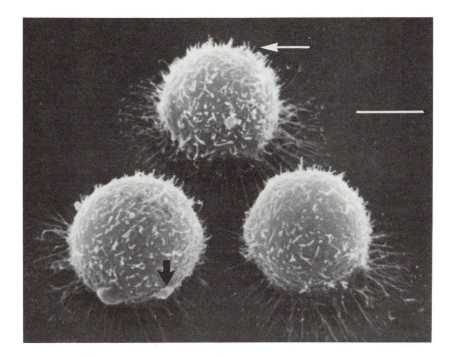

Figure 13.18. SEM photomicrograph of HeLa tissue culture cells showing microvilli and attachment fibrils. Bar = 5.0 μm. Micrograph courtesy of Ann Mullinger, Department of Zoology, University of Cambridge.

Results. Figure 13.18 shows the microvilli and attachment fibrils of the HeLa tissue culture cells. The fixation procedure has resulted in excellent preservation of these extremely delicate cellular structures, and demonstrates the importance of maintaining the correct osmotic balance of the fixative vehicle. The cells retain their round shape and the microvilli stand free of the cell surface. The polylysine-coated cover-glass provides smooth featureless support for the cells and enables one to clearly see the mode of attachment of the finely elongated fibrils. The coating layer is quite thick, but this is necessary in situations where the cells have a relatively small proportion of their surface area in contact with the underlying substrate. This micrograph (Figure 13.18) is one of a stereo pair, and an examination of the stereo pairs allows the investigator to count the number of microvilli on the cell surface and express the density of these features as a numerical value. The attachment can be seen radiating out from the underside of each cell and the actual point of attachment to the substrate is characterized by a small swelling of the fibril. Some of the attachment fibrils cross each other, and a number appear branched. One of the cells has a number of small blebs at the under surface. These blebs are unlikely to be a fixation artifact because they are devoid of microvilli. The bright region on the upper cell is not charging but a consequence of enhanced signal collection due to the position of this part of the specimen relative to the electron beam and secondary-electron collector.

14

Data Base

In order to calculate basic range equations (Chapter 3) and various quantitative x-ray correction schemes (Chapters 7 and 8), the analyst must have at hand the necessary input parameters. This chapter contains various tables (listed below) which include such data as atomic number, atomic weight, density, common oxides (atomic and weight fraction), mass absorption coefficients, and characteristic x-ray line energies. Also included are the J, ω, and R factors for quantitative ZAF corrections. To aid in the selection of coating materials, a compilation of important properties of selected elements is included.

Table 14.1. Atomic Number, Atomic Weight, and Density of Metals

Element	Atomic number	Atomic weight	Density of selected Metals (g/cm^3)
H	1	1.008	—
He	2	4.003	—
Li	3	6.941	0.534
Be	4	9.012	1.848
B	5	10.81	2.5
C	6	12.01	2.34 (amorphous C)
			2.25 (graphite)
N	7	14.01	—
O	8	16.00	—
F	9	19.00	—
Ne	10	20.18	—
Na	11	22.99	0.97
Mg	12	24.31	1.74
Al	13	26.98	2.7
Si	14	28.09	2.34
P	15	30.97	2.20 (red)
S	16	32.06	—
Cl	17	35.45	—
Ar	18	39.95	—
K	19	39.10	0.86
Ca	20	40.08	1.54
Sc	21	44.96	—
Ti	22	47.90	4.5
V	23	50.94	6.1
Cr	24	52.00	7.1
Mn	25	54.94	7.4
Fe	26	55.85	7.87
Co	27	58.93	8.9
Ni	28	58.71	8.9
Cu	29	63.55	8.96
Zn	30	65.37	7.14
Ga	31	69.72	5.91
Ge	32	72.59	5.32
As	33	74.92	—
Se	34	78.96	—
Br	35	79.90	—
Kr	36	83.80	—
Rb	37	85.47	1.53
Sr	38	87.62	2.6
Y	39	88.91	—
Zr	40	91.22	6.49
Nb	41	92.91	8.6
Mo	42	95.94	10.2
Tc	43	98.91	—
Ru	44	101.1	12.2
Rh	45	102.9	12.4
Pd	46	106.4	12.0

Table 14.1. *cont.*

Element	Atomic number	Atomic weight	Density of selected Metals (g/cm^3)
Ag	47	107.9	10.5
Cd	48	112.4	8.64
In	49	114.8	7.3
Sn	50	118.7	7.3
Sb	51	121.8	6.68
Te	52	127.6	6.24
I	53	126.9	—
Xe	54	131.3	—
Cs	55	132.9	1.87
Ba	56	137.3	3.5
La	57	138.9	—
Ce	58	140.1	6.75
Pr	59	140.9	—
Nd	60	144.2	—
Pm	61	(145)	—
Sm	62	150.4	—
Eu	63	152.0	—
Gd	64	157.3	—
Tb	65	158.9	—
Dy	66	160.5	—
Ho	67	164.9	—
Er	68	167.3	—
Tm	69	168.9	—
Yb	70	173.0	—
Lu	71	175.0	—
Hf	72	178.5	13.1
Ta	73	180.9	16.6
W	74	183.9	19.3
Re	75	186.2	21.0
Os	76	190.2	22.5
Ir	77	192.2	22.4
Pt	78	195.1	21.45
Au	79	197.0	19.3
Hg	80	200.6	13.55
Ti	81	204.4	11.85
Pb	82	207.2	11.34
Bi	83	209.0	9.8
Po	84	(210)	—
At	85	(210)	—
Rn	86	(222)	—
Fr	87	(223)	—
Ra	88	226.0	—
Ac	89	(227)	—
Th	90	232.0	11.5
Pa	91	231.0	—
U	92	238.0	19.05 (α), 18.9 (β)
Np	93	237.0	—

Table 14.1. *cont*.

Element	Atomic number	Atomic weight	Density of selected Metals (g/cm^3)
Pu	94	(244)	—
Am	95	(243)	—
Cm	96	(247)	—
Bk	97	(247)	—
Cf	98	(251)	—
Es	99	(254)	—
Fm	100	(257)	—
Md	101	(257)	—
No	102	(254)	—
Lr	103	(257)	—

Table 14.2. Common Oxides of the Elements

Atomic No.	Oxide	Wt. fraction of element	Wt. fraction of oxygen	Atomic No.	Oxide	Wt. fraction of element	Wt. fraction of oxygen
Na 11	Na_2O	0.7419	0.2581	Cd 48	CdO	0.8754	0.1246
Mg 12	MgO	0.60317	0.3968	In 49	In_2O_3	0.8271	0.1729
Al 13	Al_2O_3	0.5292	0.4708	Sn	SnO	0.8812	0.1188
Si 14	SiO_2	0.4675	0.5326		SnO_2	0.7876	0.2124
	SiO	0.6371	0.3629	Sb 51	Sb_2O_4	0.7918	0.2081
P 15	P_2O_5	0.4363	0.5636		Sb_2O_3	0.8353	0.1647
K 19	K_2O	0.8301	0.1699	Te 52	TeO_2	0.7995	0.2005
Ca 20	CaO	0.7147	0.2853	Ba 56	BaO	0.8110	0.1889
Sc 21	Sc_2O_3	0.6519	0.3481	La 57	La_2O_3	0.8973	0.1027
Ti 22	TiO_2	0.5995	0.4005	Ce 58	CeO_2	0.8141	0.1859
	TiO	0.7496	0.2504	Pr 59	PrO_2	0.8149	0.1851
	Ti_2O_3	0.6519	0.3481	Nd 60	Nd_2O_3	0.8573	0.1427
V 23	V_2O_3	0.6797	0.3203	Sm 62	Sm_2O_3	0.8623	0.1377
	V_2O_5	0.5601	0.4399	Eu 63	Eu_2O_3	0.8636	0.1364
Cr 24	Cr_2O_3	0.6842	0.3158	Gd 64	Gd_2O_3	0.8676	0.1364
	CrO_3	0.5199	0.4800	Tb 65	Tb_4O_7	0.8688	0.1312
Mn 25	Mn_3O_4	0.7203	0.2797	Dy 66	Dy_2O_3	0.8598	0.1402
	MnO_2	0.6319	0.3681	Ho 67	Ho_2O_3	0.8730	0.1270
Fe 26	Fe_2O_3	0.6994	0.3006	Er 68	Er_2O_3	0.8745	0.1255
	FeO	0.7773	0.2227	Tm 69	Tm_2O_3	0.8756	0.1244
Co 27	CoO	0.7865	0.2135	Yb 70	Yb_2O_3	0.8782	0.1218
Ni 28	NiO	0.7858	0.2142	Lu 71	Lu_2O_3	0.8794	0.1206
Cu 29	CuO	0.7988	0.2012	Hf 72	HfO_2	0.8479	0.1520
	Cu_2O	0.8882	0.1118	Ta 73	Ta_2O_5	0.8829	0.1171
Zn 30	ZnO	0.8034	0.1966	W 74	WO_3	0.7930	0.2070
Ga 31	Ga_2O_3	0.7439	0.2561	Re 75	Re_2O_7	0.7688	0.2312
Ge 32	GeO_2	0.8194	0.1806		ReO_3	0.7950	0.2050
As 33	As_2O_5	0.6519	0.3481	Os 76	OsO_4	0.7482	0.2518
	As_2O_3	0.7574	0.2426	Ir 77	IrO_2	0.8573	0.1427
Se 34	SeO_2	0.7116	0.2884		Ir_2O_3	0.8886	0.1114
Rb 37	Rb_2O	0.9144	0.0856	Pt 78	PtO_2	0.8591	0.1409
Sr 38	SrO	0.8456	0.1544	Hg 80	HgO	0.9261	0.0739
Zr 40	ZrO_2	0.7403	0.2597	Tl 81	Tl_2O_3	0.8949	0.1051
Nb 41	Nb_2O_5	0.3495	0.6505	Pb 82	PbO_2	0.8662	0.1338
Mo 42	MoO_3	0.6665	0.3335		PbO	0.9283	0.0717
Ru 44	RuO_4	0.6124	0.3876	Bi 83	Bi_2O_3	0.8970	0.1030
	RuO_2	0.7595	0.2405	U 92	UO_2	0.8815	0.1185
Rh 45	Rh_2O_3	0.8109	0.1819		U_3O_8	0.8480	0.1520
Pd 46	PdO	0.8693	0.1307				
Ag 47	Ag_2O	0.9309	0.0690				
	AgO	0.8708	0.1292				

Table 14.3. Mass Absorption Coefficients for $K\alpha$ Lines (μ/ρ in Units of 10^3 cm^2/g)^a

Absorber \ Emitter: λ (Å):	B 67.6	C 44.7	N 31.6	O 23.62	F 18.32	Ne 14.61
3 Li	31.59	10.31	3.812	1.602	0.735	0.366
4 Be	60.56	22.00	8.875	3.922	1.878	0.963
5 B	3.353	37.02	15.81	7.416	3.677	1.947
6 C	6.456	2.373	25.49	12.38	6.366	3.457
7 N	10.57	3.903	1.637	17.31	9.161	5.069
8 O	16.53	6.044	2.527	*1.200	12.39	6.961
9 F	23.88	8.730	3.602	1.688	*0.867	8.647
10 Ne	35.76	13.57	5.617	2.582	1.301	*0.715
11 Na	46.01	18.46	7.796	3.645	1.823	0.983
12 Mg	58.17	24.90	10.95	5.174	2.615	1.411
13 Al	65.17	30.16	13.83	6.715	3.407	1.848
14 Si	74.18	36.98	17.69	8.790	4.543	2.484
15 P	66.17	41.28	20.57	10.54	5.526	3.041
16 S	74.18	47.94	24.93	13.01	7.006	3.888
17 Cl	7.576	*50.76	27.61	14.79	8.047	4.522
18 Ar	9.369	45.58	29.51	16.16	8.924	5.070
19 K	11.28	6.315	35.90	19.37	10.62	6.091
20 Ca	13.01	6.838	35.59	22.03	12.37	7.163
21 Sc	14.19	7.461	3.988	23.59	13.30	7.754
22 Ti	15.28	8.094	4.364	22.14	14.54	8.557
23 V	16.71	8.840	4.790	24.25	15.75	9.370
24 Cr	20.67	10.59	5.630	3.143	15.91	10.84
25 Mn	22.16	11.47	6.160	3.468	17.02	11.56
26 Fe	25.78	13.30	7.121	4.001	2.328	13.21
27 Co	28.34	14.73	7.888	4.407	2.568	12.34
28 Ni	33.09	17.27	9.235	5.245	2.989	1.801
29 Cu	36.06	18.69	9.997	5.920	3.159	1.968
30 Zn	40.14	21.71	11.70	6.548	3.717	2.237
31 Ga	42.03	23.33	12.71	7.086	4.072	2.450
32 Ge	45.05	25.52	14.04	7.867	4.539	2.737
33 As	44.45	27.90	15.63	8.805	5.081	3.067
34 Se	45.60	30.21	17.00	9.633	5.585	3.373
35 Br	42.70	32.55	18.76	10.71	6.259	3.784
36 Kr	40.20	33.28	20.69	11.89	6.960	4.212
37 Rb	35.88	35.52	21.96	12.82	7.551	4.610
38 Sr	*27.34	35.31	23.31	13.76	8.210	5.025
39 Y	*30.00	31.38	*23.84	15.14	9.049	5.565
40 Zr	*38.41	31.13	25.03	16.14	9.743	6.035
41 Nb	4.417	*33.99	27.19	17.85	10.77	6.676
42 Mo	4.717	*32.42	23.22	18.66	11.36	7.087
43 Tc	5.053	*28.96	23.31	18.12	11.85	7.448
44 Ru	5.583	4.240	25.18	19.68	12.96	8.130
45 Rh	5.977	4.634	*26.49	19.67	13.92	8.779
46 Pd	6.204	5.250	*26.15	18.33	14.80	9.400
47 Ag	6.462	5.507	*3.900	19.26	14.73	10.01
48 Cd	6.567	5.793	4.152	20.19	14.87	10.44
49 In	6.732	6.135	4.476	22.22	*15.97	11.07

Table 14.3. *cont.*

Absorber \ λ (Å):	Emitter: B 67.6	C 44.7	N 31.6	O 23.62	F 18.32	Ne 14.61
50 Sn	6.771	6.332	4.760	*23.09	13.76	10.85
51 Sb	6.848	6.601	5.015	3.418	14.68	11.30
52 Te	6.756	6.666	5.148	3.566	16.40	11.43
53 I	*5.880	7.094	5.613	3.906	*17.24	10.17
54 Xe	5.895	7.403	5.990	4.200	*2.848	10.79
55 Ca	5.996	7.650	6.238	4.443	3.030	11.50
56 Ba	3.677	7.772	6.306	4.560	3.140	11.60
57 La	3.826	7.894	6.410	4.690	3.277	*12.34
58 Ce	6.811	8.102	7.254	5.207	3.584	2.455
59 Pr	8.704	8.909	7.717	5.503	3.783	2.592
60 Nd	10.39	9.543	8.026	5.691	3.909	2.679
61 Pm	12.43	10.53	8.532	5.964	4.081	2.796
62 Sm	14.56	11.49	9.101	6.314	4.284	2.934
63 Eu	15.62	12.19	9.583	6.627	4.495	3.080
64 Gd	14.89	*11.97	9.553	6.641	4.546	3.120
65 Tb	18.74	12.41	10.07	7.444	4.996	3.393
66 Dy	20.54	13.59	11.39	7.877	5.284	3.572
67 Ho	21.84	14.53	11.46	8.340	5.593	3.758
68 Er	23.60	15.72	12.28	8.915	5.950	3.999
69 Tm	24.79	16.93	13.13	9.561	6.383	4.269
70 Yb	23.18	17.71	13.77	10.00	6.682	4.482
71 Lu	21.89	17.41	13.88	10.27	6.897	4.624
72 Hf	21.64	18.03	12.91	10.13	7.227	4.859
73 Ta	20.82	18.39	13.42	10.56	7.588	5.128
74 W	19.66	18.75	13.88	10.99	7.928	5.358
75 Re	17.98	18.71	14.12	11.29	8.227	5.568
76 Os	16.06	17.06	14.13	11.33	8.439	5.757
77 Ir	14.00	16.44	14.68	11.70	8.433	6.109
78 Pt	11.27	16.05	15.21	11.34	8.893	6.452
79 Au	8.952	15.21	15.44	11.76	9.289	6.762
80 Hg	6.180	14.43	15.92	12.44	9.732	7.184
81 Tl	*4.885	13.03	14.26	12.21	9.690	7.304
82 Pb	*4.265	11.35	14.02	12.47	9.039	7.122
83 Bi	*3.136	9.533	13.81	12.69	9.285	7.380
84 Po	2.580	8.243	13.68	13.45	9.749	7.738
85 At	2.238	6.583	13.31	12.53	10.29	8.187
86 Rn	2.429	4.173	12.26	11.76	10.24	8.316
87 Fr	2.617	*3.058	11.11	11.89	10.63	7.582
88 Ra	2.667	2.444	9.201	11.87	10.56	7.779
89 Ac	*1.588	2.630	7.989	11.68	10.32	7.910
90 Th	1.879	2.120	6.293	11.30	8.942	7.886
91 Pa	2.104	2.391	3.461	11.51	9.533	8.399
92 U	2.247	2.317	*2.417	11.06	9.492	8.420
93 Np	2.478	2.431	2.654	*9.825	9.836	8.658
94 Pu	2.936	2.416	2.770	*6.381	10.27	*8.398

[a]Henke and Ebisu (1974); values marked with asterisks are uncertain.

Table 14.4. Mass Absorption Coefficients for $L\alpha$ Lines (μ/ρ in Units of 10^3 cm^2/g)^a

Absorber \ Emitter: λ (Å):	Ti 27.42	Cr 21.64	Mn 19.45	Fe 17.59	Co 15.97	Ni 14.56	Cu 13.34	Zn 12.25
3 Li	2.509	1.233	0.892	0.655	0.487	0.366	0.279	0.214
4 Be	5.967	3.064	2.257	1.684	1.268	0.963	0.745	0.578
5 B	11.00	5.868	4.393	3.314	2.531	1.947	1.510	1.188
6 C	17.96	9.902	7.495	5.786	4.451	3.457	2.714	2.147
7 N	24.83	14.00	10.74	8.344	6.496	5.069	4.022	3.205
8 O	1.756	18.74	14.50	11.30	8.877	6.961	5.601	4.476
9 F	2.490	1.344	1.021	13.86	10.91	8.647	6.941	5.620
10 Ne	3.845	2.044	1.541	1.178	0.917	*0.715	9.190	7.463
11 Na	5.429	2.884	2.166	1.647	1.269	0.983	0.775	*0.608
12 Mg	7.643	4.121	3.105	2.364	1.818	1.411	1.110	0.880
13 Al	9.720	5.368	4.040	3.081	2.379	1.848	1.453	1.152
14 Si	12.70	7.034	5.379	4.112	3.188	2.484	1.959	1.558
15 P	14.97	8.468	6.527	5.007	3.902	3.041	2.405	1.917
16 S	18.33	10.60	8.198	6.377	4.950	3.888	3.079	2.459
17 Cl	20.57	12.10	9.367	7.366	5.750	4.522	3.596	2.876
18 Ar	22.27	13.27	10.37	8.155	6.453	5.070	4.049	3.247
19 K	26.88	15.83	12.35	9.710	7.682	6.091	4.887	4.109
20 Ca	30.34	18.16	14.31	11.35	9.007	7.163	5.771	4.685
21 Sc	28.42	19.50	15.38	12.20	9.724	7.754	6.250	5.068
22 Ti	3.318	21.19	16.74	13.37	10.67	8.557	6.930	5.627
23 V	3.646	19.92	18.16	14.47	11.63	9.370	7.617	6.206
24 Cr	4.256	2.612	18.67	16.73	13.43	10.84	8.836	7.215
25 Mn	4.673	2.897	2.317	15.61	14.44	11.56	9.473	7.767
26 Fe	5.400	3.332	2.678	2.149	14.14	13.21	10.69	8.775
27 Co	5.961	3.672	2.939	2.371	1.924	12.34	11.66	9.436
28 Ni	7.036	4.383	3.468	2.745	2.268	1.801	11.51	10.90
29 Cu	7.825	4.880	3.752	2.883	2.537	1.968	1.586	10.37
30 Zn	8.882	5.399	4.277	3.424	2.788	2.237	1.816	1.520
31 Ga	9.617	5.861	4.667	3.756	3.043	2.450	1.989	1.629
32 Ge	10.64	6.545	5.216	4.182	3.385	2.737	2.232	1.828
33 As	11.88	7.325	5.838	4.681	3.788	3.067	2.506	2.057
34 Se	12.96	8.036	6.415	5.146	4.161	3.373	2.759	2.261
35 Br	14.36	8.955	7.173	5.774	4.671	3.784	3.101	2.556
36 Kr	15.88	9.973	7.989	6.415	5.191	4.212	3.452	2.837
37 Rb	17.02	10.73	8.627	6.978	5.672	4.610	3.776	3.105
38 Sr	18.14	11.59	9.361	7.594	6.170	5.025	4.130	3.403
39 Y	19.84	12.77	10.32	8.371	6.821	5.565	4.579	3.778
40 Zr	21.21	13.67	11.09	9.024	7.390	6.035	4.968	4.108
41 Nb	21.87	15.14	12.28	9.968	8.169	6.676	5.499	4.548
42 Mo	22.15	15.84	12.90	10.54	8.655	7.087	5.848	4.847
43 Tc	22.04	16.48	13.42	11.01	9.047	7.448	6.173	5.113
44 Ru	20.63	16.86	14.67	12.04	9.897	8.130	6.739	5.606
45 Rh	21.55	17.60	15.73	12.94	10.66	8.779	7.277	6.041
46 Pd	22.34	18.64	15.81	13.83	11.41	9.400	7.810	6.515
47 Ag	*25.18	16.46	16.49	13.76	12.10	10.01	8.328	6.928
48 Cd	*26.58	17.19	15.72	14.29	12.71	10.44	8.722	7.282
49 In	*30.61	18.42	14.85	14.89	12.46	11.07	9.199	7.703

Table 14.4. *cont.*

Absorber \ Emitter: λ (Å):	Ti 27.42	Cr 21.64	Mn 19.45	Fe 17.59	Co 15.97	Ni 14.56	Cu 13.34	Zn 12.25
50 Sn	3.942	*19.36	15.64	12.75	12.71	10.85	9.623	8.057
51 Sb	4.189	*21.79	17.03	13.47	11.03	11.30	9.509	8.447
52 Te	4.341	3.140	22.49	13.90	11.39	11.43	9.722	8.659
53 I	4.747	3.441	*20.35	15.63	12.42	10.17	10.71	8.879
54 Xe	5.088	3.704	*3.152	22.76	13.37	10.79	8.921	9.310
55 Cs	5.344	3.926	3.349	2.854	15.41	11.50	9.422	7.838
56 Ba	5.442	4.045	3.464	2.962	2.529	11.60	9.744	8.103
57 La	5.559	4.184	3.604	3.095	2.649	*12.34	10.16	8.442
58 Ce	6.233	4.618	3.954	3.379	2.882	2.455	*11.49	9.385
59 Pr	6.601	4.877	4.174	3.567	3.044	2.592	2.219	10.05
60 Nd	6.842	5.039	4.312	3.687	3.144	2.679	2.294	8.683
61 Pm	7.217	5.265	4.500	3.849	3.283	2.796	2.392	2.048
62 Sm	7.673	5.552	4.730	4.038	3.446	2.934	2.508	2.145
63 Eu	8.061	5.835	4.968	4.235	3.617	3.080	2.633	2.252
64 Gd	8.042	5.875	5.020	4.286	3.660	3.120	2.673	2.291
65 Tb	9.099	6.539	5.542	4.698	3.996	3.393	2.896	2.474
66 Dy	9.700	6.920	5.863	4.967	4.220	3.572	3.040	2.592
67 Ho	10.30	7.308	6.194	5.260	4.442	3.758	3.201	2.728
68 Er	11.02	7.827	6.617	5.587	4.726	3.999	3.404	2.898
69 Tm	11.02	8.380	7.088	5.997	5.064	4.269	3.626	3.092
70 Yb	11.55	8.781	7.427	6.276	5.314	4.482	3.799	3.225
71 Lu	11.72	9.020	7.648	6.484	5.481	4.624	3.929	3.348
72 Hf	12.15	9.456	8.005	6.797	5.738	4.859	4.142	3.527
73 Ta	12.61	9.927	8.397	7.142	6.048	5.128	4.368	3.712
74 W	13.02	9.737	8.778	7.461	6.335	5.358	4.556	3.883
75 Re	12.14	10.01	9.156	7.724	6.565	5.568	4.751	4.059
76 Os	12.17	10.16	8.786	7.915	6.767	5.757	4.920	4.194
77 Ir	12.74	10.65	9.270	8.400	7.161	6.109	5.231	4.463
78 Pt	13.32	11.15	9.772	8.403	7.592	6.452	5.511	4.705
79 Au	13.69	10.52	10.19	8.789	7.954	6.762	5.793	4.967
80 Hg	14.32	11.13	10.46	9.292	7.943	7.184	6.115	5.232
81 Tl	13.71	11.03	9.620	9.330	8.013	7.304	6.276	5.457
82 Pb	13.74	11.30	9.878	9.592	8.292	7.122	6.466	5.561
83 Bi	13.43	11.55	10.13	8.810	8.533	7.380	6.375	5.773
84 Po	13.01	12.14	10.62	9.259	8.958	7.738	6.683	6.061
85 At	13.13	12.24	11.18	9.781	8.445	8.187	7.072	6.098
86 Rn	12.47	11.79	11.12	9.748	8.454	8.316	7.059	6.091
87 Fr	12.16	10.98	11.55	10.11	8.787	7.582	7.350	6.361
88 Ra	11.38	11.04	10.82	10.33	8.999	7.779	6.728	6.567
89 Ac	10.53	10.97	9.886	10.42	9.119	7.910	6.870	6.674
90 Th	9.334	10.67	9.643	9.980	8.978	7.886	6.868	5.960
91 Pa	6.925	11.18	10.27	9.112	9.219	8.399	7.325	6.350
92 U	5.784	11.12	10.16	9.094	8.904	8.420	7.359	6.396
93 Np	*5.087	*10.67	10.49	9.445	8.282	8.658	7.703	6.725
94 Pu	*3.236	*9.278	10.79	9.931	8.708	*8.398	8.070	7.035

[a] Henke and Ebisu (1974); values marked with asterisks are uncertain.

Table 14.5a. Selected Mass Absorption Coefficients of the Chemical Elements for $K\alpha$ Lines (Heinrich, 1966)

i. Elements Na to Ar

Absorber \ λ (Å):	Emitter: Na 11.909	Mg 9.889	Al 8.337	Si 7.126	P 6.155	S 5.373	Cl 4.728	Ar 4.193
3 Li	169.4	99.2	60.7	38.6	25.3	17.1	11.8	8.4
4 Be	417.9	245.6	150.7	96.2	63.3	42.9	29.8	21.1
5 B	861.9	507.5	312.0	199.5	131.4	89.2	62.0	44.0
6 C	1534.0	904.8	557.2	356.8	235.4	160.0	111.3	79.1
7 N	2449.7	1447.7	893.0	572.7	378.4	257.6	179.4	127.7
8 O	4109.1	2432.8	1503.3	965.6	638.9	435.5	303.7	216.4
9 F	5169.0	3066.0	1897.7	1220.9	809.0	552.2	385.5	275.1
10 Ne	6966.9	4140.1	2566.9	1654.1	1097.6	750.2	524.4	374.7
11 Na	− 0	5409.1	3359.4	2168.1	1440.8	986.2	690.2	493.7
12 Mg	770.1	463.6	4376.5	2824.6	1877.0	1284.8	899.2	643.2
13 Al	1021.0	614.7	385.7	3493.2	2324.7	1593.4	1116.6	799.7
14 Si	1332.5	802.2	503.4	327.9	2840.1	1949.3	1367.8	980.8
15 P	1695.9	1021.0	640.6	417.4	279.8	2370.7	1663.6	1192.8
16 S	2102.5	1265.8	794.3	517.5	346.9	239.4	1965.5	1411.0
17 Cl	2578.4	1552.3	974.0	634.6	425.4	293.6	207.1	1656.6
18 Ar	3132.2	1885.7	1183.2	770.9	516.8	356.6	251.5	181.2
19 K	3729.2	2245.1	1408.8	917.8	615.3	424.6	299.5	215.8
20 Ca	4412.8	2656.7	1667.0	1086.0	728.1	502.4	354.4	255.3
21 Sc	5182.8	3120.3	1957.9	1275.5	855.1	590.1	416.2	299.9
22 Ti	6056.7	3646.4	2288.0	1490.6	999.3	689.6	486.4	350.4
23 V	6939.3	4177.7	2621.4	1707.8	1144.9	790.1	557.2	401.5
24 Cr	7943.0	4782.0	3000.5	1954.8	1310.5	904.3	637.8	459.6
25 Mn	9041.8	5443.6	3415.6	2225.3	1491.8	1029.5	726.1	523.1
26 Fe	10166.6	6120.7	3840.6	2502.1	1677.4	1157.5	816.4	588.2
27 Co	11464.5	6902.1	4330.8	2821.5	1891.5	1305.3	920.6	663.3
28 Ni	12805.6	7709.5	4837.5	3151.6	2112.8	1458.0	1028.3	740.9
29 Cu	12165.4	8569.0	5376.8	3503.0	2348.3	1620.5	1143.0	823.5
30 Zn	9690.8	9506.7	5965.1	3886.3	2605.3	1797.9	1268.0	913.6
31 Ga	− 0	8985.8	6602.5	4301.5	2883.7	1990.0	1403.5	1011.2
32 Ge	− 0	9871.3	7239.9	4716.7	3162.0	2182.1	1539.0	1108.8
33 As	2683.6	1655.2	6782.3	5163.9	3461.8	2388.9	1684.9	1214.0
34 Se	2940.7	1813.7	7370.6	5621.8	3768.7	2600.7	1834.3	1321.6
35 Br	3216.6	1983.9	1272.7	5238.5	4111.4	2837.2	2001.1	1441.7
36 Kr	3511.3	2165.6	1389.4	5685.7	4461.1	3078.5	2171.3	1564.4
37 Rb	3824.8	2359.0	1513.4	1006.3	4111.4	3339.6	2355.4	1697.0
38 Sr	4150.8	2560.1	1642.4	1092.1	4454.0	3595.7	2536.1	1827.2
39 Y	4501.9	2776.6	1781.4	1184.5	809.3	3300.2	2730.6	1967.4
40 Zr	4878.2	3008.7	1930.2	1283.4	876.9	3561.2	2939.1	2117.5
41 Nb	5260.6	3244.6	2081.6	1384.1	945.7	664.2	2702.8	2277.7
42 Mo	5674.5	3499.8	2245.3	1493.0	1020.1	716.5	2035.2	2393.4
43 Tc	6107.1	3766.6	2416.5	1606.8	1097.9	771.1	553.0	2196.0
44 Ru	6564.8	4049.0	2597.6	1727.2	1180.2	828.9	594.4	1682.8
45 Rh	7047.6	4346.7	2788.6	1854.2	1266.9	889.9	638.2	467.0
46 Pd	7543.0	4652.2	2984.6	1984.6	1356.0	952.4	683.0	499.8
47 Ag	8069.6	4977.1	3193.0	2123.1	1450.7	1018.9	730.7	534.7
48 Cd	8627.7	5321.3	3413.8	2270.0	1551.0	1089.4	781.6	571.7

Table 14.5a. *cont.*

i. Elements Na to Ar

Absorber \ Emitter: λ (Å):	Na 11.909	Mg 9.889	Al 8.337	Si 7.126	P 6.155	S 5.373	Cl 4.728	Ar 4.193
49 In	9204.5	5677.0	3642.1	2421.7	1654.7	1162.2	833.5	609.9
50 Sn	9812.7	6052.2	3882.8	2581.7	1764.0	1239.0	888.5	650.2
51 Sb	10446.0	6442.7	4133.3	2748.4	1877.9	1319.0	945.9	692.2
52 Te	11110.7	6852.7	4396.3	2923.2	1997.4	1402.9	1006.1	736.3
53 I	11800.4	7278.1	4669.2	3104.7	2121.4	1490.0	1068.5	782.0
54 Xe	12521.4	7722.8	4954.5	3294.4	2251.0	1581.0	1133.8	829.7
55 Cs	11411.6	8183.0	5249.8	3490.7	2385.1	1675.2	1201.4	879.2
56 Ba	10032.2	7463.7	5559.9	3696.9	2526.0	1774.2	1272.3	931.1
57 La	10596.5	7889.1	5880.0	3909.7	2671.4	1876.3	1345.6	984.7
58 Ce	9718.7	6922.3	6214.9	4132.4	2823.6	1983.2	1422.2	1040.8
59 Pr	10220.3	7309.0	5656.7	4366.7	2983.6	2095.6	1502.8	1099.8
60 Nd	0	6651.6	5979.2	4607.5	3148.2	2211.2	1585.7	1160.5
61 Pm	0	8121.1	6301.7	4856.7	3318.4	2330.8	1671.5	1223.2
62 Sm	0	7386.3	5507.8	5115.7	3495.4	2455.1	1760.6	1288.4
63 Eu	3387.6	7773.1	0	4635.6	3678.0	2583.3	1852.6	1355.8
64 Gd	3497.7	8198.5	5259.7	4883.0	3867.4	2716.3	1948.0	1425.6
65 Tb	3620.0	0	5532.6	4256.2	4064.6	2854.9	2047.3	1498.3
66 Dy	3742.3	0	5805.5	4454.1	3674.6	2997.4	2149.5	1573.1
67 Ho	3876.8	2566.1	6078.4	0	3866.2	3147.8	2257.4	1652.0
68 Er	3986.9	2638.9	6376.2	4239.7	4046.6	3300.6	2367.0	1732.2
69 Tm	4121.4	2728.0	0	4454.1	3516.8	2984.7	2480.5	1815.3
70 Yb	4268.1	2825.1	1933.9	4668.6	3685.9	3135.2	2599.8	1902.5
71 Lu	4402.7	2914.1	1994.9	4883.0	3336.5	3269.7	2720.7	1991.0
72 Hf	4537.2	3003.2	2055.8	5097.5	3483.0	2834.3	2846.2	2082.9
73 Ta	4659.5	3084.1	2111.2	0	3652.1	2968.9	2566.3	2178.8
74 W	4794.0	3173.2	2172.2	1533.1	3809.9	3095.6	2679.8	2276.5
75 Re	4916.3	3254.1	2227.6	1572.2	3978.9	2794.7	2322.1	2377.0
76 Os	5001.9	3310.8	2266.4	1599.6	0	2921.4	2418.6	2139.8
77 Ir	5222.1	3456.5	2366.1	1670.0	1206.4	3040.1	2526.5	2231.2
78 Pt	5368.8	3553.6	2432.6	1716.9	1240.3	3174.7	2276.7	1927.9
79 Au	5503.3	3642.7	2493.6	1759.9	1271.4	3309.3	2373.2	2006.8
80 Hg	5650.1	3739.8	2560.1	1806.9	1305.3	0	2475.4	2094.1
81 Tl	5821.3	3853.1	2637.7	1861.6	1344.8	994.6	2577.6	2185.5
82 Pb	5968.1	3950.3	2704.2	1908.6	1378.7	1019.7	2685.5	1965.3
83 Bi	6139.3	4063.6	2781.7	1963.3	1418.3	1049.0	0	2048.4
84 Po	6310.5	4176.9	2859.3	2018.1	1457.8	1078.2	811.7	2143.9
85 At	6506.2	4306.5	2948.0	2080.7	1503.0	1111.6	836.9	2239.5
86 Rn	6701.8	4436.0	3036.6	2143.2	1548.2	1145.1	862.1	0
87 Fr	6922.0	4581.7	3136.4	2213.6	1599.1	1182.7	890.4	682.0
88 Ra	0	4792.1	3280.5	2315.3	1672.5	1237.0	931.3	713.3
89 Ac	0	4986.4	3413.4	2409.2	1740.4	1287.2	969.0	742.3
90 Th	0	0	3468.9	2448.3	1768.6	1308.1	984.8	754.3
91 Pa	0	0	3590.8	2534.3	1830.8	1354.0	1019.4	780.8
92 U	0	0	3723.8	2628.2	1898.6	1404.2	1057.1	809.7
93 Np	0	0	0	2729.9	1972.0	1458.5	1098.0	841.1
94 Pu	0	0	0	2847.2	2056.8	1521.2	1145.2	877.2

Table 14.5a. *cont.*

ii. Elements K to Fe

Absorber \ Emitter: λ (Å):	K 3.742	Ca 3.359	Sc 3.032	Ti 2.748	V 2.504	Cr 2.291	Mn 2.103	Fe 1.937
3 Li	6.0	4.4	3.3	2.5	1.9	1.5	1.1	0.9
4 Be	15.2	11.2	8.4	6.3	4.8	3.7	2.9	2.3
5 B	31.8	23.4	17.5	13.2	10.1	7.9	6.2	4.9
6 C	57.3	42.1	31.5	23.8	18.3	14.2	11.1	8.8
7 N	92.5	68.2	51.0	38.6	29.7	23.1	18.1	14.4
8 O	157.0	115.8	86.7	65.7	50.6	39.4	30.9	24.5
9 F	199.8	147.5	110.6	83.9	64.6	50.3	39.6	31.4
10 Ne	272.4	201.4	151.2	114.8	88.5	69.0	54.3	43.1
11 Na	359.4	265.9	199.8	151.9	117.2	91.4	72.0	57.2
12 Mg	468.2	346.4	260.3	197.9	152.7	119.1	93.8	74.6
13 Al	582.8	431.7	324.7	247.0	190.8	149.0	117.4	93.4
14 Si	715.6	530.6	399.5	304.3	235.2	183.8	145.0	115.5
15 P	870.3	645.3	485.9	370.0	286.0	223.6	176.4	140.5
16 S	1030.7	765.1	576.7	439.6	340.1	266.1	210.1	167.4
17 Cl	1210.1	898.3	677.1	516.1	399.3	312.4	246.7	196.6
18 Ar	1390.2	1033.0	779.4	594.7	460.5	360.7	285.0	227.3
19 K	158.1	1189.8	897.7	685.0	530.4	415.4	328.2	261.8
20 Ca	187.1	139.4	1011.1	772.2	598.6	469.2	371.0	296.2
21 Sc	219.8	163.7	123.8	879.1	681.4	534.1	422.2	337.2
22 Ti	256.8	191.3	144.6	110.6	85.8	597.0	472.5	377.5
23 V	294.3	219.1	165.7	126.7	98.3	77.1	531.1	424.3
24 Cr	336.8	250.8	189.7	145.0	112.5	88.2	69.9	474.2
25 Mn	383.4	285.5	215.9	165.1	128.1	100.5	79.5	63.5
26 Fe	431.1	321.1	242.7	185.6	144.0	113.0	89.4	71.4
27 Co	486.2	362.1	273.7	209.3	162.4	127.4	100.8	80.6
28 Ni	543.0	404.4	305.8	233.8	181.4	142.3	112.6	90.0
29 Cu	603.6	449.5	339.8	259.8	201.6	158.1	125.2	100.0
30 Zn	669.6	498.7	377.0	288.3	223.6	175.4	138.9	110.9
31 Ga	741.2	552.0	417.3	319.1	247.5	194.2	153.7	122.8
32 Ge	812.7	605.2	457.6	349.9	271.4	212.9	168.5	134.7
33 As	889.8	662.6	501.0	383.0	297.2	233.1	184.5	147.4
34 Se	968.7	721.4	545.4	417.0	323.5	253.8	200.9	160.5
35 Br	1056.7	786.9	595.0	454.9	352.9	276.9	219.2	175.1
36 Kr	1146.6	853.9	645.6	493.6	382.9	300.4	237.8	190.0
37 Rb	1243.9	926.3	700.4	535.4	415.4	325.9	258.0	206.1
38 Sr	1339.3	997.3	754.1	576.5	447.3	350.9	277.7	221.9
39 Y	1442.0	1073.9	811.9	620.7	481.6	377.8	299.0	238.9
40 Zr	1552.1	1155.8	873.9	668.1	518.3	406.6	321.9	257.2
41 Nb	1669.5	1243.3	940.0	718.7	557.5	437.4	346.2	276.6
42 Mo	1756.2	1309.3	990.9	758.3	588.9	462.4	366.3	292.9
43 Tc	1886.6	1406.5	1064.5	814.6	632.6	496.7	393.5	314.7
44 Ru	2013.3	1500.9	1136.0	869.4	675.1	530.1	420.0	335.8
45 Rh	1835.9	1600.8	1211.6	927.2	720.0	565.4	447.9	358.1
46 Pd	1401.4	1703.4	1289.2	986.6	766.2	601.6	476.6	381.1
47 Ag	397.8	1533.5	1359.9	1041.7	809.7	636.3	504.6	403.8
48 Cd	425.3	1162.8	1434.6	1099.0	854.2	671.3	532.3	426.0

Table 14.5a. *cont.*

ii. Elements K to Fe

Emitter: Absorber λ (Å):	K 3.742	Ca 3.359	Sc 3.032	Ti 2.748	V 2.504	Cr 2.291	Mn 2.103	Fe 1.937
49 In	453.7	342.7	1303.3	1168.6	908.3	713.9	566.0	453.0
50 Sn	483.7	365.3	986.1	1232.1	957.7	752.6	596.8	477.6
51 Sb	514.9	388.9	298.0	1102.5	1003.7	789.6	626.6	501.8
52 Te	547.7	413.6	316.9	832.0	0	829.9	658.6	527.5
53 I	581.7	439.3	336.6	260.7	957.9	881.4	699.5	560.2
54 Xe	617.2	466.1	357.2	276.6	712.3	781.2	726.1	582.0
55 Cs	654.0	493.9	378.5	293.1	230.1	822.1	763.8	612.2
56 Ba	692.6	523.1	400.8	310.4	243.7	622.1	688.1	645.4
57 La	732.5	553.2	423.9	328.2	257.7	647.4	0	672.9
58 Ce	774.2	584.7	448.0	346.9	272.4	216.2	539.6	602.9
59 Pr	818.1	617.9	473.4	366.6	287.9	228.5	182.9	452.9
60 Nd	863.3	652.0	499.5	386.8	303.8	241.1	193.0	473.3
61 Pm	909.9	687.2	526.5	407.7	320.2	254.1	203.4	164.2
62 Sm	958.5	723.8	554.6	429.5	337.2	267.6	214.2	173.0
63 Eu	1008.5	761.7	583.6	451.9	354.9	281.6	225.4	182.0
64 Gd	1060.5	800.9	613.6	475.2	373.1	296.1	237.0	191.4
65 Tb	1114.5	841.7	644.9	499.4	392.2	311.2	249.1	201.2
66 Dy	1170.2	883.7	677.1	524.4	411.7	326.8	261.6	211.2
67 Ho	1228.9	928.1	711.1	550.7	432.4	343.2	274.7	221.8
68 Er	1288.6	973.1	745.6	577.4	453.4	359.8	288.0	232.6
69 Tm	1350.4	1019.8	781.4	605.1	475.2	377.1	301.8	243.7
70 Yb	1415.3	1068.8	819.0	634.2	498.0	395.2	316.3	255.5
71 Lu	1481.1	1118.6	857.1	663.7	521.2	413.6	331.1	267.3
72 Hf	1549.4	1170.2	896.6	694.3	545.2	432.7	346.3	279.7
73 Ta	1620.8	1224.1	937.9	726.3	570.3	452.6	362.3	292.5
74 W	1693.4	1278.9	979.9	758.8	595.9	472.9	378.5	305.7
75 Re	1768.2	1335.4	1023.2	792.3	622.2	493.8	395.2	319.2
76 Os	1845.2	1393.5	1067.8	826.8	649.3	515.3	412.4	333.1
77 Ir	1924.6	1453.5	1113.7	862.4	677.2	537.5	430.2	347.4
78 Pt	2006.5	1515.4	1161.1	899.1	706.0	560.3	448.5	362.2
79 Au	1801.9	1578.9	1209.8	936.8	735.6	583.8	467.3	377.3
80 Hg	1879.2	1646.8	1261.8	977.1	767.3	608.9	487.4	393.6
81 Tl	1625.8	1482.2	1316.4	1019.3	800.4	635.2	508.5	410.6
82 Pb	1690.7	1542.9	1370.7	1061.4	833.5	661.5	529.5	427.6
83 Bi	1764.8	1332.9	1429.7	1107.2	869.4	690.0	552.3	446.0
84 Po	1845.2	1393.5	1287.7	1156.6	908.2	720.8	576.9	465.9
85 At	1665.9	1454.2	1345.0	1208.7	949.1	753.2	602.9	486.9
86 Rn	1740.1	1521.9	1166.1	1263.2	992.0	787.2	630.1	508.8
87 Fr	1820.5	1589.6	1218.0	1138.5	1037.0	823.0	658.7	531.9
88 Ra	1900.8	1435.6	1273.4	1191.1	1084.3	860.5	688.8	556.2
89 Ac	0	1503.3	1332.5	1031.8	977.7	900.2	720.6	581.9
90 Th	0	1575.6	1396.8	1081.7	1025.6	944.2	755.8	610.3
91 Pa	606.5	1650.3	1264.5	1132.9	1073.4	988.3	791.0	638.8
92 U	629.0	0	1332.5	1193.9	937.5	897.6	833.9	673.4
93 Np	653.3	0	1404.0	1257.6	0	945.1	878.1	709.1
94 Pu	681.4	536.2	1495.2	0	1051.7	834.6	804.8	755.4

Table 14.5a. *cont.*

iii. Elements Co to Zr

Emitter: Absorber λ (Å):	Co 1.790	Ni 1.659	Cu 1.542	Zn 1.436	Ga 1.341	Ge 1.255	As 1.177	Zr 0.787
3 Li	0.7	0.6	0.5	0.4	0.3	0.3	0.2	0.1
4 Be	1.9	1.5	1.2	1.0	0.8	0.7	0.6	0.2
5 B	3.9	3.1	2.5	2.1	1.7	1.4	1.2	0.4
6 C	7.1	5.7	4.6	3.8	3.1	2.6	2.1	0.7
7 N	11.5	9.3	7.5	6.2	5.1	4.2	3.5	1.1
8 O	19.6	15.8	12.9	10.5	8.7	7.2	6.0	1.9
9 F	25.2	20.3	16.5	13.5	11.2	9.3	7.7	2.5
10 Ne	34.6	27.9	22.8	18.6	15.4	12.8	10.7	3.5
11 Na	45.9	37.2	30.3	24.8	20.5	17.1	14.3	4.6
12 Mg	59.8	48.4	39.5	32.4	26.7	22.2	18.6	6.0
13 Al	75.0	60.7	49.6	40.7	33.6	28.0	23.4	7.6
14 Si	92.8	75.2	61.4	50.4	41.7	34.7	29.1	9.5
15 P	112.9	91.4	74.7	61.3	50.7	42.2	35.3	11.6
16 S	134.7	109.2	89.2	73.3	60.7	50.5	42.3	13.9
17 Cl	158.1	128.2	104.8	86.1	71.2	59.3	49.7	16.4
18 Ar	183.0	148.5	121.4	99.8	82.7	68.9	57.8	19.1
19 K	210.7	171.0	139.8	115.0	95.2	79.4	66.5	22.0
20 Ca	238.6	193.7	158.6	130.5	108.1	90.2	75.6	25.1
21 Sc	271.6	220.6	180.5	148.5	123.1	102.7	86.1	28.6
22 Ti	304.4	247.3	202.6	166.8	138.3	115.4	96.9	32.3
23 V	342.1	278.0	227.7	187.5	155.5	129.8	108.9	36.3
24 Cr	382.3	310.7	254.4	209.5	173.8	145.0	121.7	40.6
25 Mn	422.5	343.6	281.6	232.0	192.6	160.8	135.1	45.2
26 Fe	57.6	379.6	311.1	256.3	212.8	177.7	149.2	49.9
27 Co	64.9	52.8	341.2	281.3	233.7	195.2	164.1	55.1
28 Ni	72.5	58.9	48.3	309.0	256.7	214.5	180.3	60.6
29 Cu	80.6	65.5	53.7	44.2	280.8	234.7	197.2	66.3
30 Zn	89.4	72.7	59.5	49.0	40.7	254.8	214.3	72.3
31 Ga	99.0	80.5	65.9	54.2	45.0	37.6	232.6	78.5
32 Ge	108.6	88.2	72.3	59.5	49.3	41.2	34.6	85.0
33 As	118.9	96.6	79.1	65.1	54.0	45.1	37.8	92.1
34 Se	129.4	105.1	86.1	70.9	58.8	49.1	41.2	99.4
35 Br	141.2	114.7	93.9	77.3	64.2	53.5	44.9	107.6
36 Kr	153.2	124.5	101.9	83.9	69.6	58.1	48.8	115.4
37 Rb	166.1	135.0	110.6	91.0	75.5	63.0	52.9	123.9
38 Sr	178.9	145.4	119.1	98.0	81.3	67.9	57.0	19.0
39 Y	192.6	156.5	128.2	105.5	87.6	73.1	61.3	20.4
40 Zr	207.3	168.5	138.0	113.6	94.2	78.6	66.0	22.0
41 Nb	223.0	181.2	148.4	122.2	101.4	84.6	71.0	23.7
42 Mo	236.3	192.2	157.5	129.8	107.7	90.0	75.6	25.3
43 Tc	253.9	206.5	169.2	139.4	115.7	96.6	81.2	27.2
44 Ru	270.9	220.3	180.6	148.8	123.5	103.1	86.6	29.0
45 Rh	288.9	235.0	192.6	158.7	131.7	110.0	92.4	30.9
46 Pd	307.5	250.0	204.9	168.8	140.2	117.0	98.3	32.9
47 Ag	326.0	265.3	217.6	179.4	149.1	124.5	104.7	35.2
48 Cd	343.9	279.9	229.6	189.3	157.2	131.4	110.4	37.1

Table 14.5a. *cont.*

iii. Elements Co to Zr

Absorber λ (Å):	Emitter: Co 1.790	Ni 1.659	Cu 1.542	Zn 1.436	Ga 1.341	Ge 1.255	As 1.177	Zr 0.787
49 In	365.7	297.7	244.1	201.3	167.2	139.7	117.4	39.4
50 Sn	385.6	313.8	257.4	212.2	176.3	147.3	123.8	41.6
51 Sb	405.5	330.3	271.1	223.7	185.9	155.5	130.7	44.1
52 Te	426.2	347.2	285.0	235.1	195.4	163.4	137.4	46.4
53 I	452.7	368.7	302.7	249.7	207.6	173.6	146.0	49.2
54 Xe	470.7	383.7	315.1	260.2	216.4	181.1	152.4	51.6
55 Cs	495.1	403.6	331.5	273.7	227.7	190.5	160.3	54.3
56 Ba	521.9	425.5	349.4	288.5	240.0	200.8	169.0	57.2
57 La	544.6	444.2	365.2	301.7	251.2	210.3	177.1	60.2
58 Ce	571.3	466.0	383.0	316.5	263.4	220.6	185.7	63.2
59 Pr	597.4	487.3	400.6	331.0	275.5	230.7	194.2	66.0
60 Nd	533.9	510.0	419.5	346.9	288.9	242.1	204.0	69.6
61 Pm	0	535.1	440.2	364.0	303.2	254.0	214.0	69.6
62 Sm	416.5	473.3	456.1	377.1	314.1	263.2	221.7	75.7
63 Eu	148.3	353.7	405.1	392.8	327.4	274.5	231.4	79.3
64 Gd	155.9	370.2	424.0	411.1	342.7	287.3	242.2	83.0
65 Tb	163.8	134.5	316.0	365.2	355.8	298.5	251.8	86.7
66 Dy	172.0	141.2	329.3	0	371.0	311.3	262.6	90.4
67 Ho	180.7	148.3	122.6	283.1	328.6	322.2	271.8	93.6
68 Er	189.4	155.5	128.5	293.7	245.2	335.1	282.9	97.8
69 Tm	198.5	162.9	134.7	0	256.0	300.5	296.0	102.3
70 Yb	208.1	170.8	141.2	117.3	263.9	221.7	305.5	106.0
71 Lu	217.7	178.7	147.8	122.8	0	229.0	270.2	109.7
72 Hf	227.8	186.9	154.6	128.4	107.5	237.5	200.8	114.3
73 Ta	238.3	195.6	161.7	134.4	112.5	0	208.4	118.5
74 W	249.0	204.3	168.9	140.4	117.5	98.9	217.3	123.6
75 Re	260.0	213.3	176.4	146.6	122.7	103.3	0	127.9
76 Os	271.3	222.6	184.1	153.0	128.0	107.8	91.2	132.5
77 Ir	282.9	232.2	192.0	159.5	133.5	112.4	95.1	137.3
78 Pt	295.0	242.1	200.2	166.3	139.2	117.2	99.2	141.4
79 Au	307.3	252.2	208.6	173.3	145.1	122.1	103.3	146.3
80 Hg	320.6	263.1	217.5	180.8	151.3	127.3	107.8	151.5
81 Tl	334.4	274.5	226.9	188.6	157.8	132.8	112.4	155.8
82 Pb	348.2	285.8	236.3	196.4	164.4	138.3	117.1	137.8
83 Bi	363.2	298.1	246.5	204.8	171.4	144.3	122.1	141.9
84 Po	379.5	311.4	257.5	214.0	179.1	150.7	127.6	105.1
85 At	396.5	325.4	269.1	223.6	187.1	157.5	133.3	108.0
86 Rn	414.4	340.1	281.2	233.7	195.6	164.6	139.3	111.3
87 Fr	433.3	355.6	294.0	244.3	204.5	172.1	145.7	114.0
88 Ra	453.0	371.8	307.4	255.4	213.8	180.0	152.3	0
89 Ac	473.9	388.9	321.6	267.2	223.7	188.3	159.3	56.0
90 Th	497.1	408.0	337.3	280.3	234.6	197.5	167.1	58.7
91 Pa	520.3	427.0	353.0	293.4	245.5	206.7	174.9	61.4
92 U	548.4	450.1	372.2	309.2	258.8	217.9	184.4	64.8
93 Np	577.5	474.0	391.9	325.6	272.6	229.4	194.2	68.2
94 Pu	615.2	504.9	417.5	346.9	290.4	244.4	206.8	72.6

Table 14.5b. Selected Mass Absorption Coefficients of the
Chemical Elements for $L\alpha_1$ Lines

i. Elements Ga to Pd

Emitter: Absorber λ (Å):	Ga 11.313	Ge 10.456	Zr 6.070	Nb 5.724	Mo 5.406	Ru 4.846	Rm 4.597	Pd 4.368
3 Li	146.1	116.4	24.3	20.5	17.4	12.7	10.9	9.4
4 Be	360.8	288.0	60.8	51.4	43.7	31.9	27.5	23.7
5 B	744.6	594.9	126.3	106.8	90.8	66.5	57.2	49.4
6 C	1325.9	1060.1	226.3	191.5	162.8	119.4	102.7	88.9
7 N	2118.4	1695.1	363.8	308.1	262.1	192.3	165.7	143.3
8 O	3555.3	2847.0	614.3	520.6	443.1	325.5	280.5	242.9
9 F	4474.6	3586.0	778.0	659.7	561.8	413.2	356.2	308.6
10 Ne	6034.0	4839.6	1055.6	895.7	763.2	561.9	484.7	420.1
11 Na	7872.8	6319.4	1385.9	1176.6	1003.2	739.4	638.2	553.4
12 Mg	669.4	539.8	1805.6	1532.8	1306.9	963.2	831.4	720.9
13 Al	887.5	715.7	2236.5	1899.8	1620.7	1195.9	1032.7	896.0
14 Si	1158.2	934.1	2732.8	2322.7	1982.6	1464.5	1265.4	1098.4
15 P	1474.1	1188.8	269.4	2824.9	2411.3	1781.1	1539.0	1335.9
16 S	1827.6	1473.9	334.0	284.5	243.4	2103.9	1818.8	1579.6
17 Cl	2241.2	1807.5	409.6	348.9	298.5	221.5	191.8	1854.6
18 Ar	2722.6	2195.7	497.5	423.9	362.6	269.0	233.0	202.6
19 K	3241.5	2614.2	592.4	504.7	431.7	320.3	277.4	241.2
20 Ca	3835.6	3093.4	700.9	597.2	510.9	379.0	328.2	285.5
21 Sc	4505.0	3633.2	823.3	701.4	600.0	445.2	385.5	335.3
22 Ti	5264.6	4245.9	962.1	819.6	701.2	520.2	450.5	391.8
23 V	6031.7	4864.5	1102.2	939.1	803.4	596.0	516.1	448.9
24 Cr	6904.1	5568.1	1261.7	1074.9	919.6	682.2	590.7	513.8
25 Mn	7859.3	6338.5	1436.2	1223.6	1046.8	776.6	672.5	584.9
26 Fe	8837.0	7127.0	1614.9	1375.8	1177.0	873.2	756.1	657.7
27 Co	9965.1	8036.8	1821.0	1551.4	1327.3	984.7	852.7	741.6
28 Ni	11130.9	8977.0	2034.1	1732.9	1482.6	1099.9	952.4	828.4
29 Cu	10574.3	9977.8	2260.8	1926.1	1647.8	1222.5	1058.6	920.7
30 Zn	11732.5	9462.2	2508.2	2136.9	1828.2	1356.3	1174.4	1021.5
31 Ga	0	10463.0	2776.2	2365.2	2023.5	1501.2	1299.9	1130.6
32 Ge	0	0	3044.2	2593.5	2218.8	1646.2	1425.4	1239.8
33 As	0	0	3332.9	2839.4	2429.2	1802.2	1560.5	1357.3
34 Se	0	0	3626.3	3091.2	2644.6	1962.0	1698.9	1477.7
35 Br	0	0	3958.2	3372.2	2885.0	2140.4	1853.3	1612.0
36 Kr	0	0	4294.9	3659.1	3130.4	2322.5	2011.0	1749.1
37 Rb	0	0	3958.2	3969.4	3395.9	2519.4	2181.5	1897.5
38 Sr	0	2959.4	4288.0	3653.2	3656.3	2712.6	2348.8	2043.0
39 Y	0	0	780.6	3922.5	3355.8	2920.7	2529.0	2199.7
40 Zr	0	0	845.8	726.1	2599.5	3143.7	2722.0	2367.6
41 Nb	0	0	912.1	783.0	674.9	2891.0	2503.3	2546.7
42 Mo	0	0	983.9	844.6	728.0	2176.4	2627.0	2286.1
43 Tc	0	0	1058.9	909.0	783.5	589.6	2024.9	2454.3
44 Ru	0	0	1138.2	977.2	842.2	633.8	552.6	1880.7
45 Rh	0	0	1222.0	1049.0	904.2	680.4	593.2	519.4
46 Pd	0	0	1307.8	1122.8	967.7	728.2	634.9	555.9
47 Ag	0	0	1399.2	1201.2	1035.3	779.1	679.2	594.7
48 Cd	0	0	1495.9	1284.2	1106.9	832.9	726.2	635.8

Table 14.5b. *cont.*

i. Elements Ga to Pd

Absorber\λ (Å):	Emitter: Ga 11.313	Ge 10.456	Zr 6.070	Nb 5.724	Mo 5.406	Ru 4.846	Rm 4.597	Pd 4.368
49 In	0	0	1595.9	1370.1	1180.9	888.6	774.7	678.4
50 Sn	0	0	1701.4	1460.6	1258.9	947.4	825.9	723.2
51 Sb	0	0	1811.2	1554.9	1340.1	1008.5	879.2	769.9
52 Te	0	0	1926.4	1653.8	1425.4	1072.7	935.2	818.8
53 I	0	0	2046.0	1756.5	1513.9	1139.2	993.2	869.7
54 Xe	0	0	2171.0	1863.8	1606.4	1208.9	1053.9	922.8
55 Cs	0	0	2300.4	1974.9	1702.1	1280.9	1116.7	977.8
56 Ba	0	8628.0	2436.3	2091.5	1802.7	1356.6	1182.7	1035.6
57 La	0	0	2576.6	2211.9	1906.5	1434.6	1250.8	1095.2
58 Ce	0	0	2723.3	2337.9	2015.0	1516.4	1322.0	1157.6
59 Pr	0	0	2877.7	2470.4	2129.3	1602.3	1397.0	1223.2
60 Nd	0	0	3036.4	2606.7	2246.7	1690.7	1474.0	1290.6
61 Pm	0	0	3200.6	2747.6	2368.2	1782.1	1553.7	1360.4
62 Sm	0	0	3371.3	2894.1	2494.5	1877.1	1636.6	1433.0
63 Eu	3022.7	8985.6	3547.4	3045.3	2624.8	1975.2	1722.1	1507.8
64 Gd	3120.9	2620.2	3730.0	3202.1	2759.9	2076.9	1810.7	1585.5
65 Tb	3230.0	2711.8	3920.3	3365.5	2900.7	2182.8	1903.1	1666.3
66 Dy	3339.1	2803.4	4116.0	3533.5	3045.5	2291.8	1998.1	1749.5
67 Ho	3459.2	2904.2	0	3710.8	3198.3	2406.8	2098.4	1837.3
68 Er	3557.4	2986.6	3902.9	3350.5	3353.6	2523.6	2200.2	1926.5
69 Tm	3677.4	3087.4	3391.9	3518.5	3032.6	2644.7	2305.8	2018.9
70 Yb	3808.4	3197.3	3555.0	3051.9	3185.5	2771.8	2416.6	2116.0
71 Lu	3928.4	3298.1	3718.1	3191.9	3322.2	2900.8	2529.0	2214.4
72 Hf	4048.4	3398.9	3359.3	3341.2	2879.8	2615.1	2645.6	2316.5
73 Ta	4157.6	3490.5	3522.4	3023.9	3016.5	2736.1	2385.5	2423.2
74 W	4277.6	3591.3	3674.6	3154.5	3145.2	2366.9	2491.0	2531.8
75 Re	4386.7	3682.9	3837.6	3294.5	2839.6	2475.8	2601.8	2278.1
76 Os	4463.1	3747.0	4011.6	3443.9	2968.3	2578.7	2248.2	2379.8
77 Ir	4659.5	3911.9	0	3583.9	3088.9	2693.7	2348.5	2056.3
78 Pt	4790.5	4021.9	1202.6	0	3225.7	2427.4	2448.8	2144.1
79 Au	4910.5	4122.6	1232.7	1082.1	3362.4	2530.3	2206.0	2231.9
80 Hg	5041.4	4232.6	1265.6	1111.0	0	2639.3	2301.0	2014.8
81 Tl	0	4360.8	1303.9	1144.6	1008.2	2748.2	2396.0	2097.9
82 Pb	0	4470.8	1336.8	1173.5	1033.7	0	2496.3	2185.7
83 Bi	0	4599.0	1375.2	1207.2	1063.3	834.1	2601.8	2278.1
84 Po	0	4727.3	1413.5	1240.8	1093.0	857.4	0	2384.4
85 At	0	4873.9	1457.3	1279.3	1126.9	884.0	786.3	0
86 Rn	0	5020.4	1501.2	1317.8	1160.7	910.5	809.9	723.1
87 Fr	0	5185.3	1550.5	1361.1	1198.9	940.5	836.5	746.8
88 Ra	0	0	1621.7	1423.6	1253.9	983.7	875.0	781.1
89 Ac	0	0	1687.4	1481.3	1304.8	1023.5	910.4	812.8
90 Th	0	0	1714.8	1505.4	1326.0	1040.1	925.2	826.0
91 Pa	0	0	1775.1	1558.3	1372.6	1076.7	957.7	855.0
92 U	0	0	1840.8	1616.0	1423.4	1116.6	993.2	886.7
93 Np	0	0	1912.1	1678.5	1478.5	1159.8	1031.6	921.0
94 Pu	0	0	1994.3	1750.6	1542.0	1209.6	1076.0	960.6

Table 14.5b. *cont.*

ii. Elements Ag to Ba

Emitter: Absorber λ (Å):	Ag 4.154	Cd 3.956	In 3.772	Sn 3.600	Sb 3.439	I 3.148	Cs 2.892	Ba 2.775
3 Li	8.2	7.1	6.2	5.4	4.7	3.7	2.9	2.6
4 Be	20.6	17.9	15.6	13.6	12.0	9.3	7.3	6.5
5 B	42.8	37.3	32.5	28.5	25.0	19.4	15.3	13.6
6 C	77.1	67.1	58.6	51.3	45.1	35.1	27.6	24.5
7 N	124.4	108.3	94.6	82.9	72.9	56.7	44.6	39.7
8 O	210.8	183.7	160.6	140.8	123.7	96.4	75.9	67.6
9 F	268.0	233.6	204.3	179.2	157.6	122.9	96.9	86.3
10 Ne	365.0	318.4	278.6	244.5	215.1	167.9	132.4	118.0
11 Na	481.0	419.8	367.5	322.7	284.0	221.9	175.1	156.1
12 Mg	626.7	546.8	478.8	420.3	370.0	289.1	228.2	203.3
13 Al	779.2	680.3	595.9	523.4	460.9	360.5	284.7	253.9
14 Si	955.7	834.8	731.6	642.9	566.4	443.3	350.5	312.6
15 P	1162.3	1015.3	889.8	781.9	688.8	539.2	426.3	380.2
16 S	1375.1	1201.7	1053.7	926.3	816.4	639.6	506.1	451.6
17 Cl	1614.5	1410.9	1237.1	1087.6	958.5	751.0	594.2	530.2
18 Ar	176.7	154.6	1421.0	1249.9	1102.1	864.2	684.4	610.9
19 K	210.3	184.1	161.6	142.3	0	995.4	788.3	703.7
20 Ca	248.9	217.8	191.3	168.4	148.6	116.7	888.2	793.2
21 Sc	292.3	255.8	224.6	197.8	174.5	137.1	108.8	97.2
22 Ti	341.6	299.0	262.5	231.1	204.0	160.2	127.1	113.6
23 V	391.4	342.5	300.8	264.8	233.7	183.6	145.6	130.1
24 Cr	448.0	392.1	344.3	303.1	267.5	210.1	166.7	148.9
25 Mn	510.0	446.3	391.9	345.0	304.5	239.2	189.8	169.5
26 Fe	573.4	501.8	440.6	387.9	342.4	269.0	213.4	190.6
27 Co	646.6	565.9	496.9	437.4	386.1	303.3	240.6	214.9
28 Ni	722.2	632.1	555.0	488.6	431.2	338.8	268.7	240.1
29 Cu	802.7	702.5	616.9	543.1	479.3	376.5	298.7	266.9
30 Zn	890.6	779.4	684.4	602.5	531.8	417.7	331.4	296.1
31 Ga	985.7	862.7	757.5	666.9	588.6	462.4	366.8	327.7
32 Ge	1080.9	946.0	830.6	731.3	645.4	507.0	402.2	359.3
33 As	1183.4	1035.7	909.4	800.6	706.6	555.1	440.3	393.4
34 Se	1288.3	1127.5	990.0	871.6	769.2	604.3	479.4	428.3
35 Br	1405.4	1230.0	1080.0	950.8	839.2	659.2	523.0	467.2
36 Kr	1525.0	1334.6	1171.9	1031.7	910.6	715.3	567.4	506.9
37 Rb	1654.3	1447.8	1271.3	1119.2	987.8	776.0	615.6	549.9
38 Sr	1781.2	1558.8	1368.8	1205.0	1063.5	835.5	662.8	592.1
39 Y	1917.8	1678.4	1473.8	1297.5	1145.1	899.6	713.6	637.5
40 Zr	2064.2	1806.6	1586.3	1396.5	1232.5	968.2	768.1	686.2
41 Nb	2220.4	1943.2	1706.3	1502.2	1325.8	1041.5	826.2	738.1
42 Mo	2333.3	2043.0	1794.8	1580.8	1395.8	1097.5	871.4	778.8
43 Tc	2140.9	2194.7	1928.0	1698.2	1499.5	1178.9	936.0	836.6
44 Ru	2285.2	2000.9	2057.5	1812.2	1600.2	1258.1	998.9	892.8
45 Rh	455.8	1533.3	1876.2	1932.8	1706.7	1341.9	1065.4	952.2
46 Pd	487.8	429.7	1432.1	1758.5	1552.7	1427.9	1133.7	1013.2
47 Ag	521.9	459.7	406.1	1329.0	1634.6	1505.5	1196.4	1069.7
48 Cd	558.0	491.5	434.2	384.6	1239.4	1357.9	1262.1	1128.5

Table 14.5b. *cont.*

ii. Elements Ag to Ba

Absorber \ λ (Å):	Emitter: Ag 4.154	Cd 3.956	In 3.772	Sn 3.600	Sb 3.439	I 3.148	Cs 2.892	Ba 2.775
49 In	595.3	524.3	463.2	410.3	364.3	0	1342.1	1200.0
50 Sn	634.6	559.0	493.9	437.4	388.4	1091.7	1208.8	1265.2
51 Sb	675.6	595.0	525.7	465.7	413.4	328.5	909.3	1132.0
52 Te	718.6	632.9	559.2	495.3	439.7	349.4	280.3	854.3
53 I	763.2	672.2	593.9	526.0	467.0	371.1	297.7	267.4
54 Xe	809.8	713.3	630.2	558.2	495.6	393.8	315.9	283.7
55 Cs	858.1	755.8	667.7	591.4	525.1	417.3	334.7	300.6
56 Ba	908.8	800.4	707.2	626.4	556.1	441.9	354.5	318.4
57 La	961.1	846.5	747.9	662.4	588.1	467.3	374.9	336.7
58 Ce	1015.8	894.7	790.5	700.2	621.6	494.0	396.2	355.9
59 Pr	1073.4	945.4	835.3	739.8	656.9	522.0	418.7	376.0
60 Nd	1132.6	997.6	881.4	780.7	693.1	550.8	441.8	396.8
61 Pm	1193.9	1051.5	929.0	822.9	730.6	580.5	465.6	418.2
62 Sm	1257.5	1107.6	978.6	866.7	769.5	611.5	490.5	440.5
63 Eu	1323.2	1165.4	1029.7	912.0	809.7	643.4	516.1	463.6
64 Gd	1391.3	1225.4	1082.7	959.0	851.4	676.6	542.7	487.4
65 Tb	1462.3	1287.9	1137.9	1007.9	894.8	711.1	570.3	512.3
66 Dy	1535.3	1352.2	1194.7	1058.2	939.5	746.6	598.8	537.9
67 Ho	1612.3	1420.1	1254.7	1111.3	986.7	784.0	628.9	564.8
68 Er	1690.6	1489.0	1315.6	1165.2	1034.6	822.1	659.4	592.3
69 Tm	1771.7	1560.4	1378.7	1221.1	1084.2	861.5	691.0	620.7
70 Yb	1856.9	1635.4	1445.0	1279.8	1136.3	902.9	724.2	650.5
71 Lu	1943.3	1711.5	1512.2	1339.4	1189.2	944.9	757.9	680.8
72 Hf	2032.9	1790.5	1581.9	1401.1	1244.0	988.5	792.9	712.2
73 Ta	2126.5	1873.0	1654.8	1465.7	1301.3	1034.1	829.4	745.0
74 W	2221.8	1956.9	1729.0	1531.4	1359.6	1080.4	866.6	778.4
75 Re	2320.0	2043.3	1805.3	1599.0	1419.7	1128.1	904.9	812.7
76 Os	2089.4	2132.3	1883.9	1668.6	1481.5	1177.2	944.3	848.1
77 Ir	2177.6	1918.0	1965.0	1740.5	1545.3	1227.9	984.9	884.6
78 Pt	1881.6	2000.1	0	1814.5	1611.0	1280.2	1026.8	922.3
79 Au	1958.7	1725.1	1839.7	1890.6	1678.5	1333.8	1069.8	960.9
80 Hg	2043.8	1800.1	1918.6	1699.4	1750.7	1391.2	1115.9	1002.3
81 Tl	2133.0	1878.7	1659.9	1774.8	1575.8	1451.3	1164.1	1045.6
82 Pb	1918.1	1953.7	1726.1	1528.9	1640.3	1511.3	1212.2	1088.8
83 Bi	1999.2	1760.8	1801.9	1596.0	1417.0	1358.7	1264.4	1135.7
84 Po	2092.5	1843.0	1628.3	1668.6	1481.5	1419.8	1320.9	1186.4
85 At	2185.8	1925.1	1700.9	1741.3	1546.0	1228.5	1189.4	1239.8
86 Rn	0	2010.8	1776.6	1573.6	1618.0	1285.7	1241.6	1115.2
87 Fr	0	2103.7	1858.7	1646.3	1461.6	1342.9	1077.1	1167.8
88 Ra	698.7	626.9	1940.7	1718.9	1526.2	1404.0	1126.1	1011.5
89 Ac	727.0	652.3	0	1800.0	1598.1	1469.1	1178.3	1058.4
90 Th	738.8	662.9	596.4	0	1675.0	1331.0	1235.3	1109.5
91 Pa	764.8	686.2	617.4	0	1754.5	1394.2	1293.8	1162.1
92 U	793.1	711.6	640.2	577.2	0	1469.1	1178.3	1224.6
93 Np	823.8	739.2	665.0	599.5	541.6	1548.0	1241.6	1289.9
94 Pu	859.2	770.9	693.6	625.3	564.9	0	1322.3	1187.7

Table 14.5b. *cont.*

iii. Elements Hf to Au

Absorber \ Emitter: λ (Å):	Hf 1.570	Ta 1.522	W 1.476	Re 1.433	Os 1.391	Ir 1.351	Pt 1.313	Au 1.276
3 Li	0.5	0.5	0.4	0.4	0.3	0.3	0.3	0.3
4 Be	1.3	1.2	1.1	1.0	0.9	0.8	0.8	0.7
5 B	2.7	2.4	2.2	2.1	1.9	1.7	1.6	1.5
6 C	4.9	4.5	4.1	3.8	3.4	3.2	2.9	2.7
7 N	7.9	7.3	6.7	6.1	5.6	5.2	4.8	4.4
8 O	13.6	12.4	11.4	10.5	9.6	8.9	8.2	7.6
9 F	17.4	16.0	14.6	13.5	12.4	11.4	10.5	9.7
10 Ne	23.9	21.9	20.1	18.5	17.1	15.7	14.5	13.4
11 Na	31.9	29.2	26.8	24.7	22.7	20.9	19.3	17.9
12 Mg	41.5	38.1	34.9	32.2	29.6	27.3	25.2	23.3
13 Al	52.1	47.8	43.9	40.4	37.2	34.3	31.7	29.3
14 Si	64.5	59.2	54.4	50.1	46.2	42.6	39.3	36.3
15 P	78.5	72.0	66.2	61.0	56.1	51.8	47.8	44.2
16 S	93.8	86.1	79.1	72.9	67.1	61.9	57.2	52.9
17 Cl	110.1	101.0	92.8	85.6	78.8	72.7	67.2	62.1
18 Ar	127.6	117.1	107.6	99.2	91.4	84.4	78.0	72.1
19 K	146.9	134.9	124.0	114.3	105.3	97.2	89.9	83.1
20 Ca	166.6	153.0	140.7	129.7	119.6	110.4	102.1	94.4
21 Sc	189.6	174.2	160.1	147.7	136.1	125.6	116.2	107.4
22 Ti	212.8	195.5	179.8	165.8	152.9	141.2	130.6	120.8
23 V	239.1	219.7	202.1	186.4	171.8	158.7	146.8	135.8
24 Cr	267.2	245.5	225.8	208.3	192.0	177.3	164.0	151.7
25 Mn	295.7	271.8	250.0	230.7	212.7	196.5	181.8	168.2
26 Fe	326.7	300.3	276.2	254.9	235.1	217.1	200.9	185.9
27 Co	358.2	329.3	303.0	279.7	258.0	238.4	220.7	204.2
28 Ni	50.7	46.6	332.9	307.3	283.5	261.9	242.4	224.4
29 Cu	56.4	51.8	47.6	43.9	40.5	286.5	265.2	245.5
30 Zn	62.5	57.4	52.8	48.7	44.9	41.5	38.4	266.5
31 Ga	69.2	63.6	58.5	53.9	49.7	45.9	42.5	39.3
32 Ge	75.9	69.7	64.1	59.1	54.5	50.4	46.6	43.1
33 As	83.1	76.3	70.2	64.8	59.7	55.1	51.0	47.2
34 Se	90.5	83.1	76.4	70.5	65.0	60.0	55.5	51.4
35 Br	98.7	90.7	83.4	76.9	70.9	65.5	60.6	56.0
36 Kr	107.1	98.4	90.5	83.4	76.9	71.0	65.7	60.8
37 Rb	116.1	106.7	98.1	90.5	83.5	77.1	71.3	65.9
38 Sr	125.1	114.9	105.7	97.5	89.9	83.0	76.8	71.0
39 Y	134.6	123.7	113.8	104.9	96.8	89.3	82.7	76.4
40 Zr	144.9	133.1	122.4	113.0	104.1	96.2	89.0	82.3
41 Nb	155.9	143.2	131.7	121.5	112.0	103.4	95.7	88.5
42 Mo	165.4	152.0	139.8	129.0	119.0	109.9	101.7	94.1
43 Tc	177.7	163.3	150.2	138.6	127.8	118.1	109.3	101.1
44 Ru	189.6	174.3	160.3	147.9	136.4	126.0	116.6	107.9
45 Rh	202.3	185.9	171.0	157.8	145.5	134.4	124.4	115.1
46 Pd	215.2	197.8	181.9	167.9	154.8	143.0	132.3	122.4
47 Ag	228.5	210.1	193.3	178.4	164.6	152.1	140.8	130.3
48 Cd	241.1	221.6	203.9	188.2	173.7	160.4	148.5	137.4

Table 14.5b. *cont.*

iii. Elements Hf to Au

Absorber \ Emitter: λ (Å):	Hf 1.570	Ta 1.522	W 1.476	Re 1.433	Os 1.391	Ir 1.351	Pt 1.313	Au 1.276
49 In	256.4	235.7	216.9	200.2	184.7	170.6	157.9	146.2
50 Sn	270.3	248.5	228.6	211.0	194.7	179.9	166.5	154.1
51 Sb	284.6	261.7	240.9	222.4	205.3	189.7	175.6	162.6
52 Te	299.1	275.1	253.2	233.8	215.7	199.4	184.6	170.9
53 I	317.7	292.2	268.9	248.3	229.1	211.8	196.1	181.5
54 Xe	330.8	304.3	280.2	258.7	238.8	220.8	204.5	189.4
55 Cs	347.9	320.0	294.7	272.2	251.2	232.3	215.1	199.2
56 Ba	366.8	337.4	310.6	286.9	264.8	244.8	226.8	210.0
57 La	383.2	352.6	324.8	300.0	277.0	256.2	237.3	219.8
58 Ce	402.0	369.9	340.7	314.7	290.6	268.7	249.0	230.6
59 Pr	420.4	386.8	356.3	329.1	303.9	281.1	260.4	241.2
60 Nd	440.2	405.2	373.3	344.9	318.6	294.7	273.1	253.0
61 Pm	461.9	425.1	391.7	361.9	334.3	309.2	286.6	265.5
62 Sm	478.5	440.5	405.8	375.0	346.4	320.4	296.9	275.1
63 Eu	424.9	458.5	422.5	390.6	360.9	333.9	309.5	286.8
64 Gd	319.7	409.6	0	408.8	377.7	349.5	324.0	300.2
65 Tb	331.5	305.3	392.8	363.2	392.0	362.9	336.4	311.9
66 Dy	345.3	318.1	293.2	378.8	350.1	378.4	350.9	325.3
67 Ho	128.5	330.2	304.4	281.5	0	335.1	363.2	336.7
68 Er	134.7	124.3	315.8	292.1	270.1	250.0	322.2	298.8
69 Tm	141.2	130.2	120.2	0	282.0	261.1	242.2	314.0
70 Yb	147.9	136.5	126.0	116.7	108.0	269.1	249.7	231.6
71 Lu	154.8	142.8	131.9	122.1	113.0	104.8	257.9	239.2
72 Hf	162.0	149.4	138.0	127.7	118.2	109.6	101.8	248.1
73 Ta	169.4	156.3	144.3	133.6	123.7	114.6	106.5	98.8
74 W	177.0	163.3	150.8	139.6	129.2	119.8	111.2	103.3
75 Re	184.8	170.5	157.4	145.8	134.9	125.1	116.1	107.8
76 Os	192.9	177.9	164.3	152.1	140.8	130.5	121.2	112.5
77 Ir	201.2	185.6	171.4	158.7	146.9	136.1	126.4	117.4
78 Pt	209.8	193.5	178.7	165.4	153.1	141.9	131.8	122.3
79 Au	218.5	201.6	186.1	172.4	159.5	147.9	137.3	127.5
80 Hg	227.9	210.3	194.1	179.8	166.4	154.2	143.2	133.0
81 Tl	237.8	219.4	202.5	187.6	173.6	160.9	149.4	138.7
82 Pb	247.6	228.4	210.9	195.3	180.8	167.6	155.6	144.4
83 Bi	258.3	238.3	220.0	203.7	188.5	174.8	162.3	150.7
84 Po	269.8	248.9	229.8	212.8	197.0	182.6	169.5	157.4
85 At	282.0	260.1	240.2	222.4	205.8	190.8	177.2	164.5
86 Rn	294.7	271.8	251.0	232.4	215.1	199.4	185.2	171.9
87 Fr	308.1	284.2	262.4	243.0	224.9	208.5	193.6	179.7
88 Ra	322.1	297.1	274.4	254.1	235.2	218.0	202.4	187.9
89 Ac	337.0	310.9	287.0	265.8	246.0	228.0	211.7	196.6
90 Th	353.5	326.1	301.1	278.8	258.0	239.2	222.1	206.2
91 Pa	370.0	341.3	315.1	291.8	270.1	250.3	232.4	215.8
92 U	390.0	359.7	332.1	307.6	284.7	263.9	245.0	227.5
93 Np	410.7	378.8	349.8	323.9	299.8	277.9	258.0	239.5
94 Pu	437.5	403.5	372.6	345.0	319.4	296.0	274.9	255.2

Table 14.5b. *cont.*

iv. Elements Hg to U

Emitter:	Hg	Pb	Bi	U
Absorber λ (Å):	1.241	1.175	1.144	0.911
3 Li	0.3	0.2	0.2	0.1
4 Be	0.6	0.6	0.5	0.3
5 B	1.4	1.2	1.1	0.6
6 C	2.5	2.1	2.0	1.0
7 N	4.1	3.5	3.2	1.7
8 O	7.0	6.0	5.6	2.9
9 F	9.0	7.7	7.2	3.8
10 Ne	12.4	10.6	9.9	5.2
11 Na	16.5	14.2	13.2	7.0
12 Mg	21.5	18.5	17.2	9.1
13 Al	27.1	23.3	21.6	11.5
14 Si	33.6	28.9	26.9	14.3
15 P	40.9	35.2	32.7	17.4
16 S	49.0	42.1	39.1	20.9
17 Cl	57.5	49.5	46.0	24.5
18 Ar	66.8	57.5	53.4	28.6
19 K	77.0	66.2	61.5	32.9
20 Ca	87.5	75.3	70.0	37.5
21 Sc	99.6	85.7	79.7	42.7
22 Ti	112.0	96.4	89.7	48.1
23 V	125.8	108.4	100.8	54.1
24 Cr	140.6	121.1	112.6	60.5
25 Mn	156.0	134.4	125.0	67.3
26 Fe	172.4	148.5	138.1	74.3
27 Co	189.4	163.3	151.9	81.9
28 Ni	208.1	179.4	166.9	90.0
29 Cu	227.6	196.3	182.6	98.5
30 Zn	247.2	213.3	198.4	107.3
31 Ga	36.4	231.5	215.4	116.5
32 Ge	39.9	34.4	32.0	126.1
33 As	43.7	37.7	35.0	136.5
34 Se	47.6	41.0	38.1	147.4
35 Br	51.9	44.7	41.6	159.5
36 Kr	56.3	48.5	45.1	24.2
37 Rb	61.1	52.7	48.9	26.3
38 Sr	65.8	56.7	52.7	28.3
39 Y	70.9	61.0	56.7	30.5
40 Zr	76.3	65.7	61.1	32.8
41 Nb	82.0	70.7	65.7	35.3
42 Mo	87.3	75.2	69.9	37.6
43 Tc	93.7	80.8	75.1	40.4
44 Ru	100.0	86.2	80.2	43.1
45 Rh	106.7	92.0	85.5	46.0
46 Pd	113.5	97.8	91.0	49.0
47 Ag	120.8	104.2	96.9	52.3
48 Cd	127.5	109.9	102.2	55.2

Table 14.5b._cont._

iv. Elements Hg to U

Emitter: Absorber λ (Å):	Hg 1.241	Pb 1.175	Bi 1.144	U 0.911
49 In	135.5	116.9	108.7	58.6
50 Sn	142.9	123.2	114.6	61.8
51 Sb	150.8	130.1	121.1	65.5
52 Te	158.5	136.8	127.3	68.8
53 I	168.4	145.3	135.2	73.1
54 Xe	175.7	151.7	141.2	76.5
55 Cs	184.8	159.6	148.5	80.5
56 Ba	194.8	168.2	156.5	84.8
57 La	204.0	176.2	164.1	89.1
58 Ce	214.0	184.9	172.1	93.5
59 Pr	223.8	193.4	180.0	97.8
60 Nd	234.9	203.0	189.0	102.9
61 Pm	246.5	213.0	198.4	108.0
62 Sm	255.4	220.7	205.5	111.9
63 Eu	266.4	230.4	214.5	117.1
64 Gd	278.8	241.1	224.5	122.5
65 Tb	289.7	250.7	233.5	127.7
66 Dy	302.1	261.4	243.5	133.2
67 Ho	312.8	270.6	252.1	137.9
68 Er	325.4	281.7	262.5	143.9
69 Tm	291.8	294.7	274.6	150.5
70 Yb	300.0	304.1	283.5	155.7
71 Lu	222.3	269.0	250.7	161.2
72 Hf	230.6	199.9	260.3	167.6
73 Ta	239.5	207.5	193.5	173.9
74 W	96.1	216.3	201.7	181.1
75 Re	100.3	223.9	208.8	187.4
76 Os	104.7	90.8	84.7	193.8
77 Ir	109.2	94.7	88.3	200.9
78 Pt	113.8	98.7	92.1	176.7
79 Au	118.6	102.9	96.0	131.2
80 Hg	123.7	107.3	100.1	135.2
81 Tl	129.0	111.9	104.4	139.2
82 Pb	134.4	116.6	108.7	144.0
83 Bi	140.1	121.6	113.4	148.1
84 Po	146.4	127.0	118.5	65.5
85 At	153.0	132.7	123.8	68.5
86 Rn	159.9	138.7	129.4	71.6
87 Fr	167.2	145.0	135.3	74.8
88 Ra	174.8	151.6	141.4	78.2
89 Ac	182.8	158.6	148.0	81.9
90 Th	191.8	166.4	155.2	85.9
91 Pa	200.7	174.1	162.4	89.9
92 U	211.6	183.6	171.2	94.7
93 Np	222.8	193.3	180.3	99.7
94 Pu	237.4	205.9	192.1	106.3

Table 14.5c. Selected Mass Absorption Coefficients of the
Chemical Elements for $M\alpha$ Lines (Heinrich, 1966)

i. Elements Hf to Au

Absorber	Emitter: Hf λ (Å): 7.539	Ta 7.251	W 6.983	Re 6.728	Os 6.490	Ir 6.261	Pt 6.046	Au 5.840
3 Li	45.4	40.6	36.4	32.7	29.5	26.6	24.0	21.8
4 Be	113.0	101.1	90.8	81.6	73.6	66.4	60.1	54.5
5 B	234.2	209.6	188.3	169.3	152.8	137.9	124.9	113.1
6 C	418.7	374.9	336.8	303.1	273.6	247.1	223.7	202.7
7 N	671.7	601.6	540.8	486.8	439.6	397.1	359.7	326.1
8 O	1131.9	1014.2	912.0	821.1	741.8	670.4	607.5	550.9
9 F	1430.4	1282.1	1153.3	1038.9	938.9	848.7	769.3	697.9
10 Ne	1936.7	1736.6	1562.8	1408.2	1273.1	1151.3	1044.0	947.4
11 Na	2537.2	2275.9	2048.9	1846.9	1670.4	1511.1	1370.7	1244.3
12 Mg	3305.4	2965.0	2669.3	2406.1	2176.1	1968.6	1785.7	1621.1
13 Al	4085.5	3666.2	3301.8	2977.4	2693.7	2437.7	2212.0	2008.8
14 Si	382.5	343.9	310.3	3634.3	3289.2	2977.7	2703.0	2455.5
15 P	486.8	437.7	394.9	356.8	323.4	293.2	266.5	242.4
16 S	603.5	542.6	489.6	442.3	400.9	363.4	330.4	300.5
17 Cl	740.1	665.4	600.4	542.4	491.6	445.7	405.2	368.6
18 Ar	899.0	808.3	729.4	658.9	597.2	541.4	492.2	447.7
19 K	1070.4	962.4	868.4	784.5	711.1	644.6	586.0	533.1
20 Ca	1266.6	1138.8	1027.6	928.3	841.4	762.8	693.4	630.8
21 Sc	1487.6	1337.6	1206.9	1090.3	988.2	895.9	814.4	740.9
22 Ti	1738.5	1563.1	1410.4	1274.2	1154.9	1047.0	951.7	865.8
23 V	1991.8	1790.9	1615.9	1459.8	1323.1	1199.5	1090.4	991.9
24 Cr	2279.9	2049.9	1849.6	1671.0	1514.5	1373.0	1248.1	1135.4
25 Mn	2595.3	2333.5	2105.5	1902.2	1724.0	1563.0	1420.8	1292.5
26 Fe	2918.1	2623.8	2367.4	2138.8	1938.5	1757.4	1597.5	1453.3
27 Co	3290.7	2958.7	2669.6	2411.8	2186.0	1981.8	1801.5	1638.8
28 Ni	3675.6	3304.8	2981.9	2694.0	2441.7	2213.6	2012.2	1830.5
29 Cu	4085.4	3673.3	3314.4	2994.3	2713.9	2460.4	2236.5	2034.6
30 Zn	4532.4	4075.2	3677.0	3321.9	3010.9	2729.6	2481.3	2257.2
31 Ga	5016.7	4510.6	4069.9	3676.9	3332.6	3021.3	2746.4	2498.4
32 Ge	5501.0	4946.1	4462.8	4031.8	3654.3	3312.9	3011.5	2739.6
33 As	6022.5	5415.0	4885.9	4414.1	4000.7	3627.0	3297.0	2999.3
34 Se	5600.3	5895.1	5319.1	4805.4	4355.4	3948.6	3589.3	3265.2
35 Br	6109.4	5493.1	4956.4	5242.3	4751.4	4307.5	3915.6	3562.0
36 Kr	1069.6	4276.2	5379.6	4860.0	4404.9	4674.0	4248.7	3865.1
37 Rb	1165.1	1052.8	954.6	3786.1	4751.4	4307.5	3915.6	4192.8
38 Sr	1264.4	1142.6	1036.0	940.5	856.4	3350.3	4241.9	3858.9
39 Y	1371.3	1239.3	1123.7	1020.1	928.9	846.1	772.6	2980.7
40 Zr	1485.9	1342.8	1217.6	1105.3	1006.5	916.8	837.1	765.0
41 Nb	1602.4	1448.1	1313.0	1192.0	1085.4	988.6	902.8	825.0
42 Mo	1728.5	1562.0	1416.3	1285.7	1170.8	1066.4	973.8	889.9
43 Tc	1860.3	1681.1	1524.3	1383.8	1260.1	1147.7	1048.0	957.7
44 Ru	1999.7	1807.1	1638.5	1487.5	1354.5	1233.7	1126.6	1029.5
45 Rh	2146.8	1940.0	1759.0	1596.9	1454.1	1324.5	1209.4	1105.2
46 Pd	2297.6	2076.3	1882.7	1709.1	1556.3	1417.6	1294.4	1182.9
47 Ag	2458.1	2221.3	2014.1	1828.4	1665.0	1516.5	1384.8	1265.5
48 Cd	2628.1	2374.9	2153.4	1954.9	1780.1	1621.4	1480.6	1353.0

Table 14.5c. *cont.*

i. Elements Hf to Au

Absorber\λ (Å):	Emitter: Hf 7.539	Ta 7.251	W 6.983	Re 6.728	Os 6.490	Ir 6.261	Pt 6.046	Au 5.840
49 In	2803.8	2533.7	2297.4	2085.6	1899.2	1729.8	1579.6	1443.4
50 Sn	2989.0	2701.2	2449.2	2223.4	2024.7	1844.1	1684.0	1538.8
51 Sb	3181.9	2875.5	2607.3	2366.9	2155.3	1963.1	1792.6	1638.1
52 Te	3384.4	3058.4	2773.1	2517.5	2292.5	2088.0	1906.7	1742.4
53 I	3594.5	3248.3	2945.3	2673.8	2434.8	2217.7	2025.1	1850.5
54 Xe	3814.1	3446.8	3125.3	2837.1	2583.5	2353.2	2148.8	1963.6
55 Cs	4041.4	3652.2	3311.5	3006.2	2737.5	2493.4	2276.8	2080.6
56 Ba	4280.1	3867.9	3507.1	3183.8	2899.2	2640.7	2411.3	2203.5
57 La	4526.5	4090.6	3709.0	3367.1	3066.1	2792.7	2550.1	2330.4
58 Ce	4784.4	4323.6	3920.3	3558.9	3240.7	2951.8	2695.4	2463.1
59 Pr	5055.6	4568.7	4142.5	3760.6	3424.4	3119.1	2848.2	2602.7
60 Nd	5334.4	4820.6	4371.0	3968.0	3613.3	3291.1	3005.3	2746.3
61 Pm	4851.2	5081.3	4607.3	4182.6	3808.7	3469.1	3167.8	2894.8
62 Sm	5099.5	4608.4	4853.0	4405.6	4011.8	3654.1	3336.7	3049.1
63 Eu	5366.9	4850.0	4397.6	4635.8	4221.4	3844.9	3511.0	3208.4
64 Gd	4679.3	5108.9	4632.3	4205.3	4438.7	4042.9	3691.8	3373.6
65 Tb	4927.6	4453.0	4867.1	4418.4	4023.4	4249.1	3880.1	3545.7
66 Dy	4469.2	4660.1	4225.5	3835.9	4217.5	3841.4	4073.8	3722.7
67 Ho	4679.3	4228.6	4444.5	4034.8	3674.1	4041.7	3690.7	3372.6
68 Er	4908.5	4435.8	4022.0	4233.7	3855.2	3511.5	3862.9	3529.9
69 Tm	5156.8	4660.1	4225.5	3835.9	4036.4	3676.4	3357.2	3706.9
70 Yb	5405.1	4884.5	4428.9	4020.6	3661.2	3853.2	3518.6	3215.3
71 Lu	5653.4	5108.9	4632.3	4205.3	3829.4	3487.9	3680.0	3362.8
72 Hf	1644.3	0	4835.8	4390.0	3997.6	3641.1	3324.9	3520.1
73 Ta	1688.6	1548.7	0	4603.1	4191.6	3817.8	3486.3	3185.8
74 W	1737.4	1593.5	1465.6	0	4372.7	3982.8	3636.9	3323.5
75 Re	1781.7	1634.1	1503.0	1383.9	0	4159.6	3798.3	3470.9
76 Os	1812.7	1662.6	1529.2	1408.0	1299.8	0	3970.5	3628.3
77 Ir	1892.5	1735.7	1596.5	1470.0	1357.0	1253.0	0	3775.8
78 Pt	1945.7	1784.5	1641.4	1511.3	1395.1	1288.2	1192.0	0
79 Au	1994.4	1829.2	1682.5	1549.1	1430.1	1320.5	1221.9	1131.4
80 Hg	2047.6	1878.0	1727.4	1590.4	1468.2	1355.7	1254.5	1161.6
81 Tl	2109.7	1934.9	1779.7	1638.6	1512.7	1396.8	1292.5	1196.8
82 Pb	2162.8	1983.7	1824.6	1680.0	1555.9	1432.0	1325.1	1226.9
83 Bi	2224.9	2040.6	1876.9	1728.1	1595.4	1473.1	1363.1	1262.1
84 Po	2286.9	2097.5	1929.3	1776.3	1639.8	1514.1	1401.1	1297.3
85 At	2357.9	2162.5	1989.1	1831.4	1690.7	1561.1	1444.6	1337.6
86 Rn	2428.8	2227.6	2048.9	1886.5	1741.5	1608.1	1488.0	1377.8
87 Fr	2508.6	2300.8	2116.2	1948.5	1798.7	1660.9	1536.9	1423.1
88 Ra	2623.8	2406.4	2213.4	2038.0	1881.4	1737.2	1607.5	1488.4
89 Ac	2730.2	2504.0	2303.2	2120.6	1957.6	1807.6	1672.7	1548.8
90 Th	2774.5	2544.6	2340.5	2155.0	1989.4	1836.9	1699.8	1573.9
91 Pa	2872.0	2634.1	2422.8	2230.8	2059.3	1901.5	1759.6	1629.2
92 U	2978.4	2731.6	2512.5	2313.4	2135.6	1971.9	1824.7	1689.6
93 Np	3093.6	2837.3	2609.7	2402.9	2218.2	2048.2	1895.3	1754.9
94 Pu	3226.5	2959.3	2721.9	2506.2	2313.6	2136.2	1976.8	1830.4

Table 14.5c. *cont.*

ii. Elements Hg to U

Ab-sorber	Emitter: Hg λ (Å): 5.666	Pb 5.285	Bi 5.118	U 3.910	Ab-sorber	Emitter: Hg λ (Å): 5.666	Pb 5.285	Bi 5.118	U 3.910
3 Li	19.9	16.3	14.9	6.9	49 In	1334.3	1113.4	1024.2	508.6
4 Be	49.9	40.9	37.3	17.3	50 Sn	1422.4	1186.9	1091.9	542.2
5 B	103.8	85.1	77.7	36.1	51 Sb	1514.2	1263.5	1162.3	577.2
6 C	186.1	152.7	139.4	64.9	52 Te	1610.6	1343.9	1236.3	613.9
7 N	299.3	245.8	224.5	104.8	53 I	1710.6	1427.4	1313.0	652.0
8 O	505.9	415.7	379.7	177.7	54 Xe	1815.1	1514.6	1393.3	691.9
9 F	641.1	527.2	481.7	226.1	55 Cs	1923.2	1604.8	1476.3	733.1
10 Ne	870.5	716.3	654.7	308.1	56 Ba	2036.9	1699.6	1563.5	776.4
11 Na	1143.6	941.8	861.1	406.3	57 La	2154.1	1797.5	1653.5	821.1
12 Mg	1489.9	1226.9	1121.8	529.3	58 Ce	2276.8	1899.9	1747.7	867.9
13 Al	1846.8	1521.9	1391.9	658.5	59 Pr	2405.9	2007.6	1846.8	917.1
14 Si	2258.1	1862.1	1703.6	808.2	60 Nd	2538.6	2118.3	1948.6	967.7
15 P	2746.4	2264.7	2072.0	982.9	61 Pm	2675.8	2232.8	2054.0	1020.0
16 S	276.7	228.8	209.6	1163.5	62 Sm	2818.5	2351.9	2163.5	1074.4
17 Cl	339.4	280.6	257.1	1366.1	63 Eu	2965.8	2474.8	2276.6	1130.5
18 Ar	412.2	340.9	312.3	149.7	64 Gd	3118.5	2602.2	2393.8	1188.7
19 K	490.8	405.9	371.8	178.3	65 Tb	3277.5	2734.9	2515.9	1249.4
20 Ca	580.8	480.3	440.0	211.0	66 Dy	3441.1	2871.4	2641.4	1311.7
21 Sc	682.1	564.1	516.7	247.8	67 Ho	3613.8	3015.5	2774.0	1377.5
22 Ti	797.2	659.2	603.9	289.6	68 Er	3263.0	3161.9	2908.7	1444.4
23 V	913.3	755.2	691.9	331.8	69 Tm	3426.6	3313.6	3048.2	1513.7
24 Cr	1045.4	864.5	791.9	379.7	70 Yb	3599.3	3003.4	3194.7	1586.5
25 Mn	1190.0	984.1	901.5	432.3	71 Lu	3108.5	3132.3	2881.4	1660.3
26 Fe	1338.1	1106.5	1013.6	486.0	72 Hf	3253.9	2715.2	3014.0	1736.8
27 Co	1508.9	1247.8	1143.0	548.1	73 Ta	2944.9	2844.1	2616.3	1816.9
28 Ni	1685.4	1393.7	1276.7	612.2	74 W	3072.1	2965.5	2728.0	1898.3
29 Cu	1873.3	1549.1	1419.1	680.5	75 Re	3208.4	2677.3	2853.5	1982.1
30 Zn	2078.3	1718.6	1574.4	754.9	76 Os	3353.9	2798.6	2574.5	2068.4
31 Ga	2300.4	1902.2	1742.6	835.6	77 Ir	3490.2	2912.4	2679.1	0
32 Ge	2522.4	2085.9	1910.8	916.3	78 Pt	0	3041.3	2797.7	1940.2
33 As	2761.6	2283.6	2092.0	1003.1	79 Au	1057.9	3170.2	2916.3	2019.9
34 Se	3006.4	2486.1	2277.4	1092.1	80 Hg	1086.1	0	3041.9	1746.2
35 Br	3279.7	2712.1	2484.5	1191.3	81 Tl	1119.1	958.8	0	1822.4
36 Kr	3558.7	2942.8	2695.8	1292.7	82 Pb	1147.3	983.0	915.4	1895.2
37 Rb	3860.5	3192.4	2924.4	1402.3	83 Bi	1180.2	1011.2	941.6	1708.1
38 Sr	3553.0	3437.2	3148.7	1509.9	84 Po	1213.1	1039.4	967.9	1787.8
39 Y	3815.0	3154.7	3390.3	1625.7	85 At	1250.7	1071.6	997.9	1867.4
40 Zr	707.1	3404.2	3118.5	1749.8	86 Rn	1288.3	1103.9	1027.9	1950.6
41 Nb	762.6	636.3	2406.8	1882.1	87 Fr	1330.6	1140.1	1061.7	2040.7
42 Mo	822.6	686.4	631.4	1979.1	88 Ra	1391.8	1192.5	1110.4	0
43 Tc	885.3	738.7	679.5	2126.0	89 Ac	1448.2	1240.8	1155.5	635.6
44 Ru	951.6	794.1	730.5	1938.3	90 Th	1471.7	1261.0	1174.2	645.9
45 Rh	1021.6	852.5	784.2	2068.8	91 Pa	1523.4	1305.3	1215.5	668.6
46 Pd	1093.4	912.4	839.3	0	92 U	1579.8	1353.6	1260.5	693.4
47 Ag	1169.8	976.1	897.9	445.9	93 Np	1641.0	1406.0	1309.3	720.2
48 Cd	1250.7	1043.6	960.0	476.7	94 Pu	1711.5	1466.4	1365.5	751.2

Table 14.6. K Series X-Ray Wavelengths and Energies[a]

Element	$K\alpha_1$		K edge	
	λ (Å)	E (keV)	λ (Å)	E (keV)
4 Be	114.00	0.109	110.0	0.111
5 B	67.6	0.183	—	—
6 C	44.7	0.277	43.68	0.284
7 N	31.6	0.392	30.99	0.400
8 O	23.62	0.525	23.32	0.532
9 F	18.32	0.677	—	—
10 Ne	14.61	0.849	14.30	0.867
11 Na	11.91	1.041	11.57	1.072
12 Mg	9.89	1.254	9.512	1.303
13 Al	8.339	1.487	7.948	1.560
14 Si	7.125	1.740	6.738	1.84
15 P	6.157	2.014	5.784	2.144
16 S	5.372	2.308	5.019	2.470
17 Cl	4.728	2.622	4.397	2.820
18 A	4.192	2.958	3.871	3.203
19 K	3.741	3.314	3.437	3.608
20 Ca	3.358	3.692	3.070	4.038
21 Sc	3.031	4.091	2.762	4.489
22 Ti	2.749	4.511	2.497	4.965
23 V	2.504	4.952	2.269	5.464
24 Cr	2.290	5.415	2.070	5.989
25 Mn	2.102	5.899	1.896	6.538
26 Fe	1.936	6.404	1.743	7.111
27 Co	1.789	6.930	1.608	7.710
28 Ni	1.658	7.478	1.483	8.332
29 Cu	1.541	8.048	1.381	8.980
30 Zn	1.435	8.639	1.283	9.661
31 Ga	1.340	9.252	1.196	10.37
32 Ge	1.254	9.886	1.17	11.10

[a] Bearden (1964).

Table 14.7. L Series X-Ray Wavelengths and Energies[a]

	$L\alpha_1$		L_3 edge			$L\alpha_1$		L_3 edge	
Element	λ (Å)	E (keV)	λ (Å)	E (keV)	Element	λ (Å)	E (keV)	λ (Å)	E (keV)
22 Ti	27.42	0.452	—	—	58 Ce	2.562	4.840	2.166	5.723
23 V	24.25	0.511	—	—	59 Pr	2.463	5.034	2.079	5.963
24 Cr	21.64	0.573	20.7	0.598	60 Nd	2.370	5.230	1.997	6.209
25 Mn	19.45	0.637	—	—	61 Pm	2.282	5.433	1.919	6.461
26 Fe	17.59	0.705	17.53	0.707	62 Sm	2.200	5.636	1.846	6.717
27 Co	15.97	0.776	15.92	0.779	63 Eu	2.121	5.846	1.776	6.981
28 Ni	14.56	0.852	14.52	0.854	64 Gd	2.047	6.057	1.712	7.243
29 Cu	13.34	0.930	13.29	0.933	65 Tb	1.977	6.273	1.650	7.515
30 Zn	12.25	1.012	12.31	1.022	66 Dy	1.909	6.495	1.592	7.790
31 Ga	11.29	1.098	11.10	1.117	67 Ho	1.845	6.720	1.537	8.068
32 Ge	10.44	1.188	10.19	1.217	68 Er	1.784	6.949	1.484	8.358
33 As	9.671	1.282	9.37	1.324	69 Tm	1.727	7.180	1.433	8.650
34 Se	8.99	1.379	8.65	1.434	70 Yb	1.672	7.416	1.386	8.944
35 Br	8.375	1.480	7.984	1.553	71 Lu	1.620	7.656	1.341	9.249
36 Kr	7.817	1.586	7.392	1.677	72 Hf	1.57	7.899	1.297	9.558
37 Rb	7.318	1.694	6.862	1.807	73 Ta	1.522	8.146	1.255	9.877
38 Sr	6.863	1.807	6.387	1.941	74 W	1.476	8.398	1.216	10.20
39 Y	6.449	1.923	5.962	2.079	75 Re	1.433	8.653	1.177	10.53
40 Zr	6.071	2.042	5.579	2.223	76 Os	1.391	8.912	1.141	10.87
41 Nb	5.724	2.166	5.230	2.371	77 Ir	1.351	9.175	1.106	11.21
42 Mo	5.407	2.293	4.913	2.523	78 Pt	1.313	9.442	1.072	11.56
43 Tc	5.115	2.424	4.630	2.678	79 Au	1.276	9.713	1.040	11.92
44 Ru	4.846	2.559	4.369	2.838	80 Hg	1.241	9.989	1.009	12.29
45 Rh	4.597	2.697	4.130	3.002	81 Tl	1.207	10.27	0.979	12.66
46 Pd	4.368	2.839	3.907	3.173	82 Pb	1.175	10.55	0.951	13.04
47 Ag	4.154	2.984	3.699	3.351	83 Bi	1.144	10.84	0.923	13.43
48 Cd	3.956	3.134	3.505	3.538	84 Po	1.114	11.13	—	—
49 In	3.772	3.287	3.324	3.730	85 At	1.085	11.43	—	—
50 Sn	3.600	3.414	3.156	3.929	86 Rn	1.057	11.73	—	—
51 Sb	3.439	3.605	3.000	4.132	87 Fr	1.030	12.03	—	—
52 Te	3.289	3.769	2.856	4.342	88 Ra	1.005	12.34	0.803	15.44
53 I	3.149	3.938	2.720	4.559	89 Ac	0.9799	12.65	—	—
54 Xe	3.017	4.110	2.593	4.782	90 Th	0.956	12.97	0.761	16.30
55 Cs	2.892	4.287	2.474	5.011	91 Pa	0.933	13.29	—	—
56 Ba	2.776	4.466	2.363	5.247	92 U	0.911	13.61	0.722	17.17
57 La	2.666	4.651	2.261	5.484					

[a] Bearden (1964).

Table 14.8. M Series X-Ray Wavelengths and Energies[a]

Element	$M\alpha_1$		M_5 edge	
	λ (Å)	E (keV)	λ (Å)	E (keV)
72 Hf	7.539	1.645	—	—
73 Ta	7.252	1.710	7.11	1.743
74 W	6.983	1.775	6.83	1.814
75 Re	6.729	1.843	6.56	1.89
76 Os	6.490	1.910	6.30	1.967
77 Ir	6.262	1.980	6.05	2.048
78 Pt	6.047	2.051	5.81	2.133
79 Au	5.840	2.123	5.58	2.220
80 Hg	5.645	2.196	5.36	2.313
81 Tl	5.460	2.271	5.153	2.406
82 Pb	5.286	2.346	4.955	2.502
83 Bi	5.118	2.423	4.764	2.603
90 Th	4.138	2.996	3.729	3.325
91 Pa	4.022	3.082	—	—
92 U	3.910	3.171	3.497	3.545

[a] Bearden (1964).

Table 14.9. Fitting Parameters for Duncumb–Reed Backscattering Correction Factor R
[See Equation (7.20), Chapter 7][a]

U	R_1^1	R_2^1	R_3^1	U	R_1^1	R_2^1	R_3^1
1	1.26	0.069	0.347	5.6	2.16	0.354	0.533
1.1	1.32	0.089	0.639	5.7	2.15	0.357	0.535
1.2	1.38	0.108	0.778	5.8	2.14	0.352	0.538
1.3	1.44	0.127	0.834	5.9	2.13	0.349	0.540
1.4	1.50	0.145	0.846	6	2.12	0.346	0.543
1.5	1.55	0.161	0.835	6.1	2.11	0.343	0.545
1.6	1.60	0.178	0.811	6.2	2.10	0.341	0.547
1.7	1.65	0.193	0.783	6.3	2.09	0.338	0.550
1.8	1.70	0.207	0.752	6.4	2.08	0.335	0.552
1.9	1.74	0.221	0.723	6.5	2.07	0.332	0.555
2	1.78	0.234	0.695	6.6	2.06	0.329	0.557
2.1	1.82	0.247	0.670	6.7	2.05	0.326	0.560
2.2	1.86	0.258	0.647	6.8	2.05	0.323	0.562
2.3	1.89	0.269	0.626	6.9	2.04	0.321	0.564
2.4	1.93	0.280	0.608	7	2.03	0.318	0.567
2.5	1.96	0.289	0.592	7.1	2.02	0.315	0.569
2.6	1.98	0.298	0.578	7.2	2.01	0.313	0.571
2.7	2.01	0.307	0.566	7.3	2.00	0.311	0.574
2.8	2.04	0.315	0.555	7.4	2.00	0.308	0.576
2.9	2.06	0.322	0.546	7.5	1.99	0.306	0.578
3	2.08	0.329	0.538	7.6	1.98	0.305	0.581
3.1	2.10	0.335	0.532	7.7	1.98	0.303	0.583
3.2	2.12	0.340	0.526	7.8	1.97	0.301	0.585
3.3	2.13	0.345	0.522	7.9	1.97	0.300	0.588
3.4	2.15	0.350	0.518	8	1.97	0.299	0.590
3.5	2.16	0.354	0.515	8.1	1.96	0.299	0.592
3.6	2.17	0.358	0.513	8.2	1.96	0.298	0.594
3.7	2.18	0.361	0.511	8.3	1.96	0.298	0.596
3.8	2.19	0.364	0.510	8.4	1.96	0.298	0.598
3.9	2.19	0.366	0.509	8.5	1.96	0.299	0.600
4	2.20	0.368	0.509	8.6	1.97	0.300	0.603
4.1	2.20	0.369	0.509	8.7	1.97	0.301	0.605
4.2	2.21	0.370	0.510	8.8	1.97	0.303	0.607
4.3	2.21	0.371	0.510	8.9	1.98	0.305	0.609
4.4	2.21	0.372	0.511	9	1.99	0.307	0.611
4.5	2.21	0.372	0.512	9.1	2.00	0.310	0.613
4.6	2.21	0.372	0.514	9.2	2.01	0.313	0.615
4.7	2.21	0.371	0.515	9.3	2.02	0.317	0.616
4.8	2.21	0.370	0.517	9.4	2.03	0.321	0.618
4.9	2.20	0.370	0.518	9.5	2.05	0.326	0.620
5	2.20	0.368	0.520	9.6	2.07	0.331	0.622
5.1	2.19	0.367	0.522	9.7	2.08	0.337	0.624
5.2	2.19	0.365	0.524	9.8	2.10	0.343	0.626
5.3	2.18	0.363	0.526	9.9	2.13	0.350	0.628
5.4	2.17	0.361	0.529	10	2.15	0.357	0.629
5.5	2.17	0.359	0.531				

[a] Duncumb and Reed (1968).

Table 14.10. J Values[a] and Fluorescent Yield, ω, Values[b]

Z	J (eV)	ω_K	ω_{L3}	Z	J (eV)	ω_K	ω_{L3}	Z	J (eV)	ω_K	ω_{L3}
1	68.2	—	—	35	371	0.628	—	68	689	—	0.221
2	70.8	—	—	36	380	0.660	—	69	699	—	0.231
3	76.7	—	—	37	390	0.680	0.01	70	709	—	0.241
4	83.9	—	—	38	400	0.702	—				
5	91.8	—	—	39	409	0.719	—	71	718	—	0.251
6	100	0.0009	—	40	419	0.737	—	72	728	—	0.262
7	108	0.0015	—					73	738	—	0.272
8	117	0.0022	—	41	429	0.754	—	74	748	—	0.284
9	126	0.004	—	42	438	0.770	—	75	757	—	0.295
10	135	0.008	—	43	448	0.785	—	76	767	—	0.307
				44	457	0.799	—	77	777	—	0.319
11	144	0.01	—	45	467	0.812	—	78	786	—	0.331
12	153	0.028	—	46	477	0.822	—	79	796	—	0.344
13	162	0.038	—	47	486	0.833	0.05	80	806	—	0.357
14	172	0.038	—	48	496	0.843	—				
15	181	0.084	—	49	506	—	—	81	815	—	0.370
16	190	0.098	—	50	515	—	0.089	82	825	—	0.384
17	200	0.113	—					83	835	—	0.398
18	209	0.130	—	51	525	—	0.095	84	845	—	0.412
19	218	0.147	—	52	535	—	0.100	85	854	—	0.427
20	228	0.166	—	53	544	—	0.106	86	864	—	0.442
				54	554	—	0.112	87	874	—	0.457
21	237	0.187	—	55	564	—	0.118	88	883	—	0.472
22	247	0.209	—	56	573	—	0.125	89	893	—	0.488
23	256	0.232	—	57	583	—	0.131	90	903	—	0.505
24	266	0.257	—	58	593	—	0.138				
25	275	0.283	—	59	602	—	0.146	91	912	—	0.522
26	285	0.310	—	60	612	—	0.153	92	922	—	0.539
27	294	0.339	—					93	932	—	0.556
28	304	0.370	—	61	622	—	0.160	94	942	—	0.574
29	313	0.402	—	62	631	—	0.168	95	951	—	—
30	323	0.436	—	63	641	—	0.177	96	961	—	—
				64	651	—	0.185	97	971	—	—
31	333	0.471	—	65	660	—	0.194	98	980	—	—
32	342	0.508	—	66	670	—	0.202	99	990	—	—
33	352	0.546	—	67	680	—	0.212	100	1000	—	—
34	361	0.586	—								

[a] Berger and Seltzer (1964).
[b] Fink et al. (1966).

Table 14.11. Important Properties

Element	Symbol	Thermal Conductivity at 300 K (W cm^{-1} K^{-1})	Resistivity at 300 K ($\mu\Omega$ cm)	Melting point (K)	Boiling point (K)	Vaporization temperature at 1.3 Pa
Aluminum	Al	2.37	2.83	932	2330	1273
Antimony	Sb	0.243	41.7	903	1713	951
Barium	Ba	0.184	50.0	990	1911	902
Beryllium	Be	2.00	4.57	1557	3243	1519
Bismuth	Bi	0.079	119.0	544	1773	971
Boron	B	0.270	1.8×10^{12}	2573	2823	1628
Cadmium	Cd	0.968	7.50	594	1040	537
Calcium	Ca	2.00	4.60	1083	1513	878
Carbon	C	1.29	3500	4073	4473	2954
Chromium	Cr	0.937	13.0	2173	2753	1478
Cobalt	Co	1.00	9.7	1751	3173	1922
Copper	Cu	4.01	1.67	1356	2609	1393
Germanium	Ge	0.599	89×10^3	1232	3123	1524
Gold	Au	3.17	2.40	1336	2873	1738
Indium	In	0.816	8.37	430	2273	1225
Iridium	Ir	1.47	6.10	2727	4773	2829
Iron	Fe	0.802	10.0	1811	3273	1720
Lead	Pb	0.353	22.0	601	1893	991
Magnesium	Mg	1.56	4.60	924	1380	716
Manganese	Mn	0.078	5.0	1517	2360	1253
Molybdenum	Mo	1.38	5.70	2893	3973	2806
Nickel	Ni	0.907	6.10	1725	3173	1783
Palladium	Pd	0.718	11.0	1823	3833	1839
Platinum	Pt	0.716	10.0	2028	4573	2363
Rhodium	Rh	1.50	4.69	2240	2773	2422
Silicon	Si	1.48	23×10^4	1683	2773	1615
Silver	Ag	4.29	1.60	1233	2223	1320
Strontium	Sr	0.353	23.0	1044	1657	822
Tantalum	Ta	0.575	13.1	3269	4373	3273
Thorium	Th	0.540	18.0	2100	4473	2469
Tin	Sn	0.666	11.4	505	2610	1462
Titanium	Ti	0.219	42.0	2000	3273	1819
Tungsten	W	1.74	5.50	3669	6173	3582
Vanadium	V	0.307	18.2	1970	3673	2161
Zinc	Zn	1.16	5.92	693	1180	616
Zirconium	Zr	0.21	40.0	2125	4650	2284

of Selected Coating Elements

Latent heat of evaporation ($kJ\,g^{-1}$)	Specific heat at 300 K ($J\,g^{-1}\,K^{-1}$)	Relative sputtering yield (atoms/600 eV Ar$^+$)	Relative cost/g highest purity	Evaporation technique
12.77	0.900	1.2	14	Ta, W; coil, basket; wets and alloys
1.603	0.205	—	1	Ta; basket, wets
—	0.193	—	1	Ta, Mo, W; basket, boats; wets
24.77	1.825	—	115	Ta, W, Mo; basket; wets, toxic
0.855	0.124	—	1	Ta, W, Mo; basket, boat
34.70	1.026	—	7	Carbon crucible, forms carbides
1.199	0.232	—	52	Ta, W, Mo; basket, boat, wets, toxic
—	0.653	—	1	W; basket
—	0.712	—	1	Pointed rods, resistive, evaporation
6.170	0.448	1.3	1	W; basket
6.280	0.456	1.4	6	W; basket, alloys easily
4.810	0.385	2.3	45	Ta, W, Mo; basket, boats, not wet easily
—	0.322	1.2	—	Ta, Mo, W; basket, boat, wets Ta, Mo
1.740	0.129	2.8	60	Mo, W, loop; basket, boat, wets, alloys Ta
2.030	0.234	—	21	W, Fe; basket; Mo, boat
3.310	0.133	—	51	Refractory; thick W loop
6.342	0.444	1.3	24	W; loop, coil, alloys easily
0.857	0.159	—	12	Fe; basket, boat, not wet W, Ta, Mo
5.597	1.017	—	1	Ta, W, Mo; basket, boat, wets
4.092	0.481	—	1	Ta, W, Mo, loop; basket, wets
5.115	0.251	0.9	3	Refractory
6.225	0.444	1.5	20	W; coil (heavy), alloys
0.370	0.245	2.4	37	W; loop, coil
2.620	0.133	1.6	26	Stranded with W, Ta; alloys
4.814	0.244	1.5	39	Sublimation from W, low pressure
10.59	0.703	0.5	1	BeO crucible, contaminates with SiO
2.330	0.236	3.4	22	Ta, W, Mo; coil, basket, boat; not wet easily
—	0.298	—	1	Ta, W, Mo; basket; wets
4.165	0.140	0.6	8	Refractory
2.340	0.133	—	52	W; basket, wets
2.400	0.222	—	18	Ta, Mo; basket, boat; wets
9.837	0.523	0.6	5	Ta, W; loop, cool, basket; alloys
4.345	0.133	0.6	6	Refractory
9.000	0.486	—	25	W, Mo; basket; alloys with W
1.756	0.389	—	1	Ta, W, Mo; loop, basket; wets
5.693	0.281	—	5	Ta; basket; forms oxide

References

Abraham, J. L. (1977). *SEM/1977/II*, IIT Research Institute, Chicago, Illinois, p. 119.

Abraham, J. L., and DeNee, P. B. (1974). *SEM/1974*, IIT Research Institute, Chicago, Illinois, p. 251.

Adachi, K., Hojou, K., Katoh, M., and Kanaya, K. (1976). *Ultramicroscopy* **2**, 17.

Albee, A. L., and Ray, L. (1970). *Anal. Chem.* **42**, 1408.

Albrecht, R. M., and Wetzel, B. (1979). *SEM/1979/III*, SEM Inc., AMF O'Hare, Illinois, p. 203.

Albright, J., and Lewis, R. (1965). *J. Appl. Phys.* **36**, 2615.

Andersen, C. A., and Hasler, M. F. (1966). *X-ray Optics and Microanalysis, 4th Intl. Cong. on X-Ray Optics and Microanalysis*, eds. R. Castaing, P. Deschamps, and J. Philibert, Hermann, Paris, p. 310.

Andersen, C. A., Keil, K., and Mason, B. (1964). *Science* **146**, 256.

Anderson, R. L. (1961). "Revealing Microstructures in Metals," Westinghouse Research Laboratories, Scientific Paper 425-C000-P2.

Anderson, T. F. (1951). *Trans. N. Y. Acad. Sci.* **13**, 130.

Appleton, T. C. (1974). *J. Microsc.* **100**, 49.

Appleton, T. C. (1978). In *Electron Microprobe Analysis in Biology*, Chap. 5, ed. D. A. Erasmus, Chapman Hall, London.

Appleton, T. C., and Newell, P. F. (1977). *Nature* **266**, 854.

Armstrong, J. T. (1978). *SEM/1978/I*, SEM Inc., AMF O'Hare, Illinois, p. 455.

Armstrong, J. T., and Buseck, P. R. (1975). *Anal. Chem.* **47**, 2178.

Arnal, F., Verdier, P., and Vincinsini, P-D. (1969). *C. R. Acad. Sci. Paris* **268**, 1526.

Asquith, M. H., and Reid, D. S. (1968). *Cryoletters* **1**, 352.

ASTM (1960). "Methods of Metallographic Specimen Preparation," ASTM STP 285, Philadelphia, Pennsylvania.

Bahr, G. F., Johnson, F. B., and Zeitler, E. (1965). *Lab Invest.* **14**, 1115.

Bambynek, W., Crasemann, B., Fink, R. W., Freund, H. U., Mark, H., Swift, C. D., Price, R. E., and Rao, P. V. (1972). *Rev. Mod. Phys.* **44**, 716.

Barber, T. A. (1976). In *Principles and Techniques of SEM: Biological Applications*, Vol. 5, ed. M. A. Hayat, Van Nostrand Reinhold, New York.

Barber, V. C. (1972). *SEM/1972*, IIT Research Institute, Chicago, Illinois, p. 321.

Barbi, N. C. (1979). *SEM/1979/II*, SEM Inc., AMF O'Hare, Illinois, p. 659.

Barbi, N. C., and Skinner, D. P. (1976). *SEM/1976/I*, IIT Research Institute, Chicago, Illinois, p. 393.

Barkalow, R. H., Kraft, R. W., and Goldstein, J. I. (1972). *Met. Trans.* **3**, 919.

Barnard, T. (1980). *J. Microsc.* **120**, 93.

649

Barnard, T., and Seveus, L. (1978). *J. Microsc.* **112**, 281.

Barnard, T., and Thomas, R. S. (1978). *J. Microsc.* **113**, 269.

Bartlett, A. A., and Burstyn, H. P. (1975). *SEM/1975*, IIT Research Institute, Chicago, Illinois, p. 305.

Bauer, E. L. (1971). *A Statistical Manual for Chemists*, 2nd ed., Academic, New York, p. 189.

Beaman, D., and Isasi, J. (1972). "Electron Beam Microanalysis," ASTM STP 506, Philadelphia, Pennsylvania, p. 27.

Bearden, J. A. (1964). "X-Ray Wavelengths," Report NYO 10586, U.S. Atomic Energy Commission, Oak Ridge, Tennessee.

Bearden, J. A. (1967). "X-Ray Wavelengths and X-Ray Atomic Energy Levels," NSRDS-NBS 14, National Bureau of Standards, Washington.

Becker, R. P., and DeBryun, P. P. H. (1976). *SEM/1976/II*, IIT Research Institute, Chicago, Illinois, p. 171.

Beeuwkes, R. (1979). *SEM/1979/II*, SEM Inc., AMF O'Hare, Illinois, p. 767.

Bence, A. E., and Albee, A. (1968). *J. Geol.* **76**, 382.

Berger, M. J. (1963). *Methods in Computational Physics*, Vol. 1, eds. B. Adler, S. Fernback, and M. Rotenberg, Academic, New York.

Berger, M. J., and Seltzer, S. M. (1964). Nat. Acad. Sci./Nat. Res. Council Publ. 1133, Washington, 205.

Bertin, E. P. (1975). *Principles and Practice of X-Ray Spectrometric Analysis*, Plenum, New York.

Bethe, H. (1930). *Ann. Phys. (Leipzig)* **5**, 325.

Bethe, H. A. (1933). In *Handbook of Physics*, Vol. 24, Springer, Berlin, p. 273.

Birks, L. S. (1971). *Electron Probe Microanalysis*, 2nd ed., Wiley/Interscience, New York.

Birks, L. S., and Brooks, E. J. (1957). *Rev. Sci. Instr.* **28**, 709.

Bishop, H. (1966). *X-Ray Optics and Microanalysis, 4th Intl. Cong. on X-Ray Optics and Microanalysis*, eds. R. Castaing, P. Deschamps, and J. Philibert, Hermann, Paris, p. 153.

Bishop, H. E. (1965). *Proc. Phys. Soc.* **85**, p. 855.

Bolon, R. B., and Lifshin, E. (1973a). "Proc. 8th Nat. Conf. Elect. Probe Anal. EPASA, New Orleans," Paper 31.

Bolon, R. B., and Lifshin, E. (1973b). *SEM/1973*, IIT Research Institute, Chicago, Illinois, p. 285.

Bolon, R. B., and McConnell, M. D. (1976). *SEM/1976/I*, IIT Research Institute, Chicago, Illinois, p. 163.

Booker, G. R. (1970). In *Modern Diffraction Techniques in Materials Science*, ed. S. Amelinclex, North-Holland, Amsterdam, p. 647.

Borile, F., and Garulli, A. (1978). *X-Ray Spectrometry* **7**, 124.

Borovskii, I., and Ilin, N. P. (1953). *Dokl. Akad. Nauk SSR* **106**, 655.

Bowen, I. D., and Ryder, T. A. (1978). "The Application of X-Ray Microanlaysis to Histochemistry," in *Electron Microprobe Analysis in Biology*, ed. D. A. Erasmus, Chapman Hall, London.

Boyde, A. (1973). *J. Microsc.* **98**, 452.

Boyde, A. (1974a). *SEM/1974*, IIT Research Institute, Chicago, Illinois, p. 101.

Boyde, A. (1974b). *SEM/1974*, IIT Research Institute, Chicago, Illinois, p. 93.

Boyde, A. (1976). *SEM/1976/I*, IIT Research Institute, Chicago, Illinois, p. 683.

Boyde, A. (1978). *SEM/1978/II*, SEM Inc., AMF O'Hare, Illinois, p. 203.

Boyde, A. (1979). *SEM/1979/II*, SEM Inc., AMF O'Hare, Illinois, p. 67.

Boyde, A. (1980). "Proc. 7th European Cong. of Elec. Microscopy," Vol. II, Den Haag, Netherlands, p. 768.

Boyde, A., and Howell, P. G. T. (1977). *SEM/1977/I*, IIT Research Institute, Chicago, Illinois, p. 571.

Boyde, A., and Maconnachie, E. (1979). *Scanning* **2**, 164.

Boyde, A., and Stewart, A. D. G. (1962). *J. Ultrastruc. Res.* **7**, 159.

Bracewell, B. L., and Veigele, W. J. (1971). *Developments in Applied Spectroscopy*, Vol. 9, Plenum, New York, p. 357.

Brätan, T. (1978). *J. Microsc.* **113**, 53.

Broers, A. N. (1969a). *Rev. Sci. Instr.* **40**, 1040.

Broers, A. N. (1969b) *J. Phys. E* **2**, 273.

Broers, A. N. (1973). *Microprobe Analysis*, ed. C. A. Andersen, J. Wiley, New York, p. 83.

Broers, A. N. (1974a) "8th Int'l. Cong. on Electron Microscopy, Canberra," Vol. 1, p. 54.

Broers, A. N. (1974b). *SEM/1974*, IIT Research Institute, Chicago, Illinois, p. 9.

Broers, A. N. (1975). *SEM/1975*, IIT Research Institute, Chicago, Illinois, p. 662.

Broers, A. N., Panessa, B. J., and Gennaro, J. F., Jr. (1975). *Science* **189**, 637.

Brown, J. A., and Teetsov, A. (1976). *SEM/1976/I*, IIT Research Institute, Chicago, Illinois, p. 385.

Brown, J. D., von Rosenstiel, A. P., and Krish, T. (1979). *Microbeam Analysis—1979*, ed. D. E. Newbury, San Francisco Press, San Francisco, p. 241.

Bruining, H. (1954). *Physics and Application of the Secondary Emission Process*, Pergamon, London.

Buchheim, W., and Welsch, U. (1977). *J. Microsc.* **111**, 339.

Bulger, R. E., Beeuwkes, R., and Saubermann, A. J. (1981). *J. Cell Biol.* **88**, 274.

Burstone, M. S. (1969). *Phys. Tech. Biol. Res.* **IIIc**, 1.

Cadwell, D. E., and Weiblen, P. W. (1965). *Economic Geology* **60**, 1320.

Cameron, I. L., and Smith, N. K. R. (1980). *SEM/1980/II*, SEM Inc., AMF O'Hare, Illinois, p. 463.

Castaing, R. (1951). Ph.D. thesis, University of Paris.

Castaing, R. (1960). In *Advances in Electronics and Electron Physics*, Vol. 13, ed. L. Marton, Academic, New York, p. 317.

Castaing, R., and Deschamps, J. (1954). *C. R. Acad. Sci. Paris* **238**, 1506.

Castaing, R., and Guinier, A. (1950). "Proceedings 1st Intl. Conf. on Electron Microscopy," Delft 1949, p. 60.

Castaing, R., and Henoc, J. (1966). *X-Ray Optics and Microanalysis, 4th Intl. Cong. on X-Ray Optics and Microanalysis*, eds. R. Castaing, P. Deschamps, and J. Philibert, Hermann, Paris, p. 120.

Catto, C. J. D. (1972). Ph.D. dissertation, Cambridge University.

Champness, P. E., Cliff, G., and Lorimer, G. W. (1976). *J. Microsc.* **108**, 231.

Chandler, J. A. (1977). "X-Ray Microanalysis in the Electron Microscope," in *Practical Methods in Electron Microscopy*, Vol. 5, ed. A. M. Glauert, North-Holland, Amsterdam.

Chandler, J. A., and Battersby, S. (1979). "X-Ray Microanalysis of Diffusible and Non-Diffusible Elements in Ultra Thin Biological Tissues Using Human Sperm Cells as a Model for Investigating Specimen Preparation," in *Microbeam Analysis in Biology*, eds. C. Lechene and R. Warner, Academic, New York.

Chang, S. H., Mergner, W. J., Pendergrass, R. E., Bulger, R. E., Berezesky, I. K., and Trump, B. F. (1980). *J. Histochem. Cytochem.* **28**, 47.

Chatfield, E. J., and Dillon, M. J. (1978). *SEM/1978/I*, SEM Inc., AMF O'Hare, Illinois, p. 487.

Clark, J., Echlin, P., Moreton, R., Saubermann, A., and Taylor, P. (1976). *SEM/1976/I*, IIT Research Institute, Chicago, Illinois, p. 83.

Clay, C. S., and Peace, G. W. (1981). "Ion Beam Sputtering: An Improved Method of Metal Coating SEM Samples and Shadowing CTEM Samples," *J. Microsc.* **122**, 281.

Cleaver, J. R. A., and Smith. K. C. A. (1973). *SEM/1973*, IIT Research Institute, Chicago, Illinois, p. 50.

Cliff, G., and Lorimer, G. W. (1975). *J. Microsc.* **103**, 203.

Coates, D. G. (1967). *Phil. Mag.* **16**, 1179.

652

Cobet, U. (1972). "3rd Int. Conf. Med. Phys. Goteborg, Sweden," p. 324.

Cobet, U., and Millner, R. (1973). *Wiss. Z. Univ. Hall* **22**, 127.

Cohen, A. L. (1979). *SEM/1979/II*, SEM Inc., AMF O'Hare, Illinois, p. 303.

Colby, J. W. (1965). National Lead Co., Ohio Report NLCO-969.

Colby, J. W. (1968). "Quantitative Microprobe Analysis of Thin Insulating Films" in *Advances in X-Ray Analysis*, Vol. 11, eds. J. B. Newkirk, G. R. Mallett, and H. G. Pfeiffer, Plenum, New York, p. 287.

Colby, J. W., Wondisler, D. R., and Conley, D. K. (1969). "Proc. 4th Nat. Conf. Elect. Prob. Anal." EPASA, Pasadena, California, Paper 9.

Coleman, J. R. (1978). *SEM/1978/II*, SEM Inc., AMF O'Hare, Illinois, p. 911.

Coleman, J. R. and Terepka, A. R. (1974). "Preparatory Methods for Electron Probe Analysis" in *Principles and Techniques of Electron Microscopy*, Vol. 4, ed. M. A. Hayat, Van Nostrand Reinhold, New York, Chap. 8.

Compton, A. H., and Allison, S. K. (1943). *X-Rays in Theory and Experiment*, Van Nostrand, New York, p. 106.

Considine, D. M. (1976). *Van Nostrand's Scientific Encyclopedia*, 5th ed., Van Nostrand, New York, p. 710.

Cooke, B. A., and Stewardson, E. A. (1964). *Brit. J. Appl. Phys.* **15**, 1315.

Cosslett, V. E. (1966). *X-Ray Optics and Microanalysis, 4th Intl. Cong. on X-Ray Optics and Microanalysis*, eds. R. Castaing, P. Deschamps, and J. Philibert, Hermann, Paris, p. 85.

Cosslett, V. E., and Duncumb, P. (1956). *Nature* **177**, 1172.

Cosslett, V. E., and Thomas, R. N. (1964a). *Brit. J. Appl. Phys.* **15**, 883.

Cosslett, V. E., and Thomas, R. N. (1964b). *Brit. J. Appl. Phys.* **15**, 1283.

Cosslett, V. E., and Thomas, R. N. (1965). *Brit. J. Appl. Phys.* **16**, 779.

Cosslett, V. E., and Thomas, R. N. (1966). In *The Electron Microprobe*, eds. T. D. McKinley, K. F. J. Heinrich, and D. B. Wittry, John Wiley, New York, p. 248.

Costello, M. J., and Cortess, J. M. (1978). *J. Microsc.* **112**, 17.

Crawford, C. K. (1979). *SEM/1979/II*, SEM Inc., AMF O'Hare, Illinois, p. 31.

Crewe, A. V. (1969). *SEM/1969*, IIT Research Institute, Chicago, Illinois, p. 11.

Crewe, A. V., Eggenberger, E. N., Wall, J., and Welter, L. M. (1968). *Rev. Sci. Instr.* **39**, 576.

Crewe, A. V., Isaacson, M., and Johnson, D. (1970). *Rev. Sci. Instr.* **41**, 20.

Criss, J. W., and Birks, L. S. (1966). In *The Electron Microprobe*, eds. T. D. McKinley, K. F. J. Heinrich, and D. B. Wittry, John Wiley, New York, p. 217.

Crissman, R. S., and McCann, P. (1979). *Micron* **10**, 37.

Curgenven, L., and Duncumb, P. (1971). Tube Investments Res. Labs. Report 303.

Cuthill, J. R., Wyman, L. L., and Yakowitz, H. (1963). *J. Met.* **15**, 763.

Dahlberg, E. P. (1976). *SEM/1976/I*, IIT Research Institute, Chicago, Illinois, p. 715.

Davies, T. W., and Morgan, A. J. (1976). *J. Microsc.* **107**, 47.

Dawson, P. H. (1966). *J. Appl. Phys.* **37**, 3644.

Deming, S. N., and Morgan, S. L. (1973). *Anal. Chem.* **45**(3), 278A.

DeNee, P. B. (1978). *SEM/1978/I*, SEM Inc., AMF O'Hare, Illinois, p. 479.

DeNee, P. B., and Abraham, J. L. (1976). in *Principles and Techniques of SEM: Biological Applications*, Vol. 5, Van Nostrand, NY, Chap. 8, p. 144.

DeNee, P. B., Frederickson, R. G., and Pope, R. S. (1977). *SEM/1977/II*, IIT Research Institute, Chicago, Illinois, p. 83.

Dengler, L. A. (1978). *SEM/1978/I*, SEM Inc., AMF O'Hare, Illinois, p. 603.

Derian, J. D., and Castaing, R. (1966). In *Optique des Rayons X et Microanalyse*, eds. R. Castaing, P. Deschamps, and J. Philibert, Hermann, Paris, p. 193.

Dingley, D. J. (1969). *Micron* **1**, 206.

Dinnis, A. R. (1971). *SEM/1971*, IIT Research Institute, Chicago, Illinois, p. 41.

Dinnis, A. R. (1972). *Electron Microscopy 1972, Proc. 5th European Cong. of Electron Microscopy*, Institute of Physics, London, p. 178.

Dinnis, A. R. (1973). In *Scanning Electron Microscopy: Systems and Applications 1973*, Institute of Physics, London, Conf. Ser., Vol. 18, p. 76.

Dolby, R. M. (1959). *Proc. Phys. Soc.* **73**, 81.

Dörge, A., Rick, R., Gehring, K., Mason, J., and Thurau, K. (1975). *J. Microsc. Biol. Cell* **22**, 205.

Dörge, A., Rick, R., Gehring, K., and Thurau, K. (1978). *Pflügers Arch. European J. Physiol.* **373**, 85.

Drier, T. M., and Thurston, E. L. (1978). *SEM/1978/II*, SEM Inc., AMF O'Hare, Illinois, p. 843.

Duerr, J. S., and Ogilvie, R. E. (1972). *Anal. Chem.* **44**, 2361.

Duncumb, P. (1971). *Proc. EMAG*, Institute of Physics, London, Conf. Ser., Vol. 10, p. 132.

Duncumb, P., and Melford, D. A. (1966). *X-Ray Optics and Microanalysis, 4th Intl. Cong. on X-Ray Optics and Microanalysis*. eds. R. Castaing, P. Deschamps, and J. Philibert, Hermann, Paris, p. 240.

Duncumb, P., and Reed, S. J. B. (1968). In *Quantitative Electron Probe Microanalysis*, ed. K. F. J. Heinrich, National Bureau of Standards Spec. Pub. 298, p. 133.

Duncumb, P., and Shields, P. K. (1966). In *The Electron Microprobe*, eds. T. D. McKinley, K. F. J. Heinrich, and D. B. Wittry, John Wiley, New York, p. 284.

Dyson, N. A. (1956). Ph.D. thesis, University of Cambridge.

Echlin, P. (1975a). *SEM/1975*, IIT Research Institute, Chicago, Illinois, p. 217.

Echlin, P. (1975b). *J. Microsc. Biol. Cell* **22**, 129.

Echlin, P. (1978). *J. Microsc.* **112**, 47.

Echlin, P., and Galle, P. (1976). *Biological Microanalysis*, Pub. Societe Francaise de Microscopie Electronique, 24 rue Lhomond 75231 Paris.

Echlin, P., and Kaufmann, R. (eds.) (1978). *Microscopita Acta Suppl.* 2, S. Hirzel Verlag, Stuttgart.

Echlin, P., and Kaye, G. (1979). *SEM/1979/II*, SEM Inc., AMF O'Hare, Illinois, p. 21.

Echlin, P., Lai, C. E., Hayes, T. L., and Hook, G. (1980). *SEM/1980/II*, SEM Inc., AMF O'Hare, Illinois, p. 383.

Echlin, P., Pawley, J. B., and Hayes, T. L. (1979). *SEM/1979/III*, SEM Inc., AMF O'Hare, Illinois, p. 69.

Echlin, P., Ralph, B., and Weibel, E. (eds.) (1978). *Low Temperature Biological Microscopy and Microanalysis*, Royal Microscopical Society, Oxford.

Echlin, P., and Saubermann, A. J. (1977). *SEM/1977/I*, IIT Research Institute, Chicago, Illinois, p. 621.

Echlin, P., Saubermann, A. J., Franks, F., and Skaer, H. leB. (1978). *Microscopica Acta Suppl.* 2, S. Hirzel Verlag, Stuttgart, p. 64.

Echlin, P., Skaer, H. leB., Gardiner, B. O. C., Franks, F., and Asquith, H. M. (1977). *J. Microsc.* **110**, 239.

Edelman, L. (1979). *Mikroskopie* **35**, 31.

Eisenstadt, L. F., Levin, B., Golomb, H. M., and Riddell, R. H. (1976). *SEM/1976/II*, IIT Research Institute, Chicago, Illinois, p. 263.

Elad, E., Inskeep, C. N., Sareen, R. A., and Nestor, P. (1973). *IEEE Trans. Nucl. Sci.* **20**, 354.

Ellisman, M. H., Friedman, P. L., and Hamilton, W. J. (1980). *J. Neurocytol.* **9**, 185.

Eshel, A., and Waisel, Y. (1978). *Micron* **9**, 155.

Evans, R. D. (1955). *The Atomic Nucleus*, McGraw-Hill, New York.

Everhart, T. E., Herzog, R. F., Chang, M. S., and DeVore, W. J. (1972). *Proc. 6th Intl. Conf. on X-Ray Optics and Microanalysis*, eds. G. Shinoda, K. Kohra, and T. Ichinokawa, Univ. of Tokyo Press, Tokyo, p. 81.

Everhart, T. E., and Thornley, R. F. M. (1960). *J. Sci. Instr.* **37**, 246.

Everhart, T. E., Wells, O. C., and Oatley, C. W. (1959). *J. Electron. Control* **7**, 97.

Exley, R. R., Butterfield, B. G., and Meylan, B. A. (1974). *J. Microsc.* **101**, 21.

Exley, R. R., Meylan, B. A., and Butterfield, B. G. (1977). *J. Microsc.* **110**, 75.

Fathers, D. J., Jakubovics, J. P., Joy, D. C., Newbury, D. E., and Yakowitz, H. (1973). *Phys. Status Solidi A* **20**, 535.

Fathers, D. J., Jakubovics, J. P., Joy, D. C., Newbury, D. E., and Yakowitz, H. (1974). *Phys. Status Solidi A* **22**, 609.

Ficca, J. F. (1968). "Proc. 3rd Natl. Conf. Elect. Prob. Anal. EPASA," Chicago, Illinois, Paper 15.

Fink, R. W., Jopson, R. C., Mark, H., and Swift, C. D. (1966). *Rev. Mod. Phys.* **38**, 513.

Fiori, C. E., and Myklebust, R. L. (1979). "Proc. American Nuclear Society Topical Conference, Mayaquez, Puerto Rico, April–May 1978," p. 139.

Fiori, C. E., Myklebust, R. L., Heinrich, K. F. J., and Yakowitz, H. (1976). *Anal. Chem.* **48**(1), 172.

Fiori, C. E., and Newbury, D. E. (1978). *SEM/1978/I*, SEM Inc., AMF O'Hare, Illinois, p. 401.

Fiori, C. E., Yakowitz, H., and Newbury, D. E. (1974). *SEM/1974*, IIT Research Institute, Chicago, Illinois, p. 167.

Fisher, D. B. (1972). *Plant Physiol.* **49**, 161.

Fisher, D. W., and Baun, W. L. (1967). *Norelco Reporter* **14**, 92.

Fisher, G., Kaneshiro, E. S., and Peters, P. D. (1976). *J. Cell Biol.* **69**, 429.

Fisher, G. L., and Farningham, G. D. (1972). "Quantitative Carbon Analysis of Nickel Steels with the Electron Probe Microanalyzer," ASM Materials Engrg. Cong., Cleveland, Ohio, October.

Fisher, R. M., and Schwarts, J. C. (1957). *J. Appl. Phys.* **28**, 1377.

Fitzgerald, R., Keil, K., and Heinrich, K. F. J. (1968). *Science* **159**, 528.

Flood, P. R. (1975). *SEM/1975*, IIT Research Institute, Chicago, Illinois, p. 287.

Flood, P. R. (1980). *SEM/1980/I*, SEM Inc., AMF O'Hare, Illinois, p. 155.

Fomenko, V. S. (1966). *Handbook of Thermionic Properties*, ed. G. V. Samsonov, Plenum, New York, p. 88.

Franks, F. (1980). *SEM/1980/II*, SEM Inc., AMF O'Hare, Illinois, p. 349.

Franks, F., Asquith, M. H., Hammond, C. C., Skaer, H. leB., and Echlin, P. (1977). *J. Microsc.* **110**, 223.

Franks, J., Clay, C. S., and Peace, G. W. (1980). *SEM/1980/I*, SEM Inc., AMF O'Hare, Illinois, p. 155.

Frasca, J. M., Carter, H. W., and Schaffer, W. A. (1978). *SEM/1978/II*, SEM Inc., AMF O'Hare, Illinois, p. 485.

Frederik, P. M., and Busing, W. M. (1981). *J. Microsc.* **122** (2), 217.

Freund, H. U., Hansen, J. S., Karttunen, E., and Fink, R. W. (1972). *Proceedings of the 1969 International Conference on Radioactivity and Nuclear Spectroscopy*, eds. J. H. Hamilton and J. C. Manthuruthil, Gordon & Breach, New York, p. 623.

Fuchs, W., Lindemann, B., Brombach, J. D., and Trösch, W. (1978). *J. Microsc.* **112**, 75.

Garland, C. D., Lee, A., and Dickison, M. R. (1979). *J. Microsc.* **116**, 227.

Garland, H. O., Brown, J. A., and Henderson, I. W. (1978). In *Electron Microprobe Analysis in Biology*, Chap. 7, ed. D. A. Erasmus, Chapman Hall, London.

Gavrilovic, J. (1980). National Bureau of Standards Spec. Pub. 533, Washington, DC, p. 21.

Gedcke, D. A., Ayers, J. B., and DeNee, P. B. (1978). *SEM/1978*, SEM Inc., AMF O'Hare, Illinois, p. 581.

Geissinger, H. D. (1972). *J. Microsc.* **95**, 471.

Geller, J. D. (1977). *SEM/1977/I*, IIT Research Institute, Chicago, Illinois, p. 281.

Gilfrich, J. V., Birks, L. S., and Criss, J. W. (1978). In *X-Ray Fluorescence Analysis of Environmental Samples*, ed. T. G. Dzubay, Ann Arbor Science Publ., p. 283.

Glaeser, R. M. (1980). In *Introduction to Analytical Electron Microscopy*, eds. J. J. Hren, J. I. Goldstein, and D. C. Joy, Plenum, New York, p. 423.

Glaeser, R. M., and Taylor, K. A. (1978). *J. Microsc.* **112**, 127.

Glauert, A. M. (ed.) (1974). *Practical Methods in Electron Microscopy*, North Holland, Amsterdam.

Goldmark, P. C., and Hollywood, J. J. (1951). *Proc. IRE*, **39**, p. 1314.

Goldstein, J. I. (1967). *J. Geophys. Res.* **72**, 4689.

Goldstein, J. I. (1979). In *Introduction to Analytical Electron Microscopy*, eds. J. J. Hren, J. I. Goldstein, and D. C. Joy, Plenum, New York, p. 83.

Goldstein, J. I., Axon, H. J., and Agrell, S. O. (1975). *Earth Planet. Sci. Lett.* **28**, 217.

Goldstein, J. I., Costley, J. L., Lorimer, G. W., and Reed, S. J. B. (1977). *SEM/1977*, IIT Research Institute, Chicago, Illinois, p. 315.

Goldstein, J. I., Hanneman, R. E., and Ogilvie, R. E. (1965). *Trans. Met. Soc. AIME* **233**, 812.

Goldstein, J. I., Hewins, R. H., and Romig, A. D., Jr. (1976). In *Proc. Lunar Sci. Conf. 7th*, Pergamon, New York, p. 807.

Goldstein, J. I. and Ogilvie, R. E. (1966). In *X-ray Optics and Microanalysis*, eds. R. Castaing, P. Deschamps and J. Philibert, Hermann, Paris, p. 594.

Gomer, R. (1961). *Field Emission and Field Ionization*, Harvard University Press, Cambridge, Massachusetts.

Green, M. (1962). Ph.D. thesis, University of Cambridge.

Green, M. (1963). *X-Ray Optics and Microanalysis, 3rd Intl. Symposium*, eds. H. A. Patee, V. E. Cosslett, and A. Engstrom, Academic, New York, p. 361.

Green, M., and Cosslett, V. E. (1961). *Proc. Phys. Soc.* **78**, p. 1206.

Green, M., and Cosslett, V. E. (1968). *J. Phys. E* **1**, 425.

Gupta, B. L., Hall, T. A., and Moreton, R. B. (1977). In *Transport of Ions and Water in Animals*, Chap. 4, eds. B. L. Gupta, R. M. Moreton, J. L. Oschman, and B. J. Wall, Academic, New York.

Haggis, G. H., and Bond, E. F. (1979). *J. Microsc.* **115**, 225.

Haggis, G. H., Bond, E. F., and Phipps, B. (1976). *SEM/1976/I*, IIT Research Institute, Chicago, Illinois, p. 281.

Haine, M. E., and Mulvey, T. (1959). *J. Sci. Instr.* **26**, 350.

Hall, C. E. (1953). *Introduction to Electron Microscopy*, McGraw Hill, New York.

Hall, T. A. (1971). In *Physical Techniques in Biological Research*, Vol. IA, ed. G. Oster, Academic, New York, p. 157.

Hall, T. A. (1975). *J. Microsc. Biol. Cell* **22**, 271.

Hall, T. A. (1979a). In *Microbeam Analysis in Biology*, eds. C. Lechene and R. Warner, Academic, New York, p. 185.

Hall, T. A. (1979b). *J. Microsc.* **117**, 145.

Hall, T. A., Echlin, P., and Kaufmann, R. (eds.) (1974). *Microprobe Analysis as Applied to Cells and Tissues*, Academic, New York.

Hall, T. A., and Gupta, B. L. (1979). In *Introduction to Analytical Electron Microscopy*, Chap. 5, eds. J. J. Hren, J. I. Goldstein, and D. C. Joy, Plenum, New York.

Hall, T. A., and Werba, P. (1969). *Proc. 5th Intl. Cong. X-Ray Optics and Microanalysis, Tubingen*, eds. G. Mollenstadt and K. H. Gaukler, Springer, Berlin, p. 93.

Halloran, B. P., and Kirk, R. G. (1979). In *Microbeam Analysis in Biology*, eds. C. P. Lechene and R. Warner, Academic, New York, p. 571.

Halloran, B. P., Kirk, R. G., and Spurr, A. R. (1978). *Ultramicroscopy* **3**, 175.

Harland, C. J., and Venables, J. A. (1980). In *Proc. 38th Ann. EMSA Meeting San Francisco*, Claitors Press, Baton Rouge, Louisiana, p. 378.

Harvey, D. M. (1980). *SEM/1980/II*, SEM Inc., AMF O'Hare, Illinois, p. 409.

Harvey, D. M. R., Hall, J. L., and Flowers, T. J. (1976). *J. Microsc.* **107**, 189.

Hayat, M. A. (1970). *Principles and Techniques of Electron Microscopy: Biological Applications*, Vol. 1, Van Nostrand Reinhold, New York.

Hehenkamp, Th., and Böcher, J. (1974). *Mikrochim. Acta Suppl.* **5**, 29.

Heinrich, K. F. J. (1966a). *X-Ray Optics and Microanalysis, 4th Intl. Cong. on X-Ray Optics and Microanalysis*, eds. R. Castaing, P. Deschamps, and J. Philibert, Hermann, Paris, p. 1509.

Heinrich, K. F. J. (1966b). *The Electron Microprobe*, eds. T. D. McKinley, K. F. J. Heinrich, and D. B. Wittry, Wiley, New York, p. 296.

Heinrich, K. F. J. (1967). National Bureau of Standards Technical Note 278.

Heinrich, K. F. J. (ed.) (1968). "Quantitative Electron Probe Microanalysis," National Bureau of Standards Spec. Pub. 298, Washington, DC, p. 5.

Heinrich, K. F. J. (1969). National Bureau of Standards Technical Note 521.

Heinrich, K. F. J. (1981). *Electron Probe Microanalysis*, Van Nostrand, New York.

Heinrich, K. F. J., Fiori, C. E., and Yakowitz, H. (1970). *Science* **167**, 1129.

Heinrich, K. F. J., Myklebust, R. L., Rasberry, S. D., and Michaelis, R. E. (1971). National Bureau of Standards Spec. Pub. 260-28.

Heinrich, K. F. J., Newbury, D. E., and Yakowitz, H. (1976). National Bureau of Standards Spec. Pub. 460, Washington, DC.

Heinrich, K. F. J., Vieth, D. L., and Yakowitz, H. (1966). In *Advances in X-Ray Analysis*, eds. G. R. Mallett, M. J. Fay, and W. M. Mueller, Vol. 9, Plenum, New York, p. 208.

Heinrich, K. F. J., and Yakowitz, H. (1968). *Mikrochim. Acta* **5**, 905.

Heinrich, K. F. J., and Yakowitz, H. (1970). *Mikrochim. Acta* **1**, 123.

Heinrich, K. F. J., and Yakowitz, H. (1975). *Analyt. Chem.* **47**, 2408.

Hendee, C. F., Fine, S., and Brown, W. D. (1956). *Rev. Sci. Instr.* **27**, 531.

Henderson, W. J., and Griffiths, K. (1972). In *Principles and Techniques of Electron Microscopy*, Vol. 2, ed. M. A. Hayat, Van Nostrand Reinhold, New York.

Henke, B. L. (1964). In *Advances in X-Ray Analysis*, Vol. 7, Plenum, New York, p. 460.

Henke, B. L. (1965). In *Advances in X-Ray Analysis*, Vol. 8, Plenum, New York, p. 269.

Henke, B. L., and Ebisu, E. S. (1974). In *Advances in X-Ray Analysis*, Vol. 17, Plenum, New York, p. 150.

Henoc, J. (1968). In "Quantitative Electron Probe Microanalysis," ed. K. F. J. Heinrich, National Bureau of Standards Spec. Pub. 298, Washington, p. 197.

Henoc, J., Heinrich, K. F. J., and Myklebust, R. L. (1973). National Bureau of Standards Technical Note 769, Washington, DC, p. 127.

Henoc, J., and Maurice, F. (1976). In "Use of Monte Carlo Calculations in Electron Probe Microanalysis and Scanning Electron Microscopy," eds. K. F. J. Heinrich, D. E. Newbury, and H. Yakowitz, National Bureau of Standards Spec. Pub. 460, Washington, DC p. 61.

Herzog, R. F., Lewis, B. L., and Everhart, T. E. (1974). *SEM/1974*, IIT Research Institute, Chicago, Illinois, p. 175.

Heuser, J. E., Reese, T. S., Dennis, M. J., Jan, Y., Jan, L., and Evans, L. (1979) *J. Cell Biol.* **81**, 275.

Hewins, R. H. (1979). *Proc. 10th Lunar Sci. Conf.*, Pergamon, New York, p. 1109.

Hewins, R. H., Goldstein, J. I., and Axon, H. J. (1976). *Proc. 7th Lunar Sci. Conf.*, Pergamon, New York, p. 819.

Hillier, J. (1947). U.S. Patent 2,418,029.

Hodges, G. M., and Muir, M. D. (1976). *Principles and Techniques of Scanning Electron Microscopy: Biological Applications*, Vol. 5, ed. M. A. Hayat, Van Nostrand Reinhold, New York.

Hodson, S., and Marshall, J. (1972). *J. Microsc.* **95**, 459.

Hodson, S., and Williams, L. (1976). *J. Cell Sci.* **20**, 687.

Hojou, K., Oikawa, T., Kanaya, K., Kimura, T., and Adachi, K. (1977). *Micron* **8**, 151.

Holbrook, K. A., Holbrook, J. R., and Odland, G. F. (1976). *SEM/1976/I*, IIT Research Institute, Chicago, Illinois, p. 266.

Holburn, D. M., and Smith, K. C. A. (1979). *Electron Microscopy and Analysis, 1979*, Institute of Physics London Conf. Ser. Vol. 52, p. 69.

Holliday, J. E. (1963). In *The Electron Microprobe*, eds. T. D. McKinley, K. F. J. Heinrich, and D. B. Wittry, Wiley, New York, p. 3.

Holliday, J. E. (1967). *Norelco Reporter* **14**, 84.

Holman, W. R. (1974). In *Principles and Techniques of Electron Microscopy*, Vol. 4, ed. M. A. Hayat, Van Nostrand Reinhold, New York.

Horisberger, M., and Rosset, J. (1977). *SEM/1977/II*, IIT Research Institute, Chicago, Illinois, p. 75.

Horl, E. M., and Mugschl, E. (1972). *Proc. 5th Cong. on Electron Microscopy*, Institute of Physics, London, Conf. Ser. Vol. 14, p. 502.

Howell, P. G. T. (1975). *SEM/1975*, IIT Research Institute, Chicago, Illinois, p. 697.

Humphreys, W. J. (1977). *SEM/1977*, IIT Research Institute, Chicago, Illinois, p. 537.

Humphreys, W. J., and Henk, W. G. (1979). *J. Microsc.* **116**, 255.

Humphreys, W. J., Spurlock, B. O., and Johnson, J. S. (1977). *Principles and Techniques of SEM: Biological Applications*, Vol. 6, ed. M. A. Hayat, Van Nostrand Reinhold, New York.

Hutchinson, T. E., and Borek, J. R. (1979). *Ultramicroscopy* **4**, 233.

Ingram, F. D., and Ingram, J. J. (1975). *J. Microsc. Biol. Cell* **22**, 193.

Ingram, F. D., and Ingram, M. J. (1980). *SEM/1980/IV*, SEM Inc., AMF O'Hare, Illinois, p. 147.

Ingram, F. D., Ingram, M. J., and Hogben, C. A. M. (1972). *J. Histochem. Cytochem.* **20**, 716.

Ingram, M. J., and Hogben, C. A. M. (1967). *Anal. Biochem.* **18**, 54.

Irino, S., Ono, T., Watanabe, K., Toyota, K., Uno, J., Takasugi, N., and Murakami, T. (1975). *SEM/1975*, IIT Research Institute, Chicago, Illinois, p. 267.

Issacson, M. S. (1979). *Ultramicroscopy* **4**, 193.

Jackman, J. (1980). *Ind. Res. Dev.* **22**, 115.

Jones, A. V., and Smith, K. C. A. (1978). *SEM/1978*, SEM Inc., AMF O'Hare, Illinois, p. 13.

Joy, D. C., and Jakubovics, J. P. (1968). *Phil. Mag.* **17**, 61.

Joy, D. C., and Maher, D. M. (1976). *SEM/1976/II*, IIT Research Institute, Chicago, Illinois, p. 361.

Joy, D. C., Newbury, D. E., and Hazzledine, P. M. (1972). *SEM/1972*, IIT Research Institute, Chicago, Illinois, p. 97.

Junger, E., and Bachman, L. (1980). *J. Microsc.* **119**, 199.

Kahn, L. E., Frommes, S. P., and Cancilla, P. A. (1977). *SEM/1977/I*, IIT Research Institute, Chicago, Illinois, p. 501.

Kanaya, K., and Okayama, S. (1972). *J. Phys. D. Appl. Phys.* **5**, 43.

Kanter, H. (1961). *Phys. Rev.* **121**, 677.

Kardon, R. H., and Kessel, R. G. (1979). *SEM/1979/III*, SEM Inc., AMF O'Hare, Illinois, p. 743.

Katoh, M. (1978). *J. Electron Microsc.* **27**, 329.

Kay, M. M. B. (1978). In *Principles and Techniques of SEM: Biological Applications*, Vol. 6, ed. M. A. Hayat, Van Nostrand Reinhold, New York.

Kazmerski, L. L., and Racine, D. M. (1975). *J. Appl. Phys.* **46**, 791.

Kehl, G. L. (1949). *Principles of Metallographic Laboratory Practice*, 3rd ed., McGraw-Hill, New York.

Kelley, R. O., Dekker, R. A., and Bluemink, J. G. (1973). *J. Ultrastruct. Res.* **45**, 254.

Kiessling, R., and Lange, N. (1964). "Non-Metallic Inclusions in Steel," Special Report 90, The Iron and Steel Institute, London.

Kimoto, S., and Hashimoto, H. (1966). In *The Electron Microprobe*, John Wiley, New York, p. 480.

Kittel, C. (1956). *Introduction to Solid State Physics*, Wiley, New York.

Kohlhaas, V. E., and Scheidling, F. (1969). *Arch. Eisenhüttenwessen* **40**, 1.

Koshikawa, T., and Shimizu, R. (1974). *J. Phys. D: Appl. Phys.* **7**, 1303.

Kramers, H. A. (1923). *Phil. Mag.* **46**, 836.

Kubotsu, A., and Veda, M. (1980). *J. Elect. Microsc.* **29**, 45.

Kumon, H., Uno, F., and Tawara, J. (1976). *SEM/1976/II*, IIT Research Institute, Chicago, Illinois, p. 85.

Kyser, D. F. (1972). In *Proc. 6th Intl. Conf. X-Ray Optics and Microanalysis*, eds. G. Shinoda, K. Kohra, and T. Ichinokawa, Univ. Tokyo Press, Tokyo, p. 147.

Kyser, D. F., and Murata, K. (1974). *IBM J. Res. Dev.* **18**, 352.

Kyser, D. F., and Wittry, D. B. (1966). In *The Electron Microprobe*, eds. T. D. McKinley, K. F. J. Heinrich, and D. B. Wittry, John Wiley, New York, p. 691.

Lafferty, J. M. (1951). *J. Appl. Phys.* **22**, 299.

Laguitton, D., Rousseau, R., and Claisse, F. (1975). *Anal. Chem.* **47**, 2174.

Lamb, J. C., and Ingram, P. (1979). *SEM/1979/II*, SEM Inc., AMF O'Hare, Illinois, p. 459.

Lander, J. J., Schreiber, H., Buck, T. M., and Mathews, J. R. (1963). *Appl. Phys. Lett.* **3**, 206.

Lane, G. S. (1969). *J. Phys. E* **2**, 565.

Lane, W. C. (1970). *Proc. 3rd Ann. Stereoscan Coll.*, Kent Cambridge Scientific, Morton Grove, p. 83.

Langmuir, D. B. (1937). *Proc. IRE* **25**, 977.

Läuchli, A. (1975). *J. Microsc. Biol. Cell* **22**, 239.

Läuchli, A., Stelzer, R., Guggenheim, R., and Henning, L. (1978). "Precipitation Techniques as a Means of Intracellular Ion Localization by Use of Electron Probe Analysis," in *Microprobe Analysis as Applied to Cells and Tissues*, eds. T. A. Hall, P. Echlin, and R. Kaufmann, Academic, New York.

Leaver, C., and Chapman, B. (1971). *Thin Films*, Wykeham, London.

Lebiedzik, J., Burke, K. G., Troutman, S., Johnson, G. G. Jr., and White, E. W. (1973). *SEM/1973*, IIT Research Institute, Chicago, Illinois, p. 121.

Lechene, C. (1978). *Microscopica Acta Supp.* **2**, S. Hirzel Verlag, Stuttgart, p. 228.

Lechene, C. P., Bonventre, J. V., and Warner, R. R. (1979). In *Microbeam Analysis in Biology*, eds. C. P. Lechene and R. R. Warner, Academic, New York, p. 409.

Lechene, C. P., and Warner, R. R. (1977). *Ann. Rev. Biophys. Bioeng.* **6**, 57.

Lechene, C. P., and Warner, R. R., eds. (1979). *Microbeam Analysis in Biology*, Academic, New York.

Liebhafsky, H. A., Pfeiffer, H. G., and Zemany, P. D. (1955). *Anal. Chem.* **27**, 1257.

Liebhafsky, H. A., Pfeiffer, H. G., and Zemany, P. D. (1960). In *X-Ray Microscopy and X-Ray Microanalysis*, eds. A. Engström, V. Cosslett, and H. Pattee, Elsevier/North-Holland, Amsterdam, p. 321.

Liepins, A., and deHarven, E. (1978). *SEM/1978/II*, SEM Inc., AMF O'Hare, Illinois, p. 37.

Lifshin, E. (1974). "Proc. 9th Annual Conf. Microbeam Analysis Society," Ottawa, Canada, Paper 53.

Lifshin, E. (1975). In *Advances in X-Ray Analysis*, eds. R. W. Gould, C. S. Barrett, J. B. Newkirk, and C. O. Roud, Vol. 19, Kendall/Hunt Pub., Dubuque, Iowa, p. 113.

Lifshin, E., and Ciccarelli, M. F. (1973). *SEM/1973*, IIT Research Institute, Chicago, Illinois, p. 89.

Lifshin, E., Ciccarelli, M. F., and Bolon, R. B. (1973). "Proc. 8th Nat. Conf. Elect. Probe Anal.," New Orleans, Louisiana, Paper 29.

Lifshin, E., Ciccarelli, M. F., and Bolon, R. (1980). *Proc. 8th Intl. Cong. on X-Ray Optics and Microanlaysis*, eds. D. R. Beaman, R. E. Ogilvie, and D. B. Wittry, Pendell, Midland, Michigan, p. 141.

Lifshin, E., and DeVries, R. C. (1972). "Proc. 7th Conf. Elect. Probe Anal. Microbeam Analysis Society," San Francisco, California, Paper 18.

Lohnes, R. A., and Demirel, T. (1978). *SEM/1978/I*, SEM Inc., AMF O'Hare, Illinois, p. 643.

Long, J. V. P., and Agrell, S. O. (1965). *Min. Mag.* **34**, 318.

Lotan, R. (1979). *SEM/1979/III*, SEM Inc., AMF O'Hare, Illinois, p. 549.

Love, G., and Scott, V. D. (1978). *J. Phys. D. Appl. Phys.* **11**, 1369.

Love, G., and Scott, V. D. (1980). *Electron Microsc.* **3**, 146.

Lytton, D. G., Yuen, E., and Richard, K. A. (1979). *J. Microsc.* **115**, 35.

MacKenzie, A. P. (1976). *Dev. Biol. Stand.* **36**, 51.

MacKenzie, A. P., Kusler, T. A., and Luyet, B. J. (1975). *Cryobiology* **12**, 427.

Maeki, G., and Benoki, M. (1977). *J. Electron Microsc.* **26**, 223.

Maissel, L. I., and Glang, R., eds. (1970). *Handbook of Thin Film Technology*, McGraw-Hill, New York.

Malick, L. E., and Wilson, R. B. (1975). *Stain Technol.* **50**, 265.

Marinenko, R. B. (1981). National Bureau of Standards Spec. Pub. 260.

Marinenko, R. B., Heinrich, K. F. J., and Ruegg, F. C. (1979). *Microbeam Analysis—1979*, ed. D. E. Newbury, San Francisco Press, San Francisco, California, p. 221.

Marshall, A. T. (1975). "Electron Probe X-Ray Microanalysis," in *Principles and Techniques of Scanning Electron Microscopy*, Vol. 4, ed. M. A. Hayat, Van Nostrand Reinhold, New York.

Marshall, A. T. (1980). *SEM/1980/II*, SEM Inc., AMF O'Hare, Illinois, p. 395.

Marshall, A. T., and Wright, A. (1973). *Micron* **4**, 31.

Martoja, R., Szollosi, A., and Truchet, M. (1975). *J. Microsc. Biol. Cell* **22**, 247.

Marton, L. L. (1941). U.S. Pat. 2,233,286.

Masters, S. K., Bell, S. W., Ingram, P., Adams, D. O., and Shelburne, J. D. (1979). *SEM/1979/III*, SEM Inc., AMF O'Hare, Illinois, p. 97.

McCrone, W. C., and Delly, J. (1973). *The Particle Atlas Two*, Vol. I, Ann Arbor Science Publishers, Ann Arbor, Michigan.

McKee, A. E. (1977). *SEM/1977/II*, IIT Research Institute, Chicago, Illinois, p. 239.

McMullan, D. (1952). Ph.D. dissertation, Cambridge University.

Meryman, H. (1966). In *Cryobiology*, Chap. 13, ed. H. Meryman, Academic, New York.

Meyer, G. W., and Stewart, W. (1977). *SEM/1977/I*, IIT Research Institute, Chicago, Illinois, p. 341.

Michaelis, R. E., Yakowitz, H., and Moore, G. A. (1964). *J. Res. Natl. Bur. Stand. (U.S.)* **68A**, 343.

Millette, J. R., Clark, P. J., and Pansing, M. F. (1978). *SEM/1978/I*, SEM Inc., AMF O'Hare, Illinois, p. 253.

Molday, R. S. (1977). *Principles and Techniques of SEM: Biological Applications*, Vol. 5, ed. M. A. Hayat, Van Nostrand Reinhold, New York.

Moll, S. H., Baumgarten, N., and Donnelly, W. (1980). *Proc. 8th Int'l. Cong. on X-ray Optics and Microanalysis*, eds. D. R. Beaman, R. E. Ogilvie, and D. B. Wittry, Pendell, Midland, Michigan, p. 87.

Moll, S. H., Healy, F., Sullivan, B., and Johnson, W. (1978). *SEM/1978*, SEM Inc., AMF O'Hare, Illinois, p. 303.

Moor, H., Kistler, J., and Mutler, M. (1976). *Experientia* **32**, 805.

Morel, F., and Roinel, N. (1969). *J. Chem. Phys.* **66**, 1084.

Moreton, R. B., Echlin, P., Gupta, B. L., Hall, T. A., and Weis-Fogh, T. (1974). *Nature* **247**, 113.

Morgan, A. J. (1979). *SEM/1979/II*, SEM Inc., AMF O'Hare, Illinois, p. 635.

Morgan, A. J., Davies, T. W., and Erasmus, D. A. (1978). In *Electron Microprobe Analysis in Biology*, Chap. 4, ed. D. A. Erasmus, Chapman Hall, London.

Moseley, H. G. J. (1913). *Phil. Mag.* **26**, 1024.

Moseley, H. G. J. (1914). *Phil. Mag.* **27**, 703.

Moza, A., Strickler, D. W., and Austin, L. G. (1978). *SEM/1978/I*, SEM Inc., AMF O'Hare, Illinois, p. 289.

Muller, M., Marti, T., and Kriz, S. (1980). *Proceedings of the Seventh European Congress on Electron Microscopy*. Seventh European Congress on EM Foundation, Leiden, The Netherlands, Vol. 2, p. 720.

Munger, B. L. (1977). *SEM/1977/I*, IIT Research Institute, Chicago, Illinois, p. 481.

Munger, B. L., and Mumaw, V. R. (1976). *SEM/1976/I*, IIT Research Institute, Chicago, Illinois, p. 275.

Munoz-Guerra, S., and Subirana, J. A. (1980). *Proceedings of the Seventh European Congress on Electron Microscopy*. Seventh European Congress on EM Foundation, Leiden, The Netherlands, Vol. 2, p. 724.

Murata, K. (1973). *SEM/1973*, IIT Research Institute, Chicago, Illinois, p. 268.

Murata, K. (1974). *J. Appl. Phys.* **45**, 4110.

Murphey, J. A. (1978). *SEM/1978/II*, SEM Inc., AMF O'Hare, Illinois, p. 175.

Murphey, J. A. (1980). *SEM/1980/I*, SEM Inc., AMF O'Hare, Illinois, p. 209.

Myklebust, R. L., Fiori, C. E., and Heinrich, K. F. J. (1979). National Bureau of Standards Tech. Note 1106.

Myklebust, R. L., Yakowitz, H., and Heinrich, K. F. J. (1973). "Proc. 8th Nat. Conf. Elect. Probe Anal.," New Orleans, Louisiana, Paper 26.

Nagel, D. (1968). National Bureau of Standards Spec. Pub. 298, ed. K. F. J. Heinrich, Washington, DC, p. 189.

Nagy, I. Z., Pieri, C., Giuli, C., Bertoni-Freddari, C., and Nagy, V. (1977). *J. Ultrastruct. Res.* **58**, 22.

National Bureau of Standards (1979). "Standard Reference Materials Catalog," National Bureau of Standards Spec. Pub. 260.

Nelder, J. A., and Mead, R. (1965). *Computer J.* **7**, 308.

Nemanic, M. K. (1979). *SEM/1979/III*, SEM Inc., AMF O'Hare, Illinois, p. 537.

Newbury, D. E. (1974). *SEM/1974*, IIT Research Institute, Chicago, Illinois, p. 1047.

Newbury, D. E., and Greenwald, S. (1980). "Observations on the Mechanisms of High Resistance Junction Formation in Aluminum Wire Connections," *J. Res. Natl. Bur. Stand.*, Washington, DC, in press.

Newbury, D. E., Myklebust, R. L., Heinrich, K. F. J., and Small, J. A. (1980). National Bureau of Standards Spec. Pub. 533, p. 39.

Newbury, D. E., and Yakowtiz, H. (1976). National Bureau of Standards Spec. Pub. 460, eds. K. F. J. Heinrich, D. E. Newbury, and H. Yakowitz, Washington, DC, p. 15.

Newbury, D. E., Yakowitz, H., and Myklebust, R. L. (1973). *Appl. Phys. Lett.* **23**, 488.

Niedrig, H. (1978). *Scanning* **1**, 17.

Nopanitaya, W., Aghajanian, J. G., and Gray, L. D. (1979). *SEM/1979/III*, SEM Inc., AMF O'Hare, Illinois, p. 751.

Nowell, J. A., and Lohse, C. L. (1974). *SEM/1974*, IIT Research Institute, Chicago, Illinois, p. 267.

Oatley, C. W. (1972). *The Scanning Electron Microscope, Part 1, The Instrument*, Cambridge University Press, Cambridge.

Oatley, C. W., and Everhart, T. E. (1957). *J. Electron.* **2**, 568.

Oatley, C. W., Nixon, W. C., and Pease, R. F. W. (1965). In *Advances in Electronics and Electron Physics*, ed. L. Marton, Academic, New York, p. 181.

Oda, Y., and Nakajima, K. (1973). *J. Jn. Inst. Met.* **37**, 673.

Okagaki, T., and Clark, B. (1977). *SEM/1977/II*, IIT Research Institute, Chicago, Illinois, p. 153.

Oron, M., and Tamir, V. (1974). *SEM/1974*, IIT Research Institute, Chicago, Illinois, p. 207.

Pallaghy, C. K. (1973). *Aust. J. Biol. Sci.* **26**, 1015.

Pameijer, C. H. (1979). *SEM/1979/II*, SEM Inc., AMF O'Hare, Illinois, p. 571.

Panayi, P. N., Cheshire, D. C., and Echlin, P. (1977). *SEM/1977/I*, IIT Research Institute, Chicago, Illinois, p. 463.

Parobek, L., and Brown, J. D. (1978). *X-Ray Spectrom.* **7**, 26.

Paulson, G. G., and Pierce, R. W. (1972). *Proc. 30th Ann. EMSA*, Claitors Publishing, Baton Rouge, Louisiana, p. 406.

Pawley, J. B. (1972). *SEM/1972*, IIT Research Institute, Chicago, Illinois, p. 153.

Pawley, J. B. (1974). *SEM/1974/I*, IIT Research Institute, Chicago, Illinois, p. 27.

Pawley, J. B., and Dole, S. (1976). *SEM/1976/I*, IIT Research Institute, Chicago, Illinois, p. 287.

Pawley, J. B., and Norton, J. T. (1978). *J. Microsc.* **112**, 169.

Pease, D. C. (1974). "Proc. 8th Intl. Cong. Electron Microscopy, Canberra," Vol. 2, p. 126.

Pease, D. C., and Peterson, R. G. (1972). *J. Ultrastruct. Res.* **41**, 133.

Pease, R. W., and Bailey, J. F. (1975). *J. Microsc.* **104**, 281.

Pease, R. F. W. (1963). Ph.D. dissertation, Cambridge University, England.

Pease, R. F. W., and Nixon, W. C. (1965). *J. Sci. Instr.* **42**, 81.

Peters, K. R. (1979). *SEM/1979/II*, SEM Inc., AMF O'Hare, Illinois, p. 133.

Peters, K. R. (1980a). *SEM/1980/I*, SEM Inc., AMF O'Hare, Illinois, p. 143.

Peters, K. R. (1980b). *J. Microsc.* **118**, 429.

Petroff, P. M., Long, D. V., Strudel, J. L., and Logan, R. A. (1978). *SEM/1978/I*, SEM Inc., AMF O'Hare, Illinois, p. 325.

Philibert, J. (1963). *Proc 3rd Intl. Symp. X-Ray Optics and X-Ray Microanalysis*, Stanford University, eds. H. H. Pattee, V. E. Cosslett, and A. Engström, Academic, New York, p. 379.

Philibert, J., and Tixier, R. (1969). *Micron* **1**, 174.

Postek, M. T., and Tucker, S. C. (1977). *J. Microsc.* **110**, 71.

Powell, C. J. (1976). National Bureau of Standards Spec. Pub. 460, eds. K. F. J. Heinrich, D. E. Newbury, and H. Yakowitz, p. 97.

Quinton, R. M. (1978). *SEM/1978/II*, SEM Inc., AMF O'Hare, Illinois, p. 391.

Rall, W. F., Reid, D. S., and Farrant, J. (1980). *Nature* **286**, 511.

Rampley, D. N. (1976). *J. Microsc.* **107**, 99.

Rao-Sahib, T. S., and Wittry, D. B. (1972). In *Proc. 6th Intl. Cong. on X-Ray Optics and Microanalysis*, eds. G. Shinoda, K. Kohra, and T. Ichinokawa, Univ. Tokyo Press, Tokyo, p. 121.

Rao-Sahib, T. S., and Wittry, D. B. (1975). *J. Appl. Phys.* **45**, 5060.

Reed, S. J. B. (1965). *Brit. J. Appl. Phys.* **16**, 913.

Reed, S. J. B. (1966). In *X-Ray Optics and Microanalysis, 4th Intl. Cong. on X-Ray Optics and Microanalysis*, eds. R. Castaing, P. Deschamps, and J. Philibert, Hermann, Paris, p. 339.

Reed, S. J. B. (1971). *J. Phys. D (Appl. Phys.)* **4**, 1910.

Reed, S. J. B. (1975). *Electron Microprobe Analysis*, Cambridge University Press, Cambridge.

Reed, S. J. B., and Ware, N. G. (1972). *J. Phys. E: Sci. Instr.* **5**, 582.

Reimer, L. (1975). In *Physical Aspects of Electron Microscopy and Microbeam Analysis*, eds. B. M. Siegal and D. R. Beamon, John Wiley, New York, p. 231.

Reimer, L., and Tollkamp, C. (1980). *Scanning* **3**, 35.

Reimer, L., and Volbert, B. (1979). *Scanning* **2**, 238.

Reuter, W. (1972). In *Proc. 6th Intl. Conf. X-Ray Optics and Microanlaysis*, eds. G. Shinoda, K. Kohra, and T. Ichinokawa, Univ. Tokyo Press, Tokyo, p. 121.

Rick, R., Dörge, A., and von Arnim, E. (1978). *Microscopica Acta Suppl.* **2**, 156.

Rick, R., Dörge, A., von Arnim, E., and Thurau, K. (1978). *J. Membrane Biol.* **39**, 313.

Robards, A. W., and Crosby, P., (1979). *SEM/1979/II*, SEM Inc AMF O'Hare, Illinois, p. 325.

Robinson, V. N. E. (1975). *SEM/1975*, IIT Research Institute, Chicago, Illinois, p. 51.

Robinson, V. N. E. (1978). *Scanning* **1**, 149.

Robinson, V. N. E. (1980). *Scanning* **3**, 15.

Robinson, V. N. E., and Robinson, B. W. (1978). *SEM/1978/I*, SEM Inc., AMF O'Hare, Illinois, p. 595.

Roli, J., and Flood, P. R. (1978). *J. Microsc.* **112**, 359.

Romig, A. D., and Goldstein, J. I. (1978). *Met. Trans.* **9A**, 1599.

Roomans, G. M. (1979). *SEM/1979/II*, SEM Inc., AMF O'Hare, Illinois, p. 649.

Roomans, G. M., and Seveus, L. A. (1976). *J. Cell Sci.* **21**, 119.

Rose, A. (1948). In *Advances in Electronics*, ed. A. Marton, Academic, New York, p. 131.

Rostgaard, J., and Tranum-Jensen, J. (1980). *J. Microsc.* **119**, 213.

Russ, J. C. (1974). In *Microprobe Analysis Applied to Cells and Tissues*, eds. T. A. Hall, P. Echlin, and R. Kaufmann, Academic, New York, p. 269.

Ryder, P. L. (1977). *SEM/1977/I*, IIT Research Institute, Chicago, Illinois, p. 273.

Ryder, T. A., and Bowen, I. D. (1974). *J. Microsc.* **101**, 143.

Sanders, S., Alexander, E., and Braylan, R. (1975). *J. Cell Biol.* **67**, 476.

Saubermann, A. J. (1978). *Microscopica Acta Suppl.* **2**, 130.

Saubermann, A. J. (1980). *SEM/1980/II*, SEM Inc., AMF O'Hare, Illinois, p. 421.

Saubermann, A. J., and Echlin, P. (1975). *J. Microsc.* **105**, 155.

Saubermann, A. J., Echlin, P., Peters, P. D., and Beeukes, R. (1981). "Application of Scanning Electron Microscopy to X-Ray Analysis of Frozen-Hydrated Sections. I. Specimen Handling Techniques." *J. Cell Biol.* **88**, 257.

Saubermann, A. J., Riley, W., and Echlin, P. (1977a). *SEM/1977/I*, IIT Research Institute, Chicago, Illinois, p. 347.

Saubermann, A. J., Riley, W., and Beeukes, R. (1977b). *J. Microsc.* **111**, 39.

Sawyer, G. R., and Page, T. F. (1978). *J. Mat. Sci.* **13**, 885.

Schamber, F. H. (1977). In *X-ray Fluorescence Analysis of Environmental Samples*, ed. T. G. Dzubay, Ann Arbor Science, p. 241.

Schiller, A., Sonnhof, U., and Taugner, R. (1979). *Mikroskopie* **35**, 23.

Schneider, G. B., Pockwinse, S. M., and Billings-Gagliardi, S. (1978). *SEM/1978/II*, SEM Inc., AMF O'Hare, Illinois, p. 77.

Seiler, H. (1967). *Z. Angew. Phys.* **22**, 249.

Sela, J., and Boyde, A. (1977). *J. Microsc.* **111**, 229.

Shimizu, R., Ikuta, T., Everhart, T. E., and Devore, W. J. (1975). *J. Appl. Phys.* **46**, 1581.

Shimizu, R., Kataoka, Y., Ikuta, T., Koshikawa, T., and Hashimoto, H. (1976). *J. Phys. D: Appl. Phys.* **9**, 101.

Shimizu, R., and Murata, K. (1971). *J. Appl. Phys.* **42**, 387.

Shiraiwa, T., Fujino, N., and Murayama, J. (1972). In *Proc. 6th Intl. Conf. on X-Ray Optics and Microanalysis*, eds. G. Shinoda, K. Kohra, and T. Ichinokawa, Univ. Tokyo Press, Tokyo, p. 213.

Shuman, H., and Somlyo, A. P. (1976). *Proc. Natl. Acad. Sci. USA* **73**, 1193.

Shuman, H., Somlyo, A. V., and Somlyo, A. P. (1976). *Ultramicroscopy* **1**, 317.

Simpson, J. A. V., Bank, H. L., and Spicer, S. S. (1979). *SEM/1979/II*, SEM Inc., AMF O'Hare, Illinois, p. 779.

Sjostrom, J., and Thornell, L. E. (1975). *J. Microsc.* **103**, 101.

Skaer, H. le B., Franks, F., Asquith, M. H., and Echlin, P. (1977). *J. Microsc.* **110**, 257.

Skaer, H. le B., Franks, F., and Echlin, P. (1978). *Cryobiology* **15**, 589.

Small, J. A., Heinrich, K. F. J., Fiori, C. E., Myklebust, R. L., Newbury, D. E., and Dilmore, M. F. (1978). *SEM/1978/I*, SEM Inc., AMF O'Hare, Illinois, p. 445.

Small, J. A., Heinrich, K. F. J., Newbury, D. E., and Myklebust, R. L. (1979). *SEM/1979/II*, SEM Inc., AMF O'Hare, Illinois, p. 807.

Smith, D. G. W. (1975). "Proc. 10th Ann. Conf. of the Microbeam Analysis Society," Las Vegas, Nevada, Paper 21.

Smith, D. G. W., Gold, C. M., and Tomlinson, D. A. (1975). *X-Ray Spectrom* **4**(3), 149.

Smith, K. C. A. (1956). Ph.D. dissertation, Cambridge University.

Smith, K. C. A., Unitt, B. M., Holburn, D. M., and Tee, W. J. (1977). *SEM/1977/I*, IIT Research Institute, Chicago, Illinois, p. 49.

Somlyo, A. V., Shuman, H., and Somlyo, A. P. (1977). *J. Cell Biol.* **74**, 828.

Spivak, G. V., Rau, E. I., Lukianov, A. E., Petrov, V. I., and Bicov, M. V. (1972). In "Proc. 5th Eur. Cong. Electron Microscopy," Manchester, England, p. 92.

Spriggs, T. L. B., and Wynne Edwards, D. (1976). *J. Microsc.* **107**, 35.

Springer, G. (1966). *Mikrochim. Acta* **3**, 587.

Springer, G. (1972). In *Proc. 6th Intl. Conf. X-Ray Optics and Microanalysis*, eds. G. Shinoda, K. Kohra, and T. Ichinokawa, Univ. Tokyo Press, Tokyo, p. 141.

Springer, G. (1973). In *Quantitative Analysis with Electron Microprobes and Secondary Ion Mass Spectrometry*, ed. E. Preuss, Kernforschung Juelich, West Germany.

Spurr, A. R. (1975). *J. Microsc. Biol. Cell* **22**, 287.

Statham, P. J. (1974). "Proc. 9th Annual Conf. Microbeam Analysis Society," Ottawa, Canada, Paper 21.

Statham, P. J. (1977). *Anal. Chem.* **49**, 2149.

Statham, P. (1979). *Mikrochim Acta, (Wien) Suppl.* **8**, 229.

Statham, P. J., and Pawley, J. B. (1978). *SEM/1978/I*, SEM Inc., AMF O'Hare, Illinois, p. 469.

Steffens, W. L. (1978). *J. Microsc.* **113**, 95.

Steinbiss, H. and Schmitz, K. (1973). *Algen Planta*, **112**, 253.

Streitwolf, H. W. (1959). *Ann. Phys. (Leipzig)* **3**, 183.

Sturgess, J. M., and Moscarello, M. A. (1976). *SEM/1976/II*, IIT Research Institute, Chicago, Illinois, p. 145.

Sumner, A. T. (1978). *J. Microsc.* **114**, 19.

Swanson, L. W., and Dickinson, T. (1976). *Appl. Phys. Lett.* **28**, 578.

Sweeney, W. E., Seebold, R. E., and Birks, L. S. (1960). *J. Appl. Phys.* **31**, 1061.

Sylvester-Bradley, P. C. (1969). *Micropaleontology* **15**, 366.

Talmon, Y., and Thomas, E. L. (1977a). *SEM/1977/I*, IIT Research Institute, Chicago, Illinois, p. 265.

Talmon, Y., and Thomas, E. L. (1977b). *J. Microsc.* **111**, 151.

Tanaka, K. (1974). In *Principles and Techniques of SEM: Biological Applications*, Vol. 1, ed. M. A. Hayat, Van Nostrand Reinhold, New York.

Taylor, C. M., and Radtke, A. S. (1965). *Economic Geology* **60**, 1306.

Tegart, W. (1959). *Electrolytic and Chemical Polishing of Metals in Research and Industry*, 2nd rev. ed., Pergamon, New York.

Thomas, P. M. (1964). U.K. Atomic Energy Auth. Rept. AERE-R 4593.

Thomas, R. S. (1969). *Adv. Opt. Electron Microsc.* **3**, 99.

Thornley, R. F. M. (1960). Ph.D. dissertation, Cambridge University.

Thorpe, J. R., and Harvey, D. M. R. (1979). *J. Ultrastruct. Res.* **68**, 186.

Thurley, K. W., and Mouel, W. C. (1974). *J. Microsc.* **101**, 215.

Tovey, N. K., and Wong, K. Y (1973). In "Proc. Intl. Symp. on Soil Structure," Swedish Geotechnical Society, Stockholm, Sweden, p. 59.

Troyon, H., Kuo, H. P., and Siegel, B. M. (1973). In *Proc. 31st Annual EMSA Meeting*, Claitors Press, Baton Rouge, Louisiana, p. 298.

Valdre, U. (1979). *J. Microsc.* **117**, 55.

Vallyathan, N. V., and Brody, A. R. (1977). *SEM/1977/II*, IIT Research Institute, Chicago, Illinois, p. 93.

Van Eekelen, C. A. G., Boekestein, A., Stols, A. L. H., and Stadhouders, A. M. (1980). *Micron* **11**, 137.

Van Essen, C. G. (1974). *J. Phys. E* **7**, 98.

Van Steveninck, R. F. M., Ballment, B., Peters, P. D., and Hall, T. A. (1976). *Aust. J. Plant Physiol.* **3**, 359.

Van Steveninck, R. F. M., and Van Steveninck, M. E. (1978). *Ion Localization in Electron Microscopy and Cytochemistry of Plant Cells*, Chap. 4, ed. J. L. Hall, Elsevier/North-Holland, Amsterdam.

Van Zyl, J., Forrest, Q. C., Hocking, C., and Pallaghy, C. K. (1976). *Micron* **7**, 213.

Von Ardenne, M. (1938a). *Z. Phys.* **109**, 553.

Von Ardenne, M. (1938b). *Z. Techn. Phys.* **19**, 407.

Walker, D. A. (1978). *SEM/1978/I*, IIT Research Institute, Chicago, Illinois, p. 185.

Ware, N. G., and Reed, S. J. B. (1973). *J. Phys. E: Sci. Inst.* **6**, 286.

Warner, R. R., and Coleman, J. R. (1973). *Micron* **4**, 61.

Warner, R. R., and Coleman, J. R. (1975). *Micron* **6**, 79.

Watson, J. H. L., Page, R. H., and Swedo, J. L. (1975). *SEM/1975*, IIT Research Institute, Chicago, Illinois, p. 417.

Wehner, G. K., and Anderson, G. S. (1970). "The Nature of Physical Sputtering," in *Handbook of Thin Film Technology*, eds. L. I. Maissel, and R. Glang, McGraw-Hill, New York.

Wells, O. C. (1960). *Brit. J. Appl. Phys.* **11**, 199.

Wells, O. C. (1974a). *Scanning Electron Microscopy*, McGraw-Hill, New York.

Wells, O. C. (1974b). *SEM/1974*, IIT Research Institute, Chicago, Illinois, p. 1.

Wells, O. C. (1977). *SEM/1977/I*, IIT Research Institute, Chicago, Illinois, p. 747.

Wells, O. C. (1978). *SEM/1978*, SEM Inc., AMF O'Hare, Illinois, p. 293.

Welter, L. M., and Coates, V. J. (1974). *SEM/1974/I*, IIT Research Institute, Chicago, Illinois, p. 59.

Wetzel, B., Cannon, G. B., Alexander, E. L., Erickson, B. W., and Westbrook, E. W. (1974). *SEM/1974/I*, IIT Research Institute, Chicago, Illinois, p. 581.

Wetzel, B., Jones, G. M., and Sanford, K. K. (1977). *SEM/1977/I*, IIT Research Institute, Chicago, Illinois, p. 545.

Witcomb, M. J. (1981). *J. Microsc.* **121**, 289.

Wittry, D. B. (1957). Ph.D. dissertation, California Institute of Technology.

Wittry, D. B. (1963). ASTM STP 349, Philadelphia, Pennsylvania, p. 128.

Wittry, D. B. (1966). In *X-Ray Optics and Microanalysis, 4th Intl. Cong. on X-Ray Optics and Microanalysis*, eds. R. Castaing, P. Deschamps, and J. Philibert, Hermann, Paris, p. 168.

Wofsy, L. (1979). *SEM/1979/III*, SEM Inc., AMF O'Hare, Illinois, p. 565.

Woldseth, R. (1973). *X-Ray Energy Spectrometry*, Kevex Corp., Foster City, California.

Woods, P. S., and Ledbetter, M. E. (1976). *J. Cell Sci.* **21**, 47.

Yakowitz, H., Fiori, C. E., and Michaelis, R. E. (1971). National Bureau of Standards Spec. Pub. 260-22.

Yakowitz, H., and Heinrich, K. F. J. (1968). *Mikrochim Acta*, p. 183.

Yakowitz, H., and Heinrich, K. F. J. (1969). *J. Res. Natl. Bur. Stand. A. Phys. Chem.* **73A**, 113.

Yakowitz, H., Michaelis, R. E., and Vieth, D. L. (1969). In *Advances in X-Ray Analysis*, Vol. 12, Plenum Press, New York, p. 418.

Yakowitz, H., Myklebust, R. L., and Heinrich, K. F. J. (1973). National Bureau of Standards Tech. Note 796.

Yakowitz, H., and Newbury, D. E. (1976). *SEM/1976/I*, IIT Research Institute, Chicago, Illinois, p. 151.

Yakowitz, H., Ruff, A. W., Jr., and Michaelis, R. E. (1972). National Bureau of Standards Spec. Pub. 260-43.

Yakowitz, H., Vieth, D. L., Heinrich, K. F. J., and Michaelis, R. E. (1965). National Bureau of Standards Spec. Pub. 260-10.

Yakowitz, H., Vieth, D. L., Heinrich, K. F. J., and Michaelis, R. E. (1966). In *Advances in X-Ray Analysis*, Vol. 9, Plenum, New York, p. 289.

Yarom, R., Maunder, C., Scripps, M., Hall, T. A., and Dubowitz, V. (1975a). *Histochemistry* **45**, 59.

Yarom, R., Hall, T. A., and Peters, P. D. (1975b). *Experientia* **31**, 154.

Yeo, A. R., Läuchli, A., Kramer, D., and Gullasch, J. (1977). *Planta* **134**, 35.

Yew, N. C., and Pease, D. E. (1974). *SEM/1974*, IIT Research Institute, Chicago, Illinois, p. 191.

Ziebold, T. O. (1967). *Anal. Chem.* **39**, 858.

Ziebold, T. O., and Ogilvie, R. E. (1964). *Anal. Chem.* **36**, 322.

Zworykin, V. K., Hillier, J., and Snyder, R. L. (1942). ASTM Bulletin 117, p. 15.

Index